U0275591

寰宇文献 Universal Library | SINOLOGY 系列

SELECTED WORKS OF BERTHOLD LAUFER

劳费尔著作集

第八卷

[美] 劳费尔 著

黄曙辉 编

中西书局

ZHONGXI BOOK COMPANY

图书在版编目(CIP)数据

劳费尔著作集 / (美) 劳费尔著；黄曙辉编. 一上
海：中西书局，2022
(寰宇文献)
ISBN 978-7-5475-2015-4

Ⅰ. ①劳… Ⅱ. ①劳… ②黄… Ⅲ. ①劳费尔 – 人类
学 – 文集 Ⅳ. ①Q98-53

中国版本图书馆CIP数据核字（2022）第207067号

第 8 卷

1

110

与栉孔扇贝和叙利亚羔羊有关的故事

THE JOURNAL OF
AMERICAN FOLK-LORE

VOLUME XXVIII

LANCASTER, PA., AND NEW YORK

𝕻𝖚𝖇𝖑𝖎𝖘𝖍𝖊𝖉 𝖇𝖞 𝖙𝖍𝖊 𝕬𝖒𝖊𝖗𝖎𝖈𝖆𝖓 𝕱𝖔𝖑𝖐-𝕷𝖔𝖗𝖊 𝕾𝖔𝖈𝖎𝖊𝖙𝖞

G. E. STECHERT & CO., AGENTS

NEW YORK: 151–155 WEST 25TH STREET   PARIS: 16 RUE DE CONDÉ

LONDON: DAVID NUTT, 57, 59 LONG ACRE

LEIPZIG: OTTO HARRASSOWITZ, QUERSTRASSE, 14

MDCCCCXV

THE JOURNAL OF

AMERICAN FOLK-LORE.

Vol. XXVIII. — APRIL–JUNE, 1915. — No. CVIII.

———•———

THE STORY OF THE PINNA AND THE SYRIAN LAMB.

BY BERTHOLD LAUFER.

THE Chinese Annals of the Later Han Dynasty (A.D. 25–220), in the account of Ta T'sin, ascribe asbestine cloth to the Roman Orient. The text then continues, "Further, they have a fine cloth said by some to originate from the down of a water-sheep, and they have also a stuff made from wild-silkworm cocoons."[1] The name of the former of these

[1] Hou Han shu, ch. 118, p. 4 b. The previous translators of this passage did not treat it with full justice. Hirth (China and the Roman Orient, p. 41) offered the rendering, "They further have 'fine cloth,' also called *Shui-yang-ts'ui* [that is, down of the water-sheep]; it is made from the cocoons of wild silkworms." G. Schlegel ("The Shui-yang or Water-Sheep," Actes du 8e Congrès des Orientalistes à Stockholm, 1889, p. 22) criticised this translation on some point, and himself proposed, "They have fine cloth which some say is made from the down of the water-sheep and the cocoons of wild silkworms." Chavannes (T'oung Pao, 1907, p. 183) translates, "They have, besides, a light cloth, of which some say that it is from the down of the aquatic sheep, but which in reality is fabricated from the cocoons of wild silkworms." M. Chavannes himself, however, contradicts this translation by his mode of interpretation: for he explains the cloth from the down of the water-sheep as the textiles made from the fibres of the pinna (the textiles are not styled "byssus," as stated by him), and in regard to the silk material refers to Aristotle's mention of "silks from wild silkworms on the island of Cos." In this case the two articles are entirely distinct, and it is clear that the above Chinese clause consists of two separate and co-ordinated parts. A stuff made from wild-silkworm cocoons is not capable of eliciting a tradition pertaining to a water-sheep. The latter, as plainly suggested by this name, is an aquatic product, while silk is not. That this view of the matter is correct, is solidly testified by the texts of the T'ang shu, and of Ma Tuan-lin quoted above, which speak of the water-sheep only, without any reference to wild silkworms. The text of the Wei lio (Hirth, *l. c.*, p. 71), however, is perfectly conclusive: "They weave fine cloth, saying that they utilize for this purpose the down of the water-sheep; this product is termed 'cloth from the west of the sea.' All domestic animals of this country are produced in the water. Some say that they make use not only of sheep's wool, but also of tree-bast [that is evidently flax] and the silk of wild silkworms in the production of textiles." Here the wild silkworms are separated from the water-sheep by two intervening sentences, and it is patent that the two subjects are not interrelated. — The passage of Aristotle in regard to the silkworm, to which Chavannes alludes, has frequently been misunderstood. Aristotle does not say that the animal was bred or the raw material produced in Cos: he merely

- 3 -

two textiles is imparted in the "Wei lio," written by Yü Huan between 239 and 265, who states, "They weave fine cloth, saying that they utilize for this purpose the down of the water-sheep; this product is termed 'cloth from the west of the sea' (*hai si pu*)." The same name appears in the Annals of the T'ang Dynasty,[1] in the account of Fu-lin (Syria), where "the wool of the water-sheep is woven into cloth." Ma Tuan-lin, in his "Wên hien t'ung k'ao," completed in 1319, has the same information; but the name is altered by him into "cloth occurring in the sea" or "cloth from within the sea" (*hai chung pu*). This was presumably effected under Arabic influence; for Ibn al-Baiṭār calls the product yielded by the *Pinna nobilis* or *P. squamosa* "wool of the sea" (*suf el-bahr*),[2] and, as will be seen, after Greek model.

The failure of previous authors to explain these accounts correctly resulted from their neglect to study the corresponding traditions of the ancients regarding this matter. Bretschneider[3] observed with reference to the passage in the Han Annals, "This is perhaps the byssus, a cloth-stuff woven up to the present time by the inhabitants of the Mediterranean coast, especially in southern Italy, from the thread-like excrescences of several sea-shells, especially *Pinna squamosa*." A modern condition of affairs is here invoked to account for a fact relating to antiquity; while the ancients find no place at all, and no attempt is made to explain the origin of the curious Chinese term "water-sheep." There is, moreover, a grave error in Bretschneider's statement when he designates this fabric as "byssus." Byssus, as everybody knows, was a fine tissue of the ancients, produced in the vicinity of Elis in Achaia.[4] It is variously interpreted as cotton or flax.[5] More probably it was the latter.[6] At any rate, it has nothing to do with the ancient

states that a woman of Cos, Pamphila by name, daughter of Plateus, is credited with the first invention of the fabric. Only subsequent authors — as Pliny (XI, 77) and Isidorus (XIX, 22, 13), who lived from 570 to 636 — mention the actual occurrence of a wild silk-worm on Cos (compare J. Yates, Textrinum Antiquorum, p. 163; Blümner, Technologie, vol. i, 2d ed., p. 202). On the other hand, the opinion is expressed that Aristotle, in this passage, does not speak at all of a silkworm (Aubert and Wimmer, Aristoteles Tierkunde, vol. i, p. 162); and E. Hahn (Haustiere, p. 563) even goes so far as to reject, with good reason, this whole text as unauthentic. At any rate, it seems doubtful that Aristotle should have written all the unintelligible absurdities of this account. Be this as it may, the notice ascribed to Aristotle's name cannot be enlisted to explain the wild silk mentioned by the Chinese Annals as having been wrought in the Roman Orient. This kind of silk has nothing to do with Cos or *vestes Coae*, but distinctly points to what was termed by the ancients *bombycinae*, — textiles manufactured in Assyria or Syria, and obtained from a wild silkworm whose cocoons could not be reeled off, but were combed and spun. This silk (in French *galette*) possessed less gloss and fineness than the Chinese material.

[1] T'ang shu, ch. 221 B, p. 8. The T'ang dynasty ruled from 618 to 906.
[2] L. Leclerc, Traité des simples, vol. ii, p. 386.
[3] On the Knowledge possessed by the Ancient Chinese of the Arabs, p. 24.
[4] Pliny, Naturalis historia, XIX, 4.
[5] Blümner, Technologie, vol. i, 2d ed., p. 192.
[6] J. Yates, Textrinum Antiquorum, pp. 267–280.

textiles obtained from the fibres of the pinna. The error of Bretschneider was caused by the fact that in our zoölogical nomenclature the filaments secreted by the foot of this animal and other bivalve mollusks, and serving for attachment to fixed objects, are styled "byssi" (plural of "byssus").[1] In this sense, however, the word was not used in the language of the ancients. Notwithstanding, we are under obligations to Bretschneider for his ingenious suggestion, as it will be seen that, as a matter of fact, he was quite correct in his presentiment.

Yule[2] connected the water-sheep of Chinese tradition with Friar Odoric's story of the vegetable lamb of the Volga. This, however, is plainly an unmethodical procedure and a chronological *saltus mortalis*, — first, as the two traditions are widely different without an attempt on the part of Yule to explain this difference; and, second, as a Chinese tradition of the third century pertaining to the Hellenistic Orient cannot be brought into direct contact with reports of mediæval European travellers, but must be correlated with coeval Hellenistic thought. Hirth[3] justly emphasized the wide gap of the chronological interval that separates the two events, but did not cope with the problem involved. Schlegel[4] attacked it in an uncritical manner, and brought new confusions into the discussion by dragging into the tangle also the camel. It is Chavannes'[5] merit to have clearly discriminated between the water-sheep and the so-called *Agnus scythicus* of mediæval travellers,[6] and to have established for the former the only correct interpretation by means of the filaments of the pinna; but, in so doing, Chavannes has recourse solely to an Arabic author, Iṣṭakhri, of the tenth century, and reconstructs from his report a legend which should have given rise to the Chinese idea of a water-sheep. It is clear, however, that the Arabic as well as the Chinese traditions must be reducible to a Hellenistic tradition; and it is obvious alike that the Chinese notion which first appears in the "Wei lio" of the third century is not due to the Arabs, but received a direct impetus from Hellenism. It is therefore imperative to go straight to headquarters, and to study what the ancients themselves have to say about the pinna and its products.

[1] This bunch of silky fibres suitable for weaving projects only from one side of the animal, near the lower pointed extremity, which is fixed perpendicularly in submarine sand or rocks, the *byssus* having the function of an anchor. *Pinna* (more correctly *Pina*) is the generic name for a large family of marine mussels (*Pinnidae*), belonging to the class of *Pelecypodes*, and occurring in the Mediterranean and Indian Oceans (see P. Fischer, Manuel de conchyliologie, p. 963, Paris, 1887; and A. Hyatt, Remarks on the Pinnidae, Proc. Boston Soc. Nat. Hist., vol. xxv, 1892, pp. 335–346). The species utilized by the ancients is known as *Pinna nobilis* or *P. squamosa*.

[2] Cathay, vol. i, p. LVII.

[3] China and the Roman Orient, p. 262.

[4] "The Shui-yang or Water-Sheep," *l.c.*, pp. 19–32.

[5] T'oung Pao, 1907, p. 183, note 4.

[6] It will be seen in the further course of this article, however, that an historical and inner connection between the two exists, nevertheless.

First of all, it is remarkable that the classical Greek and Roman authors, while thoroughly acquainted with the pinna as a species of edible mollusk, are entirely reticent about the employment of its filaments for textiles. This industry is foreign to the classical epoch, and does not appear before the second century A.D.; it is an offshoot of Hellenistic, not of Greek culture. Aristotle, in his treatise on zoölogy (v, 15), describes the pinna as follows: "With regard to the *limnostreae,* or lagoon oysters, wherever you have slimy mud, there you are sure to find them beginning to grow. Cockles and clams and razor-fishes and scallops grow spontaneously in sandy places. The pinna grows straight up from its tuft of anchoring fibres in sandy and slimy places. These creatures have inside them a parasite nicknamed the 'pinna-guard,' — in some cases a small carid, and in other cases a little crab. If the pinna be deprived of this pinna-guard, it soon dies." Again he says, "Some shift about from place to place, others remain permanent on one spot. Of those that keep to one spot, the pinnae are rooted to the ground. The razor-fish and the clam keep to the same locality, but are not so rooted; but still, if forcibly removed, they die." Of special importance for a consideration of the legend of the vegetable lamb (to be discussed farther on) is another passage in the same work of Aristotle (VIII, 1): "Nature proceeds little by little from things lifeless to animal life in such a way that it is impossible to determine the exact line of demarcation, nor on which side thereof an intermediate form should lie. Thus next after lifeless things in the upward scale comes the plant; and of plants, one will differ from another as to its amount of apparent vitality; and, in a word, the whole genus of plants, while it is devoid of life as compared with an animal, is endowed with life as compared with other corporeal entities. Indeed, as we just remarked, there is observed in plants a continuous scale of ascent towards the animal. So, in the sea, there are certain objects concerning which one would be at a loss to determine whether they be animal or vegetable. For instance, certain of these objects are fairly rooted, and in several cases perish if detached. Thus the pinna is rooted to a particular spot, and the solen (or razor-shell) cannot survive withdrawal from its burrow. Indeed, broadly speaking, the entire genus of testaceans have a resemblance to vegetables, if they be contrasted with such animals as are capable of progression." [1]

Theophrastus [2] speaks of certain animals living only in others like those existing in the pinna; in another passage [3] he compares the pearl-oyster of India and the Red Sea to the pinna of the Mediterranean.

[1] Smith and Ross, Works of Aristotle: vol. iv, Historia animalium, by D'Arcy W. Thompson, pp. 547 b, 548 a, 588 b (Oxford, 1910). Aubert and Wimmer, Aristoteles Tierkunde, vol. i, p. 155; vol. ii, pp. 112–115.

[2] De causis plantarum, II, 17, 8 (Opera, ed. Wimmer, p. 215).

[3] De lapidibus, 36 (*Ibid.*, p. 345).

Pliny[1] describes the animal in the manner of Aristotle, emphasizing its parasite (*comes*) called the *pinoteres* or *pinophylax*, — a crustacea that really lives in shells,[2] and, according to the naïve notions of the ancients, helped the pinna toward its food-supply. The pinna, which is without eyesight, opens its shells, which are soon filled by small fish; the vigilant pinoteres gives notice to the pinna at the right moment by a gentle bite; the bivalve closes its shell, kills the captives by this pressure, and divides its booty with the companion. Aristophanes, in his "Wasps" (v. 1511), alluded to this fable; and Aelian[3] reiterates the same as a good story.[4] Neither Pliny nor Aelian, however, alludes to any textile product obtained from the pinna; and the silence of Pliny, who is well informed on the subject of textiles, is particularly significant and conclusive. The origin of pinna textiles is therefore suspected to have taken place, not in the classical world, but in the Hellenistic Orient. The "Periplus Maris Erythraei," written between A.D. 80 and 89, lends color and support to this opinion. This Greek work mentions five times under the name πινικόν the textile obtained from the pinna. It must be remembered that the pinna belongs to the mussels that furnish the genuine pearl; and it is my impression that the same people who were engaged in the business of the pearl-fishery in the Persian Gulf and around Ceylon also hit upon the idea of making the best possible use of the by-product of the filaments. The technique of byssus textiles grew as a side-issue out of the pearl-industry. This is confirmed by the data of the "Periplus," which mentions the pearl-oyster of the Persian Gulf as πινίκιος κόγχος ("pina conch"),[5] and the byssus textiles as πινικόν, being exported from the place, styled the "emporium of Apologus," and from Ommana to Barygaza, the important trading-port in the Dekkan, but inferior to those of India.[6] The πινικόν is likewise a product of Taprobane

---

[1] IX, 42, § 142.

[2] Compare O. Keller, Antike Tierwelt, vol. ii, p. 488. According to Isidorus of Charax (in a fragment preserved by Athenæus, III, 46), this parasite lives also in the mouth of the oyster-shell (see text and translation in W. H. Schoff, Parthian Stations by Isidore of Charax, pp. 10–11, Philadelphia, 1914).

[3] Hist. anim., III, 29.

[4] Also Cicero (De finibus, III, 19; and De natura deorum, II, 48) and Horapollo (Hieroglyphica, II, 108) have noted it (compare J. Beckmann, De historia naturali veterum, p. 239).

[5] § 35 (ed. of Fabricius, p. 74); compare also § 59 (p. 102).

[6] It is wrong, as translated by W. H. Schoff (Periplus of the Erythraean Sea, pp. 36, 46, 47), to speak in this case of "pearls;" for the pearl is called μάργαρον, μαργαρῖτις, etc., and the Periplus itself (§ 56) styles the pearl μαργαρίτης. The word πινικόν, however, is a derivative from πῖνα designating the animal as a species, not any part of it. Certainly the total animal itself was not subjected to exportation, but only those portions useful in mercantile enterprise; that is, the pearls and the byssi or filaments. Consequently the term can but refer to the latter, and denotes either the raw material destined for weaving

(Ceylon; § 61), and the product is traded to a port on the Ganges (§ 63). When and exactly in what locality these textiles were first made, we have no means of ascertaining precisely; but the "Periplus," written at Alexandria toward the end of the first century, contains the earliest conspicuous allusion to their existence, and in general determines their geographical area in the Oriental sphere along the lines of Indo-Persian commerce.[1]   I would not emphasize so strongly, however, the point that fine cloths of this substance were made exclusively in India, as has been done by J. Yates.[2]   Without invalidating or corroborating this inference, we should keep in mind that nothing about such a textile is known to us from India, ancient or modern; and, in view of the deep-rooted Hindu aversion to the taking of animal life, I even have the feeling that a textile secured from an animal, whose death for this purpose was necessarily involved, could not well have been an Indian idea, at least in its origin not a Hindu invention. The unknown author of the "Periplus," not having himself visited India, can hardly be regarded as an authority on Indian subjects, unless his statements may be checked or confirmed by other sources; also his text has been handed down to us in a bad condition, and in many cases is open to doubt and conjecture.   The question of the local origin must therefore be held in abeyance; and its definition, as stated, is to be restricted to the maritime expanse of the Erythrean Sea (bordered by the littorals of Arabia, Persia, and India) rather than extended to any particular territorial or ethnical group.[3]   It is therefore

or the ready-made woven product. Lassen (Indische Altertumskunde, vol. iii, p. 46), Fabricius (p. 77), and Blümner (Technologie, vol. i, 2d ed., p. 204), have decided in favor of the latter, and I concur with them in this opinion.   Fabricius, it is true, is not wholly consistent in his interpretation, for in § 59 he renders κολύμβησις τοῦ πινικοῦ by "capture of pearls," and at the end of this chapter πινικόν by "*Steckmuschel*," whereas J. Yates (in his classical work Textrinum Antiquorum, An Account of the Art of Weaving among the Ancients, p. 158, London, 1843) upholds the meaning of byssus textiles for this very chapter (his interpretation of σινδὼν ἐβαργαρείτις as "fine cloth obtained from shells yielding pearls," of course, is untenable [see Fabricius, p. 104, note 1]). — In the British Museum there are two Greek bronze figures with the head of an Ethiopian, or negro, clasping a pinna which they have just brought up (H. B. Walters, Cat. of the Bronzes in the Dept. of Greek and Roman Antiquities, Brit. Mus., p. 269, Nos. 1674, 1675).

[1] In the Greek papyri the byssus textiles have not yet been pointed out.   We owe to Th. Reil (Beiträge zur Kenntnis des Gewerbes im hellenistischen Ägypten, pp. 116–122) a complete list of the textiles mentioned in the papyri, among which no reference to the pinna occurs.

[2] Textrinum Antiquorum, pp. 157–159, with reference to § 59 of the Periplus.

[3] In fact, none of the Greek writers to be cited presently mentions a locality where the weaving of pinna fibres was carried on.   It has commonly been said that the manufacture took place at Tarentum in southern Italy; but this statement is advanced for no other reason than that the pinna is obtained, and the fabrication principally conducted, at Taranto in modern times.   There is, however, no direct evidence that this place was the seat of the ancient industry.   On the contrary, as set forth above, the evidence available points to the Orient.   There are now two processes of catching the pinna in the Gulf of

logical that we find the first knowledge of this material in Hellas during the second century, where it had meanwhile apparently arrived from the Orient.

The first Greek author to testify to the fabrication of textiles from the pinna fibres is the sophist Alciphron of the second century, who, in the collection of his letters,[1] styles them "woollen stuffs out of the sea" (τὰ ἐκ τῆς θαλάσσης ἔρια).[2] The principal wool-furnishing animal of the ancients was the sheep; and the term used by Alciphron is either the index of a belief existing at that time in a marine sheep that furnished the wool of the pinna, or directly responsible for the formation of such a notion. The same idea turns up in Father Tertullian (born about A.D. 160; according to others, *circa* 155–*circa* 222), who, speaking of the materials used in weaving, observes, "Nor was it enough to comb and to sow the materials for a tunic. It was necessary also to fish for one's dress; for fleeces are obtained from the sea where shells of extraordinary size are furnished with tufts of mossy hair."[3] The Chinese terms "water-sheep" and "cloth from the west of the sea" (or "cloth from within the sea") and the Arabic designation "wool of the sea"[4] are immediately to be connected with the descriptions of Alciphron and Tertullian, and present the outflow of that Hellenistic tradition which inspired their statements. The water-sheep of the

Taranto, — by diving and by fishing. The latter method is performed by means of the *pernonico*, which consists of two semicircular bars of iron fastened together at the ends. At one end is a wooden pole; at the other end, a ring and cord. The fishermen bring their boat over the place where the pinna is seen through the clear water, let down the *pernonico*, and, having loosened the pinna by embracing it with the iron bars and twisting it round, draw it up to the boat (compare J. Yates, Textrinum Antiquorum, pp. 152–154). According to P. Petròcchi (Novo dizionario della lingua italiana, vol. ii, p. 316, Milano, 1902), large quantities of the filaments are gathered on the coasts of Sardinia, under the name *nàcchera* or *pelo di nàcchera*.

[1] *Epistolae*, I, 2, 3 (Hercher, Epistolographi Graeci, p. 44, Paris, 1873).

[2] Compare Blümner, Technologie, vol. i, 2d ed., p. 204; and O. Keller, Antike Tierwelt, vol. ii, p. 549. In the ancient Glossaries we find πίννινον in the sense of *marinum*, that is, *vellus marinum* ("sea-wool"); and *pinnosum* =*laniosum*.

[3] Nec fuit satis tunicam pangere et serere, ni etiam piscari vestitum contigisset: nam et de mari vellera, quo mucosae lanusitatis plautiores conchae comant (Liber de pallio, III, Patrologia latina, ed Migne, vol. ii, col. 1093). I have adopted the translation of J. Yates, Textrinum Antiquorum, p. 155. Tertullian's treatise De pallio contains a defence of his wearing the pallium instead of the toga, and belongs to the group of his works which were written later than the year 208.

[4] This term is certainly older than the time of Ibn al-Baiṭār (1197–1248), who merely was a compiler and translator, and who derives his notes on the pinna from "the book called 'er-Rihla.'" Rihla (that is, "The Voyage") was the work of al-Baiṭār's teacher, Abu'l Abbās, styled en-Nebāti ("the Botanist"), born in Sevilla, where he died in 1239. He traversed Spain as a collecting botanist, extended his excursions into Arabia, Syria, and Irak, and laid down the results of his explorations in the work mentioned, which is unfortunately lost, and only preserved in the citations compiled from it by al-Baiṭār (see the introduction of L. Leclerc, Traité des simples, vol. i, p. v).

Chinese records is by no means a Chinese invention, but the spontaneous reproduction of a popular term current in the Hellenistic Orient. It was there that the raw material employed in the textile products yielded by the pinna filaments was styled "water (or marine) sheep," or "marine wool," — a mental process suggested by the same spirit that nicknamed "goats" the close-textured sponges which are particularly hard and rough.[1] The Italians still call the fibres *lana pesce* or *lana penna;* that is, "fish wool," or "pinna wool."

Basilius the Great (Basilios Megas, 329 or 331–379), Bishop of Caesarea in Cappadocia, in one of his homilies, dilates on the wonders of the sea, pointing to the coral which grows in the water as an herb, but, taken up into the air, assumes the solidity of stone; and to the pearl which is hidden in an animal of low order, yet is craved by the treasuries of kings, the oyster-shells being scattered around along coasts and rough rocks. On this occasion he speaks also with admiration of "the pinna's raising a golden fleece which none of the dyers was hitherto able to imitate."[2] Another Greek ecclesiastic writer[3] even says that the product of the pinna is superior to sheep-wool. The Byzantine historian Procopius of the sixth century, in his work "Ctismata," dealing with the buildings executed or restored by the Emperor Justinian,[4] informs us that Armenia was governed by five hereditary satraps, who received their insignia from the Roman Emperor. Among these was a chlamys made from wool, — not from the wool, however, obtained from sheep, but from wool gathered out of the sea. The animals in which the outgrowth of the wool originates are usually styled *pinnoi*.[5] Accordingly the notion of marine fleece, and comparison of it with sheep-wool, were constantly awake in the minds of Greek authors. The description of the wool as "gold-colored" by Basilius answers the facts.[6]

Of Arabic authors, we owe the most interesting description of the pinna to Abu'l Abbās, to whom reference has already been made. This author, though to a certain extent under the influence of Greek tradition, as shown by his term "marine wool," evidently speaks from personal observation enriched by information gathered during his travels. We shall revert to his account later, in another connection.

---

[1] Aristotle, Hist. anim., v, 16 (fol. 548 b).

[2] Πόθεν τὸ χρυσοῦν ἔριον αἱ πίνναι τρέφουσιν, ὅπερ οὐδεὶς τῶν ἀνθοβαφῶν μέχρι νῦν ἐμιμήσατο (Homilia VII in Hexaemeron; Patrologia, ed. Migne, vol. xxix, col. 161).

[3] Cited by Blümner, Technologie, vol. i, 2d ed., p. 204, note 8.

[4] Περὶ κτισμάτων, III, I (written after 558).

[5] Χλαμὺς ἡ ἐξ ἐρίων πεποιημένη, οὐχ οἷα τῶν προβατίων ἐκπέφυκεν, ἀλλ' ἐκ θαλάσσης συνειλεγμένων· πίννους τὰ ζῷα καλεῖν νενομίκασι, ἐν οἷς ἡ τῶν ἐρίων ἔκφυσις γίνεται.

[6] There is a muff of dark gold color, made from byssus-fibres at Taranto, in the collections of the Field Museum; also a pair of gloves and a cap knitted from the same material. The latter specimens have a dull cinnamon-brown color, without gloss.

The oldest Arabic account of byssus textiles, already pointed out by Chavannes, is that of Iṣṭakhri, who wrote about 951. His story, according to M. Reinaud's[1] translation, is worded as follows: "At a certain period of the year an animal is seen running out of the sea and rubbing itself against certain stones of the littoral, whereupon it deposes a kind of wool of silken hue and golden color.[2] This wool is very rare and highly esteemed, and nothing of it is allowed to waste. It is gathered and serves for the weaving of tissues that are now dyed in various tinges. The Ommayad princes who then ruled at Cordova reserved for themselves the use of this wool; only surreptitiously a small portion of it may be abstracted. A robe made of this wool costs more than a thousand gold-pieces." The same story is repeated by Qazwīnī (1203–83), who localizes it at Santarem, a city in Spain on the Tajo, near Bāga on the coast of the sea: "One of the wonders of this sea is what is told regarding a certain animal which there comes out of the water to rub itself on the shore, whereby its hair falls out; these have the color of gold and the softness of *khezz*.[3] These are rare and highly esteemed, for which reason the people gather them and weave them into clothes. The kings prevent their exportation, which can be done but secretly. The value of a garment amounts to more than a thousand gold-pieces owing to its beauty and rarity." Maqdisī has exactly the same notice as Qazwīnī, but adds a new name for the animal in the form *abū qalamūn*, which is derived from Middle-Greek ὑπο-κάλαμον, and says that the garments glitter in different colors on the same day.[4]

The most curious development of the Arabic notions regarding byssus textiles was that these were ultimately taken for the plumage of a bird,

[1] Géographie d'Aboulféda, vol. ii, pt. 2, p. 242. The text is in De Goeje, Bibl. Geogr. Arab., p. 42; it has been translated also by Dózy, Supplément des dict. arabes, p. 853.

[2] This, of course, is a fabulous story, the *raison d'être* of which will be discussed below. In fact, the shells must be opened, and the filaments are cut off from the gland. When the bottom of the sea is sandy, the shell with its bunch of silky fibres may easily be extracted; but in rushy and muddy sea-bottoms they stick so fast as to be generally broken in being drawn up. In Italy the "wool" is twice washed in tepid water, once in soap and water, and again in tepid water, then spread on a table to dry. While yet moist, it is rubbed and separated with the hand, and again spread on the table. When quite dry, a wide comb of bone is drawn through it; afterwards this process is repeated with a narrow comb. The material destined for very fine work is combed also with iron combs called *scarde* (cards). It is then spun with a distaff and spindle. The threads are now almost universally knit, a technique unknown to the ancients (compare J. Yates, Textrinum Antiquorum, pp. 154–155).

[3] According to G. Jacob (Handelsartikel, pp. 45–47), furs of the beaver, and also the name for a silken material. It seems to me that this word is the result of a fusion of two originally different words (compare Hindustani *kesh* ["hair"] and *khaz* ["filoselle silk"] and H. Blochmann's note in his translation of Ain I Akbari [vol. i, p. 92]).

[4] G. Jacob, Studien in arabischen Geographen, vol. ii, pp. 60, 61. The Arabic-Greek word is evidently connected with the name "chameleon."

and that a bird species was construed which was alleged to yield the product of the pinna. Qazwīnī opens his chapter on ornithology with the description of a bird, styled *abū barāqish*, "being of fine shape, of long neck and feet, with a red bill, and of the size of a stork; every hour its plumage glitters in another color, — red, yellow, green, blue. In imitation of the color of this bird are woven garments styled *abū qalamūn* and exported from the land of the Romaei. Only for its color and shape this bird is noteworthy; of its functions and the medical properties of its parts nothing has come to my knowledge."[1] It is no wonder that, as said by Jacob, even Damīrī did not know what kind of bird should be understood by *abū barāqish;*[2] for, in my opinion, this bird is plainly fictitious, and reconstructed on the basis of real and alleged byssus textiles. How and why this was accomplished is obvious also. There are linguistic and commercial reasons for this metamorphosis. The word *pinna* (properly *pina*), the name for the bivalve in question, is likewise the classical Latin form for the subsequent word *penna* ("feather"),[3] and this ambiguity may have given rise among the Arabs to the conception of the filaments of the pinna as bird-plumage, — a conception easily furthered by the strong mutual resemblance of the two substances. Abu'l Abbās,[4] in his description of the pinna, says that it terminates in a point resembling the beak of a bird. On the other hand, as stated by Qazwīnī, textiles obtained from the pinna were exceedingly scarce, made stealthily, and were a sort of royal prerogative. Their exorbitant price was prohibitive to the masses. Feather fabrics were accordingly passed off as byssus weavings, and a wonderful bird was invented to boom the sale of this product. The real existence of such feather fabrics in western Asia is attested by Chinese sources.[5] Such makeshifts must have been in vogue as

---

[1] G. Jacob, Studien in arabischen Geographen, vol. ii, p. 97.

[2] Damīrī says that it is a certain bird like the sparrow, assuming various colors, and that it is applied to a changing and variable disposition (A. S. G. Jayakar, Ad-Damīrī's Hayāt al-Hayawān, vol. i, p. 352, Bombay, 1906). This description is difficult to reconcile with Qazwīnī's stork.

[3] In modern Italian the words *penna* and *pinna* are interchangeable.

[4] L. Leclerc, Traité des simples, vol. i, p. 387.

[5] The Arabic word *suf* ("wool" or "down") that we met in the term *suf el-bahr* ("marine wool") for the byssus of the pinna, passed from the days of the Mongol period into the Chinese language in the form *su-fu* or *so-fu* (variously written; see Watters, Essays on the Chinese Language, p. 355). In the Annals of the Yüan Dynasty (Yüan shi, ch. 78) it is mentioned as the cloth worn by the grandsons of the sovereign, and described as the finest of the woollen fabrics of the Mohammedans. The Geography of the Ming (Ta Ming i t'ung chi, ch. 89, fol. 24 a, ed. of 1461) defines *so-fu* as a textile made from bird's-down with designs as found in open-work, variegated silk (compare Bretschneider, Mediæval Researches, vol. ii, p. 258). An author, Chu Tsê-min, ascribes *so-fu* also to the country Fu-lin (Syria), saying that it is made from twisted hair which is dyed a dull green, and that on being washed it does not fade out (Ko chi king yüan, ch. 27, p. 16 b). *So-fu* was sent to China from Samarkand in 1392, from Ispahan in 1483, and from Lu-mi (Rum, Byzance)

early as the ninth century, in the time of Iṣṭakhri; for this author's statement that the pinna textures were then dyed in various colors is highly suspicious. A genuine pinna stuff would most assuredly not have been subjected to this vandalizing process, apt to destroy its original appearance. The Greek authors insist on the golden color and the silky quality of the byssus of the pinna, and these properties constituted the merit of the fabric for the sake of which it was craved. Basilius the Great, Bishop of Caesarea in the fourth century, accentuated the fact that none of the dyers could imitate the golden wool raised by the pinna; and a Syriac work wrongly ascribed to Aristotle, dealing with objects of natural history and partially based on Basilius' writings, says still more explicitly that "there are no dyers so clever in their work that they could accomplish something similar after the model of the colors of the pinna."[1] These passages show that from

in 1548 and 1554 (Bretschneider, *l.c.*, pp. 258, 291, 308). The feather fabrics *suf*, therefore, seem to have been in vogue in the Byzantine Empire and Persia. Dr. A. Yohannan, lecturer at Columbia University (a Persian by birth), told me that he himself had seen in Persia the manufacture of these textiles from bird's-down. The same industry is met with among the tribes of the Hindu Kush. We owe this information to J. Biddulph (Tribes of the Hindoo Koosh, p. 74, Calcutta, 1880): "A curious kind of cloth is sometimes woven out of bird's-down. That of wild fowl and of the great vulture is most generally used. The down is twisted into coarse thread, which is then woven like ordinary cloth. Robes made of it are very warm, but always have a fluffy uncomfortable look, suggestive of dirt. They are made only in the houses of those in good circumstances." It should not be supposed, however, that the Chinese made the first acquaintance with feather fabrics in consequence of their trade with Arabs and Persians. Such were indeed manufactured in China from ancient times, though we are ignorant of the technique employed, which may have been different from that practised in western Asia. In a study of asbestos and the salamander (to be published in the T'oung Pao) the writer has shown that this industry played a signal *rôle* also among the aboriginal tribes of southern China. In view of the fact that it is widely distributed in ancient America, it would be an important task to study in detail the exact history and the geographical and ethnographical diffusion of the industry in Asia (my reference, of course, is strictly applied to the use of feathers for weavings, not for mosaics or any other ornamental purpose). For the benefit of Orientalists not familiar with the literature on America, the following brief indications may serve as an aid to preliminary information. Franz Boas (Second General Report on the Indians of British Columbia, p. 14, in Sixth Report on the North-Western Tribes of Canada, 1890) states, in regard to the Lkuñgen tribe on Vancouver Island, "Blankets are woven of mountain-goat wool, dog-hair, and duck-down mixed with dog-hair. The downs are peeled, the quill being removed, after which the downs are mixed with dog-hair. A variety of dogs with long white hair was raised for this purpose; it has been extinct for some time. The hair which is to be spun is first prepared with pipe-clay." W. H. Holmes (Prehistoric Textile Art, Thirteenth Annual Report, Bureau of Ethnology, p. 27) observes, "Feather work was one of the most remarkable arts of the natives of Mexico and other southern countries at the period of the conquest. The feathers were sometimes woven in with the woof and sometimes applied to a network base after the fashion of embroidery. Rarely, it may be imagined, were either spun or unspun fabrics woven of feathers alone." Compare further W. Hough, Culture of the Ancient Pueblos of the Upper Gila River Region (U. S. Nat. Mus. Bull. 87, pp. 71–72, Washington, 1914).

[1] Syriac *pūnōs*. See K. Ahrens, Buch der Naturgegenstände, p. 75.

the fourth century onward dyers had indeed attempted to produce imitation pinna stuffs, but that their efforts were unsuccessful; certainly they did not utilize byssus in these experiments, but some other inferior fabric of a similar appearance. In the ninth century these reproductions had evidently advanced beyond the experimental stage, and deluded the public. The dyed byssus fabrics mentioned by Iṣṭakhri, indeed, are makeshifts, and as shown by Qazwīnī, in all likelihood, must have been textiles woven from bird's feathers. This is borne out also by Maqdisī's statement that the garments glitter in different colors on the same day, which is true only of feather fabrics, not, however, of byssus textiles. The latter do not glitter at all, but have a uniform gold-brown or dull-cinnamon hue. The fact that woven bird's-plumage represents a very close resemblance to pinna tissues may be gauged from Chinese descriptions of feather weavings, in which almost the same descriptive elements are used as by the Arabic authors in their references to pinna. A few examples may be cited from Chinese records. In the period Shang-yüan (674–676) of the T'ang dynasty, the Princess Ngan-lo[1] had two skirts made in the Shang-fang.[2] They were woven from the down of various kinds of birds. When viewed in front, the weaving presented a definite color; when viewed sideways, another color; when viewed in the sunlight, again another color; and when viewed in the shade, again a diverse color; while the forms of the various birds were visible in the skirts. One of these she presented to the Empress Wei.[3] The "Lang hüan ki," a work of the Mongol period, contains the following: "Phœnix-feather gold (*fêng mao kin*) means the feathers growing beneath the neck of the phœnix; they are like ribbons and glittering like gold, being matchless and as fine and soft as silk floss. In the spring the feathers drop to the foot of the mountains. The people gather them and weave them into gold brocade that bears the name 'phœnix-feather gold.' At the time of the Emperor Ming (713–755) people of the country brought such feathers as tribute, and many garments were adorned with them in the palace; at night they emitted a brilliant light. Only Yang Kuei-fei[4] was presented with a sufficient quantity to have them made into a dress and a screen, dazzling like sunlight."[5]

[1] A daughter of the Emperor Chung-tsung; she died in 710 (Giles, Biographical Dictionary, p. 3).

[2] The imperial factories supplying the wants of the reigning house.

[3] Kiu T'ang shu, ch. 37, p. 13.

[4] The favorite court-lady of the Emperor Ming, who died in 756 (Giles, Biographical Dictionary, p. 708).

[5] The text is in T'u shu tsi ch'êng, IV, 197, kung hien pu ki shi 3, p. 1 b. — D. J. Macgowan (American Journal of Science and Arts, 2d ser., vol. xviii, 1854, p. 156) mentions women's jackets composed of the feathered head-skins of peacocks, made in Shen-si. He describes the prevailing tints of these garments as green and blue, of resplendent metallic lustre, of varying intensity, mutually changing into each other, or shotted according as the light falls upon them in different directions.

In the Annals of the T'ang Dynasty (618–906) we meet another tradition, which at first sight is widely different from the older story of the water-sheep, but on closer examination proves to be an interesting continuation or further development of it. This new tradition hailed from the country of Fu-lin (Syria, with the probable inclusion of Byzance), as the former came from Ta Ts'in, the Hellenistic Orient, and is worded as follows: "There are lambs engendered in the soil. The inhabitants wait till they are going to sprout, and then build enclosures around as a preventive measure for wild beasts that might rush in from outside to devour them.[1] The umbilical cord of the lambs is attached to the soil, and when forcibly cut off, they will die. The people donning cuirasses and mounted on horseback beat drums to frighten them. The lambs shriek from fear, and thus their umbilical cord is ruptured. Thereupon they set out in search of water and pasture."[2]

Chavannes[3] has been so fortunate as to discover an earlier version of this legend in the commentary which Chang Shou-tsie published in 737 on the historical memoirs of Se-ma Ts'ien. This author cites the "I wu chi" of Sung Ying as follows: "In the north of Ts'in, in a small canton dependent on it, there are lambs spontaneously engendered in the soil. Awaiting the moment when they are ready to sprout, the people build enclosures around them, for fear lest they might be devoured by wild beasts. Their umbilical cord is attached to the ground, and its forcible cutting will cause the animal's death. Instruments are therefore beaten to frighten the lambs which shriek in terror, so that the umbilical cord breaks. Thereupon they set out in search of water and pasture, and form herds." This version has doubtless emanated from the same source as that of the Old T'ang Annals, with which it substantially agrees, except that the equestrian

[1] "Shepherds in the East lead a lonely and romantic life. They wander with their flocks far from human habitations, in order to bring them to pasture, and also because it is necessary for them to watch over them by night, to protect them from wild beasts. The sheep are usually on these occasions driven into a fold which is merely a space enclosed with a loose stone wall. Sometimes, where possible, a cave is selected. A doorway is formed in the boundary wall where one exists" (H. C. Hart, Animals mentioned in the Bible, p. 196). In the same manner the sheepcotes of the ancient Israelites appear to have been open enclosures walled round, in which the sheep were guarded from the scorching heat at noon and from beasts of prey at night (Numbers XXXII.16; 2 Samuel VII.8; Jeremiah XXIII.3; John X.1–5).

[2] Kiu T'ang shu, ch. 198, p. 12. In the New Annals of the T'ang (Sin T'ang shu, ch. 221 B, p. 8) the following version is given: "In the northern districts there are sheep growing in the soil, their umbilical cord rooting in the ground and causing their death when cut. It is therefore the practice to gallop around on caparisoned horses and to frighten the animals by beating drums. Their umbilical cord is thus ruptured, and they set out in search of water and pasture, without being able, however, to form flocks (or, they are not gregarious)."

[3] T'oung Pao, 1907, p. 183.

feat of the armored shepherds is lacking. Further, the locality is not laid in Fu-lin, but in the north of Ts'in. Obviously we have to make a slight emendation in the text, and to read "Ta Ts'in" in lieu of plain "Ts'in," which would consequently carry this version also into western Asia. That this conjecture is correct, is visible from two other texts. Ma Tuan-lin has reproduced the passage of Chang Shou-tsie, and arrayed it in the chapter on Ta Ts'in:[1] consequently Ma Tuan-lin must have encountered the reading "Ta Ts'in" in the edition of Chang which was before him.[2] Further, the "Pei hu lu," written by Tuan Kung-lu about 875,[3] explicitly naturalizes the same story in Ta Ts'in.[4] It is therefore possible that the oldest version of the legend, when it first penetrated into China, was labelled as originating from Ta Ts'in; that is to say, that it was transmitted to China before the beginning of the sixth century, when the name "Fu-lin" made its début.

I propose to examine this curious legend without any bias toward speculations which have previously been advanced. It is obvious that any rationalistic explanation evolved from our mind cannot render it justice, but that it must be explained from the thought developments of Ta Ts'in and Fu-lin. The failure of the former efforts is chiefly due to the neglect of this regard to cultural environment. The understanding of an idea generated in Ta Ts'in or Fu-lin cannot be approached by having recourse to a rumor of mediæval travellers, or still more recent authors, pertaining to totally different localities.

The student of folk-lore and the trained observer will be conscious of two points, — first that the germ of a fact or observation relative to natural history underlies the legend; and, second, that, as not all its constituents can satisfactorily be explained from natural events, it must have been construed with a certain end in view, which may have an allegorical purport or religious cause. Let us first discuss the zoölogical background. It is the question of a certain peculiar kind of

---

[1] Hirth, China and the Roman Orient, pp. 79, 115.

[2] As Ma Tuan-lin joined this story to his chapter on Ta Ts'in, he naturally suppressed the addition "Ta Ts'in" in the beginning of the story, but otherwise opened it exactly in Chang's words, — "in a small canton dependent on it in the north." The only divergences in Ma Tuan-lin's text are the omission of the phrase that the lambs shriek in terror, and the alteration at the end, "they do not form herds," — the latter point in agreement with the text in the Sin T'ang shu.

[3] Pelliot, Bull. de l'Ecole française, vol. ix, p. 223.

[4] The version of this work, which is in T'u shu tsi ch'êng (section on sheep, *hui k'ao* 2, p. 16 b), has heretofore not been utilized for the study of the legend. Besides the specific definition of Ta Ts'in, it has another interesting feature, inasmuch as it entitles the animal "earth-born sheep" (*ti shêng yang*) from which the lamb originates. The text runs thus: "In Ta Ts'in there is the earth-born sheep. Its lamb is born in the earth. The inhabitants build enclosures all around the lambs. Their umbilical cord is attached to the soil, and when forcibly cut, the animal will die. By means of equestrian stunts and drum-beating they frighten the lambs, that shriek from fear when their umbilical cord breaks off. Thereupon they set out in search of water and pasture."

lamb[1] (the word is used advisedly) characteristic of Fu-lin (Syria), and formerly also of Ta Ts'in (the Roman or Hellenistic Orient). The growth of this lamb is described in terms referring partly to a plant and partly to an animal. The primordial generation in the soil evidently is derived from the planting of a seed.[2] The word *mêng* ("to sprout, shoot forth") used in the Chinese text is exclusively employed in regard to vegetation, never to fauna. Fields as well as flocks may be safeguarded by fences, but only the latter for protection from raids of wild beasts, that as a rule are not interested in the crops. Again, the umbilical cord is an animal organ, and plants are not impressed by the beating of drums. From that act of release onward, the creature retains its pure animal character to the end. We need not for a moment trouble our thoughts about the question of the "to be or not to be" in nature, of such a being. This point of view is immaterial; while the issue at stake is whether a zoöphyte of this peculiar character and description existed in the scientific knowledge or popular lore of the Hellenistic Orient. Indeed, it existed, and has already been introduced to us by Aristotle, in his "History of Animals" (VII, 1) quoted *in extenso* on p. 106. In this passage the father of all zoölogical science dilates on the boundary-lines between plant and animal life, where the plant ascends toward the animal, and the animal descends toward the plant. At this point, according to Aristotle, it is difficult to discriminate with absolute certainty between animal and plant; and he cites as illustration of this doctrine the example of the pinna, which, devoid of motion, is rooted like a plant to a fixed spot, and must perish when detached from its intrenchment. That the pinna was conceived during the Hellenistic epoch as a wool-furnishing sheep, has already been demonstrated with sufficient evidence from both the Hellenic and Chinese camp. Thus we are enabled to grasp an essential point of our legend: the lamb engendered in the soil and firmly attached to it by means of its umbilical cord, which when forcibly cut off will cause the animal's death, represents a metamorphosis of the biological condition of the pinna, as described by Aristotle, — the umbilical cord which befits a mammal taking the place and being the transformation of the byssus.[3] It is needless to insist on the fact that Aristotle was the great

[1] Only the Sin T'ang shu speaks of sheep.

[2] The verb *shêng* of course is not conclusive, as it is used with reference to both plants and animals. Hirth and Schlegel take it in the sense of "to grow," which is not necessary; Chavannes more correctly translates "*naissent dans le sol.*" The word plainly refers to the very initial stage in the formation of the organism; Pliny would say in this case "*nascuntur in terra.*"

[3] There is accordingly a positive historical interrelation of the water-sheep of old and the vegetable lamb, which Chavannes (T'oung Pao, 1907, p. 183) has denied, merely on the ground that in the case of the latter the question is never of water. The lack of the attribute "water," however, does not constitute a fundamental or characteristic diver-

universal teacher of natural history to all subsequent generations, and that his works translated into Arabic were worshipped like a fetich in the Orient.[1] How the further elements of the legend were formed we are allowed to recognize from the accounts of the Arabs. We remember that Iṣṭakhri and Qazwīnī relate the story regarding the pinna, that at a certain time of the year it comes out of the sea and deposits its wool by rubbing itself against the rocks of the shore. Consequently the belief prevailed that the pinna was not deprived of its

gence, but is merely a chronological difference due to the further development of the legend. In the Hellenistic stage of development correlative with the Han epoch the matter was still fairly rational, the pinna being regarded as the water-sheep, in the manner rather of a metaphorical expression than of a palpably convincing notion of reality. Yet beliefs spread and grow, and in the fifth or sixth century the basic origin was forgotten; the water-sheep, owing to its equipment with a navel, the seat of its life, then could no longer be believed to exist in the sea, but was wrested from the watery element to be transplanted into solid land and to grow into a veritable, full-fledged ovine species equipped with phenomena of plant-growth. According to the nature-philosophy of the ancients, there was no difficulty in associating an umbilical cord with the life of plants: not only was this organ compared with the root of a plant, but also the stalks of tree-fruits, particularly the figs, and the germs of seeds were straightway called ὀμφαλός or *umbilicus* (the evidence is collected by W. H. Roscher, "Omphalos," pp. 7–8, Abhandl. sächs. Ges. d. Wiss., vol. xxix, No. 4, 1913; and R. Meringer, Wörter und Sachen, vol. v, 1913, p. 63; compare also the same journal, vol. vi, 1914, p. 144; both Roscher and Meringer, in their admirable studies of Omphalos, have neglected the legend in question, which we trust will furnish them with additional material in the prosecution of their highly interesting researches). On the other hand, Aristotle (Hist. anim., 1, 54) designates the animal Omphalus as the "root of the abdomen" (μετὰ δὲ τὸν θώρακα ἐν τοῖς προσθίοις γαστήρ, καὶ ταύτης ῥίζα ὀμφαλός). There is a still deeper reason to be discussed below as to why the water-sheep was ultimately transformed into a land-animal.

[1] In general, compare the interesting essay of William M. Sloane, "Aristotle and the Arabs" (Classical Studies in Honor of Henry Drisler, pp. 257–268, New York, 1894). It has already been demonstrated by H. E. Stapleton, one of the most successful students of Arabic alchemy (in his treatise "Sal-Ammoniac: a Study in Primitive Chemistry," Mem. As. Soc. of Bengal, vol. i, 1905, pp. 28, 36), that one of the essential features of this science, inherited from Greek alchemy, was the re-establishment of a belief in the strong interrelation of animals, plants, and minerals, in the paramount unity of the world of nature. "No strict line of demarcation separated plants and minerals from animals and man; all were looked upon as closely related units of a single whole." Stapleton quotes two characteristic examples from Berthelot's La Chimie au moyen âge. A Syro-Arabic text of the tenth or eleventh century says, "We can bring it about that a vegetable turns into an animal, and that an animal produces another animal. Take, for example, hair. When human hair putrefies, after a time it becomes a live snake. In the same way, the flesh of an ox changes into bees and hornets; an egg beomes a dragon; the raven engenders flies. Many things, by the process of putrefaction and transformation, engender animal species. From the putrefaction of plants originate certain animals." According to the Arabic alchemist Tughrāī, who died in 1121, seeds are produced by planting the horns of hoofed animals. Still older examples are found in the Kitāb al-Hayawān of al-Gāhiz, who died in 869; he discussed the origin of flies from beans, vermin from ordure, wasps from the marrow of palms, etc. (E. Wiedemann, "Zur Alchemie bei den Arabern," Journal für praktische Chemie, vol. 76, 1907, p. 73).

byssi through human agency, but voluntarily abandoned them, thus saving its own life. For another and still more specific statement of the case we are indebted to the Arabic botanist and traveller Abu'l Abbās, who died in 1239 at Sevilla, and who says in his work "Rihla," [1] "The inhabitants of the shores where the pinna is caught told me that a marine animal, a crustacea, captures this mollusk; that it spies the latter in the low water as soon as the pinna lets its wool escape; that it then pounces down upon the pinna and subsists on it to the exclusion of every other animal." This story opens our eyes to another feature of the Chinese legend: the frightening of the lamb on the part of men who don cuirasses with the intention of enforcing the rupture of the lamb's umbilical cord through a psychological process operating in the lamb's mind. In the original animal fable these cuirassed men were crustacea, the shelly crusts of which were subsequently transmuted into cuirasses; they terrified the pinna, which, taken aback at the sight of the enemy, dropped its byssi. These byssi drifted ashore, where they were picked up by men for the purpose known to us. The essence of the Chinese story, as far as it is originally founded on a pure animal fable, is therefore not difficult to reconstruct: it is based on the alleged struggle between pinna and crab, combined with Aristotle's discussion of the pinna's biological functions. In the Chinese version, moreover, the idea crops out that the wool of the dead lamb is useless, that while alive the lamb must be shorn. The story as recorded by the Chinese, certainly, — and in view of the accuracy of the Chinese we have no reason to question this point, — is an exact reproduction of the legend as it was current in the Orient. If the pinna was there identified with a sheep or lamb, it was entirely natural that the belief should develop that byssus-wool, in like manner as sheep-wool, could not be secured from the slain animal; and the animal, to the way of thinking in that community, would have been killed by the act of depriving it of its wool, the wool being the same as the byssus identified with an umbilical cord. For this reason it was necessary to devise a process by which the creature could be induced to give up the prized wool of its own accord; and this *rôle*, in popular imagination, was assigned to the crab. The Chinese legend, as recorded in the T'ang Annals, is therefore capable of the following retranslation or re-interpretation: "A peculiar animal of Fu-lin is the pinna (lamb), whose life is bound to the soil. The inhabitants wait till the animal, which has the nature of a plant and is devoid of motion, is going to sprout, and guard it by enclosures from attacks of rapacious beasts. The byssus (umbilical cord) of the pinna (lamb) is firmly rooted in the ground; and when forcibly detached, the animal will die. It is much terrorized by the crab, which hunts it for food. At the sight of this armored adversary,

[1] Quoted by Ibn al-Baiṭār (L. Leclerc, Traité des simples, vol. ii, p. 386).

the pinna, stricken with fright, sheds its byssi, which in this manner do not lose their vitality. The byssus-wool thus drifts ashore, where it is gathered by men to be woven into cloth." Now, the further development was that the pinna-lamb, when once rescued from the sea, was finally landed as a realistic lamb, whose wool was directly craved by men: so man remained no longer a mere looker-on, but actively took a hand in the game and elicited the wool. Our Chinese version of course is incomplete, or perhaps merely forgetful, in not alluding to the utilization of the wool; but this is certainly the purport of the musical performance. The animal is liberated from its vegetal existence and becomes a live lamb able to roam about for water and pastures; and then, certainly, man would shear it to secure its wool.

We have noted that the pinna of old was transformed into a sheep, a lamb, and even a bird; but this is not all. It was even conceived as a human being, and an intimation to this effect is given in the Talmud.[1] In the Mishna *Kilaim* (VIII, 5), a portion of the Talmud, we meet the passage, "Creatures called *adne sadeh* ('lords of the field') are regarded as beasts." Rabbi Simeon, who died about 1235, comments on this statement as follows: "It is asserted in the Jerusalem Talmud that this creature is the 'man of the mountain.' It draws its food out of the soil by means of the umbilical cord: if its navel be cut, it cannot live. Rabbi Meir, the son of Kallonymos of Speyer, has added these remarks: 'There is an animal styled *Yedua*,[2] with the bones of which witchcraft is practised. It issues from the earth like the stem of a plant, just as a gourd. In all respects, the *yedua* has human form in face, body, hands, and feet. No creature can approach within the tether of the stem, for it seizes and kills all. As far as the stem (or umbilical cord) stretches, it devours the herbage all around. Whoever is intent on capturing this animal must not approach it, but tear at the cord until it is ruptured, whereupon the animal soon dies.'" The coincidence of this legend with that of the Chinese is very striking, but the novel feature cropping out in the Palestinian Talmud is the identification of the strange creature with a human being, the "man of the mountain." Who is this mysterious man of the mountain?

The Chinese version of the legend hailed from Syria (Fu-lin). At the time when it was learned by the Chinese, Syria was a Christian country, and the guess therefore is plausible that the old Hellenistic story of the water-sheep had been modified there under the influence of Christian allegory. The most surprising alteration of the Syrian

---

[1] The Talmudic texts, on the ground of information furnished by H. Adler, have been reproduced by H. Lee (The Vegetable Lamb of Tartary, pp. 6–8, London, 1887), to whose work we shall come back. The same material had already received intelligent discussion from L. Lewysohn (Zoologie des Talmuds, pp. 65, 356–358, Frankfurt, 1858).

[2] According to the nature of Hebrew writing, in which only the consonants are fixed, the vocalization of this word, of course, is uncertain.

redaction is the substitution of the lamb for the sheep; and the Chinese term *yang kao* is so specific and intentionally chosen, that the Chinese without any doubt have reproduced correctly and exactly what Syrian tradition intended. The lamb among Christendom was the symbol of the Savior, Agnus Dei (John 1.29); and the lamb that according to the Talmud is the "man of the mountain" unquestionably represents an allusion to the "Divine Lamb standing on Mount Sion" (καὶ εἶδον, καὶ ἰδοὺ ἀρνίον ἑστηκὸς ἐπὶ τὸ ὄρος Σιών. — Revelation XIV.1). Thus the Lamb is represented in Christian art from the fourth century onward.[1] While this symbolism may well be hidden under the story of the Syrian Lamb, it is obvious, on the other hand, that it is incapable of explaining in full the whole gist of the legend. It is inconceivable that Christ should have been conceived as a lamb immovably rooting in the soil, and liberated by the action of the mounted shepherds. It remains to be considered that prior to the fourth century it was not the person of the Savior who was represented under the figure of the lamb, but that it was the faithful who were thus depicted,[2] either as the retinue of the Good Pastor, or enjoying the delights of Paradise after their salvation. This affords a satisfactory clew to the understanding of the Christian symbolism associated with our legend in Syria. The lambs attached with their umbilical cord to the ground are Christian devotees who still cling to earthly pleasures, Christians during their temporary passage or pilgrimage through this world. They are threatened by rapacious beasts, wolfish devils of temptation. The good shepherd guards his lambs by a protecting wall, but their final salvation must come through their own will and effort. The mounted and armored horsemen awakening and rousing them symbolize the Last Judgment.[3] The connection of the lambs with this earth is severed, their earthly existence ceases, to be crowned by their resurrection and ultimate redemption in the Heav-

---

[1] M. Laurent, L'Art chrétien primitif, vol. i, p. 152; vol. ii, p. 162, and Plate LXIV, Fig. 3. A. N. Didron, Christian Iconography, vol. i, pp. 318–344. The Sixth Council of Constantinople forbade the representation of Christ as a lamb (O. M. Dalton, Byzantine Art, p. 158).

[2] Matthew XXV.32; John X.1–5. The notion is traceable to the Old Testament, where the people of God are styled his "sheep" (1 Kings XXII.17; Psalms LXXIX.13; LXXX.1).

[3] Compare Revelation IX.17 (the armored horsemen) and VIII. 6 (the trumpet-blowing angels). The concatenation of the lambs with Judgment was presumably elicited or at least supported by the passage in Jeremiah (XXIII.3–5): "And I will gather the remnant of my flock out of all countries whither I have driven them, and will bring them again to their folds; and they shall be fruitful and increase. And I will set up shepherds over them which shall feed them: and they shall fear no more, nor be dismayed, neither shall they be lacking, saith the Lord. Behold, the days come, saith the Lord, that I will raise unto David a righteous Branch, and a King shall reign and prosper, and shall execute judgment and justice in the earth."

enly Kingdom. "They set out for water and pasture"[1] is the symbolical expression for the salvation, the water in the Christian sense denoting the communion of faith and the eternal kingdom of God. It is not known to me whether a Christian tradition of such a form really existed in Syria;[2] but the reconstruction here attempted is justifiable in itself, in order to do full justice to the Chinese version of the story. The Christian element and tendency are a necessary postulate, without which its fundamental features cannot be understood. It is most striking that this story opens in a sober manner, as though it were its only purport to describe a useful domestic animal of Fu-lin; not a word, however, is said about the utilization of any product of this animal, and we should certainly expect to hear at least what is done with the wool. Consequently the question is not here of a commercial proposition; at least, the Syrians who transmitted the tradition to the Chinese were not interested in this side of the matter, but solely in the peculiar life-story of the lambs, so that we are fully entitled to regard it as an allegory, and to seek its origin in the tenets of their Christian creed. The modification of the sheep into lambs, the cuirassed cavaliers, the water and pasture, and the Talmudic "man of the mountain," are unmistakable features characteristic of Christian notions. There is, further, a negative criterion pointing in the same direction: there was a sentence closing the story, the significance of which was either variable or vacillating in Syria, or not fully grasped by the Chinese interpreter. The recension of Chang Shou-tsie makes the lambs form a gregarious company after their release. In the redaction of the New T'ang Annals, compiled by Ngou-yang Siu in 1060, it is denied that they are able to form herds; while Tuan Kung-lu in his "Pei hu lu" (875), and Liu Hü in the "Old History of the T'ang Dynasty" (934), apparently embarrassed over this dilemma, dodged this point. Sheep are naturally gregarious animals; but for this very

---

[1] Compare Psalms XXIII. 1, 2: "The Lord is my Shepherd; I shall not want. He maketh me to lie down in green pastures: He leadeth me beside the still waters."

[2] Such allegories, however, were quite in keeping with the spirit of that time. Basilius the Great, whom we cited on the pinna, for instance, illustrated the doctrine of resurrection from the life-story of the silkworm: "What have you to say, who disbelieve the assertion of the Apostle Paul concerning the change at the resurrection, when you see many of the inhabitants of the air changing their forms? Consider, for example, the account of the horned worm of India, which, having first changed into a caterpillar, then in process of time becomes a cocoon, and does not continue even in this form, but assumes light and expanded wings. Ye women, who sit winding upon bobbins the produce of these animals, namely, the threads, which the Seres send to you for the manufacture of fine garments, bear in mind the change of form in this creature; derive from it a clear conception of the resurrection; and discredit not that transformation which Paul announces to us all" (J. Yates, Textrinum Antiquorum, p. 215). Again, it is interesting that Basilius, who appears to have known the silkworm only from books and by report, copied his description of it chiefly from Aristotle's account (Hist. anim., v, 19).

reason I am not inclined to believe that the Syrian original version, with its wondrous and supernatural tendency, should have terminated in such a platitude.   On the contrary, it is my impression that the Syrians did say that these extraordinary lambs, quite at variance with the common kind, did not assemble into flocks; that means, in Christian speech, the self-responsibility of the individual, and the obligation to his personal endeavor toward the path of redemption.

In the Mongol period we have a much debased version of our story from Ch'ang Tê, who was sent by Mangu Khan in 1259 to his brother Hulagu, King of Persia, and who describes the "sheep planted on hillocks" (*lung chung yang*) as a product of the countries of the Western Sea (*Si Hai*) as follows:[1] "The umbilical cord of a sheep is planted in the soil and watered.   At the time of the first thunder-peals it begins to grow, while the cord still remains connected with the ground.   When full-grown, they are frightened by the sounds of wooden instruments: the cord breaks off, and the animal roams around to feed on the herbage.   In autumn the sheep can be eaten, and there are seeds, to be used for planting, contained in its navel."   Ch'ang Tê must have overheard this story in Persia.   Certainly it is not a further Chinese development, but one of Arabo-Persian origin; certainly, also, it does not refer to any product, animal or vegetal, of western Asia, but merely represents a literary outgrowth of the older Fu-lin legend sensually deteriorated in the popular mind.

The section of the cyclopædia T'u shu tsi ch'êng entitled "Earth-Born Sheep" (already quoted) gives the following extract from the "Wu ts'ê yüan ying tsi:"[2] "As regards the earth-born sheep of the Western Regions, a vertebra of the neck is taken and planted in the soil.   On hearing the sounds of thunder, the kid is generated out of this bone.   When frightened by horsemen, its umbilical cord is severed. Its skin can be utilized as a mattress.   Another account has it that the people north of Mo[3] plant the horns of sheep, whereby is engendered an animal of the size of a hare, fat and beautiful.   The report is rather strange, and it is not ascertained what kind of fruit it is which is planted by those people.   Though what Liu Yu[4] relates may be correct, yet it remains a mystery.   Indeed, it is a marvel and subtlety of nature." It is evident that in the Mongol period the interest shifted in a certain measure and largely centred around the cause leading to the germination of the curious zoöphyte.

[1] Compare Bretschneider, Mediæval Researches, vol. i, p. 154.

[2] Apparently identical with the Yüan ying tsi, — writings of Wu Lai of the Yüan period (Bretschneider, Bot. Sin., pt. i, p. 214, No. 1125).

[3] *Mo pei jên.*   We have to read perhaps "Sha-mo" (the desert of Gobi), or, as another text cited by Schlegel (*l.c.,* p. 25) has it, "Ta-mo."

[4] Editor of the Si shi ki, — the memoirs of the journey of Ch'ang Tê, whose account has been given above.

During the fourteenth century the legend of the Syrian Lamb appeared in the diaries of European travellers. Odoric of Pordenone, who started on his journey between 1316 and 1318 and returned in 1330 (he died in January of the ensuing year), tells of very large melons growing in the Caspean Mountains in the kingdom Cadeli; and when these be ripe, they burst, and a little beast is found inside like a small lamb, so that they have both melons and meat.[1] Sir John Mandeville (or Maundeville), who travelled in Asia from 1322 to 1356, has the same report about gourd-like fruits which when ripe are cut, and disclose within a little beast in flesh, bone and blood, as though it were a little lamb. Men eat both the fruit and the beast, and this is a great marvel. The traveller assures us that he himself has eaten of this fruit. These trivial and puerile stories gave rise in Europe to numerous wild speculations in regard to a Scythian lamb of vegetal origin, growing on trees, as may be read in the monograph of H. Lee, "The Vegetable Lamb of Tartary: a Curious Fable of the Cotton Plant" (London, 1887). This work, though of considerable merit and not devoid of critical ability, is a failure in its main tendency, which is to prove that it was the cotton-plant which caused the origin of the story of the vegetable lamb.[2] True it is that in the European versions (and only these are taken into account by Lee) a reminiscence of cotton-pods bursting forth and laying bare the white cotton wool is alive; this, however, is not the origin, but the ultimate result, the most recent adjustment of the story, the antecedents of which must be connected with the Fu-lin traditions of the earth-born lamb. Even without the knowledge of these, Lee's conclusion could not be upheld. Years ago, when I first read his treatise without having access to the chain of Chinese texts, it did not prove convincing to me. It is inconceivable that in the fourteenth century, when cotton and the manner of its production were perfectly known in Asia and Europe, any such abstruse fable should have arisen in regard to cotton. The Indian cotton-plant became intimately familiar to the classical world, thanks to Alexander's campaign;[3] and I do not know that it ever became the object of fables in India, China, Greece, or Rome,[4] or in Syria, or among the Arabs.

[1] Yule, Cathay (new ed. by H. Cordier, vol. ii, p. 240).

[2] Lee was not the first to make this suggestion; for Yule, in a note of his Cathay (vol. ii, p. 242), remarks that Erman thinks the whole story a mythical view of the cotton-plant.

[3] Compare H. Bretzl, Botanische Forschungen des Alexanderzuges, pp. 136–139.

[4] H. Lee (*l.c.*, p. 46) makes a case of the passage in Herodotus (III, 106), who is the first Western author to mention Indian cotton, and says, "There are trees growing wild there, the fruit of which is a wool exceeding in beauty and goodness that of sheep." This certainly means nothing at all, particularly not with reference to the story of a vegetable lamb appearing in Europe as late as the fourteenth century. Herodotus, who merely compares cotton with sheep's wool, cannot be made responsible for a legend that is brought home in the middle ages from some dark corner of Asia. It is the history and the transformation of this legend which must be studied with critical methods. No philologist,

The Chinese of the sixth century, and assuredly of the T'ang period, knew very well what the cotton-plant and its products were;[1] and neither is there in the Chinese documents regarding cotton any reference to lambs, nor is there the slightest allusion to cotton in the Ta Ts'in and Fu-lin texts regarding the water-sheep and the earth-born lamb. The two groups of traditions are most clearly differentiated, and offer absolutely no point of contact.

The European mediæval fables are intelligible only when we read them together with the earlier traditions of the Chinese. Both Odoric and Mandeville reported their stories as coming from a certain part of Asia, and the mutual resemblance of these is close enough to arouse the suspicion that one copied the other; but this point is not of importance to me. The point to be emphasized is that their stories are the worthy counterpart of those prosaic and grossly materialized versions which we encountered among the Chinese of the Mongol period, and which are contemporaneous with Odoric and Mandeville, when the spiritual drift of the sacred Syrian allegory had long sunk into oblivion. Of course, the Chinese are not guilty of this sacrilege, but Persians and Turks, and that host of minor tribes composing the Western empire of the Mongols. Yule has identified the Caspean Mountains of Odoric with Mount Kasbin, about eighty miles due south of the Caspian Sea, in Persian territory near Teheran. Ch'ang Tê, as noticed, recorded his version of the story in Persia on his mission to Hulagu. Odoric's agreement with Ch'ang Tê proves that both have reproduced with tolerable correctness a bit of folk-lore picked up by them on Persian soil. The Persians were interested in the edibility of the lamb, and are duly seconded by Odoric and Mandeville, who have both lamb and fruit consumed. These people were interested in the material birth of the lamb, which they explained as growing from a seed planted in the ground. Accordingly it was a cultivated plant, bearing the lamb as a fruit, and raised anew every year; and this tradition again is echoed by the European mimics. The only novel features reported by the latter, and not yet revealed by a Chinese or other Oriental text, are the identification of this fruit with a melon, and the lamb harbored behind its rind. [2] Maybe both Odoric and Mandeville overheard the story from their informants in this manner; maybe they themselves

either, will subscribe to Lee's hypothesis (p. 50) that the word μῆλον used by Theophrastus for the capsule of the cotton-plant, because it means also "apple" and "sheep," might have contributed to convey, many centuries later, to readers of a dead language, an erroneous idea of fleeces that grow on trees.

[1] Compare the valuable notes of Hirth and Rockhill, Chau Ju-kua, p. 218; and Watters, Essays on the Chinese Language, p. 439.

[2] The strange combination of melon and cotton-plant may have as its *raison d'être* the phonetic similarity of the Persian words *kharbuz* or *kharbuza* ("water-melon") and *karbās* or *kirbāsa* ("cotton, muslin;" derived from Sanscrit *karpāsa*).

are responsible for this assimilation having a remote flavor of the cotton-pod; but, on this assumption, we are forced to admit that one was forestalled by the other. The traditions of the Chinese have enabled us to study the development of the story in its various stages, from the beginning of the Christian era down to the thirteenth century, and to recognize its origin, growth, and significance. We have seen that it takes its birth from the pinna, and that the Aristotelian doctrine of the fusion of vegetal and animal characteristics, applied to the life-habits of the pinna, is the very germ, the protoplasm, so to speak, which has called into existence the West-Asiatic notion of a vegetal lamb. This vegetal lamb therefore was evolved from a marine mollusk, never from a plant, and least of all from the cotton-plant. For this reason Yule[1] was misguided in seeking for "the plant about which these fables have gathered," and in regarding it as a certain genus of fern. Animal figures shaped by the Chinese from the rhizome of a fern greatly stirred the imagination of scholars in the eighteenth and nineteenth centuries, and were believed to have yielded the basis for the so-called "Scythian lamb." It is the uncontested and great merit of H. Lee[2] to have utterly destroyed these scientific fables, which, as usual, are more colossal and more baffling than the fables themselves, whose mystery they try to solve.

Entirely baseless is the opinion of G. Schlegel,[3] who, "after more than two years' study of the subject," as he avers, arrived at the result that "the Chinese have confounded two quite distinct things, — the cultivation of the cotton-plant[4] and the training of the camel, — from both of which fine stuffs can be fabricated." I am unable to see the justification of either point. There is in the Chinese records no trace that could lead to the one or the other supposition. On the one hand, according to Schlegel, "the Chinese accounts of that part of western Asia are peculiarly exact, though often seemingly shrouded in ambiguous and vague descriptions." On the other hand, he asserts,[5] "That the Chinese mistook the young dromedary or one-humped camel for a sheep, is not unnatural." The way in which Schlegel got at the camel from the sheep is a somewhat unusual one. There is no necessity of criticising it in detail, as no apprehension of an imitation of such methods need be entertained in our day.[6]

[1] Cathay, vol. ii, p. 241.

[2] *L.c.*, pp. 24–44.

[3] "The Shui-yang," *l.c.*, p. 20.

[4] This result he adopted from the work of Lee.

[5] *L.c.*, p. 32.

[6] The sinological reader, however, should be aware of the fact that the germ of Schlegel's erroneous argumentation rests on a misunderstanding of a passage in Ma Tuan-lin (p. 30 of his paper), though he had the correct translation of Hirth (China and the Roman Orient, p. 80; but see p. 255; it is certainly impossible to make rugs from pinna fibres) before his

The case presented in the preceding investigation may offer several points of general interest to the scientific student of folk-lore. We are allowed to pursue the history of the legend of the pinna-lamb through the interval of a millennium and a half from the dormant, embryonic beginning of a seemingly unimportant natural fact to a full-fledged, complex wonder-story, making all Europe talk for many centuries, and keeping scientists and learned societies on the trot in search of the secret of the marvellous lamb. The theatre of action on which the development of the story was staged is western Asia, chiefly Syria. The irony of fate, however, has ruled that the principal documentary evidence in the case enabling us to trace the real history of the story is preserved in the records of the Chinese, whose masterly historical sense permits us to establish the accurate chronology in the various phases which the story has adopted within the course of a long run. Without this solid staff we should presumably, like blind men, grope in the dark. We clearly recognize three principal stages of development, — first, the nature-philosophical stage inaugurated by the submarine life of the pinna and the conception of its byssus as marine wool, which idea reacted on the mollusk and resulted in the construction of a water-sheep; second, the mystic and allegoric stage, introduced by the Aristotelian doctrine of floristic and faunistic intermediate forms, and shaped and consecrated by the symbolism of Christian philosophy; and, third, the degenerate, materialized, in the true sense of the word animalized, form of the story, turning up in China and Europe simultaneously in the thirteenth century. Greek sources were enlisted to corroborate and to substantiate the basis of the first stage; and they were found equally effectual in accounting for the primeval foundation of stage second. In other words, the accounts of the Chinese, which simply reproduce Western folk-lore

eyes. Schlegel understood that rugs, mats, carpets, and curtains were made of the wool of the water-sheep; and by assuming that the latter refers to cotton, and by wrongly arguing that rugs may be made of hair or wool but can hardly be made of cotton, he finally hits upon Persian stuffs of camel-hair, and lands from this airship ascent upon the camel itself. Ma Tuan-lin, of course, does not say that rugs are made of the wool of the water-sheep; but the matter relative to the rugs is a new paragraph and entirely distinct from the former. Very strange, also, is the objection of Schlegel (p. 29) raised to Lee's theory that "the cotton-plant was not cultivated in the country where the vegetable lamb grew, on the west side of the Volga, neither was it grown in Persia." If this be true, it would not speak against Lee's view, but, on the contrary, in favor of it; for if such a legend, as erroneously assumed by Lee, should ever have originated around the cotton-plant, it could most certainly have started only in a region where the cotton product was but dimly known and the plant itself was not cultivated. Contrary to the opinion of Schlegel, carpets and rugs can certainly be made of cotton, and in fact are so made, for instance, in India: the so-called Suttringee are manufactured entirely of cotton; in another kind the warp is of cotton, the woof is of wool (J. F. Watson, Textile Manufactures and Costumes of India, p. 143).

and bear no relation whatever to genuine or indigenous Chinese thought, are perfectly matched and elucidated by the analogous traditions cropping out in the West. In one important respect, however, the preceding investigation remains deficient: I have not been able to point out an exact Western parallel of the Christian parable, as which I endeavored, on strong internal evidence, to prove the Syrian version of the vegetal lamb. At this point I have to ask the friendly co-operation of scholars versed in Syriac or Arabic Christian literature, a field foreign to me, and I trust that the prototype of our legend will some day be discovered there. Any search in this direction was heretofore precluded at the outset, since the history of the legend had not yet adequately or correctly been represented. Indeed, the subject had been dealt with only within the narrow boundaries of sinology, and had never been brought to the attention of Semitists. If these students will become aware of the fact that it very properly belongs to their domain, the day will not be distant when we may hope for the ultimate solution of that single point which still remains to be settled.

Field Museum of Natural History,
        Chicago.

111

作为文化历史问题的因纽特人螺丝

AMERICAN ANTHROPOLOGIST

NEW SERIES

ORGAN OF THE AMERICAN ANTHROPOLOGICAL ASSOCIATION,
THE ANTHROPOLOGICAL SOCIETY OF WASHINGTON,
AND THE AMERICAN ETHNOLOGICAL
SOCIETY OF NEW YORK

PUBLICATION COMMITTEE

F. W. HODGE, Chairman ex-officio; PLINY E. GODDARD, Secretary ex-officio;
HIRAM BINGHAM, STEWART CULIN, A. A. GOLDENWEISER,
GEORGE BYRON GORDON, WALTER HOUGH, A. L. KROEBER,
BERTHOLD LAUFER, EDWARD SAPIR, MARSHALL H. SAVILLE,
JOHN R. SWANTON, ALFRED M. TOZZER.

PLINY E. GODDARD, *Editor*, New York City
JOHN R. SWANTON and ROBERT H. LOWIE, *Associate Editors*

(F. W. Hodge, Editor for pp. 1–221)

VOLUME 17

LANCASTER, PA., U. S. A.
PUBLISHED FOR
THE AMERICAN ANTHROPOLOGICAL ASSOCIATION

1915

DISCUSSION AND CORRESPONDENCE

THE ESKIMO SCREW AS A CULTURE-HISTORICAL PROBLEM

THE highly interesting article of Mr Morten P. Porsild, *The Principle of the Screw in the Technique of the Eskimo*,[1] is fully entitled to the serious attention of all students interested in technology and the general development of human culture. The positive conclusion that "the principle of the screw is an old and original invention of the Eskimo, executed by a method of his own," indicates a problem of so great importance, that it cannot be left unnoticed or viewed with indifference by the historian of civilization. It is from the latter's point of view that I wish briefly to discuss the problem raised by Mr Porsild, and the remarkable conclusions reached by this scholar. By way of apology, I feel I have to state at the outset that I thoroughly disclaim any special knowledge of the Eskimo and their culture, and that the insignificant knowledge I may have of the subject is solely based on accessible literature and museum material. In anticipation, however, of criticism of my opinion on the part of students of real Eskimo life, which I should certainly be pleased to receive, I may say that my sympathy for this admirable people is no less than their own, and that my views are not biased by any sentiment in the matter. The principal question which I endeavor to answer for myself, and for others similarly inclined, is, Will the universal history of human culture be able to accept Mr Porsild's result as a definite fact, and how will it be reconciled or assimilated to other allied facts known to us? In other words, what is the position of the Eskimo screw in the general history of this mechanical device? And if it be true that the screw is an original and independent contrivance of the Eskimo, how will the previous views held concerning the development of the screw be affected or modified by the amazing thesis of Eskimoan originality? In using the word "amazing," I strike a keynote untouched by Mr Porsild. Amazing it is, because the Eskimo are the only known primitive tribes in the world acquainted with the application of the screw; and not only that, even the most advanced civilizations of antiquity—like those of ancient Egypt, Mesopotamia, India, and China—have never as yet yielded the slightest trace that would reveal any familiarity with

[1] *American Anthropologist*, N. S., vol. 17, pp. 1–16.

396

a screw device. The Chinese, we know positively, never were acquainted with it, but learned it only in the possession of Europeans, and have not yet adopted it in their home industries. The screw is an exclusively and decidedly European invention, peculiar to the Mediterranean culture area, in all probability first conceived by Greek mechanicians approximately in the second or first century B.C. It represents one of the most recent, if not the very latest, of all human implements. In view of these facts, which are here cited merely for preliminary information, the problem of the Eskimo screw does not appear to me so simple and so easy of solution, but, on the contrary, extremely complex; and the whole question merits renewed and more profound discussion in the light of the history of the screw.

Before entering on this subject, I may be allowed to add some bibliographical references which seem to have escaped the attention of Mr Porsild. In 1901, E. Krause, at the Museum für Völkerkunde in Berlin, published a brief article on the Eskimo screw.[1] He spontaneously discovered the screw in three arrow-heads with points of reindeer-antler, collected in 1882 and 1883 by Captain J. A. Jacobsen at Singrak, Alaska. These are well figured and described in his article. Krause made a plea for the native origin of the screw, though the interrogative form of the title of his article shows that he was not fully confident of his proposition. He advanced two reasons which induced him to ascribe the invention to the Eskimo: if the Eskimo screw had originated in imitation of an introduced screw, it would certainly have received more threads, in harmony with the latter, instead of only the one which it has; there are further rudimentary screws forming a missing link between the pegs with a single thread and the usually conical or also double-conical peg of the arrow-heads. Krause's opinion was immediately antagonized by K. von den Steinen,[2] who most emphatically denied that the Eskimo ever invented the screw. This ethnologist adopted an historical point of view in the study of the subject, without, however, citing positive and specific historical facts which would substantiate his theory that the Eskimo screw is the result of attempts to reproduce imported European models. He generally invoked the commercial intercourse of the Alaskan Eskimo with Chukchee and Russians, and, among other things, appealed to a deserted and well-preserved telegraphic station on Bering Strait, erected in 1867, and visited by Jacobsen.

[1] "Die Schraube, eine Eskimo-Erfindung?" *Globus*, Vol. LXXIX, 1901, pp. 8–9 (7 figs.).

[2] "Die Schraube, keine Eskimo-Erfindung," *Globus*, Vol. LXXIX, 1901, pp. 125–127 (9 figs.).

The fact that the Eskimo opposite the Asiatic East Cape, twenty years ago, had ample opportunity of acquainting themselves with European screws, cannot be contested, Von den Steinen concludes. This possibility nobody will deny, but possibilities are not historical facts. The question first to be answered would be, Were screws of European or American origin ever actually found among the Alaskan or other Eskimo? The present evidence, as conscientiously gathered by Porsild, surely points to a far greater age of the screw than a mere score of years. It was the merit of Von den Steinen at that time, however, to have increased the material available for discussion by thirteen interesting objects. Among these are arrow-heads from graves in the western part of Greenland, showing well-developed screws. This afforded Von den Steinen occasion for the conclusion that, if it is difficult to believe in the Eskimo invention of the screw, it is indeed impossible to assume that it should have been invented by them twice,—once in the west, and once more in the east,—and in both cases in those localities where they were well acquainted with European iron-ware. The left-handedness of the Eskimo screw was explained by Von den Steinen in the same manner as by Ryder and Porsild. He figured also a wooden pole in which a right-handed screw of fourteen rather regular grooves is carved,—very similar to the object illustrated by Porsild in Fig. 7. In conclusion, he generalized that the Eskimo were particularly gifted at learning technical stunts from Europeans, as emphasized by observers; the screw is merely an occasional and sporadic application of an introduced technique, whose influence on the Eskimo we are perhaps inclined to undervalue also in other lines. None of these arguments, to which we shall revert further on, really hits the case, and none is convincing to me.

Half a year later Miss H. Newell Wardle of Philadelphia took the platform.[1] She reviewed Eskimo material in the Museum of the Academy of Natural Sciences, Philadelphia, and rejected Von den Steinen's views, for the reason that the same ideas must originate in the same states of culture or in similar environments, intellectual or economic. "The intellectual abilities of the Eskimo cannot be doubted, their inexhaustible ingenuity has astounded all observers, and even the Aryan did not think it below his dignity to appropriate the more perfect harpoon of his 'barbaric' brother whale-hunter." Miss Wardle tried to explain the left-handed Eskimo screw from the spirally twisted tusk of the narwhal, which the people should somehow have attempted to imitate. This suggestion, however ingenious, is not plausible; we may even say it is rather improbable.

---

[1] "Die Eskimos und die Schraube," *Globus*, Vol. LXXX, 1901, pp. 226–227 (3 figs.).

If the Eskimo are acquainted with the screw, the main question that arises is, What relation does it bear to the European screw? And what is the history of the latter? We have no opportunity of studying the gradual development of this device, like that of so many other implements. It all of a sudden appears as an accomplished fact, scientifically understood and applied. Its first description and illustration, at once intelligent and scientific, we owe to the genius of Heron of Alexandria, instructor of a school for mechanicians and surveyors, and the most prominent physicist produced by the ancients. His lifetime is not exactly ascertained; but, in the judgment of the latest and most competent scholars, he lived in the second half of the second century B.C., while others assign him to the first century A.D.[1] The Greek original of Heron's work on mechanics is lost; but it is preserved in an Arabic translation, ordered by the Caliph Abu'l Abbās Ahmed Ibn al-Mutasim (833–842), and extant in four manuscripts (Leiden, British Museum, Constantinople, and Cairo).[2] The screw,[3] according to Heron, forms the fifth of the six mechanical powers, and is theoretically explained from a cylinder moving over a plain. A side of the cylinder is supposed to be in motion, and a point to move on this side from its extremity; this point runs through the whole side in the same interval as it takes this side to turn once around the surface of the cylinder and to come back to its starting-point. The curve described by this point on the cylindrical surface is the thread of a screw, designated as the screw. Then Heron proceeds to describe minutely the making, use, and properties of screws, single, and in combination with other mechanical powers.[4] The most interesting of these instruments is the screw-press employed in the production of olive-oil.[5] Oil being one of the chief articles of food among the ancients, the mechanical improvement of the oil-presses was a matter of large bearing upon economic progress. Pliny[6] informs us that

---

[1] W. Schmidt, *Heronis opera*, Vol. I, p. XXIV (Leipzig, 1899).

[2] First edited by Carra de Vaux, *Les mécaniques ou l'élévateur de Héron d'Alexandrie publiées pour la première fois sur la version arabe de Qostà ibn Luqâ et traduites en français.* Paris, 1894 (Extrait du *Journal asiatique*). Best edition by L. Nix in Nix and Schmidt, *Heronis opera*, Vol. II, fasc. 1 (Leipzig, 1900).

[3] Greek κόχλος or κοχλίας (Latin *cochlea*), which originally means "a snail with a spiral shell." The female screw is περικόχλιον; Latin has no word for the latter.

[4] Carra de Vaux, *l. c.*, pp. 101–106; Nix, *l. c.*, pp. 104–112. A small portion of Heron's work on mechanics is preserved in the "Synagoge," the mathematical collectanea of Pappus of Alexandria, who lived under the Emperor Diocletian (285–305). Among these fragments is also the description of the screw (Nix, *ibid.*, pp. 283–291).

[5] Nix, *ibid.*, pp. 240–252 (with two illustrations).

[6] *Nat. hist.*, XVIII, 74, § 317 (ed. Mayhoff, Vol. III, p. 231).

the ancients used to hold down the press-boards with ropes and leather thongs, wrought by levers; that within the last hundred years the Greek press (*Graecanica*) had been invented, in which the grooves of the male screw pass through and around the female screw;[1] and that only within the last twenty-two years had a less unwieldy press with smaller boards been contrived. In this apparatus the stem of the screw was placed in the center. The screw was directly utilized in the act of pressing, the whole pressure being concentrated upon broad planks placed over the olives or grapes. Representations of presses have come down to us on wall-paintings of Pompeii; for instance, a clothes-press, which is worked by two upright screws, precisely in the same manner as our own linen-presses.[2]

Besides the screw-press, the ancients were acquainted likewise with the employment of screws for magnifying a motion and rendering it easily manageable and measurable,—the same principle as we still apply in the screw-feet of instruments of precision. The vaginal specula of the Roman surgeons were provided with screws to open or close the bows of these instruments according to need. Three specimens discovered in Pompeii are in the Naples Museum. They are illustrated and described by J. S. Milne.[3] In one of these, 23 cm. long, the blades are at right angles to the instrument, and when closed form a tube the size of the thumb. On turning the screw, a cross-bar forces the two upper blades outward till sufficient dilation is secured for operative purposes. In

---

[1] This passage offers many technical difficulties, the discussion of which would be out of place here. I have adopted the interpretation of H. Blümner (*Technologie und Terminologie der Gewerbe und Künste bei Griechen und Römern*, Vol. I, 2d ed., p. 348). According to Nix and Schmidt (*Heronis opera*, Vol. II, p. 388), who reproduce the text of Pliny, this kind of press survived in the Canton of Graubünden till the seventeenth and even the nineteenth century.

[2] Figured by Blümner, *l. c.*, p. 188; Overbeck-Mau, *Pompeji*, p. 393; W. Smith, *Dictionary of Greek and Roman Antiquities*, Vol. I, p. 464. Aside from Pliny, the screw is mentioned by the architect and engineer Vitruvius (*De architectura*, VI, 7, 3), who lived in the first century B.C.; it was therefore known to the Romans before Pliny's lifetime. In the translation of M. H. Morgan (Vitruvius, *The Ten Books on Architecture*, p. 184, Cambridge, 1914) the passage runs thus: "The pressing room itself, if the pressure is exerted by means of levers and a beam, and not worked by turning screws, should be not less than forty feet long, which will give the lever man a convenient amount of space." E. Krause, accordingly, in his article in *Globus* cited above, was entirely misguided in asserting that "the Eskimo had advanced much further than the highly cultivated and over-refined Romans in the age of the imperium; for despite their progressive culture they did not know the screw." K. von den Steinen, in his reply, quoted this observation without adding any comment.

[3] *Surgical Instruments in Greek and Roman Times*, pp. 151, 152 (Oxford, 1907).

another, on turning the screw, the lower blades could be drawn downward, at the same time separating slightly, while the upper blades diverged also. In one instrument the screw is left-handed; in another, right-handed. Paulus of Aegina, a celebrated physician of the seventh century A.D., in his description of the speculum, does not fail to call attention to the screw, which he says is to be turned by the assistant, while the speculum itself is to be held by the operator.

As to the further history of the screw,[1] it may suffice for the purpose in view to emphasize two important facts,—first, that the Romans passed the principle of the screw on to the peoples of central Europe, to whom it was a foreign affair; in short, that the modern development of the screw is an inheritance of classical antiquity; and, second, that the Arabs derived their knowledge of it from the Greeks. Reference has already been made to the Arabic translation of Heron's fundamental work on mechanics, in which the principle and application of the screw are described theoretically and practically. The most prominent of the native Arabic works on technology is the *Mafātīh al-ulūm* (*Keys of the Sciences*), by al-Khārizmī, a mathematician who lived about 820 under the caliphate of Mamun. He describes the screw (*al-laulab*) as a well-known contrivance employed by carpenters and architects in laying foundations; and he mentions the oil-press in conformity with Heron, his term for the latter, *al-gālājarā*, being reproduced from the Greek γαλεάγρα.[2] The history of the screw, accordingly, presents an unbroken chain of development which is well determined within the area of Mediterranean

---

[1] A history of the screw has not yet been written. What F. M. Feldhaus (*Technik der Vorzeit*, col. 981, Leipzig, 1914) offers on this subject can hardly be looked upon as a useful or trustworthy contribution, as it suffers, like many articles in this technological dictionary, from serious defects, lack of criticism and historical sense, misinformation and inaccuracy. The so-called Archimedean screw (see Blümner, *l. c.*, Vol. IV, pp, 121–126; and C. G. de Montauzan, *Essai sur la science et l'art de l'ingénieur aux premiers siècles de l'empire romain*, pp. 90–91, Paris, 1908), which is a spiral pump for raising water, with a pipe coiled like a screw, is not to be connected with the history of the screw proper, as asserted by Feldhaus; still less so what he styles the peculiar thread of a screw in the spindle whorls of the Himalaya. These whorls are quite familiar to me, and are decorated with grooves of concentric circles which have nothing to do with the threads of a screw. According to Feldhaus, the screw-line occurs as ornament as early as the bronze age, particularly in the large rings for the neck. Such twisted metal rings, as occur also in China in the form of bracelets or in the handles of vessels, are modelled after ropes, and, at any rate, bear no relation to a screw. The fundamental passages of Heron's work on mechanics are not pointed out or discussed, while reference is made to an incidental mention of the screw in his *Pneumatica*.

[2] E. Wiedemann, "Zur Mechanik und Technik bei den Arabern" (*Sitzungsber. phys.-med. Soz. Erlangen*, Vol. 38, 1906, p. 21).

civilization. The screw is not an invention of primitive man: in the works of ethnologists (for instance, in those of Lubbock, E. B. Tylor, O. T. Mason, H. Schurtz, etc.) it is conspicuous by its absence. Neither is it a prehistoric invention: prehistoric archeology is reticent about it. It is a comparatively late invention of historical times, appearing under the full searchlight of history. The ancient Greeks and Romans, as well as the ancient Egyptians, were still in complete ignorance of it, till it unexpectedly came to light in the Alexandrian epoch. We cannot say exactly when, where, and under what circumstances, the idea sprang up. Certain it is that the first invention was conceived on Egyptian soil, perhaps within the precincts of Alexandria, by men of thorough mechanical training, presumably those attached to or in close touch with the scientific schools of the great Hellenistic emporium. We hear nothing of experiments gradually leading to the idea: it was born like Pallas Athene from Jupiter's head,—a clear, self-conscious, and accomplished fact. Whether those mechanics were Egyptians or Greeks (Heron was probably an Egyptian) cannot be decided, either. The only safe formula to express the event is that the invention was accomplished within the time and boundary of Hellenistic civilization in Egypt, during the second or first century prior to our era. In the first century B.C. it spread from Egypt or Hellas to Italy. We further note that the screw then was not an accidental or passing toy, but, as a matter of necessity, was applied with perfect logic and volition to machines or instruments, which without it were clumsy and inefficient, and which in combination with it were suddenly turned into highly useful and time-saving devices, denoting a rapid step of technical and economic progress.

In applying this lesson to the Eskimo screw, we fail to see this idea of absolute necessity, essential function, and technical progress, in the adoption of the screw, however manifest its purpose may be. The Eskimo implements provided with this device are not only conceivable without it, but, in the case of the arrow-heads and plugs, in fact are so made. There is no Eskimo implement in which the screw fulfills so indispensable a function that it would not be equally effectual if it were devoid of it. In the mechanical contrivances of the ancients we readily recognize the motive prompting the introduction of the screw, while the motive is not conspicuous in the Eskimo screw devices; at least, students of Eskimo culture have not yet discussed this side of the question. Another point to be noticed is that Eskimo knowledge of the screw is somewhat limited or one-sided, inasmuch as only the male screw, not the female screw, is known to them; and in case the theory of indigenous origin can be upheld, more stress should be laid on the study of the diversity of Eskimo screws from our own. Our inquiry bears out the fact that the Eskimo screw is an isolated and exceptional case in ethnic life, outside of our own culture sphere; and this is the very reason why the case is so intensely interesting and sympathetic. If the originality of this invention among the Eskimo could unobjectionably be proved, the case would certainly be remarkable, and would considerably affect and influence our opinions regarding analogous phenomena. It would stand out as an illustration of the first order, of independent invention. In view of this significance, however, we must insist on intrinsic and clean-cut evidence which is acceptable to the thinking majority; while at the outset such broad generalizations, based on theorizing speculations, as those advanced by Von den Steinen and Miss Wardle, should be strictly barred. The conclusion, for instance, that the Eskimo could not have invented the screw, because many other peoples of a higher degree of culture never invented it, and because this supposition would seriously disturb the circle constructed by us for the history of "our" screw, might prove also a fallacy, although the extraordinary isolation of the Eskimo screw remains a fact not easily to be disregarded. Everything is possible, and nothing is impossible. The possibility that the Eskimo might have conceived the principle of the screw independently must plainly be admitted. I for my part not only admit this, but even wish that definite proof might soon be established beyond any doubt. For the present, however, I have to confess that, despite all the meritorious efforts hitherto made, exact evidence in favor of Eskimo originality does not yet seem to me to be satisfactory or positively assured. We have had expressions of opinion from our prominent Eskimo scholars —opinions highly valued, as they are based on solid information—but we have not yet heard from the Eskimo themselves, who should be induced to take the witness-stand in their own defence. May I venture to suggest to Mr Porsild, who apparently is stationed among the people in Greenland, that he question his friends as to their knowledge of the bolder man, who, as the first among them, ventured on the manufacture of a screw, and what reasons prompted him to embark on this scheme? Should the folklore of the Eskimo not make any allusion to this device? Also they may be able to throw some light on the possible interrelation of it and the narwhal-tooth suggested by Miss Wardle, though this idea appears somewhat romantic. The evidence, as summed up by Mr Porsild on p. 15, is not wholly convincing; at least, not to an outsider, who, like the writer, is willing to be convinced. The utmost concession I can make to him is, that his conclusion is a theory—a very attractive and suggestive one, to be sure—but certainly it is very far, as asserted by him, from becoming a fact which might readily be accepted by science. The "facts which lead to the belief that screw-bearing implements were made before the advent of Europeans into West Greenland," strictly speaking, are not facts, but suppositions elicited from certain observations, and formulated in a somewhat dogmatic manner. The linguistic evidence gives rise to suspicion. The lack of a proper word for the screw, though not conclusive, augurs a serious defect; at least, it is disappointing. If it existed, if it were named, for instance, "narwhal-tooth," in analogy to the Greek "snail,"[1] my faith in the theory of indigenous invention would be considerably strengthened. It is certainly possible that the lacunae still existing in our knowledge of the subject (lack of necessity and motive, absence of direct Eskimo information, deficiency of circumstantial evidence), and militating against the opinion of native invention, will be filled in the course of time; but, as long as these postulates are not satisfied, it will be safer to hold judgment in abeyance.

While, on the one hand, I am of the opinion that the case of those pleading the aboriginal invention of the Eskimo screw is not yet substantially backed up, I must confess, on the other hand, that the argumentation of Von den Steinen, who, as far as I know, is the only one to take the opposite stand, is weak, and does not at all carry conviction with me. Von den Steinen would have us believe that the Eskimo of Alaska and those of Greenland, independent of each other, constructed the screw in imitation of foreign products; but he omits to tell us why this double attempt at imitation yielded exactly the same results, and, while there is no other example of a primitive tribe which made a similar effort, the Eskimo in two widely different regions and at different times should have successfully performed the same experiment twice. This point of view seems to me more miraculous than the supposition of the native origin of the screw; it even seems to me to be inspired by poetic rather than scientific insight. It is inconceivable to me that this process of imitation should have taken place at various times and in various localities. If imitation it is, this act must be reduced to the experience of a single primeval occasion, which takes the responsibility for all subsequent instances. With Boas and Porsild, I concur in the opinion that

---

[1] This, of course, would not mean that Miss Wardle's theory has the right of existence. The ancients did not learn the principle of the screw from the snail, but merely termed it for the snail by way of comparison.

we have to do with a uniform phenomenon affecting all Eskimo tribes alike; and these scholars are certainly justified in regarding its wide geographical distribution and its occurrence in isolated places and ancient sites as symptoms of a certain antiquity.

Thus far we have been confronted with two opinions diametrically opposed,—the declaration of independence and the imitation theory. There remains, however, a third point of view to be taken into consideration. Let us start from an analogous case. The wood-carved snow goggles of the Eskimo with narrow horizontal slit, commanding a sufficient range of vision, without any doubt, represent spontaneous and indigenous invention of this people. Besides this type of goggles, there is another one with two large apertures assuming the outline of the eyeballs, and covered on the inner side with sheets of American or European glass.[1] The latter type, in my opinion, is the result of an adaptation (probably recent) to European-American spectacles or goggles. Accordingly we here face a foreign idea grafted upon an object of old native manufacture, and seemingly producing a net result which might lead the casual observer to condemn the whole affair with the sweeping judgment of being due to an outside impetus. But we have only to study carefully the large varieties and variations of slit-goggles in order to become convinced of the utter baselessness of such a grotesque generalization. This example illustrates that native inventiveness does not preclude the reception of foreign technical features, if these are apt to improve the object or to be pleasing to its makers. An allied process may have been in operation in the case of the screw. Granting the possibility of the comparatively great antiquity of the latter, two characteristics are discernible which in my opinion would speak in favor of a germ of native pre-invention. These are the simple single-threaded screw first emphasized by Krause, for which, as far as I know, no analogy exists in our technique, and the screw-like designs cut into the wedge-shaped plugs for closing the wounds of the seal. In the variations of the latter implement, the history of the development of the technique is perhaps preserved. We note plain pegs, those provided with horizontal, parallel grooves, and those with grooves cut spirally.[2] Did the screw of the latter, perhaps, result from the design of grooved circles? This primitive

---

[1] This observation refers to three goggles in the collections of the Field Museum, Chicago (Cat. Nos. 13820, 20288, and 53295), from the Alaskan Eskimo. Also O. T. Mason ("Primitive Travel and Transportation," *Report of U. S. Nat. Mus. for 1894*, pp. 293, 294, 296) has pointed out this type as a "modern adaptation."

[2] Compare Boas, "Eskimo of Baffin Land and Hudson Bay" (*Bull. Am. Mus. Nat. Hist.*, Vol. XV, 1901, p. 21); and Porsild, *l. c.*, p. 12.

foundation being given, foreign ideas could have been adapted to or transplanted on it. After all, it remains an open question whether the Eskimo screw, technically speaking, may not have had an origin different from that of our own. The fact that it is this very origin which is still shrouded in mystery accounts for the obscurity from which the whole subject labors, and for the variation of possible opinions.

BERTHOLD LAUFER

FIELD MUSEUM OF NATURAL HISTORY,
   CHICAGO, ILL.

THE "RED PAINT PEOPLE" OF MAINE.

In justice to New England archeology I cannot let pass unnoticed the communication of your correspondent on page 207 of the preceding number of the *American Anthropologist*. Mr Bushnell says: "The graves discovered by Mr Moorehead on the coast of Maine differ in no respect from those rifled by the Pilgrims near the present Provincetown." This is a statement which facts do not justify. The grave at Cape Cod to which Mr Bushnell especially refers is described by the Pilgrims as follows:[1]

Wee found a place like a graue, but it was much bigger and longer than any that we had yet scene. It was also covered with boords, so as we mused what it should be, we resolved to digge it vp, where we found, first a matt, and vnder that a fayre Bow, and there another Matt, and vnder that a boord about three quarters long, finely carued and paynted, with three tynes, or broches on the top, like a Crowne; also betweene the Matts we found Boules, Trayes, Dishes, and such like Trinkets; at length we came to a faire new Matt, and vnder that two Bundles, the one bigger, the other lesse, we opened the greater and found in it a great quantitie of fine and perfect red Powder, and in it the bones and skull of a man. The skull had fine yellow haire still on it, and some of the flesh vnconsumed; there was bound vp with it a knife, a pack-needle, and two or three old iron things. It was bound vp in a Saylers canvas Casacke, and a payre of cloth breeches; the red Powder was a kind of Embaulment, and yielded a strong, but no offensiue smell; It was as fine as any flower. We opened the lesse bundle likewise, and found of the same Powder in it, and the bones and head of a little childe; about the leggs, and other parts of it was bound strings, and bracelets of fine white Beads; there was also by it a little Bow, about three quarters long, and some other odd knackes; we brought sundry of the pretiest things away with vs, and covered the Corps vp againe. After this, we digged in sundry like places, but found no more Corne, nor any things els but graues: There was varietie of opinions amongst vs about the embalmed person; some thought it was an *Indian* Lord and King: others sayd, the *Indians* haue all blacke hayre and never

---

[1] *Journal of the Pilgrims at Plymouth*, Cheever reprint, p. 38.

112

印支语系中的前缀 A-

1915   **1915**

THE

JOURNAL

OF THE

ROYAL ASIATIC SOCIETY

OF

GREAT BRITAIN AND IRELAND

FOR

1915

PUBLISHED BY THE SOCIETY

22 ALBEMARLE STREET, LONDON, W.

M DCCCC XV

MISCELLANEOUS COMMUNICATIONS

THE PREFIX A- IN THE INDO-CHINESE LANGUAGES

Under the modest title "Sur quelques textes anciens de chinois parlé"[1] M. Henri Maspero has made a most remarkable contribution to the history of the Chinese language, which takes the same high rank as T. Watters' *Essays on the Chinese Language*, published in 1889 and, like the latter, will form a fundamental basis for the elaboration of a real life-history of Chinese speech to be written in the future. The careful study of M. Maspero, based on the critical analysis of five Chinese Buddhist works traceable to the ninth century, discloses for the first time what has come down to us of the remains of the colloquial language of those days,[2] the rich material being well arranged under convenient grammatical categories. It is, however, not only for the historical grammar of Chinese, but also for the comparative study of Indo-Chinese languages that M. Maspero's researches are of prime importance ; and for this reason I take the liberty to draw the attention of students of Indo-Chinese philology to the interesting work of this scholar, simultaneously demonstrating through a practical example how his studies may be rendered useful and fertile in their application to cognate languages. It is as a tribute to M. Maspero's scholarship, and an expression of my grateful recognition

---

[1] Bulletin de l'Ecole française d'Extrême-Orient, vol. xiv, No. 4, 1914.

[2] Attempts in this direction have not been many in the past. We have from the pen of C. de Harlez a study on "Le chinois parlé au vi⁰ siècle A.C., d'après l'I-li" (*T'oung Pao*, vol. ix, pp. 215-25, 1898), and from M. Jametel, "L'argot pékinois et le Kin-ping-mei" (reprint from Mém. Soc. sinico-japonaise, vol. vii, Paris, 1888, pp. 18) ; but de Harlez and Jametel, on the ground of their literary sources, treat only of the style and phraseology of colloquial speech, not, as M. Maspero does, of its structure. His work, therefore, is entirely original.

of the high character of his researches, that the following observations have been penned.

On p. 13 of his treatise M. Maspero notes that in the Chinese oral language of the ninth century, terms denoting relationship are generally preceded by a sort of prefix, *a* 阿 : as, for instance—

> *a-ye* 阿 爺, father.
> *a-fu* 阿 父, father.
> *a-niang* 阿 娘, mother.
> *a-hiung* 阿 兄, elder brother.
> *a-ši* 阿 師, master, monk.
> *a-kiu* 阿 舅, uncle ⎫
> *a-yi* 阿 姨, sister ⎭ in Buddhist texts.[1]

This phenomenon, however, is older even than the ninth century, for in a work of the epoch of the Six Dynasties, the *Han wu ku shi* 漢 武 古 事, we meet the form *a-kiao* 阿 嬌 ("lass").[2] Moreover, we encounter the same prefix in combination with the interrogative pronoun *a-šei* or *a-sui* 阿 誰,[3] and in the interrogative adjectives *a-na* 阿 那 and *a-na-ko* 阿 那 箇 ("quel?").[4] Again, the great antiquity of this case is borne out by an instance of the use of the pronoun *a-šei* in the *San kuo chi* 三 國 志.[5]

An ancient (perhaps dialectic) demonstrative pronoun seems to be preserved in *a-tu* 阿 堵, chiefly used with reference to money;[6] and since Tibetan and other Indo-Chinese languages combine the prefix *a-* with the

---

[1] Maspero, loc. cit., p. 34.

[2] Maspero, loc. cit., p. 34.

[3] The interesting coincidence with Lo-lo *a-sa* (Nyi dialect) and *a-so* (A-hi dialect), "who?" old Burmese *a-su* (see B. Houghton, JRAS.. 1896, p. 33), Lo-lo *a-mi*, "what?" (Burmese *a-be*, "what?"), Mo-so *a-ne* ("who?"), and *a-tse* ("what?"), may be pointed out right here, as well as the apparent relationship of Chinese *šui*, Tibetan and Newari *su*, and Lo-lo *so*, *sa*.

[4] Maspero, loc. cit., pp. 25–6.

[5] Maspero, loc. cit., p. 34.

[6] Giles, *Chinese-English Dictionary*, No. 12045 (likewise in Palladius).

demonstrative pronoun, it is justifiable to regard the element *a* in *a-tu* as the same prefix. What is still more curious is, that the same expression *a-tu* in a passage of the *Tsin shu*, according to a gloss in the *P'ei wên yün fu*, assumes the significance "eyes".[1] Again, as numerous related languages affiliate the prefix *a*- with names of bodily parts (see below), we are entitled to identify the element *a* in the word *a-tu* (" eye ") with the prefix, and to explain *tu* as a dialectic word for "eye", which has its counterpart in the second element of the Lo-lo-p'o compound *me-du* (" eye ").[2]

It was known, of course, that similar formations occur in the present Chinese dialects : thus in Ning-po,

> *a-tia* 阿 爹 (likewise in Amoy), father.
> *a-niang* 阿 娘 or *a-m* 阿 媽, mother.
> *a-tsi* 阿 姐, elder sister.
> *a-me* 阿 妹, younger sister.
> *a-ko* 阿 哥, elder brother.[3]
> *a-di* 阿 弟, younger brother.
> *a-bo* 阿 婆, mother-in-law.[4]

Likewise it was known that, particularly in Cantonese, it frequently occurs in connexion with proper names ; and

---

[1] Pelliot, Bull. de l'Ecole française, vol. ix, p. 573, n. 1, 1909.

[2] A. Liétard, Bull. de l'Ecole française, vol. ix, p. 553, 1909. The first element, *me*, in the Lo-lo word, is identical with Mo-so *mo* (*m'o*), Si-hia *mei*, Tibetan *mig*, Burmese *myak*, Gešitsa dialect of Tibetan *muk*, Chinese *muk* 目. Lepcha *a-mik* and Southern Chin *a-mi* are analogous in form to ancient Chinese *a-tu*. A further relationship of the word *tu* (*du*) might possibly be given in the series *ta* of the T'ai languages.

[3] Watters (*Essays on the Chinese Language*, p. 366) states that the expression *a-ko* is in very common use among the Chinese as a respectful mode of address. He is quite correct in assuming that, if the same word was chosen for the rendering of Manchu *age* or *agu* (not, as written by him, *agù*), this was partly due to the meaning of the Chinese term. *A-ko*, accordingly, in this case, is not the Chinese transcription of a Manchu word, but the assimilation of a pre-existing Chinese term to the latter.

[4] C. Arendt, *Handbuch der nordchinesischen Umgangsprache*, pt. i, p. 282.

such names are met with also in Chinese literature.[1] But Maspero is the first to show on the basis of documentary evidence that this phenomenon is an old constituent of the Chinese language, which, thanks to his investigations, is now well traceable almost to the beginning of our era. The case is the more interesting, as this *a*- is the only survival of a prefix in Chinese, and as the same feature is found in a great number of Indo-Chinese languages; and the essential point is that it represents a prominent characteristic of this widely distributed family, and bears witness to the phonetic and morphological relationship of its members.

Burmese was the language in which this peculiar trait was first pointed out by W. v. Humboldt.[2] He observed that Burmese is capable of forming nouns by the addition of a prefixed *a*-. Schleiermacher, in his *Grammaire barmane*,[3] likewise observed the employment of *a*- in the formation of nouns and adverbs.[4] A. Schiefner[5] studied the question somewhat more profoundly by drawing upon

[1] For example, *A-yü* 阿玉 (E. Rocher, *T'oung Pao*, vol. x, p. 347, 1899), *A-hêng* 阿衡 (Chavannes, *Mémoires historiques de Se-ma Ts'ien*, vol. i, p. 178 ; vol. v, p. 196) ; *A-jung* 阿戎, mentioned in *Tsin shu* (see Pétillon, *Allusions littéraires*, p. 274) ; *A-po* 阿波, mentioned in *Sui shu*, Ch. 51 (Chavannes, *Dix inscriptions chinoises de l'Asie centrale*, p. 28, n. 3). Giles, in his Dictionary, cites *A-hiang* 阿香, as the name of a fairy who assists the God of Thunder by pushing his car. In Mayers (*Chinese Reader's Manual*, p. 1) we read of *A-kiao* 阿嬌, *A-man* 阿瞞 and *A-to u* 阿斗. *Tao-yun* 道韞 elegantly spoke of her uncle as *A-ta-chung-lang* 阿大中郎 (Lockhart, *Manual of Chinese Quotations*, p. 130). Compare also *a-weng* 阿翁 ("grandfather" or "father-in-law") in Pétillon (loc. cit., pp. 126, 259), *a-kia* 阿家 ("mother-in-law," ibid., p. 259), and *a-p'o* 阿婆 ("vieillard,' ibid., p. 418).

[2] *Einleitung in die Kawi Sprache*, p. cccliv.

[3] Inserted in his work *De l'influence de l'écriture sur le langage* (Darmstadt, 1835).

[4] Loc. cit., pp. 144, 244, 256.

[5] *Mélanges asiatiques*, vol. i, pp. 361–3.

Abor, Luhuppa, Manipurī, and Tangkhul, in which, besides
a, he met an alternation with the vowels *o*, *u*, *i*, and *e*
(*o-mit*, "eye"; *ummah* [Tib. *me*], "fire"; *i-pa*, "father";
e-lag, "hand"). It was his further merit to call attention
to Tibetan where the prefixed forms chiefly belong to the
vernacular, the written language usually being destitute
of them—

> *a-p'a*, father (written language *p'a*).
> *a-ma*, mother (w.l. *ma*).
> *a-žaṅ*, maternal uncle (w.l. *žaṅ-po*).
> *a-sru*, maternal aunt (w.l. *sru-mo*).
> *a-ne*, paternal aunt (w.l. *ne-ne-mo*).
> *a-bo*, elder brother (w.l. *p'u-bo*).
> *a-rgya*, daddy.
> *a-yas*, dear mother.
> *a-č'e* (*a-je, a-že*), elder sister.

The relation of these *a*-formations to the written
language, accordingly, is the same in Tibetan as in
Chinese.

Besides terms of relationship, Schiefner recorded the
following words:—

> *a-čug*, ankle-bone.
> *a-dogs*, table.
> *a-luṅ*, clasp.
> *a-lon*, circle, ring (Burmese *a-lunh*).
> *a-rog*, companion (from *grogs, rogs*).
> *a-re*, a little (also interjection).
> *a-p'rag*, arm (Jäschke: bosom of a garment), from *p'rag-pa*,
>   shoulder.

At present this list may be considerably increased; and
we may distinguish the following categories of words into
which the prefix enters:—

Nouns denoting relationship

The following additions may be made to Schiefner's list:

> *a-tā* (Balti and Purig), father.
> *a-k'u*, paternal uncle (w.l. *k'u-bo*).
> *a-jo*, elder brother (w.l. *jo-bo*).

a-p'yi, grandmother (w.l. *p'yi-mo*).

a-p'yim, old woman.

a-baṅ, husband of the father's or the mother's sister.

a-mes, grandfather (w.l. *mes-po*).

a-po, junior husband.

a-k'a, infant (w.l. *k'a-ba*).

Nouns denoting professions

a-č'os-pa (colloquial of Eastern Tibet), a religious man, monk (written language *č'os-pa*, from *č'os*, religion).

a-mč'od, one who recites prayers for compensation outside of the Lamaist ceremonies (from *mč'od-pa*, to offer, to sacrifice).

a-druṅ, groom.

Nouns denoting bodily parts

a-čug, ankle-bone.

a-ts'om, beard (popularly for *ag-ts'om*; compare A-hi Lo-lo *ni-ts'o*).

a-ra, beard (in Tsang).

a-ku (written *a-sku*), body; used in the Kuku-nōr dialect in the phrase *a-ku de-mo* (*bde-mo*), "How are you?"[1]

Nouns denoting animals

a-lü (written *a-lus*), in Sikkim *a-li*, or *a-liu*, cat.

a-li-k'ug-ta, swallow.

a-gas or *a-ges*, a fabulous animal.

a-pra, earless marmot.

Nouns denoting plants

a-kar (Sikkim), red pepper.

a-kroṅ, an Alpine plant.

a-ya-zva-ts'od, dead-nettle.

a-rum, a species of garlic.

a-byag, *a-ạbras*, *a-ạbre*, names of medicines.

---

[1] See Rockhill, *Diary of a Journey through Mongolia and Tibet*, p. 88. T. de Lacouperie (*Les langues de la Chine avant les Chinois*, p. 71) noted that in certain dialects of the Miao tribes of Southern China a prefix *a-* is joined to terms of bodily parts (*a-pu*, hand: *a-t'au*, foot; *a-biu*, ear).

Nouns denoting articles of food

a-log, ball of dough.

a-sla, cake.

a-sam, thick broth.

a-sbyar, thin broth.

Other nouns

a-po, building (only in dBus).

a-ñar, mirror.

a-kam, firewood.

a-sgor, ear-ring.

The prefix *a*- is not so frequent in connexion with proper names as in Lo-lo and Mo-so, but such names occur also in Tibetan. In historical records we meet A-šo-legs, A-rog-lde, A-so-lde, A-tʻog rkod-btsan. Well-known is the name A-nu, the father of Tʻon-mi, who was sent by King Sroṅ-btsan sgam-po to India in order to frame an alphabet for Tibetan.[1] A tribe of the Golog is known as A-chü.[2]

The prefix *a*-, further, serves for the formation of adverbs and a few adjectives—

a-tsʻad, a little, as to quantity (from *tsʻad*, measure).

a-tsʻe, a little, as to time (from *tsʻe*, time).

a-li, a little.

a-tsʻod, at present (from *tsʻod*, measure, time-measure).

a-ta, now.

a-mtsʻar, admirably (from *mtsʻar*, wonderful).

a-gsar, newly (from *gsar*, new).

---

[1] The name *A-nu* appears as that of an author of two grammatical works in the *Tanjur* (vol. cxxiv, Nos. 2, 3). Schiefner supposed the identity of this A-nu with Anubhūti, which is purely a conjecture unsupported by evidence. True it is that there is also a Sanskrit proper name Anu; but it does not follow therefrom that a Tibetan name A-nu is borrowed from India. On the contrary, it appears as a genuine Tibetan word, being derived from *nu-bo* (" younger brother ") in the same manner as, for instance, *a-jo* from *jo-bo* (" elder brother "). There seems also good reason to believe that the Tibetan tradition ascribing the authorship of these two treatises to Tʻon-mi A-nu (Tʻon-mi assuming his father's name) is well founded (see Huth. ZDMG., vol. xlix, p. 284, 1895).

[2] W. W. Rockhill, *Land of the Lamas*. p. 189.

a-gsal-la, openly, manifestly (from *gsal*, clear).

a-riñ, to-day (from *riñ*, long, with reference to time: compare *de-riñ*, to-day).

a-na, *a-k'a*, or *a-rtsa*, here.

a-ạdra (pronounced *andra*), thus.

a-sbyar (Old Tibetan), afterwards.

a-ruñ, once more.

a-čañ, of course, very.

a-bo-tse, good, tolerable.

a-t'o-ba, beautiful, good.[1]

a-yu, hornless (of cattle).

In Eastern Tibet the prefix is combined with the demonstrative pronouns *ạdi* and *de*: *a-ạdi* ("this one"), *a-de* ("that one"); also with *ya* ("one of a pair"). A single case is known where *a-* is combined with a verbal form which is an imperative: *a-gyis*. Jäschke interprets this expression only as an interjection—a word of caress used by mothers — but the Tibetan-French Dictionary intimates that it means *veuillez faire*: accordingly *gyis* is the imperative form of *bgyid-pa* ("to do"). This point of view is confirmed by the existence of the prohibitive form, *a-ma-gyis* ("oh! ne faites pas!").[2]

[1] Jäschke has queried this word with an interrogation-mark. It was derived by him from the dictionary of I. J. Schmidt, who on his part culled it from the Tibetan-Mongol dictionary *rToq-par sla-ba*, where it is indeed thus given and rendered by Mongol *sain*. The word presumably belongs to a dialect of Amdo.

[2] Chandra Das (*Tibetan-English Dictionary*, p. 1342) states that *a-gyis* is "an interrogative pleonastic term signifying: have you done it or done so?" This is possible; in this case, however, *a* has nothing to do with the prefix here in question, but is the interrogative particle *a* largely employed in Eastern Tibet (cf. A. Desgodins, *Essai de grammaire tibétaine*, p. 26). This phrase, accordingly, has no right to be in the lexicon any more than several others listed by Das under separate headings: as, for instance, *a-t'ul*, explained as "a colloquial expression of doubt as to whether an enemy would be vanquished": *a-ñan*, "hesitation to listen to one's advice"; or *a-drag*, "doubt as to whether a thing is good or bad." In these examples we simply encounter a verbal form or adjective prefixed by the interrogative particle *a*: and they should have been entered under the latter as catchword, which, however, is not given under *a*.

The prefix *a*- is finally employed to a large extent in association with interjections. These will be enumerated below when we come to discuss the meaning of the prefix.

Also in Tibetan the use of the prefix is not a matter of yesterday, but an old affair organically inherent in the language. The wide expansion of the prefix over all dialects, and its frequent occurrence in ancient works of literature (the cycle of Padmasambhava and in Mi-la-ras-pa), plainly mark it as an ancient and genuine component of the Tibetan language.

There can be no doubt as to the identity of the Chinese and Tibetan prefixes *a*-. First, they are physically identical, inasmuch as the two have the same tone in common, which is the even high tone. Secondly, their application is the same, both being prefixed to terms of relationship, proper names, and pronouns. Hence we may infer that the origin and the inward significance of the prefix are one and the same in the two languages. This side of the subject will be discussed farther on. In Tibetan the prefix is utilized to a much wider extent than in Chinese; but other languages of the same family, again, by far outrank Tibetan. It seems premature to conclude that in a former period Chinese might have made a proportionately larger use of the prefix and subsequently restricted it; the development may have worked in the opposite direction as well, Chinese approximately representing the original state of affairs, while further progress was gradually effected in Tibetan and allied languages.

In Lepcha we meet a prefix *a*-, the grammatical functions of which are quite apparent.

First, to use the language of our Lepcha grammarians, it is prefixed to verbal roots in order to form substantives and adjectives. It must be observed, however, that the noun character is essentially caused by the affixes -*m*, -*n*, and -*t*.

zo, to eat; *a-zo-m*, food (compare Burmese *tsāḥ*, to eat, and *a-tsāḥ*, food).

ryu, to be good: *a-ryu-m*, good.

čor, to be sour: *a-čor*, sour.

nan, to sit; *a-nan*, dwelling.

ti, to be great; *a-ti-m*, large.

t'i, to come; *a-t'i-t*, arrival.

śu, to be fat; *a-śu-m*, fat (adj.); *a-śu-t*, fat (noun).

dyu, to fight; *a-dyu-t*, war, battle.

tu, to be ominous: *a-tu-m*, evil effect of the omen.

hru (Tib. *dro*), to be hot: *a-hru-m*, hot.

kul, to be encircled with: *a-kul*, girdle.

Second, *a*- is prefixed to substantives to form others of a more specified notion or of diminutive comparison.

yel, to be beautiful; *a-yel*, beauty of plumage in cocks and game-birds.

uṅ, water; *a-uṅ*, water in which meat has been boiled.

vi, blood; *a-vi*, menses.

kuṅ, tree; *a-kuṅ*, bush.

rip, flower; *a-rip*, flower of cloth.

vyeṅ, door; *a-vyeṅ*, pass.

mon, medicine: *a-mon*, grain.[1]

Aside from these two cases in which a grammatical function is as conspicuous as in Burmese, there are other word formations in Lepcha with the prefix *a*- which do not permit an association with a grammatical category, and which are identical with what is found in Chinese and Tibetan. Thus *a*- appears in connexion with names of bodily parts—

a-fyam, hump.

a-mik (Tib. *mig*), eye.

a-boṅ, mouth.

---

[1] Cf. Col. G. B. Mainwaring, *Grammar of the Róng (Lepcha) Language*, pp. 111–12, Calcutta, 1876, and Mainwaring-Grunwedel, *Dictionary of the Lepcha Language*, p. 439. Grunwedel compares with the first category Burmese *c'a* ("to be hungry")—*a-c'a* ("hunger"); with the second, Burmese *im* ("house")—*a-im* ("sheath"). See also L. Vossion, *Grammaire franco-birmane d'après A. Judson*, p. 22.

a-fo, tooth.

a-ñor (Tib. *s-ñan*, Gurung *nha*, Sunwār *nophā*, Tōṭo *nānu*), ear.

a-t'yak, head.

a-ka (beside *ka*), hand.

a-tsom, hair (compare Tib. *a-ts'om*, beard).

a-li (Gurung, Murmi, and Sunwār *le*, Magar *let*, Tōṭo *lebe*, Tib. *lče*),[1] tongue.

a-lim, spleen.

Further, in names of animals :

a-lyu (Tib. *a-lü*, written *a-lus*), cat.

a-lok-fo, raven.

Also in names of plants :

a-tok, a rhododendron.

a-pyoñ } ear of corn.
a-gi }

a-mon, grain in the husk.

a-pi, bark.

a-kok, bark of bamboo.

a-bor, flower.

Even in abstract nouns :

a-pum, origin.

a-pryom, solution of a riddle or problem.

Or in others, like—

a-tit, egg.

a-gli, barrel.

a-nyol, cooking-vessel.

a-fup, crust.

a-lap, carpet.

Finally, as in other Indo-Chinese languages, in terms of relationship :

a-bo, father.

a-mu, mother.

a-num, elder brother (from *nu*, younger brother).

---

[1] Tibetan *lče*, accordingly, is composed of two elements. *l* + *če*, the former being preserved by Gurung, etc., and Si-hia *la* (glossary of Ivanov), the latter corresponding to Chinese *she* 舌.

> *a-nom* (beside *nom*), elder sister.
> *a-yu*, wife (from *yu*, woman).
> *a-kup*, child.
> *a-joṅ*, maternal uncle.
> *a-ku*, paternal uncle.
> *a-nyu*, aunt.
> *a-nop*, sister-in-law, etc.
> *a-vo*, husband.
> *a-zoṅ*, husband of paternal aunt.

The same phenomenon is met in the language of the Gurung in Nepal :[1]

> *a-ba*, father.
> *a-ma*, mother.
> *a-li*, younger brother.
> *a-ṅa*, younger sister.
> *a-gu*, companion.

In Kanāwarī we note *a-te* (" elder brother "), frequently used by the people in addressing one another. *a-yo* (" great grandmother "), *a-i* (" grandmother ").[2]

In regard to the languages in Assam, Sten Konow[3] states in general that the prefixes *a-*, *e-*, *i-*, etc., are used in the same way as the corresponding prefix *a-* in Tibetan and most of the Tibeto-Burman languages of Assam, while the peculiar use of the prefix *a-* in Kachin and Burmese seems to be foreign to them. In the *Linguistic Survey of India*,[4] however, it is stated that the prefix *a-* in Miri and Dafla " is connected with the Burmese prefix *a* which is used in the formation of nouns and adjectives, and with the Tibetan prefix *a* in words such as *a-ma*, mother ". Again, on p. 616 of the same publication, we read in

---

[1] Cf. Grierson, *Linguistic Survey of India*, vol. iii, pt. i, p. 183. I do not believe that the *a*, as here marked, is long (cf. also the editor's remark on pronunciation on p. 182)

[2] Joshi, *Grammar and Dictionary of Kanāwari*, pp. 29, 32, Calcutta, 1909.

[3] " Note on the Languages spoken between the Assam Valley and Tibet " (JRAS., 1902, p. 134).

[4] Vol. iii, pt. i, p. 589.

regard to Digaru, that "the prefix *a-* does not appear to be used in the same way as the prefix *a-* in Kachin and Burmese, in order to form nouns from verbs". On the other hand, again, it is asserted with reference to the language of the Chulikata, a division of the Mishmi, that "a prefix *a* plays a great rôle in the formation of nouns and adjectives" (p. 614). A principal difference between the application of this prefix in the Assam languages and that in Kachin and Burmese, however, can hardly be discovered. In my estimation it is exactly the same, and the latter languages may only claim a higher degree of intensity or a wider extension in its use.

In the language of the Lo-lo we observe the same phenomenon. Father P. Vial[1] has well studied it in the dialect Nyi or Nyi-p'a. According to him it is prefixed to monosyllabic nouns serving to call somebody; for instance, *a-ba* (" father"), *a-ma* ("mother"), *a-pu* ("elder brother"), *a-j'a* ("elder sister," Tib. *a-je*), *a-bu* ("grandfather"), *a-p'i* ("grandmother," Tib. *a-p'yi*), *a-ñi* ("aunt," Tib. *a-ne*). Vial terms this prefix *appellatif*, as it is likewise placed before the names applied to children, when consisting of a single syllable; for example, *A-šle, A-t'o*.[2] When such names are composed of two syllables, the prefix *a-* disappears, "parce qu'elle perd son utilité qui est d'appuyer la voix"; for instance, *Mu-šle, Ts'i-pu*. The latter rule, however, does not apply to terms of relationship, as shown by *a-pu-šle* ("second elder brother"), *a-ba-giai* ("uncle, father's elder brother"). This *a* is further found, as observed by Vial, in connexion with certain other words

---

[1] *Dictionnaire français-lolo*, p. (21).

[2] *A-bi* 阿 毗 is known as the inventor of Lo-lo script (Devéria, "Les Lolos et les Miao-tze," p. 7, extrait du *Journal Asiatique*, 1891). The Lo-lo adopt as personal name also the terms of the zoö-zodiac under which they have been born, this term being linked with the prefix *a-*; for example, A-nu (" born in the year, month, or day of the monkey "), A-ie ("Mr. Rat-Year"). A-jo (" Mr. Sheep-Year "); see P. Vial, *Les Lolos*, p. 37 (Shanghai, 1888, publication of Siccawei).

which are outside of the two categories laid down by him ;
for example, *a-ne-ma* ("raven"),[1] *a-šla-ma* ("hare"),
a-nu-ma ("a kind of bean"). In A-hi, another dialect of
Lo-lo, *a* is prefixed to adjectives : *a-t'o* ("white"), *a-nye*
("black"), side by side with *t'o* and *nye*.[2] In the same
dialect the prefix *a-* alternates with *i-*: in the place of
a-ba, a-mo ("mother"), *i-ba* and *i-mo* may be said as well,
and the employment of these vocalic prefixes is optional.[3]
In his study of the Lo-lo-p'o dialect A. Liétard[4] likewise
drew attention to the same grammatical feature, giving an
abundant selection of examples, in which not only terms
of relationship but also names of animals, professions,
utensils, etc., appear: *a-no* ("dog"), *a-di* ("worm"), *a-pi-p'o*
("sorcerer"), *a-šo-p'o* ("bonze"),[5] *a-t'o* ("knife"), *a-tso*
("axe"), *a-to* ("fire"), *a-mu* ("heaven"), *a-do* ("door").
This *a*, he adds, is never suppressed, except in *a-bo*
("father"), *a-mo* ("mother"), and *a-pa* ("cake"). It is
employed also in the name of girls ; as, for instance, *A-sö*
("the fourth"), *A-lu* ("the sixth"). Finally, the word

[1] A comparative series of this word in the various Lo-lo dialects is
given by A. Liétard (Bull. de l'Ecole française, vol. ix, p. 552, 1909).

[2] A. Liétard, Bull. de l'Ecole française, vol. ix, pp. 290, 294, 555, 1909.
In Kachin also the prefix *a-* enters into the formation of colour adjectives ;
for instance, *a-èyan* ("black"). Likewise in Thādo, which belongs to
the group of Northern Chin languages, and is spoken in southern
Manipur : *a-rom* ("black"), *a-yen* ("green, yellow"; cf. *a-yen*,
"turmeric," in the same language), *a-bon* ("white"), *a-wa* ("bright,
light"; from *wat*, "to shine"), *a-yin* ("dark, dense"); see T. C. Hodson,
Thādo Grammar, pp. 61, 64 (Shillong, 1905). The same feature occurs
in Lepcha : *a-nok* or *a-tyan* ("black"), *a-bok* or *a-don* ("black and
white"), *a-dum* ("white"), *a-fon* ("green"), *a-hyir* ("red").

[3] A. Liétard, ibid., p. 289.

[4] *Au Yun-nan. Les Lo-lo P'o*, p. 217 (Munster, 1913). The premature
death of Father Liétard, who died on July 5, 1912, in Chao-t'ung,
Yun-nan, before the publication of his important work, is an irreparable
loss to science.

[5] The word *šö* apparently is a Chinese loan-word, derived from *ši*
師 ; and it is particularly interesting that M. Maspero (loc. cit., p. 13)
has discovered the Chinese counterpart of the above Lo-lo term in the
form *a-ši* 阿 師 ("maître, moine"). The affix *p'o* means "male".

a-hi ("man") as the designation of the tribe belongs to this category.[1]

A rather extensive use of the prefix *a-* is made also in the Cho-ko dialect, of which we owe to Father Liétard a very valuable word-list in comparison with other Lo-lo dialects and Tibetan.[2] In Cho-ko we meet the prefix in many words where the other languages are lacking in it—

> *a-si*, gold (literally, "the yellow one," from *si*, "yellow," corresponding to Nyi Lo-lo *śe* and A-hi Lo-lo *śa*, "yellow" and "gold"; compare Tibetan *g-ser* from *ser*, "yellow," developed from *ge* (*ke*) + *ser*; Mo-so *ke-se*; Miao-tse *ko*; Si-hia *k'ä* 皆).
> *a-ko*, silver (compare Lepcha *kom*, Mo-so *de-gu*).
> *a-k'u*, iron.
> *a-ko*, fruit.
> *a-ka*, leaf.
> *a-suṅ*, onion (analogous to Nyi Lo-lo *a-ts'e* and Lepcha *o-tsoṅ*; compare Tibetan *b-tsoṅ*, Burmese *krak-swan*,[3] Southern Chin *kwet-šon*, Chinese *suan* 蒜, "garlic," and *ts'uṅ* 葱, "onion").
> *a-ñi*, cat.
> *a-lom*, horse (A-hi Lo-lo *a-lo-mo*; *lo* for *ro*; compare Jyaruṅ *mo-rŏ*, Si-hia *riṅ-ro*, Tibetan *r-ta* from *rŏ-tá*).
> *a-i*, rat.
> *a-si*, monkey.
> *a-na*, raven (A-hi Lo-lo *a-ñe*).
> *a-ni-ku*, beak.

---

[1] More examples will be found in A. Liétard's "Essai de dictionnaire lo-lo français" (*T'oung Pao*, 1911, pp. 17–21). Lolo *a-nò* ("milk") is comparable with Tibetan *nu-ma*, Chinese *nou* 孛殳: Siamese *nom*: *a-ño* ("fish"), with Tibetan *ña*. Lepcha *ño*, Mo-so *ni*, Chinese *ñu, ñ* 魚.

[2] Bull. de l'Ecole française, vol. ix, pp. 549–56. 1909.

[3] Burmese *krak* is doubtless related to Tibetan (written language) *syog-pa* ("garlic"), Yun-nan Tibetan *gau-pa*. Sung-pan Tibetan *čon-grog* (Potanin, *Tanguto-Tibetan Border-Land of China*, vol. ii. p. 395, in Russian). Cf. further Lepcha *suṅ-gu* ("garlic"). A-hi Lo-lo *šo*, Lo-lo-p'o *śu*, Nyi Lo-lo *śe-ma*.

a-čo, man (P'u-p'a *čo*, A-hi Lo-lo *ts'u*, Lo-lo-p'o *ts'a*, Nyi
   Lo-lo *ts'o*, Mo-so *zu-ču*, Leṅ-ki Miao *tsi-ne*, Si-hia *tsu-ni*,
   Manyak *č'o*, Tibetan *ts'o* suffix of plural).

a-sɩ, blood (A-hi Lo-lo *se*, P'u-p'a *su*, Chinese 血).

a-tsɩ, oil.

a-dsɩ, lamp (P'u-p'a *a-teṅ* is a Chinese loan-word : and the
   prefix in connexion with a loan-word is interesting, as it
   shows that the speakers are perfectly conscious of its use).

a-löm-je, to ride on horseback (compare *a-lom*, horse).

a-si-bya, yellow (compare, above, *a-si*, gold).

a-n-zya, green.

a-n-bya, red.

As already mentioned, *a-* appears also in combination
with the interrogative pronouns, *a-su* (Nyi dialect) and
a-so (A-hi dialect), in the same manner as in Chinese
a-šui. The same phenomenon occurs in the language of
the Mo-so : *a-ne* (" who ? "), *a-tse* (" what ? ").[1] In Mo-so,
of which we unfortunately possess only scanty vocabularies, we find, moreover, *a-pa* (" father "), *a-me*
(" mother "), *a-p'u* (" grandfather "), *a-tsö* (" grandmother "),
a-bu (" elder brother ") ;[2] further, *a-me* (" chicken ") and
a-jo or *a-yü* (" monkey ") ; and adverbial formations like
a-ñi (" yesterday ") from *ñi* (" day "). The interesting
documents relative to Li-kiang, translated by Chavannes,
supply us with a great many Mo-so names preceded by the
prefix *a-* : A-ti, A-ch'ao-tso, A-shui-ch'eng, A-ku, A-tsung,
A-liang, A-hu, A-lie, etc.[3] The same element enters into

[1] J. Bacot, *Les Mo-so*, p. 59. Mo-so *ne* apparently is identical with
Chinese *na* 那, and Mo-so *tse* with Tibetan *ci*.

[2] Cf. H. Cordier, *T'oung Pao*, 1908, p. 683.

[3] *T'oung Pao*, 1912, pp. 611, 614 et seq. It is a feature of particular
interest that in the " History of the Yüan " (*Yüan shi*, chap. 61, p. 4)
the name A-liang is written Me-liang, and the name A-hu Me-wu (ibid.,
p. 569). M. Chavannes explains this word Me as a tribal name, the
Chinese being in the habit of prefixing to the name of a chief that of his
tribe, which was gradually looked upon by them as his family-name.
Thus Me was exchanged for Mu, the name of the Mo-so chieftains of
Li-kiang, in 1382, when the latter themselves adopted the Chinese
custom of family-names. This point of view is confirmed by Sü

the formation of proper names among the Shan tribes ; as, for instance, A-k·ing, A-jung-ho, A-li, A-fang, A-k·o.[1] T. de Lacouperie[2] has called attention to the fact that a prefix *a*- is linked to proper names in the language of the Chung-kia-tse 犭 家 子 or Chung Miao 犭 苗, which belongs to the northern branch of the T·ai group.[3]

Hung-tsu (1586–1647), who says that all the chiefs of Li-kiang bore the family-name Me from the time of the Han down to the Ming, when T·ai-tsu altered it into Mu. Of course, neither the Mo-so nor the Lo-lo or Tibetans ever had family-names ("Les Lolos n'ont pas de noms patronymiques, ils n'ont qu'un nom personnel," says P. Vial, *Les Lolos*, p. 37) ; and it is solely Su Hung-tsu's personal viewing of the matter when he takes Me for a family-name. But is Me really a designation of ethnic value which as such ever had currency among the Mo-so themselves ? I venture to doubt it, despite the alternating forms Me-ch·a and Mu-ch·a pointed out by M. Chavannes What is certain to me is, that the element *me*-, as shown by the *Yüan shi*, is a prefix on exactly the same footing as *a*- ; Me-liang is identical with and the equivalent of A-liang. The labial prefixes, *ma*-, *me*-, and *m*- are very frequent in numerous Indo-Chinese languages (*ma*-, for instance, in Lepcha, and in Chulikata and Digaru Mishmi [*Linguistic Survey of India*, vol. iii, pt. i, pp. 614, 616], *me*- in Miri [ibid., p. 589], *m*- in Miju Mishmi [ibid., p. 619], Tibetan, Mo-so, and Lolo) ; and what is particularly notable, the prefix *ma*- is interchangeable with *a*- in Kachin (Sten Konow, ZDMG., vol. lvi, p. 493, 1902). In this language nearly all personal names are combined with the prefix *ma*- (H. F. Hertz, *Handbook of the Kachin or Chingpaw Language*, p. 37, 2nd ed., Rangoon, 1902). Cf. also A-hi Lo-lo *me-ne* ("cat") with Cho-ko *a-ñi* ("cat"). For this reason it is equally probable that at the time of the Mongols there was a period of the Mo-so language when the prefix *a*- could alternate with the prefix *me*-.

[1] C. Sainson, *Histoire particulière du Nan-tchao*, pp. 116, 125, 145, 247, 258 ; see also the Index on p. 277.

[2] *Les langues de la Chine avant les Chinois*, p. 63.

[3] P. Vial (*Les Lolos*, p. 33) states that the proper mode of writing Chung-kia is 重 甲 : that is, "heavy cuirasses." This is somewhat more sensible : but the chances are that Chung-kia originally had no meaning in Chinese, but that it is the indigenous designation of the tribesmen in question, which the Chinese, *tant bien que mal*, attempted to reproduce in their writing.—It is very curious that a prefix *a*- in connexion with proper names occurs also in Khmer, which belongs to the Mon-Khmer family of languages, that is not morphologically related to Indo-Chinese. M. Moura (*Vocabulaire cambodgien-français*, p. 33) states, "*a*, devant un nom propre d'homme indique la familiarité, s'il s'agit d'un enfant ; il marque le mépris, s'il s'agit d'une personne âgée." K. Himly ("Bemerkungen über die Wortbildung des Mon": Sitzungsber. bayer. Akad., 1889, p. 274) has drawn attention to this prefix in Mon,

Various theories have been brought forward to explain the significance of this prefix *a*-. Steinthal supposed that *a*- of Burmese is identical with Siamese *an* ("matter, something"), as in *an tī* ("what is good"). F. Müller[1] acceded to this opinion; but it is decidedly untenable. Aside from the impossibility of interpreting a Burmese prefix through the medium of a material word of Siamese, this theory does not account for the extensive diffusion of the same prefix in a large number of other Indo-Chinese languages.

It would be very tempting to regard the prefix *a*- in its origin as a demonstrative pronoun. Indeed, it has been taken in this sense by several students of those languages in which a pronoun *a* of that valuation actually occurs. Thus Grierson[2] observes in regard to Abor-Miri that the prefix *a*- in the demonstrative pronouns *a-da* ("that") and *a-la* ("that portion or thing in sight but not near") is apparently an independent pronoun. He refers for comparison to *a-la* ("there"; literally, "that-in") and *a-lokka* ("therefrom"), adding that a corresponding pronoun *a* occurs in many other connected dialects. As regards Tibetan, Sten Konow[3] has advanced the opinion that the prefix *a*-ཨ., which is written with the consonantic letter transcribed by us ,*a*, seems to be identical with the demonstrative pronoun ,*a* ཨ of Ladākh and K'ams. However plausible at first sight this view may be, it seems to me, nevertheless, to be exposed to grave objections.

Stieng, and Khmer: Mon *a-kruim* ("boastful") from *kruim* ("to boast"); *a-čak* ("link") from *čak* ("to unite"); *a-gah* ("that one") from *gah* ("this one"). In view of the profound historical influence of Burmese upon Mon, the formation of nouns from verbal roots by means of *a*- in Mon might be ascribed to an impetus received from that quarter; for the rest, however, the entire question requires a special investigation in the Mon-Khmer group.

[1] *Grundriss der Sprachwissenschaft*, vol. ii, p. 352.

[2] *Linguistic Survey of India*, vol. iii, *Tibeto-Burman Family*, pt. i, p. 595 (Calcutta, 1909).

[3] ZDMG., vol. lvi. p. 493. 1902.

The pronoun ,*a* in the dialects of Ladākh and K'ams is an isolated occurrence, and therefore can hardly account for a phenomenon that affects all Indo-Chinese languages from the mountain valleys of Nepal and Assam to the plains of China. Moreover, the existence of this alleged pronoun ,*a*, as already intimated by Jäschke,[1] is at least doubtful, and in all probability it rests only on a mishearing for *a* ཨ⋅ It is well known that Tibetan has two different signs for the expression of *a*, which correspond to two strictly differentiated sounds ; the one, ,*a* ས⋅, is produced by the opening of the glottis, like the Greek spiritus lenis or the Arabic aliph ; the other, *a* ཨ⋅, is a pure vowel, without any admixture of a consonant. The old and regular demonstrative pronoun of the Tibetan language, however, is always *a* ཨ⋅ with the variants *e* ཨེ⋅, *u* ཨུ⋅, *o* ཨོ⋅ which I have discussed on a former occasion,[2] but is never ,*a* ས ⋅ This being the case, it is improbable, nay, impossible, that the Tibetan prefix ,*a* should have been derived from the pronoun *a*. Not only are the two physically distinct sounds, represented in writing by two diverse letters, but they differ also in tone, ,*a* being high-toned and *a* being deep-toned : and high-toned and deep-toned words are not comparable. These two words, accordingly, bear no relation whatever to each other. For this reason I am not ready to accept Sten Konow's theory. It is insufficient also for other reasons : while it would be plausible, for instance, that the prefix *a*- in connexion with terms of relationship might be a demonstrative pronoun, this is not evident in other groups of words, as, for example, in the colour adjectives met in Lo-lo and in Kuki-Chin languages. The theory, consequently, is too narrow, and fails to cover the entire psychology of the case.

In the same article Sten Konow[3] develops another

---

[1] *Tibetan-English Dictionary*, p. 603.

[2] *T'oung Pao*. 1914, p. 56, note.

[3] ZDMG., vol. lvi, pp. 513–4, 1902.

theory as to the origin of the prefix *a*- in the Kuki-Chin languages. In these, terms of relationship are never used in a general, abstract sense, but are always correlated with a distinct individual; that is to say, they are regularly combined with the possessive pronouns: thus in Thādo *ka-pa* ("my father"), *na-nu* ("thy mother"), *a-k'ut* ("his hand"). This pronominal *a* of the third person then became an integral component of the word-stem, and this process was facilitated by the pre-existence of another prefix *a*- identical with the corresponding prefix in Burmese. Sten Konow, accordingly, assumes two different prefixes *a*-, one of pronominal character, and another of word-forming tendency.[1] All this may well hold good for the group of languages visualized by the author, but again we are at a loss as to how to apply this explanation to other Indo-Chinese languages showing the same formative principle. Chinese, Tibetan, and Lo-lo lack a possessive pronoun *a*; and the possible conclusion that these languages might have lost it, merely in view of the fact that it exists in the Kuki-Chin group, would not seem to me to be justified. In consequence of our still imperfect knowledge of Indo-Chinese languages, etymological speculations will remain at present somewhat hazarded, and in the case under review it appears advisable to restrict our attention to the psychological significance of the case rather than to endeavour to unravel the etymological origin.

In examining carefully the list of words which in the various languages are capable of assuming the prefix *a*-, we note that they have a very specific relation, first of all, to the social life of the community, and, second, to the

---

[1] Examples of the latter kind are cited on p. 492 of his article. Cf. also T. C. Hodson, *Thādo Grammar*, pp. 9, 13. In the *Linguistic Survey of India* (vol. iii, pt. i, p. 575) it is observed that the prefixes *a*-, *e*-, or *u*- in Aka are probably identical with the possessive pronoun of the third person, while the prefix *na*- is explained as being perhaps that of the first person.

speaker. They are equivalents of social importance, in which every individual has his share, and which for this reason are somehow emphasized by the speakers. They are (in a different sense, however, from the usually accepted one) terms of endearment; that is, words for persons or objects which have endeared themselves to the individual as a member of the social unit. The terms of relationship are of prime importance in this psychological category. Here it is notable that the *a*-forms are principally employed in addressing persons. *Ma*, in Tibetan, is "mother", anybody's mother, but *a-ma* is exclusively the speaker's individual mother, thus addressed by him in the sense of "dear mother". It is of especial interest to note how the social horizon of these expressions is widened by gradually embracing larger social bodies. Thus *a-jo* (literally "elder brother") becomes the general address for every gentleman ("sir"), and *a-če* ("elder sister") assumes the broader significance of "mistress, madam". *A-ne* ("aunt") widens into an address for nuns. As to their grammatical form, these terms are vocatives; and as to their significance, they imply a tinge of an affectionate state of mind on the part of the conversationalist. This fully explains also why we find the same element in proper names of Chinese, Lo-lo, Mo-so, and Tibetan. It is the name by which a person is called by others to whom he or she is dear (*A-ho*, for instance, "my dear Ho"). Father Vial was guided by a correct feeling when conferring upon the prefix the designation *appellatif*; this word exactly describes what it is. A similar sentiment is evinced by the Tibetan or Lepcha when he calls the cat *a-lü* or *a-lyu* ("kitty"), and here our own diminutive formations spontaneously loom up in our mind. Along this line the Lepcha have further developed the application of the prefix. This foundation being inferred, it is conceivable to me that from the very beginning the element *a* might have had this vocative meaning of endearment, and that the assumption of a possible interrelation with a pronominal element, be it demonstrative or possessive, at least, is not imperative. I am very much inclined to regard it as an original, spontaneous, emotional nature-sound, very much like our own interjection *ah!* formed with or without a vocative; and the same interjectional *a!* naturally is found also in the Indo-Chinese languages.[1] As such it is met in Tibetan, and also in compounds of interjectional value in which *a* visibly plays the rôle of a prefix:

a-k'a, *a-k'a-k'a*, or *a-k'ag*, exclamation expressive of contempt, detestation, or bereavement (also in the written language). The opposite is

a-la, or *a-la-la*, expressive of joyful surprise (*a-la-la-ho* frequently occurs in the Ge-sar Saga to introduce songs).

a-gyis, a word of caress used by mothers in soothing their babies.

a-tsa, or *a-ts'a*, expressive of pain by touching fire or hot objects (hence *tsa* and *ts'a* in *ts'a-ba*, "heat, hot"; *a-tsa*, "ah, how hot!"), and in general, an interjection of sudden fright or profound regret. The opposite is

a-č'u, expressive of pain from cold, hence name of one of the Cold Hells.

a-na, expressive of grief.

a-ma-ńa, same.

a-ra-ra, cry of anguish.

a-tsi-ts'i, expressive of regret or repentance.

a-tsi, expressive of wonder.

a-mts'ar, oh dear, what wonder!

a-pi, expressive of wonder on making a new experience.

a-re, well, then!

In these examples the purely interjectional character of the element *a* is plainly obvious. It is equally manifest also that it has the function of a prefix, and that in its quality as a prefix it cannot be identified with a demonstrative pronoun.

[1] In regard to Chinese see Watters, *Essays on the Chinese Language*, p. 136.

A similar series of interjections with *a-* is found in Lepcha,[1] and doubtless occurs in other Indo-Chinese languages.

The same interjectional force of meaning is valid also for the interrogative and demonstrative pronouns. When *a-šui* or *a-su* are said in the place of the plain *sui* or *su* ("who?"), it is evident that a high degree of emphasis is laid on the pronoun; and the same stress we feel in *a-adi* and *a-de* as compared with *adi* and *de*. Thus I am inclined to think that the original vocative and interjectional significance of *a* further developed into that of a strongly emphatic word which in due course became available as a useful vehicle in the formation of words and in expressing certain shades of meaning. At this point we naturally come down to the period when the various speech-groups of the family were separated and scattered over an immense area of land. The word-forming tendency of the prefix *a-* was set in operation after the separation of the various members of the Indo-Chinese, and consequently assumed individualistic features in the single branches of the stock. The interjectional value of *a-*, being in common to all languages, must have been in force prior to the time of separation, and presents the primeval archetype which is responsible for the subsequent, separate developments in the single languages. The formation of the mechanical rule that in Lepcha, Kachin, and Burmese a prefix *a-* forms nouns and adjectives is erroneous. This merely means to view foreign languages in the light of Latin grammar, and to read our own grammatical notions into them, according to the method in which most grammars of Indo-Chinese languages have unfortunately been written. That rule may offer a certain vantage-point to the practical student of a language, but it has nothing to do with scientific observation. The distinction of Lepcha *a-hru-m* from *hru*, for instance,

[1] See the Dictionary of Mainwaring-Grunwedel, p. 440.

does not lie in the alleged fact that the former is a noun or adjective and the latter a verb. These terminological categories are merely artificial constructions of our mind suggested by the word-conditions of Indo - European languages, but it does not mean at the outset that they necessarily exist also in other languages than our own. Certainly, the word - categories in Indo - Chinese are fundamentally different from ours; and the differentiation of noun, adjective, verb, etc., does not by any means reach there that degree of essential importance that it has in Indo - European. The case of *a-hru-m* means to the philologist a word-formation from the stem *hru* by means of the prefix *a-* and the affix -*m*, expressing with greater emphasis and intensity the idea conveyed by the stem *hru*. It could not be immediately inferred from the very character of the language that the one is a noun and the other a verb. This distinction, in fact, is not made by the language itself, but only by us who read the effects of our Latin school-training into a language developed outside of our culture-sphere. Our procedure is the same as though we were to treat the manners and institutions of the Lepcha from the standpoint of Roman law, taking up section by section of the latter and religiously checking it up with what is offered by the Lepcha. He who can break away from the slave-fetters of our grammar and think objectively in a foreign language, will easily recognize that the theories previously advanced on the nature of the prefix *a-* are imaginary, and that the development, as here outlined, so far as this is possible in the present state of our knowledge, truly corresponds with observable data. There are not two prefixes *a-*, as assumed by Sten Konow, but there is only one, reducible to a monophyletic origin, and appearing as a uniform manifestation throughout the group.

B. LAUFER (Chicago).

113

哈刺章

KARAJANG

As is well known, Marco Polo (bk. ii, chap. 48) applies
to the province of Yün-nan the name Karajang. Henry
Yule, in his classical edition of Polo's travels (vol. ii, p. 72),
correctly analysed this word into *kara-jang*, explaining
the first element as the Mongol or Turkish word *kara*
(" black "), and referring to the " White Jang " (*Tsaghan-
Jang*) mentioned by Rashid-eddīn. As to the second
element of the name, Yule annotated, "*Jang* has not been
explained; but probably it may have been a Tibetan
term adopted by the Mongols, and the colors may have
applied to their clothing." M. Pelliot (*Bull. de l'Ecole
française*, vol. iv, 1904, p. 159) proposed to regard the
unexplained name *Jang* as the Mongol transcription of
Ts'uan, the ancient Chinese designation of the Lo-lo,
taken from the family name of one of the chiefs of the
latter; he gave his opinion, however, merely as an
hypothesis which should await confirmation. I now
believe that Yule was correct in his conception, and that,
in accordance with his suggestion, *Jang* indeed represents
the phonetically exact transcription of a Tibetan proper
name. This is the Tibetan *ɑJaṅ* ᘔᘔᘔ or *ɑJaṅs* ᘔᘔᘔ
(the prefixed letter ɑ and the optional affix -*s* being silent,
hence pronounced *Jang* or *Djang*), of which the following
precise definition is given in the *Dictionnaire tibétain-
latin-français par les Missionnaires Catholiques du Tibet*
(p. 351): "Tribus et regionis nomen in N.-W. provinciæ
Sinarum Yun-nan, cuius urbs principalis est Sa-t'am
seu Ly-kiang-fou. Tribus vocatur Mosso a Sinensibus
et Nashi ab ipsismet incolis." In fact, as here stated,
Jaṅ or *Jang* is the Tibetan designation of the Mo-so
and the territory inhabited by them, the capital of which
is Li-kiang fu.[1] This name is found also in Tibetan

---

[1] The Tibetan name for it, *Sa-t'am*, is entered also on p. 1010 of the
same dictionary. M. Bacot, in his attractive work *Les Mo-so*, p. 3,

literature; for instance, in the Tibetan historical work *dPag bsam ljon bzaṅ* (p. 4, l. 7). In the valuable index to this book prepared by the editor, Chandra Das (p. xxxvi), it is explained as the "name of a place on the border of China, part of eastern or ulterior Tibet". It is less obvious why the same author, in his *Tibetan-English Dictionary* (p. 452), with reference to the same Tibetan work edited by him, insists on *Jang* being a "place in N.W. Tibet which once formed the kingdom of *Jang*".[1] A conspicuous rôle is played by the country *Jang* in the Tibetan epic romances of King Ge-sar, which are divided into three parts, dealing with Tibetan wars against the Chinese, the Turkish tribes (*Hor*), and the Mo-so, styled *aJaṅs-gliṅ* ("country of Jang").[2] In a printed edition as

transcribes the name *Sa-dam*. It is an interesting case that M. Bonin (*Les royaumes des neiges*, p. 281) heard and rendered it in the form *Sdam*, as this illustrates the formation of a prefixed consonant, a phonetic phenomenon so peculiarly Tibetan (gradual elision of the vowel in the first element of a compound under the influence of a strong accentuation of the ultima).

[1] Again (JASB., 1904, extra No., p. 98), he explains the country *Jang* as Kuku-nōr region and Amdo. He cites also *Sa-t'am* as the "name of a place in K'am" from the writings of the Lama Kloṅ-rdol, so that there is no doubt that, as correctly stated by the French missionaries, *Sa-t'am* is the adopted form of the Tibetan written language. It is thus written also in the Tibetan Ge-sar Saga.

[2] This information is usually credited to E. C. Baber (for instance, by T. de Lacouperie, *Beginnings of Writing*, p. 43; G. Devéria, *Frontière sino-annamite*, p. 164; H. Cordier, *T'oung Pao*, 1908, p. 670); but Baber (*Travels and Researches in Western China*, p. 88, in Supplementary Papers, Roy. Geogr. Soc., vol. i, London, 1886) honestly acknowledged that he received this information from Mgr. Biet, Apostolic Vicar of Tibet in Ta-tsien-lu. Indeed, it is forestalled by Desgodins in *Annales de l'Extrême Orient*, vol. ii, p. 133, Paris, 1880, and hence repeated in his book *Le Tibet d'après la correspondance des missionnaires*, 2nd ed., Paris, 1885, p. 369, where the curious transcription *Guiong* for the Tibetan name *Jang* appears, while Baber and his numerous followers wrote *Djiung*. T. de Lacouperie (loc. cit.) stated that *Guiong* is identical with *Djiung*, and from M. Bacot's book (*Les Mo-so*, p. 13) it appears that *Djung*, as he writes, is merely a variant of *aJang* (*Hdjang*). The French missionaries themselves have never explained this diversity of names; in their dictionary (p. 253) they give with reference to the Ge-sar Saga

well as in a manuscript of this work, the name is written with a prefixed *l* (*lJań*), which leads to the pronunciation *Jang* also. This mode of writing was presumably inspired by assimilation to the word *ljań* ("green").

I am under the impression that the Tibetan tribal and geographical term *Jań* (*aJań*) meets all requirements of the case, and that we are justified in identifying with it the second element of the Mongolized formation *Kara-jang* transmitted by Rashid-eddīn and Marco Polo. As pointed out by M. Chavannes (*T'oung Pao*, 1912, p. 615; or in Bacot, *Les Mo-so*, p. 177), the Chinese have conveyed to us the transcription *Ch'a-han Chang* 茶罕章, answering to Mongol *Tsaghān Jang* ("the White Jang"). This name we may carry back to the Yüan period, for in the *Yüan shi*, as likewise observed by Chavannes (l.c., p. 603), it is encountered in the form 察罕章. The Mongol transcription *Jang* of the same name is preserved by Sanang Setsen (I. J. Schmidt's edition, p. 238), who calls the king of this tribe (the time refers to the end of the sixteenth century) *Sitam*, presumably a rendering of the name of the Mo-so capital, *Sa-t'am*. The character 章 is well chosen as the instrument to transcribe the Tibetan word *Jań* or *Jang*: both words perfectly agree in the tone, which is the high tone. If *j* were the initial letter of the Tibetan word, it would naturally be deep-toned: it is protected, however, by the prefix *a*, which renders it high-toned; and the insurance of this result is the essential function of this prefix, which in this case

only the form of the written language, *aJańs*. The attention of future travellers in those regions may be called to this point; it would be interesting to see this possible dialectic change of *a* into *u* confirmed. Independent of the Tibetan studies carried on at Ta-tsien-lu, K. Marx, a Moravian missionary stationed at Leh, Ladāk, read in the Tibetan epic about Ge-sar's wars against the *Jang* (see JASB., vol. lx, p. 116, n. 13, 1891).

marks the tone and precludes the aspiration of the initial media.[1]

B. LAUFER (Chicago).

THE INDIAN ORIGIN OF THE GREEK ROMANCE

By far the ablest attempt to establish the Indian origin of the Greek Romance is that made by F. Lacôte in a paper included in the *Mélanges d'Indianisme offerts par ses élèves à M. Sylvain Lévi*,[2] and it is well worth while investigating the case made out for the thesis by this scholar. His paper contains, besides direct reference to his theory, much other valuable matter which admits of less doubt and which need not be considered here.

The distinctive feature of the argument for derivation is the view that the Greek romance borrowed its form from the Indian. It is shown that the Indian Kathā is essentially narrated, not of, but by, the parties to the action, and that this rule produces a curiously involved form of narrative such as may be seen in any of the famous Indian Kathās. These, however, in their elaboration are only developments of the simpler Ākhyāyikā, which can be traced from its simplest form in the *Jātakas* through

---

[1] At first sight I was almost tempted to recognize the name *Jang* also in the *T'ang Annals*, as T. de Lacouperie (*Beginnings of Writing*, p. 41), following d'Hervey St.-Denys, mentions a tribal name *Mo-tchang Man* (thus also Cordier, *T'oung Pao*, 1908, p. 664). In looking up the text in the *T'ang shu* (Ch. 222A, p. 4b) I find, however, that this alleged *tchang* corresponds to 裳, which reads *shang*; and *shang* could hardly be taken as the transcription of Tibetan *aJaṅ*. As regards the phonetic relation of the latter to Chinese *chang*, compare the Tibetan transcriptions *ajam-mo* ("post-stage") and *aja-sa* ("edict"), reproducing the Mongol words *jam* and *jasa* respectively, which were transcribed by the Chinese *chan* 站 and *cha-sa* 札 撒 (see Pelliot, *Journal asiatique*, 1913, Mars-Avril, p. 458). The Tibetan transcriptions, presumably modelled directly after the Mongol forms, are again shielded by the prefix *a*, which has neither a grammatical nor a graphic function, but a purely phonetic rôle, safeguarding the initial *j* from aspiration.

[2] Paris, 1911, pp. 249–304.

114

光学镜片：一、中国与印度的聚焦点火透镜

T'OUNG PAO

通報

OU

ARCHIVES

CONCERNANT L'HISTOIRE, LES LANGUES,
LA GÉOGRAPHIE ET L'ETHNOGRAPHIE
DE
L'ASIE ORIENTALE

Revue dirigée par

Henri CORDIER
Membre de l'Institut
Professeur à l'Ecole spéciale des Langues orientales vivantes
ET
Edouard CHAVANNES
Membre de l'Institut, Professeur au Collège de France.

VOL. XVI.

LIBRAIRIE ET IMPRIMERIE
CI-DEVANT
E. J. BRILL
LEIDE — 1915.

OPTICAL LENSES.

BY

BERTHOLD LAUFER.

————·<⊃≡›·.———

I. BURNING-LENSES IN CHINA AND INDIA.

FIRE-PRODUCTION BY MEANS OF OPTICAL LENSES AMONG THE ANCIENTS. — Crystal lenses, wherever employed in ancient times, served for one main purpose exclusively, — the optical method of fire-making. This method is not found among any primitive tribes of the world, but it is restricted to the highly advanced nations settled around the Mediterranean and to the peoples of India and China. W. HOUGH, in his interesting study *The Method of Fire-Making*, [1] has justly observed, "Among the several ways of producing 'pure' fire the mirror and lens presented a worthy method to those ancient cultured nations possessing instruments for focussing light. It can scarcely be said that this was a wide-spread and popular plan for producing fire, but probably was a thing known to priests and scientific men of the day, and viewed as a mystery or curiosity."

The centre of gravity of the following inquiry lies in a new research of this interesting subject, as far as China and India are concerned. [2] China and India, however, were not isolated in the age

---

[1] *Report of National Museum*, Washington, 1890, p. 408.

[2] This study owes its origin to a suggestion received from Dr. Frank Brawley and Dr. Emory Hill, two prominent oculists of Chicago, who are about to issue a comprehensive cyclopædia of ophthalmology, and desire to obtain reliable information on the history of optical lenses in Asia. The second part of this essay will deal with the history of spectacles.

when the utilization of lenses loomed up on their horizons, but partook of the blessings of that great world civilization inspired and diffused by Hellenism. This subject therefore, like all other culture-historical problems, must be visualized within the frame of universal history; and it will hence not be amiss first to pass in review what we know of burning-lenses among the ancients in the western part of the world.

The peoples of classical antiquity were acquainted with two optical instruments for the production of fire, — concave burning-mirrors and convex burning-lenses focussing the sunlight. The question as to whether these are to be attributed to the inventive genius of the Greeks, or were modelled by them on the basis of previous achievements of Mesopotamian civilization, cannot be decided in our present state of knowledge. H. LAYARD [1] (1845) discovered in the palace of the Assyrian King Ashur-naṣir-pal (885 — 860 B.C.) at Nineveh a rock-crystal lens of plano-convexity, $1\frac{1}{2}$ inches in diameter, with a focus of $4\frac{1}{2}$ inches, cut much like our own burning-glasses, though somewhat crude in its workmanship. It may well have performed the function of a burning-lens, as admitted by modern technologists; [2] but we should await more evidence before crediting the first invention of burning-lenses to the nations of the Euphrates Valley.

The earliest well-authenticated literary testimony for the use of burning-lenses remains the famous scene in Aristophanes' (c. 450 — c. 385 B.C.) comedy *The Clouds* (Νεφέλαι), written in 423 B.C., where the following dialogue ensues between Strepsiades and Socrates (I quote from T. Mitchell's rendering). [3]

---

[1] *Discoveries among the Ruins of Nineveh and Babylon*, p. 197.

[2] NIEMANN and DU BOIS (in KRAMER, *Der Mensch und die Erde*, Vol. VII, p. 162); and FELDHAUS, *Technik der Vorzeit*, col. 667.

[3] The situation is this: Strepsiades, who has run up a debt of five talents, wants to dodge his obligation by destroying the bill of complaint recorded in wax by operating on it a burning-lens.

STREPSIADES.	I've hit the nail
	That does the deed, and so you will confess.
SOCRATES.	Out with it!
STREPSIADES.	Good chance but you have noted
	A pretty toy, a trinket in the shops,
	Which being rightly held produceth fire
	From things combustible —
SOCRATES.	A burning-glass,
	Vulgarly call'd —
STREPSIADES.	You are right; 'tis so.
SOCRATES.	Proceed!
STREPSIADES	Put the case now your bailiff comes,
	Shows me his writ — I, standing thus, d'ye mark me,
	In the sun's stream, measuring my distance, guide
	My focus to a point upon his writ,
	And off it goes in fumo'
SOCRATES.	By the Graces!
	'Tis wittingly devis'd.

This translation is somewhat free, and does not bring out the technical points which are of importance for a consideration of the burning-lens. Strepsiades describes it as a beautiful and diaphanous stone (λίθος διαφανής ἀφ' ἧς τὸ πῦρ ἅπτουσι); and what Socrates in the above translation calls a burning-glass is in the Greek *hyalos* (ὕαλος). It is presumed that this word here appears for the first time in Greek literature in the sense of "glass," [1] and accordingly that Aristophanes speaks of burning-lenses made from glass. [2] The reasons given in support of this opinion, however, are by no means convincing. The first Greek author with a distinct mention of glass is Herodotus (II, 69), who terms it "molten stone" (λίθος χυτή) with reference to the ear-rings placed by the Egyptians in the ears of their tame crocodiles. Herodotus (III, 24) likewise is the first to use the word ὕαλος in the description of the coffins of the Ethiopians, where it most evidently has the significance of "rock-crystal" or some other

---

[1] BLÜMNER, *Technologie*, Vol. IV, p. 384.

[2] M. H. MORGAN, *De ignis eliciendi modis apud antiquuos* (*Harvard Studies in Classical Philology*, Vol. I, 1890, p. 46). This is the most complete study of Greek and Roman methods of fire-making, inclusive of burning-lenses and burning-mirrors.

transparent stone; [1] for "they put the prepared body in a crystal pillar hollowed out for this purpose, crystal being dug up in great abundance in their country." [2] If ὕαλος has in Herodotus, as shown by the inward evidence of the passage, the meaning of "rock-crystal," I see no reason why the same meaning should not be attributed to it in Aristophanes. Besides the passage cited, there is but one other in which the great writer of comedy makes use of the word: in *The Acharnians* the Greek ambassadors, returning from a mission to the King of Persia, report,

> "At our reception we were forced to drink
> Strong luscious wine in cups of gold and crystal," [3]

as J. H. Frere translates with perfect correctness; where Blümner, Morgan, and others, however, see the first mention of glass vessels in Greek records. [4] It seems to me more probable that gold and crystal vessels are here spoken of. In order to succeed in making the burning-lenses mentioned in *The Clouds* of glass, Morgan is obliged to have recourse to two theories which are unsupported by evidence. We see plainly from the words of Aristophanes, he observes, that glass was very rare in his time (while two pages ahead glass utensils were then at Athens), since he calls it a precious stone (*gemma*); and, as it is said that this stone is for sale in the shops of the pharmacists (*pharmacopola*), it is proved by this very fact that the matter was regarded as a miracle. This "miracle" will fade away, if we adopt the reasonable and natural interpretation of taking ὕαλος in this passage as "rock-crystal" with the specific sense of "burning-

---

[1] Some authors take it for Oriental alabaster or arragonite, which is transparent when cut thin.

[2] Thus also Achilles Tatius calls rock-crystal ὕαλος ὀρωρυγμένη.

[3] Ἐξ ὑαλίνων ἐκπωμάτων.

[4] MORGAN (*l. c*, p. 44) says with regard to this passage that glass utensils were at Athens as early as in Aristophanes' times; the passage, in my opinion, would allow only of the inference that they were at the Court of Persia, and dimly known to Aristophanes.

lens of crystal;"[1] and we are thus released from the necessity of making Aristophanes speak of glass as a precious stone. Strepsiades' description fits "crystal" very well indeed. There are other, historical reasons which warrant the belief that the first burning-lenses were cut from crystal, not from glass, as will be shown by a study of this subject from Chinese and Sanskrit sources.

M. H. MORGAN,[2] it is true, makes the point that rock-crystal became known only at a late period in classical antiquity, shortly before Augustus; and he reveals the Roman poet Helvius Cinna, and Strabo, who mentions the occurrence of crystals in India, as the earliest authorities. This opinion, however, is not correct. Rock-crystal (ἡ κρύσταλλος) is distinctly alluded to by Theophrastus (372—287 B.C.)[3] as a translucent stone together with anthrax, omphax, and amethyst, all of which can be turned into signet-rings.

More important than the material of which the burning-lenses of the Greeks were made is the question as to their purpose and mode of use. The scene in Aristophanes' comedy enlightens us in this respect on two points. The effect of a burning-lens was perfectly known. The legal document of which Strepsiades speaks was certainly draughted on a tablet of wax, and related to a debt which he contracted; he intends to foil his creditors by melting the wax by

---

[1] This interpretation is adopted by LIDDELL and SCOTT in their *Greek-English Lexicon*.

[2] *Harvard Studies in Classical Philology*, Vol. I, pp. 44, 48—49.

[3] *De lapidibus*, V, 80 (opera ed. WIMMER, p 345, Paris, 1866). This fact is indicated also by KRAUSE (*Pyrgoteles*, p. 16) and SCHRADER (*Reallexikon*, p. 152). Theophrastus is the first Greek author to speak of rock-crystal. As is well known, the word κρύσταλλος occurs in Homer, but has the significance "ice" (derived from κρύος, "chill, frost"); an analogous example is presented by Hebrew *qerah* meaning "ice" and "rock-crystal." The actual utilization of the mineral is certainly much older than the allusions to it in literature It occurs among the material listed for cylinder-seals in Mesopotamia (HANDCOCK, *Mesopotamian Archæology*, p. 287) and among the intaglios of the Minoan, Mycenæan, and archaic Greek periods (D. OSBORNE, *Engraved Gems*, pp. 25, 283). On rock-crystal among the ancients, in general compare L. DE LAUNAY, *Minéralogie des anciens*, Vol. I, pp. 22—28; and C. W. KING, *Antique Gems*, pp. 90—97.

means of a burning-lens, and thus to escape judicial proceedings. Such action was not the order of the day, but the specific witty thought sprung by Strepsiades, at which Socrates laughs. The destruction of writs, therefore, was not the real object of burning-lenses; what they really were intended for we may infer from the allusion that they were kept in the shops of the pharmacists. At this point Morgan went somewhat astray by neglecting the statement of Pliny, quoted below, who assures us that crystal lenses were employed in medical practice for cauterizing the skin; and if the Chinese adopted this very same process, the chances are that also the druggists of Athens in the fifth century B.C. kept burning-lenses in stock, not for any fanciful, miraculous purpose, but with a somewhat realistic end in view, — to sell them as instruments useful in certain surgical operations. Cauterization was practised to a large extent in ancient times; and many forms of the cautery were devised, numerous specimens of which have survived. [1]

THEOPHRASTUS, in his treatise on fire, mentions crystal, bronze, and silver, when wrought in a certain manner, as means of igniting fire. [2]

PLINY (23 — 79), in his *Natural History*, makes two references to burning-lenses, both of crystal and glass. In his chapter on crystal he says, "I find it stated in medical authors that crystal balls placed opposite to solar rays are the most useful contrivance for cauterizing the human body." [3]   It will be noticed that the Chinese physicians

---

[1] J. S. MILNE, *Surgical Instruments in Greek and Roman Times*, pp. 116—120. Milne (p. 5) asserts, "The writings of Pliny contain little information of any kind and are absolutely of no use for our purpose," but Pliny's references to burning-lenses, quoted above, would have found a suitable place in his chapter on cauteries, and assisted in enlightening the text of Hippocrates on p 120.

[2] Ἐξάπτεται δὲ ἀπό τε τῆς ὑέλου καὶ ἀπὸ τοῦ χαλκοῦ καὶ τοῦ ἀργύρου τρόπον τινὰ ἐργασθέντων (*De igne*, 73; opera ed WIMMER, p. 363). Others cancel the words ἀπό τε τῆς ὑέλου and interpret the instruments as concave mirrors (MORGAN, *l. c.*, p. 52).

[3] Invenio apud medicos, quae sint urenda corporum, non aliter utilius uri putari quam crystallina pila adversis opposita solis radiis (XXXVII, 10, § 28).

made use of crystal lenses for exactly the same purpose. In the other passage it is remarked, "If glass balls filled with water are exposed to sunlight, they produce such a vigorous heat that they will ignite clothes." [1]

LACTANTIUS, the eminent Christian author of the third and fourth centuries, apparently under Pliny's influence, writes that when a glass globe full of water is held in the sun, fire will spring from the light reflected from the water, even in the severest cold. [2]

ISIDORUS, the learned Bishop of Sevilla (570 -- 636), observes that crystal opposed to solar rays attracts fire to such a degree that it ignites arid fungi or leaves. [3] His knowledge is evidently based on Pliny.

Besides the passages in Pliny we find a clear mention of crystal lenses in the *Orphica*, or Λιθικά of Orpheus, — a Greek poem wrongly associated with the name of Orpheus, and describing the magical properties believed to be inherent in stones, and revealed by the seer Theodamas to Orpheus. It is not, as formerly assumed, a work coming down from around 500 B.C., [4] but it manifestly bears the ear-marks of the late Alexandrian epoch, and is a production of post-Christian times. Crystal opens the series of stones dealt with in this work (Verses 170 –184). The deity cannot resist the prayers of him who, bearing in his hand a refulgent and transparent crystal, betakes himself into a temple: his wish will surely be granted. When crystal

---

[1] Cum addita aqua vitreae pilae sole adverso in tantum candescant, ut vestes exurant (XXXVI, 67, § 199)

[2] Orbem vitreum plenum aquae si tenueris in sole, de lumine quod ab aqua refulget ignis accenditur etiam in durissimo frigore (*De ira Dei*, x).

[3] Hic (crystallus) oppositus radiis solis adeo rapit flammam ut aridis fungis vel foliis ignem praebeat (*Origines*, XVI, 13, 1). Fungi used in cauterization are mentioned by Hippocrates and Paul.

[4] KRAUSE, *Pyrgoteles*, p. 6. The exact date of this work is not satisfactorily established (compare BERNHARD, *Grundriss d. griech. Lit*, Vol. II, pt. 1, p 359; and SUSEMIHL, *Gesch. d. griech. Lit. in der Alexandrinerzeit*, Vol. I, p 866).

is placed on dry wood-shavings, while the sun-rays strike it, smoke will soon arise, then fire, and at last a bright flame, regarded as sacred fire. No sacrifice is more pleasing to the gods than when offered by means of such fire.

The ancients, accordingly, employed optical lenses in medicine for cauterizing the skin, and in the religious cult for securing sacred fire. The opinion has been expressed also that they served the purpose of magnifying objects, with reference to a passage in SENECA, that letters, however minute and indistinct, appear larger and clearer through a glass ball filled with water.[1] LESSING[2] has ingeniously and conclusively demonstrated that there is a wide step from a magnifying-sphere to a magnifying-lens, and that the causes of the enlargement were sought by the ancients, not in the spherical shape of the glass, but in the water with which it was filled. Moreover, the passage of Seneca proves nothing beyond a personal experience of that author; and there is, in fact, no ancient tradition regarding specular or magnifying lenses. In Pompeii, Nola, and Mainz, lenses have been excavated, of which J. MARQUARDT[3] says that they could have been nothing but magnifying-lenses. I am unable to admit the force of this conclusion, and think that these lenses were simply burning-lenses.[4]

BURNING-LENSES IN THE MIDDLE AGES AND AMONG THE ARABS.— The European middle ages are doubtless indebted to the ancients for whatever knowledge of this subject then existed. The mineralogical knowledge of this period is mainly based on the important work of

---

[1] Litterae quamvis minutae et obscurae per vitream pilam aqua plenam maiores clarioresque cernuntur (*Quaestiones naturales*, 1, 6, 5).

[2] *Briefe, antiquarischen Inhalts*, No. 45

[3] *Privatleben der Römer*, p. 752

[4] M H. MORGAN (*Harvard Studies in Classical Philology*, Vol. I, 1890, p. 46) sides with Marquardt and Sacken against Lessing, but on insufficient grounds, and evidently without taking serious notice of Lessing's forcible arguments.

the French Bishop of Rennes, MARBODUS (1035 — 1123), entitled *De lapidibus pretiosis*, and written in Latin hexameters. This poem, largely founded on Pliny, Solinus, and the Orphica, conveyed the classical traditions regarding stones to mediæval Europe, became the direct source of at least four French *Lapidaires*, and successfully maintained its place as the great pedagogical manual on precious stones and as the classical handbook of the schools of pharmacy down to the end of the sixteenth century. [1] In § 41 of his work, Marbodus makes the following observation on crystal lenses:

> "But true it is that held against the rays
> Of Phœbus it conceives the sudden blaze,
> And kindles tinder, which, from fungus dry
> Beneath its beam, your skilful hands apply." [2]

As regards the further development of this matter, suffice it for our purpose to quote from KONRAD VON MEGENBERG's (1309 — 78) *Book of Nature*, — "If the sun shines on a round crystal, it ignites tinder in like manner as the beryl does; if it is round like an apple, and if it is exposed to the sun while it is moist, it ignites extinguished coal," — and to refer to the *Opus maius* of ROGER BACON (1240 — 92), [3] who attempted to analyze the operation of a burninglens. But Bacon's essay is dependent on that of the Arabic physicist Ibn al-Haiṭam (or Alhazen, 965 — 1039), who treated the problem much more profoundly and scientifically. [4]

---

[1] Compare the interesting discussion of L. PANNIER, *Lapidaires français du moyen âge*, pp. 15 *et seq* (Paris, 1882).

[2] Translation of C. W. KING, *Antique Gems*, p. 411. In the earliest French translation (PANNIER, *l c.*, p 61) this passage runs thus: "Ceste conceit le fou vermeil, | Ki la tient el raí del soleil, | E de cel fou li tondre esprent | S'il i tuchet alqes sovent."

[3] *The "Opus maius" of Roger Bacon*, ed. by J. H BRIDGES, Vol I, p. 113 (Oxford, 1897).

[4] Compare S. VOGL, *Physik Roger Bacos*, p. 80 — In regard to the more recent employment of burning-lenses, it is said that some Old-English tobacco-boxes have a lens in the lid for use on emergency; and naturalists still make occasional use of their pocketlenses as a substitute for a match (*Horniman Museum and Library, Handbook on Domestic Arts*, I, p. 35).

Arabic knowledge of crystal lenses, again, is founded on that of classical authors, and mainly linked with the name of Dioscorides. In the Arabic version of the Materia Medica of this Greek author, compiled by Ibn al-Baiṭār (1197—1248), we find it stated that rock-crystal struck by hardened iron yields abundant sparks;[1] that a piece of black linen subjected to the rays emitted by this stone, when it is exposed to solar light, will be ignited and consumed; and that it may be employed in this manner in order to obtain fire.[2] The Arabic *lapidarium* of the ninth century, traditionally but wrongly ascribed to Aristotle, mentions the sparks of crystal in the same manner, but omits the reference to lenses, which, however, occurs in the Hebrew and Latin translations of the same work.[3] Qazwīnī, the Arabic encyclopædist of the thirteenth century (1203—83), observes, "If rock-crystal is placed opposite the sun, and if a black rag or a flake of cotton is brought near it, the latter will catch fire, and objects may be lighted with such fire. There is still another kind of rock-crystal, less pure than the former, but harder; whoever beholds it, takes it for salt. If struck with hardened steel, however, sparks will easily spring from it; hence it serves as strike-a-light for the men of the kings."[4]

---

[1] The ancient Laplanders made ample use of rock-crystal in the place of flint, and an eye-witness who tried the experiment assures us that rock-crystal struck by the steel yields more sparks than flint (J. SCHEFFER, *Lappland*, p. 416, Frankfurt, 1675). Also in the prehistoric ages of northern Europe, quartzites served for the production of fire (compare the interesting study of G. F·L SARAUW, *Le feu et son emploi dans le nord de l'Europe aux temps préhistoriques*, in *Annales du XXe Congrès archéol et hist de Belgique*, Vol I, Gand, 1907, pp. 196—226, chiefly, pp. 213 et seq.).

[2] L. LECLERC, *Traité des simples*, Vol. III, p. 342.

[3] RUSKA, *Steinbuch des Aristoteles*, pp. 170, 171. The Latin text runs thus: "Bonitas huius lapidis est quod quando exponitur soli rotundatus ut radii solares penetrent ipsum erit ignis ab eo" (*ibid.*, p. 207) The word *rotundatus* denotes a burning-lens.

[4] RUSKA, *Steinbuch aus der Kosmographie des al-Qazwīnī*, p 9. E. WIEDEMANN (*Sitzungsberichte der phys.-med. Soz. Erlangen*, Vol 36, 1904, p. 332) remarks that the Arabic author omitted the word "globe" after "rock-crystal;" and he thinks it notable that Qazwīnī expressly speaks of rock-crystal

Likewise in their knowledge of burning-mirrors, the Arabs depend upon the science of the Greeks, as shown in their discussions of this subject by references to Anthemius and Diocles. [1]

REFUTATION OF THE THEORIES THAT THE ANCIENT CHINESE WERE ACQUAINTED WITH BURNING-LENSES. — In passing on to China, we face a bewildering jungle of speculations and opinions as to our subject; and only after clearing this jungle will it be possible to discuss the real facts in the case. If Dr. E. HILL [2] recently stated that "it is said that a Chinese emperor used lenses as early as 2283 B.C. to observe the stars," we here find expression of that popular opinion which credits the Chinese with lenses prior to the Greeks, — an invention which, as will be seen, was never made by the Chinese themselves. A lens could not have been manufactured at that time, as the materials required for it, glass or rock-crystal, were then unknown in China. Moreover, the Chinese in this case lay no claim whatever to a lens. The text from which this alleged lens (I do not know by whom) has been distilled is contained in the oldest historical record of the Chinese, the *Shu king* (II, 5), in which the astronomical activity of the Emperor Shun is spoken of: he is said to have availed himself of an instrument of jade, the description of which is not given in the text, but only by the late commentators. [3] Whatever this instrument of hard, untransparent stone may have been, it surely has nothing in common with a lens.

Even professional sinologues, like SCHLEGEL, [4] and quite recently FORKE, [5] have asserted that burning-lenses were known to the Chinese

---

[1] WIEDEMANN, *Sitzungsberichte der phys.-med. Soz Erlangen*, Vol. 37, 1905, p. 402.

[2] *Ophthalmic Record,* Vol. 23, 1914, p. 504

[3] See LEGGE, *Chinese Classics*, Vol. III, p 33; COUVREUR, *Chou king*, p 14; CHAVANNES, *Mémoires historiques de Se-ma Ts'ien*, Vol. I, pp. 58—59; and the writer's *Jade*, pp. 104 *et seq*.

[4] The views of Schlegel are discussed farther on.

[5] *Lun-hêng*, pt. 2, pp. 496—498.

in pre-Christian times long before they were known to the Greeks. Their conclusions, however, rest on a fallacy due to misunderstandings of the texts. We shall closely examine these, and see how those scholars were prompted to their opinions. It will be demonstrated at the same time that optical lenses of crystal or glass were absolutely unknown in China prior to our era.

Se-ma Chêng of the eighth century A.D. records, in his *Memoirs of the Three Early Sovereigns* (*San huang ki*), the following legend regarding the mythical being Nü-kua or Nü-wa, conceived as a serpent with a human head: [1] "He fought with Chu-yung [the regent of fire] and failed in victory. Flying into a rage, he butted with his head against Mount Pu-chou and brought it down. The pillar of heaven was broken, and the corners of earth were bursting. Nü-kua then fused five-colored stones to repair the firmament, and cut off the feet of a marine tortoise to set up firmly the four extremities of earth. He gathered the ashes of burnt reeds to stop the inundation, and thus rescued the land of Ki. Thereupon the earth was calm, the sky made whole, and the old order of things remained unchanged." [2] The same tradition is contained in the book going under the name of the alleged philosopher Lie-tse, [3] the present recension of which, in all probability, is not earlier than the Han period; likewise in the book of Huai-nan-tse of the second century B.C., [4] and in the *Lun-héng* of Wang Ch'ung. [5] The latter philosopher points it out as a very ancient tradition believed by most people.

---

[1] Originally a male sovereign, but from the second century A D. represented on the bas-reliefs of the Han period as a woman.

[2] Compare CHAVANNES, *Mémoires historiques de Se-ma Ts'ien*, Vol 1, pp. 11, 12; II J. ALLEN, *Ssūma Ch'ien's Historical Records* (*Journ. Roy As Soc*, 1894, p 274); MAYERS, *Chinese Reader's Manual*, p 162, HIRTH, *Ancient History of China*, p. 11.

[3] Ch. 5, *T'ang wén* (compare E. FABER, *Naturalismus bei den alten Chinesen*, p 104; L GILES, *Taoist Teachings from the Book of Lieh Tzǐ*, p. 85, L. WIEGER, *Les pères du système taoiste*, p. 131).

[4] *P'ei wén yün fu*, Ch. 21, p 217.

[5] A. FORKE, *Lun-héng*, pt. 1, p. 250; pt. 2, p. 347.

Every unbiased student will recognize in this legend concerning Nü-kua a genuine myth, in which a cosmological catastrophe is hinted at, the havoc wrought to heaven and earth being repaired with realistic expedients contrived by a primitive and naïve imagination. He whose trend of mind is bent on interpretation may fall back on the phenomenon of the rainbow, which may have impressed a primitive mind as consisting of stone-like patches for mending the sky after the destructive force of a rainstorm; and the brilliant colors of a quartz or agate may have intimated an association of ideas between the hues of a stone and those of the iris. The composite coloration of a stone may have suggested the effect of a smelting-process; at all events, the molten stones of a legend cannot be taken literally; the casting of metal is naïvely transferred to stones. Be this as it may, or whatever our interpretation of the myth may drive at, it is obvious to every sober mind that the elements of a fantastic myth, which is not reducible to an analysis of actual reality, cannot be utilized as the foundation of far-reaching conclusions as to industrial achievements of the Chinese. Some of our sinologues, however, were of a different opinion. The melting of the five-colored stones ascribed to that fabulous being was a rather tempting occasion for the exercise of ingenious speculations. MAYERS [1] championed the idea that the stone of five colors is coal, the useful properties of which Nü-kua was the first to discover; and T. DE LACOUPERIE, [2] in a very interesting article, took great pains to demonstrate that the legend has nothing to do with the introduction of glass and the discovery of mineral coal, though by no means himself arriving at any positive result.

Wang Ch'ung, [3] in connection with a fire-making apparatus for

---

[1] *Notes and Queries on China and Japan*, Vol. II, p. 99

[2] *T'oung Pao*, Vol II, 1891, pp 234—243.

[3] *Lun héng*, Ch. 16, p. 2 (ed. of *Han Wei ts'ung shu*) FORKE, *Lun-héng*, pt 2, p. 351.

drawing fire from the sky, mentions the practice, that "on the day *ping-wu* of the fifth month, at noon, they melt five stones to be cast into an instrument that is capable of obtaining fire." According to FORKE, [1] Wang Ch'ung speaks of burning-glasses as, "The material must have been a sort of glass, for otherwise it could not possess the qualities of a burning-glass. [2] Flint glass, of which optical instruments are now made, consists of five stony and earthy substances, — silica, lead oxide, potash, lime, and clay. The Taoists, in their alchemical researches, may have discovered such a mixture." By interpreting the terms *yang sui* 陽燧 or *fu sui* 夫遂 as "burning-glass," Forke reads of burning-glasses even in the *Chou li*, and is finally carried to this conclusion: "Burning-reflectors were known to the Greeks. Euclid, about 300 B.C., mentions them in his works; and Archimedes is believed to have burned the Roman fleet at Syracuse in 214 B.C. with these reflectors, — probably a myth. Plutarch, in his life of Numa, relates that the Vestals used to light the sacred fire with a burning-speculum. As the *Chou li* dates from

---

[1] *Ibid.*, p. 496.

[2] It will be seen below that this conclusion is a fallacy, and is in fact inadmissible; but, granting for a moment its *raison d'être*, the technical point is not so easily settled, as represented by Forke. Wang Ch'ung does not speak of five different stones, but, as demonstrated farther on, indeed speaks of five-colored stones with a distinct allusion to the Nü-kua legend; his term *wu shi* 五石 in this passage being merely a loose expression or abbreviation for *wu sé shi* 五色石. If, then, a multi-colored stone is here in question, and if this stone could be identified with a kind of quartz, Forke's opinion, from a technical point of view, would not be utterly wrong; for it is technically possible to make glass from quartz. This experiment was successfully carried on about a decade ago by C. Heræus in Hanau: the quartz utilized was melted in vessels of pure iridium, which melts at 2000°, while the melting-point of quartz is at 1700°. After exceeding its melting-temperature, the quartz becomes glassy. The process itself is difficult and complex, and it would be unreasonable to suppose that a technical manipulation which has succeeded only in our own time should have been familiar to the ancient Chinese, who derived from the West whatever knowledge of glass they possessed. If, however, the "five-colored stone," as shown below, was a variety of agate or soapstone (and this opinion is highly probable), nothing remains of Forke's theory.

the eleventh century B.C. (?), it is not unlikely that the Chinese invented the burning-reflector independently, and knew it long before the Greeks."

TH. W. KINGSMILL once remarked, [1] "Myths have been not inaptly described by Max Müller as a disease of language; and to this category we may perhaps relegate the group of modern myths which have grown up in and around our descriptions of China and its arts." I apprehend that the assigning to the ancient Chinese of burning-lenses belongs to this category of modern myths based on misinterpretation of terms. BIOT, [2] SCHLEGEL, [3] HIRTH, [4] and CHAVANNES [5] have clearly shown that the fire-apparatus spoken of in the *Chou li* was a metal mirror, and the Chinese commentators claim no more for it; even Forke cites their opinion, yet mechanically clings to his idea of burning-glasses. Unfortunately, he omits to tell us how the Chinese of the Chou period — when even a word for "glass," and certainly the matter itself, were unknown to them — should have obtained glass. And if the molten stones of Wang Ch'ung, in Forke's opinion, are glass, the molten colored stones of Nü-kua would be entitled to the same consideration; and thus the baffling result would be attained that not only burning-glasses, but also glass in general, are truly Chinese inventions, the latter going back to the dim past of prehistoric ages.

An intimation that the five-colored or variegated stone is a reality, is first given by Li Tao-yüan 麗道元, who died in A.D. 527, in his commentary on the *Shui king* 水經注, a book on the rivers of China: [6] "On the northern side of the Hen Mountains, along the

---

[1] *Chinese Recorder*, Vol. VII, 1876, p. 43.

[2] *Le Tcheou-li*, Vol. II, p 381.

[3] *Uranographie chinoise*, p 612.

[4] *Boas Anniversary Volume*, pp. 226—227.

[5] *Le Tai Chan*, pp 188—189.

[6] Compare CHAVANNES, *T'oung Pao*, 1905, p. 563.

Ki River, the rocky hills border the river so closely that there is no space for flat beaches; in places where the water is shallow there is plenty of five-colored stones." [1]   In another passage he refers to carvings from the stone of the same name, which served for the decoration of a palace of the Emperor Wên of the Wei dynasty in A.D. 220.

The *Yün lin shi p'u* 雲林石譜 by Tu Wan 杜綰 of 1133 [2] likewise makes mention of five-colored stones 五色石 in the Ki River 溪水 near Sung-tse 松滋, in the prefecture of King-nan 荊南府 (now King-chou), in the province of Hu-pei. Among these are some almost transparent, intersected by numerous lines that are straight like the fibres of a brush, and not different from the agate of Chên-chou 眞州. [3]

Another tradition crops out in the Gazetteer of Lai-chou 萊州志, [4] according to which the district of Ye 掖縣, forming the prefectural city of Lai-chou on the northern coast of Shan-tung, would produce five-colored stones made into vessels and dishes, and asserted to be identical with the "strange stones" (*kuai shi* 怪石) mentioned in the Tribute of Yü. [5]   This stone of Lai-chou is well

---

[1] 佷山北溪水所經皆石山略無土岸。其水淺處多五色石 (*P'ei wên yün fu*, Ch 100 A, p. 16).

[2] Ch. B, p 5^b (edition of *Chi pu tsu chai ts'ung shu*).

[3] The latter is found in the water or sandy soil of the district Liu-ho 六合, in the perfecture of Kiang-ning, province of Kiang-su According to Tu Wan's description, this agate is either a pure white or five-colored, the latter variety being characterized by the same attributes as the stone of Sung-tse, it is locally used for the carving of Buddhist images.

[4] *P'ei wên yün fu*, Ch. 100 A, p. 16.

[5] LEGGE, *Chinese Classics*, Vol. III, pp. 102, 104; COUVREUR, *Chou king*, p 67; compare *Ts'ien Han shu*, Ch. 28 A, p 1_b. Legge remarks that the "strange stones" are very perplexing to the commentators, and that Ts'ai gets over the difficulty by supposing they were articles indispensable in the making of certain vessels, and not curiosities, merely to be looked at. The above identification seems to me very plausible; on account of its numerous shades and curious designs, in which the imagination of the Chinese sees grotesque scenery, the soapstone of Lai-chou could well have merited the name "strange or supernatural stone."

known to us; it is a variety of agalmatolite or soapstone which is still carved by the Chinese into a hundred odds and ends and worked up into soap, the stone being powdered, and the powder being pressed into forms. [1] Its tinges are manifold and very pleasing, and are therefore capable of artistic effects. The Field Museum owns several albums of the K'ien-lung period, containing pictures (人物) entirely composed of Lai-chou stone of diverse colors, neatly cut out and mounted. The stone being very soft, carving is comparatively easy. [2]

We accordingly note that in post-Christian times the "five-colored stone" has been identified by the Chinese with a variety of either agate or soapstone. This certainly does not mean at the outset that the stone of the same designation attributed by tradition to times of great antiquity must be identical with one or the other; the ancient name *wu sè shi*, whatever it may have conveyed in its origin, may simply have been transferred to certain kinds of agate and soapstone in comparatively recent periods. This stricture being made, however, there remains a great deal of probability that the five-colored stone of Nü-kua, after all, was nothing else; there is, at least, no valid reason why it should have been something else. [3] To this interpretation, Forke might object that in the aforesaid passage of Wang Ch'ung the question is not of the melting of five-colored stones, as in the tradition of Nü-kua, but of the melting of five

[1] F v. Richthofen, *Schantung*, pp. 199—200; A. Williamson, *Notes on the Productions of Shan-tung* (*J. China Branch R As. Soc*, Vol. IV, 1868, p. 69); Becher, *Notes on the Mineral Resources of Eastern Shan-tung* (ibid., Vol. XXII, 1888, p. 37); A. Fauvel, *The Province of Shantung* (*China Review*, Vol. III, 1875, p. 375).

[2] It is described in the *Yun lin shi p'u*, Ch. B, p. 1ᵇ.

[3] T. de Lacouperie (*Toung Pao*, Vol. II, p. 242) based his theory of five-colored stones on certain geological conditions of Shan-si Province, where, according to A. Williamson, the strata of some hillsides are clearly marked from base to summit, the many-colored clays presenting all the hues of the rainbow. This would not be so bad if the Chinese accounts really spoke of clay, but they obstinately insist on stones, and stone and clay were strictly differentiated notions also to the ancient Chinese.

13

single stones, and that consequently the aspect of the problem is thus modified; this objection, however, could not be upheld. The solution of the problem is furnished by Wang Ch'ung himself. In two passages of his work, as already pointed out, he himself narrates the tradition regarding Nü-kua, and his mending of the sky by means of five-colored stones. At the end of the chapter, [1] in which he subjects the story to a lengthy discussion, scorning it with ruthless sarcasm, he suddenly changes his phraseology, and speaks of "the repairing of the sky by means of five kinds of stones, which may have worked like medicinal stones in the healing of disease." [2] Consequently in the diction of the author the two terms "five-colored stone" (*wu sé shi*) and "five stones" (*wu shi*) are interchangeable variants relating to the same subject-matter. It is therefore evident beyond cavil that the passage concerning the fire-apparatus, where the fusing of five stones is mentioned, likewise implies a literary allusion to the Nü-kua legend, and refers to exactly the same affair. If glass is not involved (nor can it be intended) in the Nü-kua legend, it cannot, accordingly, be sought for either, as alleged by Forke, in this passage of Wang Ch'ung.

The question now remains to be answered, Why does Wang Ch'ung bring stones on the tapis to describe an instrument which, judging from all other Chinese records, was a metal mirror? We know that the ancient Chinese possessed mirrors of stone. HIRTH [3] has indicated a jade mirror found in A.D. 485 in an ancient tomb near Siang-yang in Hu-pei Province, which the polyhistor Kiang Yen (443—504) stated to date from the time of King Süan (827—782 B.C.). The *Yün lin shi p'u* [4] mentions two localities where stone material fit for mirrors was quarried, — Mount Wu-ki 浯溪山, in the district

---

[1] FORKE, *Lun-héng*, pt. 1, p. 252.
[2] This passage is quoted also in *P'ei wén yün fu*, Ch. 100 A, p. 16
[3] *Chinese Metallic Mirrors* (Boas Anniversary Volume, p. 216).
[4] Ch. c, p. 9.

of Kʻi-yang 祁陽, prefecture of Yung-chou 永州, province of Hu-nan, the stone slabs of which, several feet wide, of deep blue (or green) hue, could reflect objects at a distance of several tens of feet: and the district of Lin-ngan 臨安, in the prefecture of Hang-chou 杭州, province of Chê-kiang. In Su-chou, such stone mirrors, usually carved from Yün-nan marble (*Ta-li shi* 大理石), are still offered for sale. When we now critically analyze the passage of Wang Chʻung, we recognize in it a fusion of three different notions, — first, the alleged melting of stones borrowed from the Nü-kua legend; secondly, a recollection of stone mirrors looming up in his mind; and, thirdly, a reminiscence of metal mirrors used in the Chou period (and also subsequently) for securing fire. In a word, his description is a downright literary concoction, pieced together from three different sources; and it is therefore impossible to regard it as an authentic and authoritative source from which any conclusions as to realities may be derived. It can prove absolutely nothing for the elucidation of facts, such as glass, burning-glasses, burning-mirrors, or anything else. Forke's thesis of the alleged priority of the Chinese in the matter of burning-glasses is untenable; and the fact remains, much more solidly founded than assumed by Forke, that the ancients were the first to make use of them. [1]

Another weapon, seemingly still more formidable, has been introduced into the discussion by Schlegel. Liu Ngan, commonly known under the name Huai-nan-tse, a member of the imperial family, philosopher and alchemist, who died in 122 B.C., is credited by SCHLEGEL [2] with the statement that "it is not absolutely necessary

[1] Forke has not clearly discriminated between burning-lenses and burning-mirrors. I hope to devote a monograph to the latter subject with particular reference to the relation of the Greek burning-mirrors to the Chinese. So much may be said here that Greek priority seems to me to be established along this line also.

[2] *Uranographie chinoise*, p. 142; and *Nederlandsch-Chineesch Woordenboek*, Vol. I, p. 674.

to employ a bright metal plaque, but that a large crystal ball like-wise, held toward the sun, can produce fire." Consequently burning-lenses should have been known to the Chinese in the second century B.C. This would indeed be very nice, were it not that Huai-nan-tse never made such an assertion, wrongly attributed to him by Schlegel. Of all that Schlegel makes him say, he has in fact said only the very first sentence, — "When the mirror is held toward the sun, it will ignite and produce fire," — while all the rest of it does not emanate from the philosopher, but from his later commentators. Schlegel, indeed, does not quote Huai-nan-tse's original text, but derives the passage from a recent work, *Liu ts'ing ji cha* 留 青 日 札.[1] We need only refer, however, to Huai-nan-tse's actual text,[2] to recognize at a glance the real state of affairs. Huai-nan-tse knew only of concave metal mirrors for the production of fire, but nothing whatever about crystal or any other lenses. He repeatedly mentions the former,[3] but never the latter, nor does any of his contemporaries, for the reason that lenses did not turn up on the horizon of the Chinese before the beginning of the seventh century A.D.[4]

BURNING-LENSES NOT A CHINESE INVENTION. DEFICIENT KNOWLEDGE OF THE SUBJECT ON THE PART OF THE CHINESE. — China has indeed known lenses, and certain optical properties of them; yet they were not invented by the Chinese, but were received and introduced by them from India. This fact will be established by the investigation to follow. The subject is somewhat complex, and has never been clearly set forth by any author, Chinese or foreign. It is indis-pensable to penetrate into the primeval sources, and to sift their

---

[1] A collection of miscellaneous essays by T'ien Yi-hêng, a writer of the Ming period.

[2] Ch. 3, p. 2 (edition of *Han Wei ts'ung shu*). In the commentary of this edition no reference is made to crystal lenses; their mention is simply an utterance of the author of *Liu ts'ing ji cha*.

[3] For instance, Ch 5, pp 11, 14; Ch. 6, p. 2ᵇ; Ch. 8, p. 1ᵇ; etc.

[4] Another argument of Schlegel in favor of early Chinese acquaintance with burning-lenses is discussed below in the paragraph on ice-lenses.

data with critical eyes, as the recent Chinese writers have been unable to cope with the matter properly; at any rate, none of their statements can be accepted without careful examination. Li Shi-chên, the great Chinese authority on physical science in the sixteenth century, who spent a lifetime on the elaboration of his praiseworthy work *Pên ts'ao kang mu*, has summarized his knowledge of optical lenses (*huo chu* 火 珠, "fire-pearls") as follows: [1] "The dictionary *Shuo wên* designates them as 'fire-regulating pearls' (*huo-ts'i-chu* 火 齊 珠). [2] The Annals of the Han Dynasty style them *mei-hui* 玫瑰, these characters having the sounds *mei hui* 枚回. The Annals of the T'ang Dynasty narrate that 'in the south-eastern ocean there is the Lo-ch'a country 羅 刹 國 producing fire-regulating pearls, the biggest of these reaching the size of a fowl's egg, and in appearance resembling crystal 水 精. They are round and white, and emit light at a distance of several feet. When exposed to the sunlight, and mugwort is placed near, the latter is ignited.' Such lenses are used in the application of moxa, which in this manner is painless. [3] At present there are such lenses in Champa (Chan-ch'êng 占 城), which are styled 'great fire-pearls of the morning dawn' (*chao hia ta huo chu* 朝 霞 大 火 珠). The *Sü Han shu* 續 漢 書 [4] says that the country of the Ai-lao barbarians [5] pro-

---

[1] *Pên ts'ao kang mu*, Ch. 8, p. 18. This notice is an appendix to his account of rock-crystal.

[2] This translation and its meaning will be explained in the following section. We have no adequate word to cover exactly the meaning of Chinese *chu* 珠, which means not only a "bead" or "pearl," but also a "gem or precious stone," usually of circular shape. Already D'HERBELOT (*Bibliothèque orientale*, Vol. IV, p. 398) has explained correctly these various shades of meaning.

[3] This sentence is not contained in the T'ang Annals, but is Li Shi-chên's own statement. For explanation see below.

[4] A continuation of the official history of the Han dynasty, written by Sie Ch'êng 謝 承 of the third century.

[5] 哀 牢 夷. These tribes (their Chinese designation is preserved in the name "Laos") formed the Shan kingdom, first appearing in history during the first century of our era, in the present territory of Sze-ch'uan and Yün-nan.

duces stones styled *huo-tsing* 火精 ('fire-essence') and *liu-li* 琉璃.
In view of this fact, the term *huo-ts'i* 火齊 is an error for *huo-tsing* 火精; the latter is correct in correspondence with the term *shui-tsing* 水精 ('water-essence,' a name for rock-crystal)." [1]   It will be seen from the following discussion that this notice is very inexact in detail, and altogether highly uncritical, — a defect for

---

[1] F. DE MÉLY (*Lapidaires chinois*, p 60), who has partially translated this text (not from the original, but from a late Japanese cyclopædia), gives wrong characters and transcriptions of the Chinese terms, — *kiu koei* instead of *mei hai* (or *mei huei*, or *mei kuei*; see farther below), and *ho chai* in lieu of *huo ts'i*. Moreover, the rendering of *huo chu* by "lupe" is inadmissible, as neither the Chinese nor the Indians have ever made use of magnifying-lenses, but both peoples were familiar only with lenses for fire-making. — The term *huo-tsing* is not an error for *huo-ts'i*, as assumed by Li Shi-chên, but denotes a red variety of rock-crystal supposed to attract fire, while the white variety of the same stone attracts water and fire at the same time (*Wu li siao shi*, Ch. 7, p. 13ᵇ); *huo-tsing* and *huo-ts'i*, in fact, refer to different minerals. In the same manner as among the ancients, the speculations of the Chinese concerning the nature of rock-crystal were divided between the opinions that, on the one hand, it was the essence of water (owing to the outward resemblance to ice) and, on the other hand, the essence of fire (because when struck with steel, it yields sparks, or when used as a lens, produces fire). HIRTH (*China and the Roman Orient*, p. 233) is quite right in deriving the former theory from classical lore. I hope to come back to this subject in detail in a series of studies dealing with Chinese-Hellenistic relations. In opposition to PLINY (XXXVII, 9, § 23), who takes crystal for a kind of ice due to excessive congelation, found only in regions where the winter snow freezes most intensely (Contraria huic causa crystallum facit, gelu vehementiore concreto. Non aliubi certe reperitur quam ubi maxime hibernae nives rigent, glaciemque esse certum est, unde nomen Graeci dedere), DIODORUS SICULUS of the first century B. C. expresses the view that crystal originates from purest water hardened into ice, not by cold, however, but through the powerful effect of solar heat (Crystallum ex aqua purissima in glaciem indurata coalescere aiunt, non quidem a frigore, sed divini ignis potentia). The celebrated French Bishop MARBODUS (1035—1123) attacked the glacial theory in his poem *De lapidibus pretiosis* (§ 41) as follows: "Crystallus glacies multos durata per annos, | Ut placuit doctis, qui sic scripsere, quibusdam, | Germinis antiqui frigus tenet atque colorem. | Pars negat, et multis perhibent in partibus orbis | Crystallum nasci, quod non vis frigoris ulla, | Nec glacialis hiems unquam violasse probatur." In China, the same theory was called into doubt by Ts'ao Chao 曹昭 in his *Ko ku yao lun* 格古要論, published in 1387: "Altough it is said that many years old ice becomes rock-crystal, this is obviously false in view of the fact that green and red crystals occur in Japan" (多年老冰爲水晶然日本國有青水晶紅水晶則水晶非冰也明矣), — an attempt at scientific thinking.

which Li Shi-chên himself is not solely responsible, but which already adheres to his uncritical predecessors. We note, first of all, that he avails himself indiscriminately of three terms, — *huo chu* ("fire-pearl"), *huo-ts'i-chu* ("fire-regulating pearl"), and *mei-hui*. On a previous occasion I ventured to express doubts of the alleged identity of the former two terms;[1] and it will now be demonstrated that they indeed relate to two different mineral substances associated by the early Chinese accounts with two different traditions. In fact, neither the *Shuo wén* nor the Han Annals speak of burning-lenses; Li Shi-chên, however, is quite correct in tracing them to the Lo-ch'a country, but cites the T'ang Annals wrongly by assigning to them the term *huo ts'i chu* instead of *huo chu*. This text of the T'ang Annals indeed is the first and earliest authentic Chinese account relative to burning-lenses. We note also that Li Shi-chên does not claim any knowledge of them on the part of Wang Ch'ung or Huai-nan-tse; and, as far as I know, there is no Chinese author who would make such a pretension. The various problems raised by the text of the *Pén ts'ao kang mu* will now be discussed in detail.

Huo-ts'i not a Burning-Lens, but Mica. — The earliest definition of the "fire-regulating pearl" (*huo ts'i chu* 火齊珠)[2] that occurs

---

[1] *Notes on Turquois in the East*, p. 28.

[2] HIRTH and ROCKHILL (*Chau Ju-kua*, p. 113) express the opinion that *huo ts'i* appears to be a foreign word, without being able, however, to indicate for which foreign word it might be intended. This supposition is hardly probable, as the phrase *huo ts'i* is good old Chinese, and yields a reasonable sense. It occurs in the ancient Book of Rites (*Li ki*, chap *Yue-ling*, ed. COUVREUR, Vol I, p. 401; LEGGE's translation, Vol. I, p. 303): "In the second month of winter, orders were given to the grand superintendent of the preparation of liquors to see that the rice and other glutinous grains be all complete, etc., that the water be fragrant, that the vessels of pottery be good, and that the *regulation of the fire* (*huo ts'i* 火齊) be right." The term *huo ts'i chu*, accordingly, is very well fitted to signify "a pearl (or gem) used in regulating fire." Indeed, the term *huo-ts'i*, as shown farther on, has been employed for a mineral indigenous in China, and belonging to the mica group, prior to her contact with India; we hear, for instance, of screens (*Shi i ki*, Ch. 5, p. 6; ed. of *Han Wei ts'ung shu*), couches, and finger-rings of *huo-ts'i*, of native manufacture (*ibid.*, Ch. 8, p. 3). This subject is not pursued here any further, as it will be treated by the writer in a special monograph on mica.

in the Annals of China is embodied in the History of the Liang
Dynasty, [1] which enumerates it among the products of Central India,
and describes it as follows: "*Huo-ts'i*, in its appearance, is like the
mica of China, [2] with a tinge like that of purple gold, and of intense
brilliancy.   Pieces split off from it are as thin as the cicada's wings;
when joined together again, they are like doubled silk gauze." [3]
This text, however, is not peculiar to the two Annals, but is

---

[1] *Liang shu*, Ch. 54, p. 7ᵇ. The Liang dynasty covers the period from 502 to 556.
Its history was compiled by Yao Se-lien in the first half of the seventh century. The
same text is found also in *Nan shi* (Ch. 78, p. 7). The latter work, comprising the
history of China from 420 to 589, was elaborated by Li Yen-shou in the seventh century.

[2] In Chinese *yün-mu* 雲母 (litterally, "cloud-mother"). On the basis of a specimen obtained from China, *yün-mu* was identified with mica by E. BIOT (in PAUTHIER-
BAZIN, *Chine moderne*, Vol. II, p. 558), who also rejected Rémusat's interpretation of this
term as "mother-o'-pearl" (this meaning is erroneously given by PALLADIUS, *Chinese-
Russian Dictionary*, Vol. II, p. 543). He pointed out seven varieties bearing different
names. Under the same name, *yün-mu*, the different varieties of mica have well been
described by GEERTS (*Produits de la nature japonaise et chinoise*, Vol. II, pp. 426—433);
while F. PORTER SMITH (*Contributions toward the Materia Medica of China*, p. 210)
mistook *yün-mu* for talc, though describing mica under that title. G SCHLEGEL (*T'oung
Pao*, Vol. VI, 1895, p. 49) has contributed to the subject a few notes which are rather
inexact; only his erroneous view that *yün-mu* is a modern term, may here be pointed out.
As in many studies of orientalists we meet the phraseology "mica or talc," it cannot be
strongly enough emphasized that mica and talc are fundamentally different minerals; and
it is even difficult to see how they could ever be confounded. The word *yün-mu* has b.en
adopted for the designation of mica in the modern scientific mineralogy of China and
Japan (see, for instance, *Journ. Geol. Soc. of Tōkyo*, Vol. XIX, 1912, p. 413), while talc
is *hua shi* 滑石 or *fei-tsao shi* 肥皂石; the identification of *yün-mu*, therefore, is absolutely certain. The Chinese name arose in consequence of the belief that this
mineral forms the basis in the origin of the clouds; that is, strictly speaking, the clouded
appearance of the mineral was instrumental in inspiring this popular belief. The Sanskrit
designattion for mica is *abhra*, a word appearing as early as the fifth century in the Bower
Manuscript (A. F. R. HOERNLE, *The Bower Manuscript*, pp 11, 117) This word means
literally "cloud, atmosphere," and thus presents a curious counterpart of the Chinese designation for the same mineral, *yün-mu* ("cloud-mother"). The Chinese alchemists took
powdered mica internally in order to insure long life; and when placed in the grave, it
was believed to have the effect of preserving the body from decay.

[3] 火齊狀如雲母。色如紫金。有光耀。別之
則薄如蟬翼。積之則如紗縠之重沓也。

encountered as early as the third century in the *Nan chou i wu chi*
南州異物志 ("Account of Remarkable Objects in the Southern
Provinces"), by Wan Chên 萬震,[1] where it is prefaced by the
statement that *huo-ts'i* comes from, or is produced in, the country
of India;[2] and it is this work which has doubtless served as a
source to the annalist. The brief description of the mineral is
perspicuous enough to enable one to recognize in it mica, — a group
of minerals that crystallize in the monoclinic system, and consist
essentially of aluminum silicate. The striking characteristic of all
species is a highly perfect basal cleavage, by which the crystals may
be split into the thinnest films (that is, the cicada wings of the
Chinese). It is to this property, and to the highly elastic nature
of the lamellæ (by which mica is distinguished from the flexible,
foliated, but inelastic mineral, talc), as well as to the fact that it
is able to withstand high temperatures and is a bad conductor of
electricity, that mica owes its commercial value.[3]

It was not in India, however, that the Chinese acquainted themselves with mica for the first time. Mica is indigenous in many
places of China; and a contemporary of Wan Chên, Chang Pu 張勃,
the author of a geographical description of the kingdom of Wu,[4]
mentions the mineral "*huo-ts'i*, which is like *yün-mu*, as occurring

---

[1] According to *Sui shu* (Ch. 33, p. 10), Wan Chên lived in the time of the Wu
dynasty (third century)

[2] 火齊出天竺國 (*T'ai p'ing yü lan*, Ch. 809, p. 2). The only variant encountered in this text is in the fourth sentence: 節如蟬翼 instead of
別之 etc., as above. The *Pên ts'ao kang mu* (Ch. 8, p. 18), in the notice of *liu-li*,
quotes the same text from the work *I wu chi*, which says that the stone is a product of
all countries of southern India.

[3] Compare the excellent article "Mica" in G. Watt's *Dictionary of Economic Products
of India*, Vol. v, pp. 509—513 (also as separate reprint), where its uses, geological and
geographical distribution, as well as mining and trade in India, are fully discussed.

[4] *Wu lu ti li chi* 吳錄地理志 (see Bretschneider, *Bot Sin*, pt. 1,
No 1043).

in the district Si-küan. [1] It is composed of many layers, and can accordingly be split. It is of yellow color, resembling gold." [2] This, again, is an unmistakable characterization of mica, and of that variety known to us as golden mica (or de chat). [3] We note that a kind of mica was known in China under the name huo-ts'i, and that the Chinese merely rediscovered this particular species in India; the term huo-ts'i, therefore, cannot be the rendering of a Sanskrit word, and such a Sanskrit name as might come into question, indeed, does not exist.

Huo-ts'i are referred by the Chinese also to some countries located in south-eastern Asia. In the year 519, Jayavarman, King of Fu-nan (Cambodja), sent an embassy to China, and offered pearls of that description, saffron (yü-kin), storax, and other aromatics. [4] In 528 and 535 two embassies arrived in China from a country called Tan-tan 丹丹, and huo-ts'i pearls or beads were included among the tribute-gifts of the latter mission. [5] Very little is known about this country, and its identification is not ascertained. At the time of the T'ang dynasty (618—906) it is mentioned again as being situated south-east of the island of Hai-nan, and west of the

---

[1] As the kingdom of Wu comprised the present territory of Kiang-su, Chè-kiang, and parts of An-hui, this locality must have been within the boundaries of these provinces.

[2] 西偈縣有火齊如雲母。重沓可開。黃似金 (T'ai p'ing yü lan, Ch. 809, p. 2). The coincidence of the terms used in this text and the Nan chou i wu chi is notable.

[3] Now termed in Chinese kin sing shi 金星石 ("gold star stone") or kin tsing shi 金精石. See GEERTS, Produits de la nature japonaise et chinoise, Vol. II, p. 430; D. HANBURY, Science Papers, p. 219; and F. PORTER SMITH, Contributions toward the Materia Medica of China, p. 148, who mentions Kiang-nan as a locality where it occurs; this is probably identical with that mentioned in the above Chinese work. The Imperial Geography (Ta Ts'ing i t'ung chi, Ch. 244, p. 11) mentions the district of Tè-hua (forming the prefectural city of Kiu-kiang, province of Kiang-si) as producing mica (yün mu).

[4] Liang shu, Ch 54, p 5ᵇ; or Nan shi, Ch. 78, p 4 (compare PELLIOT, Bull de l'Ecole française, Vol. III p. 270)

[5] Liang shu, ibid.

country To-lo-mo 多羅磨, which is otherwise unknown to us. [1]
G. SCHLEGEL, [2] in a discussion of this passage of the Liang history,
without adducing any evidence, rendered the term *huo-ts'i* by
"Labrador feldspat," which is an arbitrary and unwarranted opinion. [3]
Both Fu-nan and Tan-tan, this much is certain, were countries in
the sphere of influence of Indian civilization; and in the same manner
as Fu-nan received diamonds in consequence of its lively intercourse
with India, [4] so also its *huo-ts'i* gems were undoubtedly derived from
the same source.

Aside from India, Fu-nan, and Tan-tan, *huo-ts'i* are listed in
the Chinese Annals also among the products of Persia; that is,
Persia in the epoch of the Sassanian dynasty. [5] Since Persia was
then in close relations with India, it is highly probable that the
huo-ts'i of Persia, like many other products attributed to the country
by the Chinese, [6] also hailed from India. We shall revert once
again to Persia when discussing the term *mei-hui*.

There is not a single ancient Chinese account that speaks of the
use of burning-lenses in regard to *huo-ts'i*. [7] The only purpose to

---

[1] *Tang shu*, Ch 222 B, p. 4 (compare PELLIOT, *Bull. de l'Ecole française*, Vol. IV,
p 284).

[2] *Toung Pao*, Vol. X, 1899, p. 460.

[3] Schlegel's view that the country Tan-tan should be sought for on the Malay Peninsula, and be identified with the mysterious Dondin, placed by Odoric of Pordenone of
the fourteenth century between Ceylon and China, has been refuted by PELLIOT (*l. c*).

[4] India traded diamonds with Ta Ts'in, Fu-nan, and Kiao-chi (*Tang shu*, Ch. 221 A,
p. 10b).

[5] *Pei shi*, Ch. 97, p. 7b, *Wei shu*, Ch. 102, p. 5b; *Sui shu*, Ch. 83, p. 7b.

[6] HIRTH and ROCKHILL, *Chau Ju-kua*, p. 16.

[7] The conclusion of some Chinese authors that *huo-ts'i* are burning-lenses may have
been prompted partially by the report of a mica mirror (*huo ts'i king*) contained in the
Shi i ki (Ch. 3, p. 6b, ed. of *Han Wei ts'ung shu*) This mirror, three feet in width,
is alleged to have been sent as a gift by a country styled K'ü-su 渠胥, at the time
of the Emperor Ling of the Chou dynasty (571—545 B C) In a dark room, objects
were visible in it as in the daytime; and when words were spoken in the direction of
the mirror, an echo sounded from it as answer. HIRTH (*Boas Anniversary Volume*, p 228)
sees in this mirror a practical demonstration of the theory of sound-reflection, coupled

which the latter was turned was for making lanterns transparent and durable. This confirms the fact that *huo-ts'i* is mica, for the earliest application of it in India and China was in windows and lanterns. [1] Muscovite, a variety of mica, is still employed for lamp-chimneys, as fire-screens in the peep-holes of furnaces, and as screens in the laboratory, for observing the processes in a highly heated furnace without suffering from the intense heat. It is thus clear why the Chinese called this mineral *huo-ts'i* "fire-regulating;" and it is also clear that, since mica cannot by any means be made into a burning-lens, the alleged identity of *huo-ts'i* with the burning-lens styled *huo-chu* is absolutely wrong. Only the fact that the word "fire" forms the first element in the names of both minerals suggested this hypothesis to the Chinese philologists. But there is a fundamental difference in characterizing the two by the attribute "fire." In mica it refers to that phenomenon known to us as asterism, — the exhibition of a starlike reflection, which occurs also in sapphire, chiefly displayed by some phlogopites when a candle-flame is viewed through a sheet of the mineral, — and the frequent use of the substance for windows, as remarked by Watt, may have facilitated the observation of this peculiar property. The fact that the Chinese were perfectly aware of it has already been demonstrated by the reference to the mica windows in the palaces of Lo-yang; and there is another similar report in the Records of Kuang-tung Province, [2] according to which the mica of

---

with that of light-reflection. The text itself, like the book from which it is taken, is apocryphal. The assigning of it to the Emperor Ling is a gross anachronism, and nothing is known about the country K'ü-sü.

[1] Windows of mica are mentioned in a *Description of the Palaces of Lo-yang* (*Lo-yang kung tien ki* 洛 陽 宮 殿 記; *T'ai p'ing yü lan*, Ch. 808). They spread a dazzling brilliancy in the sunlight. Also fans were made from the same substance by Shi Hu 石 虎 (mentioned in his work *Ye chung ki* 鄴 中 記; see BRETSCHNEI-DER, *Bot. Sin.*, pt 1, No. 1079).

[2] *Kuang chou ki* 廣 州 記, by P'ei Yüan 裴 淵, who lived under the Tsin dynasty (265—419), see BRETSCHNEIDER, *Bot. Sin.*, pt. 1, No. 377.

the district of Tsêng-ch'êng, when struck by the sunlight, emits a brilliant light. [1]

LIU-LI AND LANG-KAN NOT BURNING-LENSES. — We find also the opinion heralded by Li Shi-chên that the stone *liu-li* 琉璃 (Sanskrit *vaiḍūrya*) is identical with the *huo-ts'i* gem. This notion goes back to Ch'ên Ts'ang-k'i 陳藏器, who lived during the first part of the eighth century at San-yüan (in the prefecture of Si-ngan, Shen-si Province), and who is the author of the *Pen ts'ao shi i* 本草拾遺. This work seems to be lost; but extracts of it are preserved in the later works on natural history, notably in the *Chéng lei pén ts'ao* 證類本草 of the year 1108, and in the *Pén ts'ao kang mu*. In both works he is quoted as saying that, according to the dictionary *Tsi yün* 集韻, *liu-li* is the same as the gem *huo-ts'i*. This work, of course, is not the *Tsi yün* which was begun in 1034 and completed in 1039, [2] but the *Tsi yün* or *Yün tsi* by Lü Tsing 呂靜 of the Tsin dynasty (265—419). [3] We are here confronted with a purely philological opinion of a lexicographer, which is hardly founded on a personal examination of the objects concerned, [4] nor is it very likely that Sanskrit *vaiḍūrya* ever referred to a variety of mica.

---

[1] 增城縣有雲母向日焃之光耀 (*T'ai p'ing yü lan*, Ch. 808). — The introduction of plate-glass has now supplanted the use of mica in Eastern Asia; but some curious survivals of it still occur in Tibet. The Tibetans manufacture an abundance of charm-boxes (*gau*), some of large dimensions in the form of shrines; a window is cut out in the metal surface to render the image in the interior visible. This window is now usually covered with European glass, but also with a transparent sheet of mica Ornaments of mica are still employed by the women in the territory of the Kuku-nōr for the decoration of their fantastic head-dresses

[2] WATTERS, *Essays on the Chinese Language*, p. 60.

[3] See the Catalogue of Sui Literature (*Sui shu*, Ch 32, p. 22; and WATTERS, *l. c.*, p. 40). *T'ai p'ing yü lan* (Ch 809, p 2) quotes the same definition from the dictionary *Yün tsa* 韻雜, which presumably is a misprint for *Yün tsi* 韻集.

[4] This discussion bears out the reasons which induced F. PORTER SMITH (*Contributions toward the Materia Medica of China*, p. 120) to identify *huo-ts'i* with lapis lazuli, as he took *liu-li* for the latter and encountered the equation of *huo-ts'i* with *liu-li*.

As the term *liu-li* refers to certain varieties of rock-crystal [1] and to certain vitreous products, it would be possible in theory that burning-lenses were made from this substance: but no such instance is on record. There is, however, an isolated case in which a specular lens of this material is in question.

In the **year 499**, the Buddhist monk Huei Shên 慧深 returned to China under the pretence that he had visited a marvellous island in the farthest east, called Fu-sang 扶桑, and made a glowing report of its wonders. It is well known that a number of European and American scholars sought this alleged country Fu-sang in Mexico or somewhere else in America, and pretended that this continent had been discovered by the Chinese nine centuries before Columbus. Others, of a more sober trend of mind, localized Fu-sang on Sachalin or on islands near Japan. But even this moderate attitude rests on a cardinal error, for Fu-sang, as described by Huei Shên, is not a real country at all, but a product of imagination, a geographical myth, composed of heterogeneous elements, as will be shown by me elsewhere. In this connection Fu-sang is of interest to us, as the earliest Chinese mention of a specular lens is associated with it. In the beginning of the sixth century envoys of Fu-sang are alleged to have appeared in China, "offering as tribute a precious stone for the observation of the sun (*kuan ji yü* 觀日玉), of the size of a mirror, measuring over a foot in circumference, as transparent as rock-crystal (*liu-li*); looking through it in bright sunlight, the palace-buildings could be very clearly distinguished." [2] The event

---

[1] It would be preferable to use the general term "quartz," as it is impossible to determine in each and every case what kind of crystal is intended.

[2] 扶桑國使使貢觀日玉。大如鏡。方圓尺餘。明澈如琉璃。映日以觀日 (variant: 見) 中宮殿皎然分明 (*T'ai p'ing yü lan*, Ch. 805, p. 10) This text is derived from the book *Liang se kung tse ki*, 梁四公子記, "Memoirs of the Four Lords of the Liang Dynasty

of the embassy here alluded to is apocryphal, for it is not on record in the official Annals of the Liang Dynasty; the country Fu-sang itself is an imaginary construction. Moreover, the work which contains this story, and which consists of conversations held by the four Lords [1] with the Emperor Wu of the Liang dynasty (502—549) has a decided tendency toward the wondrous, and teems with fables derived from the West. Notwithstanding, all this does not detract from the value of this first account of a specular lens, through which objects could plainly be beheld. I think that SCHLEGEL [2] was not so very wrong in lending expression to the opinion that this "precious stone for the observation of the sun" was a rock-crystal.

In his book (happily now forgotten) *Fusang or the Discovery of America by Chinese Buddhist Priests in the Fifth Century* (1875) CH. G. LELAND has utilized also this notice in support of his Fusang-American hypothesis, and has tried to establish an analogy between the observation glass of the Chinese account and the burning-mirrors of metal which the ancient Peruvians are alleged to have employed for kindling their sacred fire. BRETSCHNEIDER [3] who banished the nightmare of Leland with as much critical acumen and as a solid fund of information refuted this particular point only by discounting the credibility of the Chinese source in question. [4]

(502—556)," written by Chang Yue 張說 (667—730), statesman, poet, and painter (GILES, *Biographical Dictionary*, p 51).

[1] They were Huei-ch'uang 蜀闛, Wan-kie 飍杰, Wei-t'uan 敫黹瑞, and Chang-ki 仉臀.

[2] *T'oung Pao*, Vol III, 1892, p 139.

[3] *Über das Land Fu Sang* (*Mitt d. Ges. Ostasiens*, Vol. II, No. 11, 1876, pp. 1—11).

[4] He erroneously styled the work "the memoirs of a certain Liang sze kung." In his *Botanicon Sinicum* (pt 1, p. 169) the title is correctly explained. In an old catalogue of books from the twelfth century, Bretschneider comments, this work is described as totally unreliable, as the author narrates mostly wondrous and incredible stories This is merely a conventional Chinese mode of literary criticism. The wondrous stories of this book are of incalculable historical value to us, as many of them are exact reproductions of western legends.

This point of view is unnecessary. We certainly do not have to believe in the embassy from Fu-sang, which is not confirmed by the Annals; the instrument, however, described in the report cannot be a personal invention of Chang Yüe, the author of that work, but surely is a reality. It doubtless was a lens which permitted to see the distant palace-buildings with greater distinction; yet it was not a burning-lens, and the comparison drawn by Leland is far from the point. Moreover, the alleged burning-mirrors of the Peruvians existed merely in the imagination of Garcilaso de la Vega, whose fantasy has already been exploded by E. B. TYLOR. [1]

It is possible to trace with some degree of probability the real origin of that lens fancifully associated with the mythical land Fu-sang. The work *Liang se kung tse ki* that contains this account offers the following interesting text: "A large junk of Fu-nan which had hailed from western India arrived (in China) and offered for sale a mirror of a peculiar variety of rock-crystal (碧玻瓈鏡), [2]

---

[1] *Researches into the Early History of Mankind*, pp. 250—253 (New York, 1878).

[2] G. PAUTHIER (*L'inscription de Si-ngan-fou*, p. 31, Paris, 1858), who first called attention to this text, was quite correct in explaining the term *p'o-li* as "rock-crystal." PELLIOT (*Bull. de l'Ecole française*, Vol. III, p. 283) accepts *p'o-li* in this passage in the sense, commonly adopted, of "glass," while admitting that it etymologically corresponds to Sanskrit *sphaṭika*. The latter, however, means "rock-crystal;" and in my opinion the Chinese word *p'o-li*, derived from it, in the greater number of ancient texts, has the same significance. Evidence based on other texts will be produced farther below; here we discuss only the text under consideration. For two weighty reasons it is impossible to regard the mirror mentioned in the *Liang se kung tse ki* as a glass mirror. First,—the story of the merchants, which is an echo of the Western legend of the Diamond Valley, reveals the fact that the question is of a precious stone, not of glass; among the numerous versions of this legend, there is not one that speaks of glass, but all of them are unanimous in mentioning hyacinths, diamonds, or precious stones in general. A plain glass mirror, most assuredly, would not have been priced so highly, nor have caused such a sensation, nor have been linked with a legend of that character. Second,—glass mirrors were not yet invented at that time in the West, and for this reason the conclusion that they should have been known in India and Fu-nan during the sixth century seems to me very hazarded. True it is that HIRTH (*Chinese Metallic Mirrors*, Boas Ann. Vol., p. 219), who also regards this mirror from Fu-nan as being of "green glass" (see, however, also the following footnote), and who wonders at the incredible price solicited for it, supports his theory by

one foot and four inches across its surface, and forty catties in weight. It was pure white and transparent on the surface and in the interior, and displayed many-colored things on its obverse. When held against the light and examined, its substance was not discernible. [1] On in-

---

the statement that the ancients were acquainted with glass mirrors. This argument, however, is not valid; we have to study only the famous and ingenious treatise of J. BECKMANN (*Beiträge zur Geschichte der Erfindungen*, Vol. III, particularly pp. 302—335; an English translation of this monumental work was published in 1814 by W. JOHNSTON) to become thoroughly convinced of the baselessness of Hirth's claim; and the result of Beckmann, who wrote in 1792, is upheld both by classical philology (MORGAN, *Harvard Studies in Classical Philology*, Vol. I, 1890, pp. 50—51) and by the modern history of technology (FELDHAUS, *Technik der Vorzeit*, col. 1044). The plain fact remains that real glass mirrors in our sense did not come up in Europe before the latter part of the thirteenth century, and that they did not exist in classical antiquity. — I do not deny, of course, that in a later period the term *p'o li* assumed the meaning of "glass;" the exact date remains to be ascertained.

[1] HIRTH and ROCKHILL (*Chau Ju-kua*, p. 228), who have translated merely the beginning of this text on the basis of an incomplete quotation in *T'u shu tsi ch'êng*, render this sentence, "Objects of all kinds placed before them [the mirrors] are reflected to the sight without one's seeing the mirror itself." Even if this translation were admissible, which I venture to doubt, I am at a loss to understand what it should mean; it even seems to convey the meaning of something that is impossible. The sentence 置五色 物於其前 (see the complete text of the passage on p. 202, note 3) cannot be linked with the following 向 etc., which is a new sentence expressing a new idea. This may be inferred also from the text, as quoted in *Pên ts'ao kang mu*, in which the sentence beginning with 置 etc. is omitted, while the sentence beginning with 向 etc. is completely reproduced. Objects are certainly not placed in front of a mirror to be seen, but man wants to behold himself or objects in a mirror. It is obvious that the objects here mentioned were natural designs formed by zones of various colors in the stone. As they were not acquainted with the complete text, as handed down in *T'ai p'ing yü lan*, Hirth and Rockhill understand that the junks of Fu-nan habitually sell such mirrors to the Chinese. Our story renders it clear that only an isolated instance comes into question, and that this particular, unusual mirror could not even be disposed of in China. The *Liang se kung tse ki* is not a work on commercial geography summarizing general data, but is a story-book narrating specific events. We have in the present case not a description, but a narrative. For the rest, however, the notes contributed by Hirth and Rockhill on the history of glass are very interesting and valuable, though many problems connected with this difficult subject still remain unsolved. Hirth's opinion, that *pi-p'o-li* should be regarded as a word-formation prompted by analogy with *pi-liu-li*, is very plausible. Our text indeed renders this conception almost necessary, as the word *pi* cannot be taken here in the sense of "green," the substance of the mirror being described as white and transparent.

14

quiry for the price, it was given at a million strings of copper coins. The Emperor ordered the officials to raise this sum, but the treasury did not hold enough. Those traders said, 'This mirror is due to the action of the Devarāja of the Rūpadhātu.[1] On felicitous and joyful occasions, he causes the trees of the gods[2] to pour down a shower of precious stones, and the mountains receive them. The mountains conceal and seize the stones, so that they are difficult to obtain. The flesh of big beasts is cast into the mountains; and when the flesh in these hiding-places becomes so putrefied that it phosphoresces, it resembles a precious stone. Birds carry it off in their beaks, and this is the jewel from which this mirror is made.' Nobody in the empire understood this and dared pay that price."[3]

The story connected in this report with the crystal mirror is a somewhat abrupt and incomplete version of the well-known legend of the Diamond Valley, the oldest hitherto accessible Western version

---

[1] 色界天王 ("the Celestial King of the Region of Forms"). The Rūpadhātu is the second of the three Brahmanic worlds. The detailed discussion of this subject on the part of O. FRANKE (*Chinesische Tempelinschrift*, pp. 47—50) is especially worth reading. The Devarāja here in question is Kubera or Vaiçravaṇa, God of Wealth, guarding the northern side of the world-mountain Sumeru and commanding the host of the aerial demons, the Yaksha.

[2] 天樹. This term corresponds to Sanskrit *devataru*, a designation for the five miraculous trees to be found in Indra's Heaven (compare HOPKINS, *Journ. Am. Or. Soc.*, Vol. XXX, 1910, pp. 352, 353)

[3] 梁四公記。扶南大舶從西天竺國來賣碧玻瓈鏡面廣一尺四寸重四十斤。內外皎潔置五色物於其前。向明視之不見其質。問其價豹錢百萬貫文。帝令有司算之以府庫當之不足。其商人言。此色界天王。有福樂事天樹大雨雨衆寶山納之。山藏取之難得。以大獸肉投之。藏中肉爛類寶一。鳥銜出而此寶焉。舉國不識無敢酬其價者 (*T'ai p'ing yü lan*, Ch. 808, p. 6).— The narrative is obscure in omitting to state that the jewels adhere to the flesh which is devoured by the birds.

of which is contained in the writings of Epiphanius, Bishop of Constantia in Cyprus (*circa* 315—403). [1] Again, it is the author of that curious work, *Liang se kung tse ki*, who has preserved to us the earliest Chinese form of this legend which strikingly agrees with the story of Epiphanius. This text is worded as follows: "In the period T'ien-kien (502—520) of the Liang dynasty, Prince Kie of Shu (Sze-ch'uan) paid a visit to the Emperor Wu, [2] and, in the course of conversations which he held with the Emperor's scholars on distant lands, told this story: 'In the west, arriving at the Mediterranean, [3] there is in the sea an island of two hundred square miles (*li*). On this island is a large forest abounding in trees with precious stones, and inhabited by over ten thousand families. These men show great ability in cleverly working gems, [4] which are named for the country Fu-lin 拂林. [5] In a northwesterly direction from

---

[1] *Epiphanii opera*, ed. Dindorf, Vol. iv, p. 190 (Leipzig, 1862). On the basis of these new Chinese sources, I have treated the history of this legend in detail in a study on the diamond (unpublished manuscript of the writer), and therefore do not pursue the subject further on this occasion.

[2] He was the first emperor of the Liang dynasty and lived from 464 to 549 (Giles, *Biographical Dictionary*, p. 285).

[3] *Si hai* 西海 (the "Western Sea"). Compare Hirth, *Journ. Am. Or Soc.*, Vol. xxxiii, 1913, p 195.

[4] This must be referred to the cutting and engraving of antique intaglios (gems in the sense of Latin *gemma*).

[5] The same mode of writing (林 instead of the later 棽) as that encountered by Chavannes (*T'oung Pao*, 1904, p. 38) in a text of 607, extracted from the *Ts'e fu yuan kuei*. The same way of writing occurs also in *Yu yang tsa tsu* and in a poem of the T'ang Emperor T'ai-tsung (*Pei wén yun fu*, Ch. 27, p. 25). As our text speaks of a forest of jewelled trees, a popular interpretation of the name Fu-lin apparently is intended here, "forest" (林) of the jewels being read into Fu-lin; as if it were "forest of Fu." We are here confronted with the earliest allusion in Chinese records to the country Fu-lin, antedating our previous knowledge of it by a century, Hirth having traced the first appearance of the name to the first half of the seventh century. The reference to the period T'ien-kien (502—520), and the mention of the Liang Emperor Wu, are exact chronological indications which now carry Chinese acquaintance with Fu-lin to the beginning of the sixth century. This result perfectly harmonizes with the view expressed by Pelliot (*Journal asiatique*, Mars-Avril, 1914, p. 498), that the name Fu-lin appears with certainty about 550, and that it is possibly still older.

the island is a ravine hollowed out like a bowl, more than a thousand feet deep. They throw flesh into this valley. Birds take it up in their beaks, whereupon they drop the precious stones. The biggest of these have a weight of five catties.' There is a saying that this is the treasury of the Devarāja of the Rūpadhātu." [1] This is not the occasion to discuss the history and development of this interesting legend in connection with its Arabic and subsequent Chinese parallels; this will be done by me in another place. Suffice it to say for the present that the Chinese version is an exact parallel to that of Epiphanius, that it antedates all Arabic versions, that it represents a purer form than the earliest Arabic text in the *lapidarium* of Pseudo-Aristotle, and that it was transmitted to China directly from Fu-lin. I have here fallen back on these two texts of the *Liang se kung tse ki* to introduce the reader to the mental horizon of its author, Chang Yüe, and thus to secure a basis for judging the *raison d'être* of the specular lens ascribed by him to an embassy from Fu-sang. It was a plausible *a priori* supposition that this instrument must have been one of Western manufacture; and being now familiar with the outfits and tools of the workshop of Chang Yüe, who absorbed traditions of Fu-nan, India, and Fu-lin, we may well infer that the alleged Fu-sang lens was really a

梁四公記。梁天監中有蜀杰公謁武帝嘗
與諸儒語及方域。西至西海海中有島方二
百里。島上有大林。林皆寶樹中有萬餘家。其
人皆巧能造寶器所謂拂林國也。島西北有
坑盤坳深千餘尺。以肉投之。鳥銜寶出大者
重五斤。彼云是色界天王之寶藏 (*T'u shu tsi ch'êng*, section on national economy 321, 寶貨, *tsung pu ki shi*, p 5) — The last sentence, of course, is not an element inherent in the story, as it came from Fu-lin, but is an interpolation of the Chinese author Chang Yüe, taken from the narrative which the traders of Fu-nan had overheard in India.

product of Syria (Fu-liŋ) and reached China possibly by way of India and Cambodja (Fu-nan), in the same manner as the costly mirror of rock-crystal. [1]

A product termed *lang-kan* 琅玕 is identified with *huo-ts'i* by Su Kung 蘇恭 of the T'ang period, [2] who, at the same time defines the former as a kind of *liu-li*. K'ou Tsung-shi 寇宗奭, in his *Pên ts'ao yen i* 本草衍義 of 1116, calls him to task for this wrong statement by observing that *liu-li* is a substance evolved by fire, while *lang-kan* is not, so that the two could not represent identical species. Su Kung's identification has indeed not been adopted by any subsequent Chinese scholar. [3]

---

[1] In the writer's proposed Chinese-Hellenistic studies will be found several interesting examples of Hellenistic folk-lore traditions looming up in Fu-nan and thence transmitted to China.

[2] *Chêng lei pên ts'ao*, Ch 5, fol. 26 Also in a commentary to the dictionary *Ki tsiu pien* 急就篇 (*P'ei wên yin fu*, Ch. 7 A, p. 106 b).

[3] *Lang-kan*, in times of antiquity, appears as a mineral, mentioned already in the earliest Chinese document, the tribute of Yu, in the *Shu king* (LEGGE, *Chinese Classics*, Vol III, p. 127), as a product of the province of Yung-chou; its exact nature cannot be determined, the commentators saying no more than that it was a stone used for beads; Legge's explanation that possibly it was lazulite or lapis lazuli, is purely conjectural. The *Shuo wên* defines *lang-kan* as a stone resembling jade; and the *Erh ya* localizes it in the K'un-lun. The *Pie lu* 別錄 assigns the stone to P'ing-tsê 平澤 in Shu 蜀 (Sze-ch'uan). *Wei liv*, *Hou Han shu*, *Liang shu*, and *Wei shu* (HIRTH, *China and the Roman Orient*, pp. 41, 47, 50, 73) mention *lang-kan* among the products of Ta Ts'in, no explanation of its significance with reference to these passages is on record. We find *lang-kan* also in Kucha (*Liang shu*, Ch. 54, p. 14), in central India (*ibid.*, p. 7 b), and generally in India (*T'ang shu*, Ch. 221 A, p. 10 b). From the T'ang period onward the Chinese naturalists or pharamacists, beginning with Ch'ên Ts'ang k'i, describe *lang-kan* as a kind of coral, growing like a tree with root and branches on the bottom of the sea, fished by means of nets, and being reddish, when coming out of the water, but subsequently turning darker. The *Yan lin shi p'u* (Ch c, p. 9 b) says that it is a stone caught in shallow places near the coast of Ning-po, resembling the genuine coral (*shan-hu*), being white, when coming out of the water, and afterwards turning purple or black. Li Shi-chên objects to the application of the term *lang-kan* to these marine products which, according to him, should be credited with the name *shan-hu*, while the former should be restricted to a stone occurring in the mountains. Compare also SCHLEGEL, *T'oung Pao*, Vol. VI, 1895, p. 58; F DE MÉLY, *Lapidaires chinois*, p. 56, HIRTH and ROCKHILL, *Chau Ju-kua*, pp. 162, 226 The word *lang-kan* seems to be an onomatopoetic formation descriptive of the

THE MINERALOGICAL TERM MEI-HUI. — Finally we have to discuss the term *mei-hui* 玫瑰, which, according to Li Shi-chên, also should refer to lenses. It first appears in the poem *Tse hiü ju* 子虛賦 of Se-ma Siang ju, who died in 117 B.C., as one of the mineral products of Sze-ch'uan.[1] Kuo P'o (275—324) explains it as a stone bead 石珠; Tsin Pao 晉灼 says that it is identical with *huo-ts'i* beads; and Yen Shi-ku (579—645) reiterates the same, adding that "is is the 'fire-pearl' coming at present from the countries of the south."[2] These definitions are vague and unsatisfactory, being made by philologists who in all probability had never seen any of the stones in question. Yen Shi-ku errs in identifying *huo-ts'i* with *huo-chu*, and therefore the identification of both with *mei-hui* is presumably wrong also. The dictionary *Shuo wên* (A.D. 123) notes *huo-ts'i* as an equivalent or synonyme of *mei-hui*; as we have shown that the former covers the group of micas, it would follow from this definition, provided it is correct, that *mei-hui* should be a variety of mica, and consequently cannot be a burning-lens.

The term *mei-hui* is listed also in the ancient vocabulary *Ki tsiu chang* 急就章, edited by Shi Yu 史游 under the reign

---

sound yielded by the sonorous stone when struck (compare the words *lang* 硠, "rumbling of stones, roll of a drum;" and *lang* 朗, "clear, as light or sound;" *lang-t'ang* 朗鎕 is used in Peking as an interjectional expression, imitative of the noise of gongs and drums; in general compare chap. IV of WATTERS, *Essays on the Chinese Language*). This point of view would account for the fact that the name *lang-kan* was transferred from a stone to a coral; for Tu Wan, in his *Yun lin shi p'u* (*l. c.*), expressly states that the coral *lang-kan* when struck develops resonant properties.

[1] *Shi ki*, Ch. 117, p 2 b; and *Ts'ien Han shu*, Ch. 57 A, p. 2 b. Yen Shi-ku defines the pronunciation of the two characters as *mei* and *hui* (or *huei*), but admits for the latter also the sound *kuei* (玫音枚。瑰音回。又音瓌).

[2] 火齊珠。今南方之出火珠也。 This clause is interesting, inasmuch as it proves the importation of lenses into China in the first half of the seventh century,—a fact which, as will be seen, is confirmed by the T'ang Annals.

of the Emperor Yüan (48—33 B.C.), [1] with reference to jars made from this stone and three others. It is simply defined as "fine jade" in the commentary. This explanation, again, would banish any idea of burning-lenses. [2]

What the mei-hui mentioned by Se-ma Siang-ju was, no Chinese commentators really knew. Their explanations are makeshifts to conceal their lack of proper knowledge of the subject. This much seems certain, that the mei-hui of Sze-ch'uan was not mica (huo-ts'i), first, because mica is not known to occur there; and, second, because the name mei-hui denotes also the rose, [3] and accordingly the mineralogical term seems to refer to a rose-colored stone. For this reason it seems out of the question also that it could have been used as a lens, and there is indeed no account to this effect, mentioning the employment of mei-hui. The case, therefore, is one of purely literary extension of significance. The original meaning of the word having fallen into oblivion, it

---

[1] Regarding this work see the important study of CHAVANNES, Documents chinois découverts par Aurel Stein, pp. 1—10. The passage referred to is in Pien tse lei pien, Ch. 70, p. 13 b

[2] The apocryphal work Shu i ki, of the sixth century, which has not come down to us in its original form, is credited with the statement, "Snake-pearls are those vomited by a snake. There is a saying in the districts of the Southern Sea (Kuang-tung, etc.) that a thousand snake-pearls are not the equivalent of a single mei-hui, which means that snake-pearls are low in price. Also mei-hui is the designation of a pearl (or bead, jewel)."

[3] Rosa rugosa, with red and pink flowers (G. A. STUART, Chinese Materia Medica, p. 381; and M. J. SCHLEIDEN, Die Rose, Geschichte und Symbolik, p 228, who enumerates several species of rose in China). The Japanese naturalist Ono Ranzan states that the precious stone mei-hui is named for the color of the flowers of Rosa rugosa, and invokes the Chinese work Tien kung k'ai wu 天工開物 by Sung Ying-sing of 1628 (2d ed., 1637), as his authority (GEERTS, Produits de la nature japonaise et chinoise, Vol. II, p. 360). I cannot trace this reference in the latter work, but find there that mei-hui is treated as a special kind of precious stone "resembling yellow or green peas; the biggest are red, green, blue, yellow, in short, occurring in all colors; and there are also mei-hui like pearls" (see Tu shu tsi ch'éng, chapter on precious stones, pao shi). Yet I am convinced that Ono Ranzan encountered this statement in some Chinese book, and may have erred only in quoting the Tien kung k'ai wu.

became free to assume the same meaning as *huo-ts'i*, in the rôle of an elegant term of the *estilo culto*. The fact that it really interchanges with the latter is manifested by the account of Persia in *Nan shi*,[1] where *mei-hui* are listed among the products of that country: while, as mentioned on p. 195, the analogous reports in *Pei shi*, *Wei shu* and *Sui shu* have the term *huo-ts'i* in the same passage. Thus the greatest probability is that also *mei-hui*, as used in this text of the *Nan shi*, denotes the mica of India. As regards other foreign countries, we find *mei-hui* mentioned in the *Wei lio*, written by Yü Huan between 239 and 265, as a product of the Roman Orient (Ta Ts'in),[2] and worn on the high head-dress of the women of the King of the Ephtalites (Ye-ta).[3]

After having overthrown the nebular hypotheses of foreign and Chinese scholars, the path is finally cleared for discussing the real thing, the history of burning-lenses in China. There is only one term in the Chinese language which may lay claim to having this significance, and that is *huo chu* 火珠 (the "fire-pearl").

INTRODUCTION OF BURNING-LENSES INTO CHINA. — The first historical mention of "fire-pearls" (*huo chu*) is made in the Annals of the T'ang Dynasty (618—906),[4] where they are connected with a tribe of Malayan or Negrito stock, styled "Lo-ch'a" 羅刹, and inhabiting an island in the Archipelago east of P'o-li 婆利 (Bali). "Their country," it is said, "produces fire-pearls in great number, the biggest reaching the size of a fowl's egg. They are round and white, and emit light at a distance of several feet. When held

---

[1] Ch. 79, p 8.
[2] HIRTH, *China and the Roman Orient*, p. 73
[3] *Lo-yang kia lan ki* 洛陽伽藍記, written in 547 by Yang Huan-chi 楊衒之 (quoted in *Tu shu tsi ch'éng, Pien i tien* 67, Ye-ta, *hui k'ao* 2).
[4] *T'ang shu*, Ch 222 c, p. 1 b.

against the rays of the sun, mugwort [1] and rushes [2] will be ignited at once by fire springing from the pearl." [3] The same text, with slightly varying phraseology, is given also in the *Old History of the T'ang Dynasty*, [4] where, however, the interesting addition occurs, that this pearl is in appearance like crystal (狀 如 水 精) Hence we may justly conclude that these fire-pearls were convex crystal lenses, whose optical properties were utilized in producing fire for the medical purpose of cauterization. [5]

---

[1] Chinese *ai* 艾, *Artemisia vulgaris*, a plant common in China and from ancient times used in cauterizing the skin (see BRETSCHNEIDER, *Bot. Sin.*, pt. 2, No 429; pt. 3, No. 72),—a process known to us by the Japanese name *moxa* (properly *mogusa*, the Jap. word for *Artemisia*). The best leaves are taken and ground up with water in a stone mortar, the coarsest particles being eliminated, and the remainder being dried. A small portion is rolled into a pellet the size of a pea, placed upon the ulcer or spot to be cauterized. The preferred method of igniting the moxa is still by means of a burning glass or mirror (compare G. A. STUART, *Chinese Materia Medica*, p. 53). The most interesting and detailed account of this practice was written by ENGELBERT KAEMPFER in the seventeenth century (*History of Japan*, Glasgow edition, Vol. III, pp. 277—292). Kaempfer states that the Japanese used burning splinters or incense-sticks to ignite the moxa.

[2] KAEMPFER (*l. c.*, p. 276) informs us that the most common caustic used by the Brahmans of India is the pith of rushes, which grow in morassy places. This pith they dip into sesamum-seed oil, and burn the skin with it after the common manner.

[3] 多火珠。大者如雞卵。圓白照數尺。日中以艾藉珠輒火出。

[4] *Kiu T'ang shu*, Ch. 197, p. 1 b.

[5] GROENEVELDT (*Notes on the Malay Archipelago*, p. 206, in *Miscell. papers relating to Indo-China*, Vol. 1), who was the first to indicate the relevant passage of the *T'ang shu* (but neglected the corresponding text of the *Kiu T'ang shu*), was therefore wrong in affirming that the fire-pearl is "evidently a kind of burning-glass, but whether of glass or crystal, and manufactured in what place, we have no means to ascertain" We have, as will be seen farther on, the means of ascertaining that these crystal lenses were manufactured in India. Another error of Groeneveldt was to assign the fire-pearls to the country of P'o-li instead of Lo-ch'a. PELLIOT (*Bull de l'Ecole française*, Vol IV, p. 283, note 3) has clearly pointed out the confusion prevailing in this chapter of the T'ang Annals, and has shown that it was the wild men of Lo-ch'a visiting the coasts of Champa in order to sell these crystal lenses, carrying on their trade at night, while hiding their faces during the day (*ibid.*, p. 281, but he too speaks of "lentilles de verre"). G SCHLEGEL (*Toung Pao*, Vol IX, 1898, p. 178; and 1901, p 334), who revealed the same text from the Chinese Gazetteer of Kuang-tung Province, offered the inadequate translation, "Their country produces car-

The crystal lenses, accordingly, were employed in the same manner as the burning-mirrors of copper or bronze in a former period. The *Ku kin chu* 古今注 [1] of Ts'uei Pao 崔豹 of the fourth century states that the latter served for the purpose of setting mugwort on fire. [2]

The Annals of the T'ang Dynasty indicate also the fact that in 630 King Fan-t'ou-li 范頭黎 sent an embassy to China to present such lenses. [3] It is this text of the T'ang Annals which gave to Li Shi-chên occasion for his general statement of the subject, as quoted above. We now observe that he has cited the text inaccurately, and has credited it with the term *huo-ts'i-chu* instead of *huo chu*. The former, however, as we have seen, denotes mica, which cannot be used for lenses; the latter relates to rock-crystal; and it is essential to discriminate between the two. Likewise it is not to the point when he asserts that the lenses now found in Champa are styled "great fire-pearls of the morning dawn." "Morning dawn" (*chao hia*) is well known to us as the designation of a specific textile fabric; [4] and in the passage of the T'ang Annals indicated it happens that the two terms "morning-dawn cloth" and "fire-pearl" (*chao hia pu huo chu* 朝霞布火珠) are closely joined, hence arose, apparently, the misunderstanding of Li Shi-chên.

---

buncles (*huo chu*) which are like crystals." Carbuncles certainly are not like crystals, nor can they be utilized as optical lenses. C. PUINI (*Enciclopedia sinico-giapponese*, p. 65, Firenze, 1877) had already indicated that *huo chu* is a species of quartz.

[1] Ch. c, p. 5 b (ed. of *Han Wei Ts'ung shu*).

陽燧以銅爲之，形如鏡。何日則火生以艾承之則得火也。

[3] The last clause in the definition of these is worded in the *Old History* thus: "When held against the sun at noon in order to ignite mugwort, the latter is consumed by fire"

(正午向日以艾蒸之卽火燃).

[4] PELLIOT, *T'oung Pao*, 1912, p. 480; GILES, *Adversaria Sinica*, p. 394; LAUFER, *T'oung Pao*, 1913, pp. 339, 340; *Ling-wai tai ta*, Ch. 6, p. 13.

A book entitled *Sui T'ang kia hua* 隋唐佳話 [1] informs us that in the beginning of the period Chêng-kuan (627—650) the country Champa (Lin-yi) offered to the Court burning-lenses (*huo chu*), in appearance like rock-crystal, stating that the people of Champa had obtained them from the Lo-ch'a country, whose inhabitants have red hair, a black skin, teeth like animals, and claws like hawks. [2]

The Lo-ch'a or Rākshasa, who, judging from the unflattering description of the Chinese, were a wretched, savage tribe (but sufficiently advanced to practise navigation and to trade with Champa),

---

[1] Quoted in *Pien tse lei pien*, Ch. 21, p. 5 b.

[2] Chinese *Lo-ch'a* is the transcription of the Sanskrit word *Rākshasa*. The latter is the designation for a class of man-devouring ogres with red neck and eyes, and protruding tusks, roaming about at night and doing mischief to mankind. It was believed by Groeneveldt and Schlegel that the country of the Lo-ch'a mentioned in the T'ang Annals is identical with the Nicobar Islands, but PELLIOT (*Bull. de l'Ecole française*, Vol. IV, p. 281) has rightly demonstrated the baselessness of this theory, with the result that the country of the Lo-ch'a in question was situated east of P'o-li, which is identical with Bali, the island east of Java. GERINI (*Researches on Ptolemy's Geography of Eastern Asia*, p. 497) likewise has antagonized that theory, arguing that Lo-ch'a refers to the more southern parts of the Malay Peninsula, and perhaps stands also for the wilder tribes of Negrito-Sakai stock populating its eastern coast; but this opinion conflicts with the Chinese accounts of Lo-ch'a In the belief of the Indians, the main abode of the Rākshasa demons was Ceylon (Langkā), which for this reason was styled also Rākshasālaya ("Abode of the Rākshasa"); and as such, Ceylon appears in the great epic poem Rāmāyana, in which King Rāma combats these fierce devils of Ceylon. A country of the Rākshasa plays a signal rôle in the Tibetan cycle of legends clustering around Padmasambhava, who lived in the eighth century (see E. SCHLAGINTWEIT, *Lebensbeschreibung von P.*, I, p. 21; and LAUFER, *Roman einer tibetischen Konigin*, p. 224). It would be tempting to regard the Lo-ch'a as a tribe like the Vedda of Ceylon, but for geographical reasons it is assuredly impossible to place the Lo-ch'a on Ceylon. Such a nickname as Rākshasa could certainly have been applied by the superior castes of India to any inferior aboriginal tribes (compare the note of YULE, in his *Marco Polo*, Vol. II, p. 312, regarding a Brahman tradition that the Rākshasas had their residence on the Andamans, and the analogous application in India of the words *Nāga* and *Piçāca*). Indian traditions referring to Rākshasa tribes, therefore, cannot assist us toward the identification of the Lo-ch'a country of the T'ang period, which, as justly upheld by Pelliot, was an island in an easterly direction from Bali. It may be supposed that it was the highly cultivated peoples of Java and Bali who conferred the name "Rākshasa" on that primitive tribe in their proximity.

certainly were themselves not able to produce fire-making lenses. [1] From what quarters was their supply derived? We are informed by the Annals of the T'ang Dynasty that in the year 641 Magadha in India sent to the Chinese Court tribute-gifts among which appeared fire-lenses (*huo chu*),[2]) and, further, that Kashmir produces fire-lenses, saffron, and horses of the dragon breed. [3] The latter notice is contained also in the memoirs written by the celebrated pilgrim Hüan Tsang in 646; [4] and his statement, based on actual observation, was doubtless the source from which the official history of the T'ang dynasty drew. The Arabic mineralogists also — as, for instance, al-Akfānī — knew Kashmir as a country producing rock-crystal. [5]

In the beginning of the period K'ai-yüan (713—742) Kashmir sent as tribute "pearls of supreme purity" (*shang ts'ing chu* 上清珠), illuminating an entire house with their splendor. [6] Possibly also in this case crystal lenses are understood.

I Tsing, the Buddhist monk and traveller, who journeyed in India from 671 to 695, observes, "It is only in China where stones are internally taken as medicine. Since rock-crystal and marble emit

---

[1] GERINI (*l. c.*, p. 491), who erroneously locates the Lo-ch'a on the east coast of the southern portion of the Malay Peninsula, conjectures with reference to these crystal lenses that rock-crystal "very likely" occurs in that region. This point of view is quite immaterial. Whether rock-crystal is found there or not, the Lo-ch'a certainly did not quarry it; and if they did, it was not wrought by them into lenses. Quartz, for instance, is common on the Andamans, but the natives make it only into chips or flakes used in shaving or tattooing, while even the art of eliciting fire from the stone by means of striking is wholly unknown to them (E. H. MANN, *Journ. Anthrop. Inst.*, Vol. XII, 1883, p. 381).

[2] *Tang shu*, Ch. 221 A, p. 11.

[3] *Tang shu*, Ch. 221 B, p. 6. Compare CHAVANNES, *Documents sur les Tou-kiue occidentaux*, p. 166.

[4] JULIEN, *Mémoires sur les contrées occidentales*, Vol. I, p. 167, who translates "glass lenses", and WATTERS, *On Yuan Chwang's Travels in India*, Vol. II, p. 261.

[5] WIEDEMANN, *Zur Mineralogie im Islam*, p. 206. Al-Akfānī died in 1348.

[6] *Tu yang tsa pien* by Su Ngo, Ch. A, p. 3 (ed. of *Pai hai*).

sparks of fire, the organs of the body, if those stones are administered, may be scorched and ripped open. Many of our contemporaries, being unaware of this fact, have suffered death in consequence of this wrong treatment." [1] In Chinese alchemy preparations made from jade and mica played a signal part, and were consumed by ambitious devotees to insure long life or immortality. [2] When crystal lenses made their appearance in China, the belief was naturally fostered that fire was a substance inherent in the stone. Fire was considered as an element belonging to the male, creative, and life-giving principle called *yang*, so that a mineral partaking of it was apt to strengthen the body and to prolong life. The evil effect of the internal application of rock-crystal, as conceived by I Tsing, thus becomes intelligible: in the same manner as a crystal lens can set fire to an object, so it may cause the human body to catch fire.

The information given in the T'ang Annals with regard to the Lo-ch'a originated from the mission which carried Ch'ang Tsiün 常 駿 in the year 607 into the country Ch'i-t'u 赤 土. On his journey he is said to have reached the country of the Lo-ch'a, while in another passage it is stated that owing to this mission the inhabitants of the Lo-ch'a country entered into relations with China. [3]

---

[1] *Nan hai ki kuei nei fa chuan*, Ch. 3, p. 20 (ed. of Tōkyo); compare J. Takakusu (*Record of the Buddhist Religion*, p. 135), who wrongly takes the term *pai shi* 白 石 (literally, "white stone") for adular, which does not occur and is unknown in China, *pai shi* repeatedly appears in the votive inscriptions on Buddhist marble sculptures of the T'ang period, and is still the current expression for "marble." It would be possible that I Tsing employed the term *pai shi* as a rendering of Sanskrit *silopala* ("white stone"), which is a synonyme of *sphaṭika* and accordingly a variety of quartz or rock-crystal (R. Garbe, *Die indischen Mineralien*, p. 87). Takakusu speaks of "the swallowing of a stone;" the stones were of course triturated and powdered, the mass was kneaded and prepared with other ingredients.

[2] Under the Sui (589—618) was still extant a treatise on the Method of Prescriptions in administering Jade (*Fu yü fang fa* 服 玉 方 法). See *Sui shu*, Ch. 34, p. 21.

[3] Pelliot, *Bull. de l'Ecole française*, Vol. IV, p. 281.

The latter statement seems to be the more probable of the two. The date 607 may thus be fixed as the time when the Chinese made their first acquaintance with burning-lenses; and during the first part of the seventh century a somewhat lively trade in the article was carried on from Champa to China. Hence Yen Shi-ku (579 – 645), as mentioned, justly points to the importation of burning-lenses from the south during his time. While, as a last resort, the Lo-ch'a lenses are traceable to India, we have as yet no means of ascertaining through what channels these lenses were transmitted from India to the Lo-ch'a. At this point there is a lacune in our knowledge which I am unable to fill; it may be supposed only that Sumatra or Java, or both countries, acted as middlemen in this traffic, but I regret having no certain facts along this line to offer.

It is curious that a tribe of such a low degree of culture as the Lo-ch'a possessed burning-lenses, and was instrumental in conveying this Indian article to Champa and China. This fact we may explain from ethnographical conditions of the present time, with which we are familiar: the Lo-ch'a, though acquainted with natural fire and its uses, must have been a tribe that did not know of any practical method of producing fire. Such a people, for example, we meet among the Andamanese, of whom E. H. MAN [1] says, "The Andamanese are unable to produce fire, and there is no tradition pointing to the belief that their ancestors were their superiors in this respect. As they live in the vicinity of two islands, one of which contains an extinct, and the other an active volcano, it seems not unreasonable to assume that their knowledge of fire was first derived from this source. Being strangers to any method of producing a flame, they naturally display much care and skill in the

---

[1] *Journ. Anthrop. Inst.*, Vol. XI, 1882, p. 272; compare also Vol. XII, 1883, p. 150.

measures they adopt for avoiding such inconvenience as might be caused by the extinction of their fires. Both when encamped and while journeying, the means employed are at once simple and effective. When they all leave an encampment with the intention of returning in a few days, besides taking with them one or more smouldering logs, wrapped in leaves if the weather be wet, they place a large burning log or faggot in some sheltered spot, where, owing to the character and condition of the wood invariably selected on these occasions, it smoulders for several days, and can be easily rekindled when required." Nothing introduced by the English so impressed this people with the extent of their power and resources as matches. It is notable also that the household fire is not held sacred by the Andamanese, or regarded as symbolical of family ties, and that no rites are connected with it; there are not even beliefs with reference to its extinction or pollution. The Lo-ch'a must have lived under exactly the same conditions when burning-lenses were first introduced among them from India. Not familiar with any practical method of fire-making or any fire-ceremonial, they readily took to this easy expedient, as the modern Andamanese did to our matches. It is still the primitive tribes spending most of their time in the open air, like the Lepcha and Tibetans (see below), who evince a predilection for the application of the burning-lens in fire-making.

Besides the name *huo chu* 火珠, the term *huo sui chu* ("fire-igniting lens") is found in the *Chêng lei pên ts'ao*, completed by T'ang Shên-wei in 1108.[1] From the same work it follows also

---

[1] 火燧珠向日取得火 (*Chêng lei pên ts'ao*, Ch. 3, fol. 44, edition of 1523). This is the concluding sentence of a brief notice on *p'o-li* (see above, p. 200). Both the *Chêng lei* and the *Pên ts'ao kang mu* accept this term in the sense of "rock-crystal" (*sphaṭika*), Li Shi-chên giving as synonyme the term *shui yü* 水玉, which appears in the *Shan hai king* and in the poem on the Shang-lin Palace 上林賦

that burning-lenses were manufactured in China under the Sung. Whether this was the case under the T'ang I am unable to say.

BURNING-LENSES IN INDIA AND SIAM. — The preceding Chinese accounts are clear enough to allow the inference that the so-called "fire-pearls" were lenses of rock-crystal cut into convex shape, that they were used for cauterization in the same manner as reported by Pliny, and that they were introduced into China, through the medium of the Lo-ch'a and of Champa, from Kashmir, or other regions belonging to the culture-zone of India. In short, what the

---

of Se-ma Siang-ju: its transparency, he says, equals that of water, its hardness that of jade, hence this term; the name "water-jade" is identical with rock-crystal (其瑩 如水。其堅如玉。故名。水玉與水精同名). The opinion of both T'ang Shên-wei and Li Shi-chên goes back to Ch'ên Ts'ang-k'i of the T'ang period, whose definition of p'o-li is as follows: "P'o-li is a precious stone of the Western countries. It belongs to the category of hard stones, and is developed in the soil. According to the opinion of some it results from the transformation of ice that is a thousand years old; but this is certainly not the case" (陳藏器曰。玻瓈 西國之寶也。玉石之類。生土中。或云千歲冰 所化。亦未必然). Nobody, as far as I know, has as yet explained the statement of Li Shi-chên that the original mode of writing is 頗黎, and that this name P'o-li is the designation of a country. T'ai p'ing yü lan (Ch. 808, p. 6) quotes a work Tien chu ki 天竺記 ("Memoirs of India") as follows: "In the Himalaya, there is the mountain of precious stones producing the complete series of the seven gems (saptaratna), all of which may be obtained. Only the p'o-li gem is produced on such lofty peaks that it is difficult to obtain" (大雪山中有寶山諸七 寶並生取可得。唯頗黎寶生高峯難得). Here we are confronted with the reproduction of an Indian notion that meets its parallel in the Ratnaparīkshā, according to which rock-crystal is a product of Nepal (L. FINOT, Lapidaires indiens, p. 56). Certainly the people of India did not hunt for glass on the heights of the Himalaya. The King of Nepal adorned himself with pearls, p'o-li, mother-o'-pearl, coral, and amber (T'ang shu, Ch. 221 A, p. 1); his p'o-li certainly were a kind of rock-crystal, as also S. LÉVI (Le Népal, Vol. I, p. 164) understands, but not glass. The Buddhist monk Huei Yuan 慧苑 of the T'ang period, in his Glossary to the Buddhā-vataṁsaka-sūtra (華嚴經音義, Ch. 1, p. 8, ed of Shou shan ko ts'ung shu, Vol. 94; see Bunyiu Nanjio, No. 1606), explains p'o-li as "to some degree resembling in appearance rock-crystal (水精: that is, the variety of rock-crystal indigenous in China), yet occurring also in red and white varieties."

Chinese received were Indian manufactures. Hence it is legitimate to conclude that the Chinese name *huo-chu*, conferred upon these lenses, represents the translation of a corresponding Sanskrit term. Such, indeed, exists in the Sanskrit compound *agnimaṇi*, the first element of which (*agni*) means "fire," answering to Chinese *huo*; and the second part of which (*maṇi*) signifies a "pearl, bead, gem, or jewel," exactly like the Chinese word *chu*. [1] Moreover, Sanskrit *agnimaṇi*, according to the Sanskrit Dictionary of Boehtlingk, is an epithet of the stone *sūryakānta*, which means "beloved by the sun," so called because it produces fire under the influence of solar rays. Other synonymes are *tapanamaṇi* ("sun jewel"), *tāpana* ("dedicated to the sun"), *dīptopala* ("refulgent stone"), *agnigarbha* ("essence of fire"), — all of these, as correctly seen by L. Finot, [2] referring to rock-crystal. A Hindu treatise on precious stones, the Navaratnaparīkshā, says, under the subject of rock-crystal, that the

---

[1] Although apparently formed in imitation of this Sanskrit expression, the term *huo chu*, notwithstanding, pre-existed in China independently of Indian influence, but in a widely different sense. The following story is on record in the Annals of the Tsin Dynasty (*Tsin shu*, Ch. 99, p. 1; biography of Hêng Huan 桓玄). His mother, née Ma 馬氏, was sitting out one night with her companions in the moonlight, and saw a shooting-star fall into a copper basin filled with water. In the water appeared what looked like a fire-pearl (*huo chu* 火珠) of two inches, diffusing a bright, clear light. Madame Ma took it out with a gourd ladle and swallowed it. When she gave birth to her son, the house was filled with effulgent light; hence the infant received the name Ling-pao 靈寶 (that is, "Supernatural Treasure"). It is evident that this "fire-pearl" was a product of meteoric origin. A similar account is found in the Bamboo Annals: Siu-ki 修己, the mother of the Emperor Yü 禹, saw a falling-star, and in a dream her thoughts were moved till she became pregnant, after which she swallowed a spirit pearl (LEGGE, *Chinese Classics*, Vol. III, Prolegomena, p. 117). The term *huo chu* appears again in *Tsin shu* (Ch. 25, p. 13 b) in connection with the description of the costume, ornaments, and paraphernalia worn by the heir-apparent. There is no explanation of its meaning in this text: perhaps it was a flaming or sparkling gem. In the latter sense I encountered the term in two passages of the *Shi i ki* (Ch. 5, p. 5 b; and Ch. 7, p. 2; ed. of *Han Wei ts'ung shu*); in one case the question is of an extraneous hairpin adorned with a fire-pearl dragon and a phœnix.

[2] *Lapidaires indiens*, p. XLVII.

15

variety of the stone which, struck by sunlight, instantaneously elicits fire, is styled *sūryakānti* by the connoisseurs. The physician Narahari from Kashmir, who wrote a small *lapidarium* in the beginning of the fifteenth century, observes in regard to the same stone, "If it is smooth, pure, without fissures and flaws in the interior, if polished so that it displays the clearness of the sky, and if from contact with solar rays fire springs from it, it is praised as genuine." [1] Narahari dilates likewise on the medical virtues of the stone, to which he lends the attribute "sacred," and which, if honored, procures the favor of the sun.

Fire-production by means of lenses was not a very ancient, or a common, or a popular, practice in India, any more than in classical antiquity. [2] In the oldest epoch of India's history, the Vedic period, we hear only of fire-making by means of friction from wooden sticks. The daily birth of Agni, the god of fire, from the two fire-sticks (*araṇī*), is often alluded to in Vedic literature.

---

[1] R. GARBE, *Die indischen Mineralien*, p. 89. Garbe commits the error of regarding this stone as the sunstone, being misguided by the Sanskrit name *sūryakānta*, and speculates that also the Indian name has come with this stone to Europe. All this is erroneous. First, the sunstone is not known to occur in India, but it occurs near Verchne Udinsk in Siberia, Tvedestrand and Hitterö in Norway, Statesville in North Carolina, and Delaware County in Pennsylvania (BAUER, *Edelsteinkunde*, 2d ed., pp. 528, 529); second, the name "sunstone" is bestowed upon this kind of feldspar by us, not by the Indians, because it reflects a spangled yellow light originating from minute crystals of iron oxide, hematite, or gothite, included in the stone, and which both reflect the light and give it a reddish color (FARRINGTON, *Gems and Gem Materials*, p. 179); this case, therefore, is totally different from that which induced the Hindu to name a certain variety of rock-crystal "sun-beloved;" third, feldspars, like the sunstone, are not made into burning-lenses, such as are described by Narahari. After arriving at his fantastic result, Garbe is forced to admit that Narahari is wrong to classify the (that is, Garbe's) "sunstone" among the quartzes, but the physician of Kashmir who does not speak of "our" sunstone is perfectly right in grouping rock-crystal among quartzes, and the blunder is solely on the part of Garbe.

[2] The utility of the burning-lens, of course, has its limitations. It is dependent upon a cloudless sky and the power of strong sunlight. At night when fire may be most needed it is put out of commission.

They are his parents, the upper being the male, and the lower the female; or they are his mothers, for he is said to have two mothers. [1] The *Vāyu Purāna*, one of the oldest of the eighteen Purānas, presumably dating in the first half of the fourth century, [2] mentions three kinds of fire, — the solar fire (*saura*), or the pure one, or the fire of the gods; fire proceeding from lightning, procured from trees ignited by a lightning-stroke; and fire obtained by friction. Whether and how the first-named was secured we do not know. It would be very tempting to believe that this celestial fire, obtained by concentrating the rays of the sun, was the result of an application of lenses, as, indeed, is still the case in Siam (see below). Such a conclusion, however, would hardly be justified. In all probability, only the divine or transcendental fire, like that in the Greek myth of Prometheus, is here intended. Also in the Avesta, the sacred writings of the ancient Iranians, in which five kinds of fire are distinguished, the fire of heaven burning in the presence of Ahura Mazda is known; [3] and there is no record of the use of burning-lenses on the part of the Iranians. [4]

---

[1] Compare A. A. MACDONELL, *Vedic Mythology*, p. 91; H. OLDENBERG, *Religion des Veda*, p. 105; R. ROTH, *Indisches Feuerzeug* (*Z. D. M. G.*, Vol. 43, pp. 590—595); F. SPIEGEL, *Arische Periode*, p. 147. The modern processes of fire-making in India are well described by E. THURSTON, *Ethnographic Notes in Southern India*, pp. 464—470 (Madras, 1906).

[2] V. A. SMITH, *Early History of India*, p. 305 (3d ed., Oxford, 1914).

[3] A. V. W. JACKSON, in *Grundriss der iranischen Philologie*, Vol. II, p. 641; W. GEIGER, *Ostiranische Kultur*, p. 253.

[4] A material difference between the fire-worship of the ancient Indians and Iranians lies in the point that fire-making ceremonies predominate with the former (a good and succinct description of these will be found in the new book of L. D. BARNETT, *Antiquities of India*, pp. 156—161), while the latter were eager to seek for the sites of natural fire (JACKSON, *Zoroaster*, pp. 98—101); so that the artificial production of fire was not part of their rites. Much valuable information relative to the Persian worship of fire has been gathered by DIEULAFOY (*Suse*, pp. 393 *et seq.*). The Avesta (Vidēvdāt, XIV, 7; F. WOLFF, *Avesta*, p. 405) mentions fire-implements without description of particulars, and we seem to have no information as to Iranian methods of fire-making. This is the more deplorable, as the Persian form of fire-worship spread into all parts of the world, — to

In Sanskrit medical literature I have not yet found any reference to burning-lenses, [1] but the employment of burning-mirrors in medical practice is well ascertained for ancient India. Such mirrors, probably made of metal, [2] are twice mentioned in the medical work *Ashṭāṅga-Hṛidaya*. [3]   In one case, certain drugs are to be ground on it; and a counterpart of this practice appears in a recipe of the famous Bower Manuscript, coming down from the middle of the fifth century: "Let long pepper and turmeric be rubbed repeatedly on a mirror, and anoint with them the eye when it suffers severe pain; it will then quickly become well." In the other case (mentioned in the above work), the wound of a person bitten by a rat is to be cured by an arrow or a mirror, and, as

---

Rome (F. CUMONT, *Mysteries of Mithra*, p. 99; and *Oriental Religions in Roman Paganism*, p. 137), to India (R. G. BHANDARKAR, *Vaishnavism*, pp. 151—155), and to China (Masūdī, in B. de MEYNARD, *Prairies d'or*, Vol. I, p. 303; J. J. MODI, *References to China in the Ancient Books of the Parsees*, in his *Asiatic Papers*, pp. 241—254; CHAVANNES, *Le Nestorianisme*, *Journal asiatique*, 1897, pp. 60, 61, 74, 75; PELLIOT, *Bull. de l'Ecole française*, Vol. III, pp. 669, 670). It could very well be conceived that the Persian Magi, who appear in India under the name Maga and in China as Mu-hu (*Mémoires concernant les Chinois*, Vol. XVI, p. 230; CHAVANNES and PELLIOT, *Traité Manichéen*, p. 170), should have had a certain share in the diffusion of burning-lenses; but this, for the time being, remains purely a matter of speculation, as we are entirely ignorant of any evidence in the case. One curious coincidence, however, deserves attention in this connection, and this is the sacred candle of the Siamese lighted with "celestial fire" by means of a burning-glass (mentioned below) and the same "celestial fire" kept constantly burning in a lamp by the Persian kings as a symbol of the perpetuity of their power, and it passed with the mystical ideas of which it was the expression to the Diadochi, and from them to Rome, where the celestial fire received as its emblem the inextinguishable fire that burned in the palace of the Cæsars, and which was carried before them in official ceremonies.

[1] Cauterization was practised by Indian physicians (see HOERNLE's translation of *Suçruta Saṃhitā*, pp. 74—80).

[2] Regarding mirrors in ancient India, see the writer's *Dokumente der indischen Kunst*, I, p. 174.

[3] That is, the "Quintessence of the Eight Parts of Medicine," ascribed to the physician Vāghbata, probably written before the eighth century (J. JOLLY, *Indische Medicin*, p. 8; the time of the work is fully discussed by JOLLY in *Z. D. M. G.*, Vol. 54, 1900, pp. 260—274).

supposed by Dr. Hoernle, by the reflection of the sun-rays focussed on it. [1]

The lack of information on objects of reality so painfully obtrusive in Indian literature, combined with the defect of a sound chronological sense, renders it impossible to trace a *terminus a quo* for the utilization of burning-lenses; and the records of the Chinese present our only reliable source in this respect. Indeed, the students of India have never taken up this problem, and may now hear for the first time that burning-lenses were ever known in India. The information coming from Chinese sources, which establish the date of the first introduction of such lenses into China in the beginning of the seventh century, allows the inference that they were made and employed in India prior to this date. This result, however trifling it may appear at first sight, is significant in bearing out the fact that long before the Arabic invasion of India (710) burning-lenses were operated there, and that the idea cannot have been imported into India by the Arabs.

Sacred fire was annually obtained from crystal lenses at the Court of the Emperor Akbar, and all the fires of the imperial household were lighted from it. His historian, Abul Fazl Allami (1551—1602), thus describes the ceremony: [2] "At noon of the day, when the sun enters the nineteenth degree of Aries, the whole world being then surrounded by its light, they expose to the rays of the sun a round piece of a white and shining stone, called in Hindi *sūrajkrānt*. A piece of cotton is then held near it, which catches fire from the heat of the stone. This celestial fire is committed to the care of proper persons. The lamp-lighters, torch-bearers, and cooks of the household use it for their office; and when the year has passed in happiness, they renew the fire. The vessel

---

[1] Compare A. F. R. HOERNLE, *The Bower Manuscript*, p. 160.

[2] H. BLOCHMANN, *Ain I Akbari*, Vol. I, p. 48 (Calcutta, 1873).

in which this fire is preserved is called 'fire-pot.' There is also a shining white stone, called *chandrkrānt*, which, upon being exposed to the beams of the moon, drips water." [1]

Burning-lenses are still employed in Siam at state ceremonies, like the New Year festival, or during the tonsure-ceremonial when Buddhist monks are ordained, for obtaining what is called the "celestial fire" (*fai fa*). The medium enlisted is a huge wax candle, styled *thien chai* (literally, "victorious taper"), which is prepared under the direction of the head priest of some royal temple. The wax employed for a single taper amounts to twenty-six pounds in weight; the wick consists of a hundred and eight cotton threads, a number sacred with the Buddhists; and the length is about five feet. Round it are inscribed the magical formulas and diagrams which are prescribed by custom. This sacred candle is usually lighted by means of celestial fire, generated from the sun by the use of a huge burning-glass (*wen fai*) mounted on a richly gilded and enamelled frame. The fire thus kindled is protected in a lamp until the auspicious moment arrives for applying it to the "torch of victory." The lamp is then brought before the king, who takes

---

[1] The Hindi word corresponds to Sanskrit *candrakānta* ("beloved by the moon"), in the same manner as does *sūryakānta* to the above Hindi name for the crystal lens. *Candrakānta* is a kind of rock-crystal, generally believed in India to shed water when the moon shines on it (FINOT, *Lapidaires indiens*, p. XLVII). The Tibetan rendering of this term is *c'u šel* ("water crystal"), explained as "a fabulous magic stone supposed to have the power of producing water or even rain" (JASCHKE, *Tibetan-English Dictionary*, p. 562). GRENARD's opinion (*Mission scientifique dans la Haute Asie*, Vol. II, p. 407), that this stone "employed by the Tibetan sorcerers who have the power of causing or stopping rain" probably is jade, is inadmissible; the Tibetan word for "jade" is *yang-ṭi* or *g-yang-ṭi* (*Polyglot Dictionary of K'ien-lung*, Ch. 22, p. 64), the history of which I hope to trace some day in another place. — Tibetan has also a term for a burning-lens, — *me šel* ("fire crystal") or *sreg byed šel* ("burning crystal"); likewise Lepcha *mi šer* or *šer mi* (MAINWARING-GRÜNWEDEL, *Dictionary of the Lepcha Language*, pp. 285, 434). According to H. VON SCHLAGINTWEIT (*Reisen in Indien und Hochasien*, Vol. II, pp. 201, 202) burning-glasses imported from China are widely used in Tibet for fire-making; he himself witnessed in Sikkim the employment of such glasses directed on tinder.

a taper, termed the "ignition candle," which he lights at the celestial fire, while reciting a prayer-formula. The king then hands the ignition candle to the head priest, who applies its flame to the *thien chai*. During this performance the attendant chapter of monks rehearses a prayer. The torch is kept lighted in a special white gauze frame. A solemn ceremony takes place also at the time when it is extinguished. [1]

ICE-LENSES. — Everybody knows that also a flake of ice, if cut into the form of a convex lens, may serve as a burning-glass with good effect. The Chinese have had this experience; and one of their books, the *Po wu chi* 博物志, a collection of notes on remarkable objects and occurrences, has it on record that "fire may be obtained by cutting a piece of ice into circular shape, holding it in the direction of the sun, and placing mugwort (*Artemisia*) behind the ice, so that it falls within the shadow." [2] It should be added that this notice figures under the title "juggler's art" 戲術; and it is from this class of performers, who swallow fire and swords, that the demonstration of such an experiment might be expected. Nevertheless, Li Shi-chên found it advisable to insert this notice in his essay on the mugwort, [3] as if it had ever been a common practice of physicians to apply the moxa to their patients by means of an ice-lens. This, however, remains open to doubt. Mugwort is said to have received the name "ice-terrace" (*ping-t'ai*) from the employment of ice-lenses. The authorship of the work above quoted is attributed to Chang Hua 張華, who lived from 232 to 300. If Chang Hua of the third century should really have written this

---

[1] After G. E. GERINI, *The Tonsure Ceremony as performed in Siam*, p. 161 (Bangkok, 1893). — Regarding crystal lenses in Japan see GEERTS, *Produits de la nature japonaise et chinoise*, p. 243.

[2] 削冰令圓舉以向日以艾於後承其影則得火 (Ch. 4, p. 4 b; edition printed in Wu-ch'ang).

[3] *Pén ts'ao kang mu*, Ch. 15, p. 3.

passage, the case would indeed be notable in establishing the fact
that four centuries prior to the first introduction of burning-lenses
from Indian regions the latter were known in China as an apparently native idea.   Indeed, this text has been accepted in this
sense, and was marched forward by G. SCHLEGEL [1] as a strong
bulwark in his argumentation for the indigenous origin of burning-
lenses in China; but this plea will melt away as easily as the bit
of ice when its function as lens was over.   Also Schlegel had access to WYLIE's *Notes on Chinese Literature*, from which we learn
(p. 192) that the work *Po wu chi*, originally drawn up by Chang
Hua, was lost in the Sung period (960—1278); that the present
book with that title was probably compiled at a later period on the
basis of extracts contained in other publications; and that there are
many quotations from it in the ancient literature which do not
appear in the modern edition.   There is, accordingly, no guaranty
whatever that any text in this work, as it is now extant, goes
back to the third century and originates from the hand of Chang
Hua.   The text in question is quoted by Li Shi-chên from the
P'i ya 埤雅, a dictionary compiled by Lu Tien 陸佃 (1042—
1102), so that from this indication we may carry it to the latter
part of the eleventh century.   It is certainly far older than that;
but it cannot have been penned by Chang Hua, and, at the very
best, cannot date back farther than the first half of the seventh
century, when burning-lenses first became known in China.   The
Anuals of the T'ang Dynasty, as we noticed, record burning-lenses
in the possession of the Lo-ch'a as an entirely novel affair, describing their use and effect, and this incontrovertibly proves that
they were unknown in times previous.   Neither do the T'ang

---

[1] *Uranographie chinoise*, p. 142; *Nederlandsch-Chineesch Woordenboek*, Vol. I, p. 674;
and *T'oung Pao*, Vol. IX, 1898, p. 179.   The allegation of Schlegel that lenses of ice
were used before the invention of glass is pure invention, being contained neither in this
nor in any other Chinese text.

authors assert that they were known at an earlier date (Yen Shi-ku, on the contrary, insists on their being imported "at present;" that is, in his own lifetime), nor is there any record in the historical annals relating to the third century to the effect that such lenses should have been in vogue at that period. Whoever reads with critical eyes the account now sailing under the false flag of the *Po wu chi* will soon notice that in its style it is worded on the basis of the text of the T'ang Annals, and also that it materially depends upon the latter, — materially, because it was only after, and in consequence of, the introduction of foreign crystal lenses, that the experiment with ice could have been conducted in China. This idea was not conceived by the Chinese as the result of a natural observation or optical study, which they never cultivated; but ice was resorted to as a makeshift, as a substitute for the costly rock-crystal, on the theory of their nature philosophy, that the latter is transformed ice: crystal and ice, being products of a like origin, were thought to be able to bring about the same effect.

Conclusions. — When we now attempt to reconstruct the general history of burning-lenses, the principal fact standing out is that China, despite the opposite contention of some enthusiasts, has not the shadow of a claim to their invention, but, on the contrary, admits her debt to Lo-ch'a and Champa; that means, to India. China received them from India in the same manner as mediæval Europe and the Arabs received them from Greece and Rome. The problem, therefore, crystallizes around the central point: In what reciprocal relation or obligation are India and Hellas? Hellas, at the outset, is entitled to the privilege of chronological priority, and can point to the well-fixed date 423 B.C., when Aristophanes wrote his *Clouds*. At that time, we may assert positively, burning-lenses were unknown in India, for which we have merely a retrospective *terminus a quo* lying backward of the seventh century A.D. Negative evidence in this particular case is somewhat conclusive: for, with all their ideas of the sacredness of fire and its prominent position in religious worship, the ancient Hindu themselves would not have allowed such an excellent contrivance to escape, — a contrivance that would have brought the realization of their dreams of celestial fire. The fact remains that none of the Sanskrit rituals ever mention such an implement, which, for this reason, cannot have been of any significance in the culture-life of the nation. It is therefore highly improbable, nay, impossible, that the Hindu should have independently conceived the invention. Even if our conclusion, based on Chinese documents, that burning-lenses were employed in India prior to the seventh century, should be substantiated in the future by the efforts of Indian research, and, for example, be carried back to a few centuries earlier, this would hardly change our result fundamentally, or overthrow the impression that the use of such lenses belongs to the mediæval epoch of Indian history. There are good reasons for upholding this opinion and for connecting their introduction with the influence upon India of Hellenistic-Roman civilization. First, we may say negatively that it was not Assyria which transmitted the idea to India. In that case, we should justly expect that it would turn up there at a much earlier date, and occur simultaneously in ancient Persia; but Zoroastrian Persia, like Vedic India, lacks them entirely. This observation justifies us in concluding also that burning-lenses played a very insignificant part, if any, in Mesopotamia; if they did, we should find them also in Greece at a much earlier date. Without pressing the question of the when and where of the original invention, we must be content at present to regard the Greeks as the people who, we know positively, made the first use of optical lenses. The second negative evidence that is impressed upon us is this, that Alexander's campaign cannot be made responsible for the transmission. It is needless to insist that the historians of Alexander are silent about it; coeval India is likewise so; and it is inconceivable that an idea, though Alexander's genius should have carried it into the borders of India, would have borne fruit on her soil only as late as the middle ages. The Arabs, as already observed, did not transfer it, either, to India. If we strictly adhere to our chronological result, we are clearly carried into the Gupta period, which, taken in a wide sense, extends from about 300 to 650 A.D., and which, particularly in the fourth and fifth centuries, was a time of exceptional intellectual activity in many fields,[1] in mathematics, astronomy, and medicine, all of which have received an appreciable stamp of Western influence.[2] Indeed, as emphasized by Smith, the eminent achievements of this period are mainly due to contact with foreign civilizations, both on the East and on the West, and the fact of India's intercourse with the Roman Empire is indisputable. The conquest of Mālwā and Surāshtra by Caudragupta II Vikramāditya toward the close of the fourth century opened up ways of communication between Upper India and Western lands which

---

[1] V. A. SMITH, *Early History of India*, 3d ed., p. 304.

[2] See particularly A. WEBER, *Die Griechen in Indien* (*Sitzungsberichte Berliner Akademie*, 1890, pp. 921—925); G. D'ALVIELLA, *Ce que l'Inde doit à la Grèce*, pp. 95—119 (Paris, 1897); G. THIBAUT, *Indische Astronomie*, pp. 43, 76.

gave facilities for the reception of European ideas. It is accordingly a reasonable conclusion that burning-lenses were transmitted to India, not from Hellas, but from the Hellenistic Orient of the Roman Empire, in a period ranging between the fourth and sixth centuries, to be passed on to China in the beginning of the seventh century. The introduction of the burning-mirrors alluded to in the Bower Manuscript, in my opinion, falls within the same epoch, emanating from the same direction.

ADDITIONAL NOTES. — P. 202, note 2. The tree in question is the *pārijāta* (see *Fan yi ming i tsi*, Ch. 25, p. 27 b, ed. of Nanking).

P. 206, note. Compare also *lang-tang* 琅璫 and 銀鐺; an interesting notice on this word is contained in the *Nêng kai chai man lu*, Ch. 7, p. 27 b (*Shou shan ko ts'ung shu*, Vol. 71).

The interesting study of Dr. M. W. de Visser (*Fire and Ignes Fatui in China and Japan*, reprint from *M.S.O.S.*, 1914, pp. 97—193) reached me only a short while ago when my manuscript was in the press. Dr. de Visser touches some questions dealt with on the preceding pages, though from a different point of view, but he accepts Schlegel's statements and the text of the *Po wu chi* without criticism.

115

印度的聚焦点火透镜

MÉLANGES.

BURNING-LENSES IN INDIA. [1]

A burning-lens is mentioned, and its utilization is demonstrated, in the story of King Virūdhaka, contained in the Tibetan biographies of Buddha. This story was first disclosed by A. SCHIEFNER [2] from the Tibetan *Life of Buddha*, compiled in 1734 by Rin-ĉen ĉos-kyi rgyal-po. When the cruel king Virūdhaka had vanquished and slaughtered the Çākyas, Bhagavat betook himself to Çrāvastī, where he dwelt in the Jetavana, and predicted that Virūdhaka in the course of seven days would be consumed by fire and be reborn in Hell. The king built a palace of several stories in the water and lived there; on the seventh day, however, the sun struck a burning-lens which belonged to the royal consort, whereupon the king and Ambarīsha were seized by the flames, with loud cries for help [perished, and] were reborn in the hell Avīci. [3] This story is embodied in the Vinaya, as translated in the Tibetan Kanjur (vol. X), where it is narrated at greater length and with more details. In the rendering of L. FEER, [4] the relevant passage runs thus: "Sur ces entrefaites, le temps s'éclaircit, les rayons du soleil donnèrent sur le verre ardent; il se produisit un feu qui gagna le coussin; du coussin, il se communiqua au pavillon" Finally we read in ROCKHILL's *Life of the Buddha*, translated from the Kanjur, as follows (p. 122): "When Virūdhaka's messenger came and told him what the Buddha had said, he was filled with trouble. Ambharīsha comforted him with the assurance that Gautama had only said this because the king had killed so many of his people. Moreover, he advised him to have a kiosque built in the water, and there to pass the seven days. The king followed his

---

[1] Compare this volume, pp. 216—223.

[2] *Tibetische Lebensbeschreibung Çâkjamuni's*, p. 59 (St. Petersburg, 1849).

[3] The Tibetan text (fol. 337 b) runs as follows: de-nas bĉom ldan ḥdas mñan-yod-du gšegs-nas rgyal-byed ts'al-na bžugs-te | ḥp'ags-skyes-po žag bdun-na mes ts'ig-ste dmyal-bar skye-bar luṅ bstan-pas | des ĉ'ui naṅ-du k'aṅ bzaṅs brtsigs-te ḥdug-pa daṅ | žag bdun-pai ts'e na btsun-moi me šel la ñi-ma p'og-pas rgyal-po daṅ ma-la gnod mes ts'ig-nas o-dod ḥbod bžin-par mnar-med-du skyes-so.

[4] *Fragments extraits du Kandjour (Annales du Musée Guimet*, Vol. V, 1883, p. 76).

advice, and retired to the kiosque with all his harem. On the seventh day, as they were preparing to return to Çrāvastī, and the women were arraying themselves in all their jewels, the sky, which until then had been overcast, cleared up, and the sun's rays falling on a burning-glass which was on a cushion, set fire to the cushion, and from that the flames spread to the whole house. The women ran away and made their escape, but when the king and Ambharīsha tried to do likewise, they found the doors shut, and with loud cries they went down into the bottomless hell."

It appears from these texts that the burning-lens was mentioned in the Sanskrit original from which the Tibetan translation was made. The lens is styled *me śel* (literally, "fire crystal"), which was indicated by the writer as the Tibetan term (this volume, p. 222). The fact that in this case a burning-lens is really understood may be proved beyond doubt from another Tibeto-Sanskrit text. The story of Virūdhaka is recorded in the Avadānakalpalatā (No. 11), and here we meet likewise the lens, called in Sanskrit *sūryakānta* (this volume, p. 217), in the Tibetan version *me śel*.[1] In the Tibetan prose edition of the same work (p. 48) it is said that the lens belonged to the ornaments of the house, that it was hit by the sunlight, that thus fire broke out in the building, and everything was burnt up (kʿaṅ-pai rgyan la me śel yod-pa-la ñi-mai mdaṅs pʿog-pas rkyen byas | kʿyim-la me śor-nas kun tsʿig-go). The versified recension is briefer and simply says that through the concentration of the solar rays in the lens the conflagration was effected (me śel ñi-mai od-dag-gi sbyor-bas me ni rab-tu ạbar). The Avadānakalpalatā was compiled by the Kashmirian poet Kshemendra, who lived around 1040 A.D., from older collections of Avadānas, and was translated into Tibetan in 1273.

Hüan Tsang, while visiting the kingdom of Çrāvastī, was shown the dried-up lake in which Virūdhaka was said to have perished. In the pilgrim's narrative no allusion is made to a lens, but according to him the waves of the lake suddenly divided, flames burst forth, and swallowed the boat in which the king was.[2]

The Sanskrit term *sūryakānta* is rendered into *me śel* also in the Tibetan translation of the Lalitavistara (chap. 15; ed. of FOUCAUX, Vol. I, p. 157, line 15; Vol. II, p. 196), but a precious rock-crystal, not a burning-lens, is here in question. Compare SCHIEFNER's remarks on this passage in *Mélanges asiatiques*, Vol. I, p. 234.   B. LAUFER.

---

[1] See the edition of Chandra Das in *Bibliotheca Indica*, Vol. I, pp. 392, 393.

[2] Compare JULIEN, *Mémoires*, Vol. I, p. 307; S. BEAL, *Buddhist Records*, Vol. II, p. 12.

116

吐火罗语杂考三篇

MÉLANGES.

THREE TOKHARIAN BAGATELLES.

1. A Chinese Loan-Word in Tokharian A.

The word for "town" in the Indo-European language designated as Tokharian A is *rī*, with short or long vowel, capable of forming a plural *ri-s*. The word was pointed out by the first decipherers of the language, E. Sieg and W. Siegling. [1] Emil Smith, in his very interesting analysis of the Tokharian vocabulary, [2] has justly observed that the word *rī* cannot satisfactorily be explained as coming from any Indo-European language, and that the alternative form with the lengthening of the vowel might speak in favor of a foreign origin,

---

[1] *Tocharisch, die Sprache der Indoskythen* (*S.B.A W.*, 1908, p. 923). I do not agree with these authors in regarding the language as that of the Indo-Scythians, but side with the conservative views expressed on the subject by A. Meillet (*Le Tokharien, Indogerm. Jahrbuch*, Vol. I, pp. 1—19). The ingenious supposition of F. W. K. Müller (*S.B.A.W.*, 1907, p. 960) still lacks the precise documentary evidence. The mere attestation of the fact that an Uigur colophon mentions the translation of a Buddhist work from an Indian language into Tokharian does not yet prove substantially that the fragments now styled Tokharian by way of convention really belong to that language, although this possibility may be admitted. The fact itself, that Buddhism and Buddhistic literature existed among the Tokharians, certainly was not novel, but previously known. Tāranātha has preserved to us the names of four members of the Buddhist clergy in Tukhāra (Tibetan T'o-gar; with popular etymology also T'o-dkar; *dkar*, "white"), — viz., Ghoshaka; the Vaibhāshika teacher Vāmana (Tibetan Miu-t'un, "dwarf," mentioned also in *dPag bsam ljon bzaṅ*, p. 88); the ācārya Vibhājyavāda; and Dharmamitra, a teacher of the Vinaya (pp. 61, 78, 198 of the translation of Schiefner), — and he twice refers to the Buddhism of Tukhāra (*ibid.*, pp. 38, 282). According to the Index of the Kanjur (ed. I. J. Schmidt, p. 78, No. 513), the original text of the Ārya-pratītya-samutpāda-hridaya-vidhi-dhāranī, from which the Tibetan translation was made, had been procured from Tukhāra by the Bhikshu Ner-ban (Nirvāna?)-rakshita.

[2] *"Tocharisch" die neuentdeckte indogerm. Sprache Mittelasiens* (*Videnskabs-Selskabets skrifter*, 1910, No. 5, p. 15, Christiania, 1911).

as the long vowels, with the exception of *a*, rarely or hardly ever occur. Smith tentatively proposed a relationship of the Tokharian word to Tibetan *ris* ("quarter"), remarking that *ri* is the present and probably very ancient pronunciation of the latter. Without discussing the possibility of a contact between Tokharian and Tibetan, this suggestion is not convincing for two main reasons. The Tibetans are an essentially nomadic group of tribes, to which the notion of a town in its origin was entirely foreign; and it may be considered as certain that at the time when the Tokharian word was in existence the Tibetans had only a few towns. The T'ang History relates that the inhabitants of Tibet roam about tending their herds, without having fixed settlements, while there are but a few walled places (其 人 或 隨 畜 牧 而 不 常 厥 居 然 頗 有 城 郭. *Kiu T'ang shu*, Ch. 196 A, p. 1 b). The Tibetan designation for a settlement of any size, though it consist of a single or several habitations, is *gron* (written language also *gron-k'yer*), but the word *ris* is never applied in this sense. It is even very far from signifying "quarter" unceremoniously, but means "part, division," usually in a figurative, not in a strictly territorial sense, and as a rule appears only as the second element of a compound. It therefore seems to me that the Tokharian word *ri* has no chance to claim its derivation from Tibetan *ris*. If, however, the former should really be a loan-word, it would appear more probable and reasonable to look to Chinese for assistance and to correlate the Tokharian word with Chinese *li* 里 (Korean and Japanese *ri*), "a village comprising twenty-five or fifty families." The Chinese, as energetic colonizers in Central Asia, may well have exerted their influence upon the native population there in this direction.

This word thus far is the only Chinese loan-word discoverable in Tokharian; in going over its vocabulary at least I could find no others. As has justly been said by A. MEILLET, [1] "Le tokharien n'est pas de ces langues qui sont fortement sujettes à l'emprunt; le vocabulaire est indigène pour la plus grande partie, autant qu'on puisse le voir par les faits déjà connus."

2. A Tokharian Loan-Word in Chinese.

The earliest (and still common) Chinese designation of asafoetida, [2] *a-wei* 阿 魏 (Japanese *agi*), traced by HIRTH to the Annals of the Sui Dynasty, [3] in which it is mentioned as a product of the Kingdom of Ts'ao 漕, has not yet been explained. Hirth observes that "*a-wei* is a foreign word, derived

---

[1] *Mémoires de la Société de linguistique de Paris*, Vol. XVII, 1912, p. 292.

[2] Belonging to the genus *Ferula*, comprising some sixty species (see WATT, *Dictionary of the Economic Products of India*, Vol. III, pp. 328—337).

[3] HIRTH and ROCKHILL, *Chau Ju-kua*, p. 225. The same text is also in *Pei shi*.

presumably from the Sanskrit or Persian name of the drug." This supposition, at the outset, is not very probable, as the Sanskrit and Persian terms have been traced in Chinese, and are indeed supplied by Hirth himself: Sanskrit *hiṅgu* is handed down in the Chinese transcriptions *hiṅg-kü* (**hiṅg-gu*) 興瞿, *hiṅg-yü* (*hiṅg-ṅu*) 形虞, and *hün-kʻü* (**hün-gu*) 薰渠,[1] and Persian *aṅguža(d)* انكژد or انكژد, in Chinese *a-yü* (*-tsie*) (**a-ṅü-zi*) 阿虞(截).[2] Watters says with reference to the *Pên tsʻao kang mu* that *a-wei* is wrongly given as the Brāhman or Sanskrit name. This statement, however, is not made by Li Shi-chén, the author of the *Pên tsʻao* (Ch. 34, p. 21). Whereas he expressly notes that *a-yü* is a Persian term, and that *hiṅg-yü* is a word used in India, he fails to state from what language the word *a-wei* is derived. He indicates that it makes its first appearance in the *Pên tsʻao* of the Tʻang period, and treats us to a wonderful etymology of the name: "The barbarians themselves style it *a*, expressing by this exclamation their horror at the abominable odor of this gum-resin."[3] This is sufficient to warrant the conclusion that Li Shi-chén was ignorant of the language from which the word had sprung. He further imparts a Mongol word *ha-si-ni* 哈昔泥,[4] and, what is more important, another transcription *yang-kuei* 央匱, not mentioned by Hirth or Watters. The Nirvāṇasūtra (*Yen pʻan king* 湟槃經) is cited by him as the source for this word, and apparently the Mahāparinirvāṇasūtra is understood.[5] *Yang-kuei*, in my opinion, is the same as *a-wei*; that is to say, the two are variants, representing transcriptions of an identical foreign prototype. This one we encounter in Tokharian B *aṅkwa*, first pointed out in the plural form *aṅkwas* by M. S. Lévi from one of the documents of

---

[1] Hirth, *l. c.*, and *J.A O.S.*, 1910, p. 18; Watters, *Essays on the Chinese Language*, p. 440

[2] Also the Sanskrit loan-word *hiṅ* occurs in Persian (L. Leclerc, *Traité des simples*, Vol. I, p. 448). The Tibetan equivalent *šiṅ-kun* must be explained from **siṅ-kun* (assimilated to *šiṅ* by way of popular etymology: *šiṅ*, "tree"), the latter from **hiṅ-kun* (= Sanskrit *hiṅgu*), derived from a mediæval vernacular of India.

[3] 夷人自稱曰阿此物極臭阿之所畏也 (*wei* 畏, pun upon *wei* 魏).

[4] This word is not listed in the Mongol dictionaries of Kovalevski and Golstunski. It is nothing but a transcription of Ghazni or Ghazna 霍悉那, the capital of Zābulistān (Chavannes, *Documents*, p. 160), which, according to Huan Tsang, was the habitat of the plant (Hirth, *l c*) According to I-tsing (Takakusu's translation, p 128), asafoetida was abundant in the western portion of India

[5] This entire foreign nomenclature is ascribed to a poem of Fan Chʻêng-ta 范成大 (1126—93) in Kʻang-hi's Dictionary (under 魏香).

medical contents secured by the Mission Pelliot. [1] The element *yan* 央, as is known, represents the syllable *an* in the Chinese transcription of Sanskrit words: for instance, in Aṅgulimālya. *Kuei* 匱 is North Chinese, as compared with an older articulation *kwai* or *kuai*, as still preserved in Cantonese: so that *yang-kuei* 央匱, read in the T'ang period *an-kwai*, is a phonetically exact transcription of a word corresponding to Tokharian *aṅkwa*. The same holds good for the transcription *a-wei*: *a* 阿 answers to Sanskrit *a* in the method of Buddhist transcriptions: the character *wei* 魏, as far as I know, has not yet been pointed out among the latter, but it had the ancient pronunciation *kwai* (鬼), *gwai*, and *ṅwai*, also *ṅui*. In this manner, also this mode of transcription leads back to Tokharian *aṅkwa*. From a phonetic point of view it is interesting to note that the pair *yang-kuei—a-wei* meets with an analogous counterpart in the name of the fig (*Ficus carica*) discussed by Hirth, [2] *ying-ji* (*aṅ-it) 映日 and *a-yi* (*a-jit) 阿驛, both answering to a West-Asiatic name of the general type *anjir*: also in this case we have a double mode of transcription following similar lines, as in the previous instance, — the nasal after the initial vowel being expressed in the one form and omitted in the other, — so that we are entitled to the conclusion that the element *a* 阿 served also for the reproduction of the initial syllable *an* or *an* in foreign words during the T'ang period.

Another Tokharian term of botanical pharmacology is of great interest to us. This is *arirāk*, the designation of the myrobalan *Terminalia chebula*. [3] First of all, we receive from it a satisfactory clew as to the mysterious Tibetan name *a-ru-ra* (corresponding in meaning to Sanskrit *harītakī*), [4] which comes nearer to the Tokharian form than to any form of other languages known to us. Second, new light falls upon the Chinese transcription *ha-li-lo* 訶黎勒, first mentioned at the end of the third century in the *Nan fang ts'ao mu chuang*. [5] This word has been brought together with Arabic *halilag* ﻠﻴﻠﺞ by T. Watters, [6] and with Aramaic *halilag* הלילג by Hirth. [7] Persian *halīlah* ﻫﻠﻴﻠﻪ, also *balil* and *balila*, should be added. As the genus

---

[1] *Journal asiatique*, 1911, Juillet-Août, p. 138.

[2] *J.A.O.S.*, 1910, p. 20.

[3] S. Lévi, *l c*, p 122.

[4] H. Laufer, *Beitr. tibet. Med.*, pp. 56—67. Garcia ab Horto (*Aromatum Historia*, 1567, p. 132) gives as Indian popular name *arare*, and as medical term *arituqui*. Compare the Anglo-Indian word hara-nut. Harītakī is transcribed in Chinese 訶梨恒雞. In Newārī the name of the plant is *halala*.

[5] See Bretschneider, *Bot. Sin*, pt. 1, p. 38.

[6] *Essays on the Chinese Language*, p. 355.

[7] *J A.O.S.*, 1910, p. 23.

Terminalia is indigenous to India, however, [1] it is manifest that the West-Asiatic names, in the same manner as the Chinese and Tibetan ones, are derived from a language of India, and that there is no necessity of resorting to Persian, Aramaic, or Arabic for an explanation of the Chinese name. The Tokharian form *arirāk* demonstrates that the prototype on which the derivations of West-Asiatic, Chinese, and Tibetan are based, indeed pre-existed somewhere on Indian soil. Chinese *-li-lo* answers to an ancient articulation *-li-lak* (*-ra-rak*), and very exactly reproduces Tokharian *-rirak*. The correctness of this point of view is corroborated by the word *p'i-li-lo* 毗黎勒, [2] corresponding to Sanskrit *vibhītaka* and to Tibetan *ba-ru-ra* (*Terminalia belerica*). Again in this case the Chinese and Tibetan forms are not actually based on Sanskrit *vibhītaka*, with which they have only the first element in common; while *li-lo* (*ri-rak*) and *ru-ra* appear as the second element in the same fashion as in the type *ho-li-lo—a-ru-ra*. Consequently the Chinese and Tibetan forms allow us to presuppose the former existence of an Indo-Tokharian form **virurāk*, from which the two were derived, and which corresponded in sense to Sanskrit *vibhītaka*. The Tokharian term *triphal* (Sanskrit *triphala*, the "three myrobalans") [3] shows that a name for this kind of myrobalan must have been known.

3. Tuman.

It is well known that in New Persian a word occurs for the designation of a "myriad," *tumān* or *toman* تومان, which with insignificant phonetic modifications, is found also in the Turkish, Mongol, and Tungusian languages of inner Asia, and which passed, most probably from Turkish, also into Magyar (*tomény, temény, tömen*; usually in the combination *tomény-ezer*, "myriad, many thousands;" *toméntelen*, "innumerable"). [4] Whereas this word in popular use refers to an indefinite high number, the figure *x*, the supposition is

---

[1] The tree is abundant in northern India from Kumaon to Bengal and southward to the Deccan tablelands, and is found also in Ceylon, Burma, and the Malay Peninsula (see WATT, *l c.*, Vol. VI, pt 4, pp. 24—36) In Ibn al-Baitar we meet the term "myrobalan of Kabul" أنيليلج الكابلي (L. LECLERC, *Traité des simples*, Vol. I, p. 131); hence our "chebuli" (YULE and BURNELL, *Hobson-Jobson*, p 136)

[2] *Pên ts'ao kang mu*, Ch. 31, p. 4. It is first mentioned under the T'ang by Su Kung 蘇恭 and Li Sun 李珣.

[3] S LÉVI, *l c.*, p. 126.

[4] Compare Z. GOMBOCZ, *Die bulgarisch-türkischen Lehnwörter in der ungarischen Sprache*, p. 131 (*Mémoires de la Société finno-ougrienne*, Vol XXX, Helsingfors, 1912). Gombocz, while pointing out the analogous Mongol, Turkish, and Tungusian forms, omits reference to Persian.

granted that in more exact manner of speech it should convey the notion of
„ten thousand." Marco Polo, who spoke the Persian language, is our witness of
the fact that in his day *toman*, as he writes, covered this numerical category. [1]
This is confirmed by the *Yüan chʻao pi shi* 元朝祕史 (Ch. 12. p. 45,
ed. of Li Wên-tʻien 李文田), where the word appears in the two transcriptions *tʻu-mien* 禿綿 (*tūmän*) and *tʻu-man* 土滿 (*tuman*), both being
said to be identical, and explained as the Mongol word expressing the numeral
"ten thousand" (譯言萬數也) and also an indefinite quantity (猶言
眾耳). The Niuči vocabulary contained in the Ming edition of the *Hua i
yi yü* likewise transcribes the Niuči word *tuman* by means of the Chinese
characters 土滿 . [2] The farther removed from the original centre of its
propagation, the more was it liable, naturally, to assume the air of a fantastic
aggrandizement. When, in the summer of 1898, I was engaged in the study of
two Tungusian dialects, Ewunki and Oročon, in the village Wal on the north-
east coast of Sachalin Island, one of my Tungusian informants gave as the
highest number known to him *tumä*, and translated it into Russian by
"million." [3]

Various opinions have been expressed in explanation of the word in
question. H. YULE [4] has taken it for granted that it is a Mongol word. The
striking fact could not escape the students of Altaic languages that, while the
cardinal numbers from 1 to 10 are different in Turkish, Mongol, and Tungusian,
a curious coincidence prevails in the designations for "thousand" (Turkish

---

[1] Ed of YULE and CORDIER, Vol II, p 192.

[2] W. GRUBE, *Sprache und Schrift der Jučen*, p. 35, No. 665. T. WATTERS (*Essays
on the Chinese Language*, p 360) gives also the transcription *tʻu-mén* 圖們 .

[3] This is the easternmost region to which the word has advanced. It is notable that
it has been adopted only by Ural-Altaic, but not by any Palae-Asiatic languages. The
Yukaghir, for instance, have no words for numbers above a hundred, and used to express
a hundred by "ten tens," while they now employ *ičtoχ* (from Russian *sto*) and also the
Russian word for "thousand" in the form *tičeče* (W JOCHELSON, *Grammar of the Yukaghir
Language*, p 115).

[4] *Hobson-Jobson*, p 928. Yule (*ibid.*) has asserted also that *toman* or *tomaun*, in the
sense of a certain coin or a certain sum of money (in Persia equal to ten şābqrāns or
crans, about 9.75 fr.; in India equal to $ 15.50 [G TEMPLE, *Glossary of Indian Terms*,
p 262]; among the Ossetians equal to 10 10 Rubels [W. MILLER, *Sprache der Osseten*,
p. 109]; among the Turks equal to 3 Rubels [RADLOFF, *Wörterbuch*, Vol III, col 1518]),
is identical with the word *tuman* ("myriad"). On the authority of Yule, this has passed
into our lexicography (for example, into the *Century Dictionary*) The number "ten
thousand" is not visible in any of the instances given ; and, in my opinion, the word in
question is entirely distinct from the numeral *tuman*, and is derived from another root
with a history of its own.

myñ, byñ, buñ: Mongol *miñgan*: Tungusian *miñan*) and "ten thousand" (Codex cumanicus *tumen*: Old Chuvaš *tümän*: Orkhon inscriptions and Uigur *tümän*: Djagatai *tumän* تومن: [1] Osmanli *tuman*: Mongol *tümän* [Old Mongol, also *tuman*]; Niŭči *tuman*: Manchu *tumen*: Tungusian dialects *tumú, tümo, tumé, tumén*; Gold *tuma, tymú*). This state of affairs must naturally raise the suspicion that these two numeral series cannot be invoked as witnesses of linguistic relationship; that, on the contrary, they are derived from a foreign source. For this reason, W. SCHOTT [2] and J. HALÉVY, [3] the two scholars who thus far have discussed the numerals of this group in the most ingenious manner, [4] have advisedly passed over the series *tuman* in silence, actuated as they were by a correct feeling that the question is of a loan-word. G. J. RAMSTEDT, in a study of the numerals of the Altaic languages, [5] justly observed that the word, both in Tungusian and in Turkish, is suspicious of a late derivation; but, although referring to Russian тьма and темникъ, yet he thought that the original might perhaps be sought for in Indo-Chinese, pointing to Chinese *wan, man* ("ten thousand") and *ti-man* ("the ten-thousandth"). This unfortunate idea was accepted by Z. GOMBOCZ (*l. c.*) who, like Ramstedt, overlooked the existence of the corresponding Persian word. Long before the discovery of Tokharian there was no doubt in my mind that *tuman* is neither Turkish nor Mongol (and least of all Chinese), but Indo-European: the Persian word and the interesting Slavic forms were sufficient to justify this opinion. M. E. BLOCHET, in a very interesting notice *Le nom des Turks dans l'Avesta*, [6] makes an incidental reference to the word *tumān*, stating that "it is a very ancient borrowing from the Chinese *to-man* 多萬 ('the ten thousand')." [7] I venture to doubt that a combination like this ever had any real existence in Chinese: it is not registered in the *P'ei wên yün fu* (Ch. 73); the notion „several or many myriads" is usually expressed by *shou wan* 數萬. The

---

[1] I. KUNOS, in his edition of Suleiman Efendi's Djagatai-Osmanli Dictionary (p. 196), transcribes *tuman*.

[2] *Das Zahlwort in der tschudischen Sprachenklasse* (*Abh. B Ak. W.*, 1853, pp 1—29)

[3] *L'étroite parenté des noms de nombre turco-ougriens* (*Keleti szemle*, Vol. II, 1901, pp. 5—18, 91—108).

[4] Despite the sweeping criticism of G. J. RAMSTEDT (*Journ. de la Soc. finno ougrienne*, Vol. XXIV, 1907, p. 2), who, as far as tangible results are concerned, has not advanced much beyond his predecessors.

[5] *L. c.*, p. 22.

[6] *J.R.A.S*, 1915, p. 307.

[7] The opinion of M. Blochet is not quite clear to me. According to him, *tumān* is the older and original form (and this is also my opinion), and Persian *tumān* is intended to transcribe the Altaic word. What I do not comprehend is whether, in M. Blochet's view, the Persians or the Turks adopted the loan from the Chinese.

ancient pronunciation of *wan* was *ban*, and a Chinese *to-wan* borrowed by Turks during or before the T'ang period would have resulted in *doban* or *duban*: whereas an ancient Turkish or Mongol *tu* or *tö*, according to the phonetic rules of transcription, would always presuppose an initial aspirate on the part of modern (that is, post-T'ang) Chinese. [1] It is not necessary, however, to expatiate on this side of the argument; in the case of borrowings we have to look for motivation which is entirely lacking, and which is not produced by the supporters of the Chinese theory.

I had expected that A. MEILLET's conclusive study of the Tokharian numerals [2] had indeed brought us the ultimate solution of the principal issue of the problem, which in my opinion should be acceptable to all. M. MEILLET points out the numeral "ten thousand" (*tmāṃ* in Tokharian A, and *tumane*, *tmāne* in Tokharian B), and discusses at length the Indo-European character of this word. [3] He strongly fortifies his opinion with an excellent etymology based on the comparative study of Indo-European philology, and emphasizes Persian *tumān* and Slavic *tǐma*. It should be added that Tokharian A *tmāṃ* phonetically is on the same level as Russian *tma* (тьма or тма), which appears as early as the time of the Slavic-Church language and Old Russian. There are, further, the following derivatives: *t'emnik* (темникъ) and *tmo-načalnik* (тмоначальникъ), "commander of ten thousand:" *t'moryi* (тьмовый), "relative to ten thousand:" *tmoritseyu* (тморицею) and *tmorično* (тморично), "many times, incessantly:" *tmoričnyi* (тморичный), *tmotmušči* (тмотмущій), and *tmo-t'omnyi* (тмотёмный), "innumerable." [4] This fact bears out the close relationship of Tokharian to Slavic insisted upon by M. MEILLET, and positively uproots the idea that the Tokharian and Slavic words have been borrowed from Turkish. The word (this fact is now well assured) is of Indo-European origin; and the Turkish word owes its existence to an Indo-European language, not *vice versa*. It should certainly be borne in mind that *tuman* belongs to the medial, not the ancient, stage of Indo-European speech-development (in regard to Tokharian M. MEILLET observes, "C'est une langue de type indo-européen moyen, et non pas du type ancien"), and that the documentary evidence thus far available

---

[1] Compare, as regards this particular case, the above Chinese transcriptions *t'u-mien* and *t'u-man*

[2] *Les noms de nombre en Tokharien B* (*Mémoires de la Société de linguistique de Paris*, Vol. XVII, 1912, pp. 281—294).

[3] *l. c.*, pp. 292, 293; and *Le Tokharien* (*Indogerman. Jahrbuch*, Vol. I, p. 19).

[4] VLADIMIR DAL, Толковый словарь живого великорусскаго языка, Vol. IV, col. 767, 773, 887. The Russian word was formerly derived from Turkish by H. YULE (*Hobson-Jobson*, p. 929), and recently by GOMBOCZ (*l. c.*) Yule pointed to Herberstein, who about 1559 reported that "one thousand in the language of the people is called *tissutze* (тысяча): likewise ten thousand in a single word tma."

strictly points to mediæval times. [1] In view of Avestan *baēvar*, Pahlavi and Persian *bēvar* ("ten thousand"), it would be interesting to have some more exact chronological indications as to the time when *tuman* springs up in Persian literature.

While I perfectly concur with M. Meillet in regarding *tuman* and its congeners as Indo-European, I venture to dissent from him in the opinion that the Turkish forms are derived from Tokharian: I am rather disposed to think that they hail straight from Persian. Phonetically, the Turkish, Mongol, and Tungusian forms are decidedly based on Persian *tumān* or *tomān*, while none of those languages exhibits a final *e* like Tokharian B *tumane*, and still less a contracted form like Tokharian B *tmāne* or Tokharian A *tmām*. There is, however, a still more weighty, culture-historical reason why the word in the languages of inner Asia should be traced to Persia as its home. The scholars hitherto engaged in the discussion of this question argued it only from the philological point of view, without accounting for the reasons of the wide expansion of the word, embracing the territory from the Baltic, the Danube, and the Black Sea as far as the north-eastern Pacific. The matter is concerned with the military history of Asia. It was not the necessity of having a word for the numeral "ten thousand," or of expressing the notion of a high indefinite number, that induced Turkish, Mongol, and Tungusian tribes to adopt the word *tuman*: it reached them in consequence of the reception, on their part, of the military organization and tactics launched in Persia. On another occasion I have explained the far-reaching influences emanating from Persia along this line, and the word *tuman* belongs to the same class. Steingass says, in his revised edition of Johnson's and Richardson's *Persian Dictionary*, that *tumān* refers to "districts into which a kingdom is divided, each being supposed to furnish ten thousand fighting men;" [2] that *tumān-dar* توماندار is the commander of a *tumān*, and *tumān-dārī* the command of a *tumān*. The same is expressed by Radloff in his *Turkish Dictionary* in assigning to Djagatai *tümän* the significance "military unit of ten thousand men." As regards the Mongols, we all have read our Marco Polo, who describes the decimal system on which the Mongol army was organized, and who says that "they call the corps of a hundred

---

[1] For this reason I should hesitate to identify the name of the Hiung-nu Khan T'ou man 頭曼, who died in 209 B. C., with Turkish *tuman* ("ten thousand"), as has been suggested by E. Blochet (*Les inscriptions turques de l'Orkhon*, p. 7, note 3). The Chinese transcription *t'ou-man* may well correspond to a Turkish *tuman*; but the latter, after all, may have had another meaning.

[2] The same definition is given under *toman* by G. Temple, in his *Glossary of Indian Terms*, p 262 (London, 1897). It was the Moghul emperors who with their army organization transplanted the matter and the term into India

thousand men a *tuc*, and that of ten thousand a *toman*" (ed. of YULE and CORDIER, Vol. I, p. 261). [1] Yule certainly is on the right track when he annotates that the decimal army-division made by Chinggis at an early period of his career was probably much older than his time, and that in fact we find the Myriarch and Chiliarch already in the Persian armies of Darius Hystaspes. According to HERODOTUS (VII, 81), the Persian army invading Greece under Xerxes was divided into tens, hundreds, thousands, and ten thousands, each of these divisions having its own leader, and the leaders being placed under the command of the Myriarch. Again, an exceptional position was taken by the Immortals, those picked Ten Thousand, who were all Persians, and were led by Hydarnes. When one of this corps died, his place was forthwith filled by another man, so that their number was never greater or less than ten thousand (VII, 83). At the root the matter was deeply associated with the territorial organization of the Old-Persian monarchy and the military conscription based thereon. Here we face truly Iranian institutions; and it is self-evident that these, together with many others, were absorbed by the Turks of inner Asia, and subsequently by the imitators of the latter, the Mongols. Hence we are driven to the conclusion that the word *tuman*, as the name of a very ancient Iranian military institution, was handed on to Turks and Mongols by the Persians: it was not mathematical, but military necessity that forced this word on its route of migration and tended to preserve its life.

There are, accordingly, good philological and historical reasons for determining the position of the word *tuman* with a fair degree of exactness. It is Indo-European in its origin, and propagated in Tokharian, Persian, and Slavic. It is a Persian loan-word in Old-Turkish; a Turkish loan-word in Magyar, on the one hand, and in Mongol, on the other hand; and a Mongol loan-word in Niüci, Manchu, and other Tungusian languages. It has nothing to do with Chinese *wan*. On the contrary, wherever our word occurs in Chinese records, it is assuredly modelled after the Turkish-Mongol equivalent. T. WATTERS [2] has already made this correct observation: "The word *tuman* in Turki means a myriad, but it has other meanings also, and it is found in other languages. Certain Chinese writers seem to have adopted it, and the word occurs frequently in their writings. It is found transcribed in several different ways [see above], and it is generally used in the sense of a myriad."   B. LAUFER.

---

[1] In like manner Ibn Batūta says that each squadron of the Khan was composed of ten thousand men, the chief of whom is styled *emir ṭūmān* امير تومان (ed. of DEFRÉ-MERY and SANGUINETTI, Vol. IV, p 300) The military division of the Mongols into *tumän* appears also from the chronicle of Sanaṅ Setsen (I. J. SCHMIDT's edition, pp. 175, 193, etc, 403); *tuman*, of course, must not be conceived, with Schmidt, as a collective name of the Mongols.

[2] *Essays on the Chinese Language*, pp. 159, 160.

117

荜澄茄

VIDAṄGA AND CUBEBS.

In their monumental work *Chau Ju-kua* (p. 224), HIRTH and ROCKHILL have acquainted us with the vegetal product derived from a creeper growing in Su-ki-tan on Java, and styled by Chao Ju-kua *pi-têng-kᶜie* 蓽澄茄. The translators of this author annotate that, according to the *Pên tsᶜao kang mu*, this is a foreign word which occurs also in the transcription *pᶜi-ling-kᶜie* 毗陵茄. This name itself, however, is not explained by them. It is, first of all, important to note from which time these transcriptions come down. The earliest author cited in the *Pên tsᶜao* as speaking of *pi-têng-kᶜie* is Chᶜên Tsᶜang-kᶜi 陣藏器, who lived during the first part of the eighth century, and who localizes the habitat of the plant on Sumatra (*Fu shi* 佛誓, Bhōja). Hence we are entitled to the inference that we face a transcription made in the style of the Tᶜang period; and, to all appearances, we are confronted with the reproduction of a Sanskrit word. The three elements of which the term is composed are well known from the nomenclature of the Chinese Buddhists: Chinese *pi* or *pᶜi* renders Sanskrit *vi* or *bi*; the alternation of *têng* and *ling* allows us to presuppose an initial cerebral in Sanskrit with the choice of a cerebral *ḷ* in Prākrit; the phonetic element *têng* 登 corresponds to ancient *taṅ* and *daṅ* (for instance, in Mātaṅga and daṁshṭra), while *ling* renders *lin*, *leṅ*, or *laṅ*; *kᶜie* 茄 ("brinjal") has only the ancient phonetic value of *ga*, being the equivalent of 伽, the classifer 十 (in the same manner as in the first character *pi*) being chosen merely in view of the botanical significance of the whole term. Thus we obtain a Sanskrit form vidaṅga, and I had indeed arrived at this restoration from a purely phonetic point of view, without knowing that such a Sanskrit word exists, or what it means. The transcription *pi-ling-kᶜie* would justify the assumption of a Prākrit form viḷaṅga or viḷeṅga, and in Bengali we have *biraṅga* (in Hindustānī *baberáṅ*, *wawruṅ*; in Puštu *bábraṅ*). An Arabic form *fileṅga* (see p. 285) likewise supports this view.

The word vidaṅga is of ancient date: it occurs in the Suçruta-saṁhitā and repeatedly in the Bower Manuscript (also in the form bidaṅga). [1] This plant has been identified with *Embelia ribes* (family *Myrsineae*), an immense climber abundant in the hilly parts of India from the Central Himalaya to Ceylon and Singapore, and occurring also in Burma. Its seeds are extensively

---

[1] HOERNLE, *The Bower Manuscript*, pp. 301, 320.

employed as an adulterant for black pepper. [1]   W. ROXBURGH [2] states more specifically, "The natives of the hills in the vicinity of Silhet, where the plants grow abundantly, gather the little drupes, and when dry sell them to the small traders in black-pepper, who fraudulently mix them with that spice, which they so resemble as to render it almost impossible to distinguish them by sight, and they are somewhat spicy withal." The seeds of another species (*Embelia robusta*) are eaten by the Paharias of the Darjeeling district. [3] This description answers well the pepper-like black seeds dried in the sun, as described by Chao Ju-kua. HIRTH and ROCKHILL, however, are perfectly correct in identifying Chao Ju-kua's viḍaṅga growing on Java with *Piper cubeba* (family *Piperaceae*). [4] It was evidently from Sumatra and Java that the term viḍaṅga was introduced into China together with the cubebs. The Sanskrit term must have been transferred to this plant autochthonous to Java, because the products of the Indian and Javanese climbers were very similar in appearance and in their properties. The word doubtless belonged to the Kawi language. Other such instances are known where the Hindu settlers on Java named indigenous products of the island with Sanskrit words designating other species. An example of this kind is afforded by the *pin-kia* 頻伽 birds sent as tribute from Kaliṅga (訶陵, Java) to the Chinese Court in the year 813. [5] The name *pin-kia* apparently is an abbreviation of Sanskrit kalaviṅka, written in Chinese 迦陵 (or 羅) 頻伽, [6] exactly corresponding with

---

[1] WATT, *Dictionary of the Economic Products of India*, Vol. III, p 242. *Embelia ribes* Burm. is stated to occur also in southern China, Hahang and the Lo-fou shan in Kuang-tung Province and Hongkong being given as localities (FORBES and HEMSLEY, *Journal of the Linnean Society*, Botany, Vol. XXVI, pp. 52, 63). According to the same authors, four other species of *Embelia* occur in southern China. It seems, however, that none of them is known by a Chinese name or is mentioned in the *Pên ts'ao* literature. *Embelia ribes* Burm is found also in the Dutch East Indies (*Encyclopædie van Neder-landsch-Indië*, Vol. II, p. 218: "De vruchtjes en een uit deze bereid werkzaam beginsel [embelia-zuur] zijn in den laatsten tijd in Europa als voortreffelijk lintworm-middel in gebruik genomen") As regards Burma, it is frequent in the tropical forests of Martaban and Upper Tenasserim (S. KURZ, *Forest Flora of British Burma*, Vol. II, p. 102).

[2] *Flora Indica*, p. 197 (Calcutta, 1874).

[3] J. S. GAMBLE, *List of the Trees, Shrubs, and Large Climbers found in the Darjeeling District*, p. 53 (Calcutta, 1896).

[4] This identification is due to D. HANBURY (*Science Papers*, p 246). It is given after the latter by S. W. WILLIAMS (*Chinese Commercial Guide*, p. 117), F. P. SMITH (*Contributions toward the Materia Medica of China*, pp 79, 83), and G. A. STUART (*Chinese Materia Medica*, p. 144, Shanghai, 1911).

[5] *T'ang shu*, Ch. 222 B, p. 3

[6] *Fan yi ming i tsi*, p. 20ᵇ (edition of Nanking). Compare EITEL, *Handbook of Chinese Buddhism*, p. 67.

the Tibetan rendering *ka-la-piṅ-ka*, the Indian cuckoo extolled for its me-
lodious voice. [1]

In regard to the adjustment which has taken place in the Archipelago
between the designations for *Embelia ribes* and *Piper cubeba*, we meet a very
interesting parallel in the materia medica of the Arabs. These have been
acquainted since the early middle ages with the product of the latter species,
known to them under the name *kabāba* كبابة, whence our word "cubeb" is
derived, [2] and discussed at length by Ibn al-Baiṭār (1197—1248). [3] One of the

---

[1] It is not known to me whether the word piṅka or viṅka is recorded in the Kawi
language of Java, but, judging from the Chinese notation of it in the T'ang Annals, I feel
certain that it must have existed there with reference to a fine song-bird indigenous to
Java. GROENEVELDT (*Notes on the Malay Archipelago*, in *Misc. Papers rel. to Indo-
China*, Vol. I, p. 140) observed that "about these birds many an hypothesis is possible,
but not one seems satisfactory" It is matter of regret that he has withheld from us his
opinion on the subject. E STRESEMANN, in a most interesting study on the historical
development of our knowledge of birds of paradise (*Novitates Zoologicae*, Vol. XXI, London,
1914, pp. 13—24), has recently offered the suggestion that the Javanese *piṅ-kia* birds of
the T'ang History possibly might have been birds of paradise. This supposition, however,
is improbable Birds of paradise do not sing at all, but are sought for only on account
of their magnificent plumage. Moreover, birds of paradise do not live on Java. The
centre of their habitat is New Guinea, where twenty-seven known species breed; while
three inhabit the northern and eastern parts of Australia, and one the Moluccas (WALLACE,
The Malay Archipelago, pp. 419—440). Accordingly, the earliest opportunity of the
Javanese to become acquainted with birds of paradise was granted at the time when the
people of Java reached the Moluccas; and this was not the case before the middle of the
fourteenth century, when King Mājapāhit extended his power into those regions, as narrated in the Old-Javanese poem Nāgarakretāgama of the year 1365 (translated by H.
KERN, *De Indische Gids*, Vol XXV, 1903, pp. 341—360) As admitted by STRESEMANN
in another article (*Novitates Zoologicae*, Vol. XXI, 1914, p 39), it was at that time that
the cassowary of Ceram was first introduced into Java (and it is Stresemann's particular
merit that he rejected the old error that the original home of the cassowary, known to
the Chinese as *huo chi* 火鷄 [see GROENEVELDT, *l. c*, pp 192, 193, 198, 253, 262]
was on Sumatra, Java, or Banda); but the same admission must hold good for birds of
paradise. Regarding the possibility of the importation of the dried skins for these birds
into China, compare F. W. K. MULLER in *T'oung Pao*, Vol. IV, 1893, pp. 82—83 (an
article not consulted by Stresemann, nor did he utilize YULE's important contribution to
the subject in his *Hobson-Jobson*, p 95), with comments by HIRTH (*T'oung Pao*, Vol. V,
1894, pp 390—391) and GROENEVELDT (*ibid.*, Vol. VII, 1896, p. 114). This subject
would be deserving of a renewed and more profound investigation: the objections raised
by Hirth and Groeneveldt to Muller's thesis are by no means convincing to me, and at
all events will not terminate the discussion.

[2] YULE and BURNELL, *Hobson-Jobson*, p 277. The introduction of cubebs into our
pharmacopœia is due to the Arabic physicians of the middle ages.

[3] L. LECLERC, *Traité des simples*, Vol III, p. 138.

earliest authors cited by him, Ibn al-Heitsem, discriminates between two varieties, a larger and a smaller one, the larger one being *habb al-a'rus* حبّ العروس, the smaller one *falinja* or *falenja* فلنجة. The latter kind is treated by Ibn al-Baiṭār, who has arranged his material in alphabetical order, under a separate entry. [1] where LECLERC, the excellent translator of the Arabic work, annotates, "Nous ignorons quelle est cette graine. Ce n'est pas le cubèbe ni la muscade. C'est la graine d'une plante qui croît dans l'Inde et atteint la hauteur d'environ une coudée," etc. Both the description given in the text and the very name *falenja* leave no room for doubt that the vegetal product in question is the viḍaṅga of India. Arabic *falenja* is merely a reproduction of this word, and the older Arabic articulation doubtless was *filenga* or *filanga*, which is in perfect harmony with the Chinese transcription *pi-liṅ (leṅ)-ga*. [2]

Hirth and Rockhill err in restricting the occurrence of *Piper cubeba* to Java only. [3] According to WATT, [4] the plant is a native of Java and the Moluccas, and is cultivated to a small extent in India (most probably due to importation from the Archipelago). The well-informed *Encyclopædie van Nederlandsch-Indië* [5] states that the creeper occurs wild in Java and Borneo, and is cultivated throughout the Dutch East Indies, being exported in large quantities to Holland, where it receives its function in the pharmacopœia. [6] Chʿên Tsʿang-kʿi, as stated, refers the plant to Sumatra; and whether it grows there or not, its ready-made product seems to have first reached the Chinese from Sumatra rather than from Java. [7] It is interesting to note that at the same time cubebs had entered India; for Ibn-Khordādbeh, who wrote between

---

[1] LECLERC, *l. c.*, Vol. III, p. 40, No. 1695.

[2] In view of the Arabic importation of both cubebs and viḍaṅga from India and of cubebs also from the Archipelago and China (see below), these two products ought to have been included by G. FERRAND (*Relations de voyages et textes géographiques arabes, persans et turks relatifs à l'Extrême-Orient*, Vol. I, p. 234) in his list of Indian and East-Asiatic products assembled from the great work of Ibn al-Baiṭār. It is gratifying, at any rate, that Ferrand calls the special attention of "indianistes, sinologues et indo-sinologues" to the translation of Leclerc, which "is not as well known as it ought to be." The writer has ploughed through Leclerc's work for the last fifteen years, and has always found it a most trustworthy, helpful, and inspiring companion.

[3] They do not refer to Marco Polo, who mentions cubebs among the products of Java (ed. of YULE and CORDIER, Vol. II, p. 272).

[4] *L. c.*, Vol. VI, pt. 1, p. 257.

[5] Vol. II, p. 255.

[6] The Dutch name *staartpeper* ("tail-pepper") presents a literal translation of Malay *lāda barekor*, or *marica buntut*.

[7] According to the *Encycl Brit.* (Vol. VII, p. 607), *Piper cubeba* is indigenous to South Borneo, Sumatra, Prince of Wales Island, and Java.

20

844 and 848, enumerates them among the export-articles of India. [1]   Li Sün 李珣, the author of the *Hai yao pên ts'ao* 海藥本草 in the second half of the eighth century, quotes a work *Kuang chou ki* 廣州記 ("Records of Kuang-tung") as saying that cubebs grow in all maritime countries and are identical with tender black pepper. [2]   Li Shi-chên comments that they are found in Hai-nan and all foreign countries (scil., of the south?). [3]   Of greater importance is the fact that under the Sung dynasty the plant was cultivated in the soil of Kuang-tung Province, as reported by Su Sung 蘇頌 in his *T'u king pên ts'ao* 圖經本草. [4]   In Persian, in Hindustāni, Bengālī, and other Indian languages, cubebs are still called *kabāb-čini* كباب چينى; that is, kabāb from China. [5]

GARCIA DA ORTA [6] supplies us with some information on this point, which is interesting enough to be cited in extenso: "Tametsi cubebis raro in Europa utamur, nisi in compositionibus: attamen apud Indos magnus earum in vino maceratarum est usus ad excitandam venerem; tum etiam in Iaoa [Java] ad excalfaciendum ventriculum. [7]   Appellatur hic fructus ab Arabibus medicis Cubebe et Quabeb: a vulgo Quababechini: in Iaoa, ubi frequens nascitur, Cumuc; [8]

---

[1] G FERRAND, *l. c.*, Vol. I, p. 31.

[2] 澄茄生諸海國乃嫩胡椒也 (according to another reading, "the tenderest of black pepper" 胡椒之嫩者).

[3] 海南諸番皆有之. — Ibn Rosteh, who wrote about 903, mentions cubebs as products of the island Salāhat in the Archipelago; Masūdī, as products of the kingdom of the Mahārāja (G. FERRAND, *l. c*, Vol I, pp 79, 99, 110).

[4] Finally the word *pi-têng* 畢澄 was transferred to a kind of wild pepper 山胡椒 growing in Kuang-si, as stated in the *Chi wu ming shi t'u k'ao* 植物名實圖考 (Ch. 25, p. 69) of 1838 (see BRETSCHNEIDER, *Bot. Sin.*, pt. 1, p 72). This work contains also an illustration of the plant; so does the *Chêng lei pên ts'ao* (Ch. 9, fol. 44), where it is entitled *"pi-têng-k'ie* of Kuang-chou."

[5] See YULE in his edition of *Marco Polo*, Vol. II, p 391.

[6] Latinized ab Horto. Garcia went to India in 1534 as physician of the Portuguese Viceroy, and during thirty years made a most thorough study of Indian drugs, products, and medicine The results of his labor were published at Goa, 1563, under the title "Coloquios dos simples, e drogas e cousas medicinais, e assi dalguas frutas achadas nella India Oriental onde se tratam algumas cousas tocantes a medicina, pratica, e outras cousas boas para saber." Only six copies of this original edition are said to be in existence. I quote from the Latin edition of C. CLUSIUS (p. 111), published at Antwerp in 1567.

[7] For the warming of the stomach. ACOSTA, who wrote a treatise on the drugs of India in 1578, as quoted by Yule, says that the Indian physicians use cubebs as cordials for the stomach.

[8] Javanese *kumukus*; Malayan *temukus*

a reliquis Indis, praeterquam in Malayo, Cubabchini. Non est autem sortitus hanc appellationem, quod in China nascatur, quandoquidem ex Cunda [1] et Iaoa, ubi plurimus est, in Chinam perferatur: sed quoniam Chinenses, qui Oceanum Indicum navigabant, hunc fructum, quem in iam enumeratis insulis emerant, cum aliis mercibus in alios maris Indici portus et emporia deferebant." Garcia, accordingly, regarded the Chinese only as the importers of the product, not as its growers: and it may be admitted that the bulk of the Chinese importation into India traced its origin to the Archipelago. Garcia, however, never visited China: and we have no reason to question the accuracy of the Chinese account claiming indigenous cultivation, which is amply confirmed by modern observers. In 1789 LOUREIRO, in his *Flora Cochinchinensis*, pointed it out as being cultivated in Indo-China. [2] F. P. SMITH refers to the probable introduction of the species from Sumatra or Java into the province of Kuang-tung. FORBES and HEMSLEY, [3] in their comprehensive work on the systematic botany of the East, state in regard to the species (named by them *Litsea cubeba*), "We have only seen the fruit as it appears in commerce, and it is similar to that of the 'mountain pepper' of Central China (*Litsea pungens*, Hemsl.), yet evidently not the same, nor even a cultivated variety of it."

In the Tibetan-Chinese List of Drugs *Fan Han yao ming* 番漢藥名 [4] we meet the Sanskrit viḍaṅga under No. 117 in the Tibetan transcription *byi-taṅka* or *byi-taṅga*, [5] explained through Chinese *man-king-tse* 曼荊子 (*Vitex trifolia*), [6] a plant growing abundantly in northern China, and furnishing a black berry which is used in medicine. Hence the adjustment with viḍaṅga was effected: indeed, Ch'ên Ts'ang-k'i remarks that the *pi-têng-k'ie* (viḍaṅga), in their appearance, resemble the seeds of the *wu-t'ung* 梧桐 (*Sterculia platanifolia*) and those of the *man-king*. On the other hand, we encounter in the same List of Drugs (No. 192) the Chinese term *pi-têng-k'ie*

---

[1] Identical with Çunda, Sunda (see YULE and BURNELL, *Hobson-Jobson*, p. 868).

[2] BRETSCHNEIDER, *Early European Researches into the Flora of China*, p. 171

[3] *Journal of the Linnean Society* (sect Botany), Vol. XXVI, p 380.

[4] See for the present BRETSCHNEIDER, *Bot. Sin.*, pt 1, p 104. I hope to give shortly a bibliographical study of this work, which would be too long to insert here. My quotations from it refer to a critical edition (in manuscript) prepared by me. The substance of the work is embodied in A. POZDNAYEV's Учебникъ тибетской медицины (Vol. I, pp 247—301). A very poor and careless edition of it was published in 1913 by HUBOTTER (*Beiträge zur Kenntnis der chin. sowie der tib-mong Pharmakologie*).

[5] Likewise in Mongol *byidaṅga* (the addition of the letter *y*, as in Tibetan, denoting palatalized *b'*). The word viḍaṅga is not contained in the Mahāvyutpatti, and it is not known to me how old the Tibetan transcription is

[6] BRETSCHNEIDER, *Bot. Sin.*, pt. 2, p. 357; STUART, *Chinese Materia Medica*, p. 457.

碧澄 [1] 茄, with a Tibetan equivalent *rin-po-č'e myag*. The first element of this compound means "precious, valuable;" the word *myag*, not recorded in our Tibetan dictionaries, still awaits explanation. It was not known heretofore that the seeds of *Piper cubeba* or *Embelia ribes* were employed in Lamaist pharmacology, but to all appearances this seems to have been (or still to be) the case.

The previous notes bear out the fact that it is not always sufficient to define pharmacological terms of East-Asiatic languages merely by way of determination of the specimens to which the technical terms at present relate, but that philological and historical researches are indispensable in order to reach a full understanding of the real facts. New associations of ideas were formed when new products turned up and crossed the experience of an earlier allied substance; new adaptations of terms were brought about, rallying most diverse species under the same flag.

B. LAUFER.

---

[1] If S. W. Williams and his successors transcribed this character *ching* and *ch'ëng*, they were, as far as the modern language is concerned, quite correct; for the Tibetan-Chinese work, in which the Chinese names are transcribed in Tibetan letters for the benefit of the Tibetans trading with Chinese in drugs, renders the character in question by *č'en*.

118

柔克义讣告

NÉCROLOGIE.

William Woodville ROCKHILL.

In memory of Mr. Rockhill, whose useful and noble career was so suddenly and lamentably ended at Honolulu on December 8, 1914, and with whom it was my privilege to have been acquainted and to have corresponded for a period extending over eighteen years, I take the liberty to add the following data to the bibliography of his works given by M. Cordier on pp. 162—164 of this volume:

> Le traité de l'émancipation ou Pratimoksha-sûtra, traduit du tibétain. (*Revue de l'histoire des religions*, Vol. IX, 1884, pp. 3—26, 167—201).
>
> The Tibetan "Hundred Thousand Songs" of Milaraspa, a Buddhist Missionary of the Eleventh Century. (*Proc. Am. Or. Soc.*, 1884, pp v—ix).
>
> Notes on some of the Laws, Customs, and Superstitions of Korea. (*American Anthropologist*, 1891, pp. 177—187).
>
> A Pilgrimage to the Great Buddhist Sanctuary of North China. (*Atlantic Monthly*, 1895, pp. 758—769). [Interesting record of a visit to the Wu-t'ai shan.]
>
> China's Intercourse with Korea from the XVth Century to 1895 London (Luzac & Co.), 1905 (60 p).
>
> Diplomatic Audiences at the Court of China. London (Luzac & Co), 1905 (54 p.)

The last of Mr. Rockhill's literary products is the edition of the *Chu fan chi* 諸蕃志 elegantly printed at Tōkyo with movable copper types in one volume, with English postscript, dated April 1, 1914. On September 8, when I conveyed to him my thanks for the copy which he had kindly addressed to me, and expressed my satisfaction at this fine example of Japanese book-making, he wrote me: "The book was published by the Kokumin shimbun Press at Tōkyo I had 250 copies struck off; some of these I had sent to Kelly & Walsh at Shanghai, others to Luzac & Co., London. I hope they reached them safely, but I have not yet heard. I am much pleased that you like the way the book was printed, I rather like it myself." Mr. Rockhill was a rare type of scholar, singularly broad-minded, and equipped with common sense and an unusually wide knowledge of all peoples of the Far East. His *Life of the Buddha* will remain a household book with all of us; and his four great works devoted to Tibet, the goal of his lifelong ambition, will continue to serve as an inexhaustible mine of valuable information, with their solid fund of geographical

and ethnological data. Besides his writings he left two lasting monuments,—a
remarkable collection of Tibetan objects housed in the U. S. National Museum;
and the nucleus of a Tibetan, Mongol, and Chinese Library, belonging to the
treasures of the Library of Congress in Washington,—the interests of which
he always furthered with a liberal spirit. Mr. Rockhill was a man of extreme
modesty, and seldom talked about himself and his achievements. He received
no honors from this country, but indeed he craved none; and it is decidedly
to his credit that he was never chosen by a university for an honorary degree.
It is painful to think that at the end of his life his diplomatic services were
valued more highly by China than by his own Government.

B. LAUFER.

John ROSS.

Le Rév. John Ross qui avait appartenu à la United Presbyterian Church
of Scotland est mort âgé de 73 ans, le 7 août 1915 à Edimbourg. Il était ar-
rivé à Tche fou à l'automne de 1872, mais les missionnaires étant fort nom-
breux dans ce port, il se rendit à Nieou tchouang où ne se trouvait qu'un agent
de la Société biblique écossaise. Ross ouvrit une école de garçons et en 1873,
une chapelle dans une boutique; plus tard il se rendit à Moukden. Il était
rentré en Europe, il y a cinq ans. Il a publié un certain nombre d'ouvrages,
la plupart relatifs à la Mandchourie et à la Corée.   H. C.

— Visit to the Corean Gate. (*Chinese Recorder*, V, 1874, pp. 347—354.)
— History of Corea Ancient and Modern with description of manners and customs, language
    and geography. Maps and Illustrations. Paisley; J. and R. Parlane, s.d. [1879], in-8,
    pp. xii—404.
— Corea. (*China Review*, XVI, pp 19—25)
— The Gods of Corea. (*Chinese Recorder*, XIX, Feb. 1888, pp. 89—92).
— The Products of Corea. (*Ibid.*, April 1888, pp 165—167)
— The Corean Language. (*China Review*, VI, pp. 395—403).
— Corean Primer, being Lessons in Corean on all ordinary subjects, transliterated on the
    principles of the "Mandarin Primer", by the same Author. Shanghai: American
    Presbyterian Mission Press, MDCCCLXXVII, in-8, pp. 89.
— Korean Speech with Grammar and Vocabulary. New Edition. Shanghai and Hongkong,
    Kelly & Walsh, 1882, in-8, pp 101.
— The Rise and Progress of the Manjows. (*Chinese Recorder*, VII, 1876, pp 155—168,
    235—248, 315—329; VIII, 1877, 1—24, 197—208, 361—380).
— The Manchus, or the reigning Dynasty of China: their Rise and Progress. Maps and
    Illustrations. Paisley: J. and R. Parlane, 1880, in-8, pp. xxxii—751.
— Cheaper Edition. London: Elliot Stock, 1891, in-8, pp xxxij—751.
— Notes on Manchuria. (*Ibid*, VI, 1875, pp. 214—221).
— History of the Manchurian Mission. (*Chinese Recorder*, XVIII, 1887, pp. 255—263)
— Missionary Arithmetic (*Ibid.*, XXIII, Dec 1892, pp 568—570)
— Manchuria. [With a Map] (*Scottish Geog. Mag.*, XI, 1895, pp 217—231).
— Mandarin Primer: being Easy Lessons for Beginners, Transliterated According to the
    European Mode of Using Roman Letters. Shanghai: American Presbyterian Mission
    Press, 1876, in-8, pp. viii—122.
— — 1877, in-8, pp. viii—122.

119

石棉和蝾螈——在中国和希腊民间传说中的研究

ASBESTOS AND SALAMANDER,

AN ESSAY IN CHINESE AND HELLENISTIC FOLK-LORE.

BY

BERTHOLD LAUFER.

————••◁⊃⊂▷••————

It is my object, not to write a history of asbestos and its application with reference to human culture, but to unravel the curious traditions entertained by the Chinese regarding this marvellous production of nature, and to correlate their notions of it with the corresponding thoughts of the ancients, the Syrians and Arabs, and of mediæval Europe. Without due consideration of the Western folk-lore, the Chinese traditions, the elements of which are thoroughly based on Occidental ideas, would forever remain a sealed book. We are indebted to A. WYLIE [1] for a most scholarly study, *Asbestos in China*, which contains an almost complete array of Chinese sources relative to the subject; in fact, without his energetic pioneer-labor, the present investigation could not have been carried to the point to which it has now attained. My obligations to him for his able research-work are acknowledged in each and every case. The present state of science, however, has permitted me to go far beyond the results which Wylie was able to reach a generation ago. WYLIE [2] merely noted in the most general way that the accounts

---

[1] *Chinese Researches*, section iii, pp. 141—154 (Shanghai, 1897).
[2] *L. c.*, p. 149.

21

of the Chinese corroborate the statements of ancient classical writers, mainly emphasizing the point that the Chinese, in the same manner as the ancients, mention handkerchiefs or napkins woven from asbestos. No attempt, however, was made by him to explain all the curious lore that was lavishly accumulated on top of this subject. Here WYLIE [1] merely offered the remark, "The speculations of native writers as to the material of which it was made will probably not be thought equally worthy of credit with the bare recital of facts which came under their notice. In early times they appear not to have suspected that it was a mineral product, but have contented themselves with applying to the animal and vegetable kingdoms respectively for a solution of the difficulty." From the viewpoint of comparative folk-lore and Chinese relations with the West, these speculative theories which partially take their root in Hellenism certainly present most attractive material for study. Further, Wylie's representation of the matter suffers from various defects. It is not well arranged in chronological or any other order, and the sources are not sifted critically. Moreover, as admitted by himself, he did not succeed in identifying most of the geographical terms to be found in the Chinese texts. [2] At present this task is greatly facilitated, chiefly thanks to P. Pelliot's learned researches, which form the basis of many an important conclusion reached on the following pages. The geographical point of view is indispensable in this case, as only in this manner is it possible to trace the routes over which ideas have wandered.

By "asbestos" we understand the fibrous varieties of tremolite, actinolite, and other kinds of amphibole, the fibres of which are sometimes very long, fine, flexible, and easily separable by the fingers,

---

[1] *L. c.*, p. 144.

[2] Also HIRTH (*China and the Roman Orient*, p. 252) confessed that he was unable at the time when he wrote (1885) to identify these names.

and look like flax. The colors vary from white to green and wood-brown. The name "amiantus" is now applied usually to the finer and more silky kinds. Much that is called asbestos is chrysotile, or fibrous serpentine. [1] Asbestos, then, is a term of generic character, applied to the peculiar fibrous form assumed by several minerals, and not a name given to any one particular species; the asbestiform condition being simply a peculiar form under which many minerals, especially serpentine, occasionally present themselves. The varieties of asbestos are very numerous. They are all silicates of lime and magnesia or alumina, and commonly occur in crystalline rocks of metamorphic origin. The most valuable property of asbestos, its infusibility, is due to the large proportion of magnesia in its composition, which, like lime, has proved absolutely infusible at the highest temperatures attainable in furnaces or otherwise. Under the blowpipe a single fibre will fuse into a white enamelled glass or opaque globule, but in the mass some varieties have been known to resist the most intense heat without any visible effect. Chrysotile, however, if exposed for some time to long-continued heat, will lose somewhat of its tenacity and silkiness, and become rough and brittle. [2] The word "asbestos," then, in its present loosely-defined significance, is rather a commercial than a mineralogical term, and covers at least four distinct minerals, having in common only a fibrous structure and more or less fire and acid proof properties. [3] It will be well to keep this in mind, as it cannot be expected that the Greek, Roman, Arabic, and Chinese writers, in their accounts of asbestos, should have in their minds a uniform and well-defined mineralogical species.

---

[1] E. S. Dana, *System of Mineralogy*, p. 389 (New York, 1893).

[2] R. H. Jones, *Asbestos, its Properties, Occurrence, and Uses*, pp. 13, 22, 23 (London, 1890).

[3] G. P. Merrill, *Notes on Asbestos and Asbestiform Minerals* (*Proc. U. S. Nat. Mus.*, Vol. XVIII, 1895, p. 281).

ASBESTOS IN CLASSICAL ANTIQUITY.—It is possible that Theophrastus (372—287 B.C.) [1] makes mention of asbestos, although this name does not appear in his writings. He states, "In the mines of Scaptesylae is found a stone, in its external appearance resembling rotten wood, which is kindled by oil poured over it; when the oil is consumed, the stone itself ceases to burn, as though it were not affected by fire." Theophrastus discusses in this connection the different effects which the action of fire may bring about upon stones; but while he may have had asbestos in mind, this conclusion is by no means forcible. Others hold, for instance, that he speaks here of bitumen, [2] and this view seems more probable.

STRABO (circa 63 B.C.—A.D. 19; x, 1, § 6) states that "in the quarries near Carystus, at the foot of Mount Ocha in Euboea, is extracted a stone which is combed like wool, and spun and woven; of this substance, among other things, are made napkins ($\chi\epsilon\iota\rho\delta\mu\alpha\kappa\tau\rho\alpha$) which, when soiled, are thrown into the fire, and whitened and cleaned, in the same manner as linen is washed." [3]

---

[1] *De lapidibus*, 17 (opera ed. F. WIMMER, p. 343).

[2] JOHN HILL, in his still very useful work *Theophrastus's History of Stones with an English Version, and Critical and Philosophical Notes* (p. 40, London, 1746), makes the following interesting comment on this passage: "It is much to be questioned whether this was the true original reading, and genuine sense of the author; in all probability some errors in the old editions have made this passage express what the author never meant to say. The substance, and indeed the only substance described by the other ancient naturalists as resembling rotten wood, is the gagates or jet before mentioned among the bitumens; but that has no such quality as the author has here ascribed to this stone of Scaptesylae. The ancients, it is to be observed, had a common opinion of the bitumens, that the fire of them was increased by water, and extinguished by oil; and very probably this was the sentiment originally delivered here by the author, however errors upon errors in different copies of his works may since have altered the sense of them. The stone itself was probably a bitumen of the lapis Thracius kind, as the place from whence it has its name was a town of that country."

[3] Compare F. DE MÉLY, *Lapidaires grecs*, p. 14. Carystus (now Castel Rosso) was a city situated at the southern extremity of the island of Euboea, south of the mountain Ocha (now St. Elias). It was there that in 490 B.C. the Persian expedition under Datis and Artaphernes landed (HERODOTUS, VI, 99). At the time of Plutarch the mine was exhausted (see below). Celebrated was the marble of Carystus (mentioned also by Strabo),

DIOSCORIDES (v, 156) of the first century A.D., who designates asbestos by the name "amiant," [1] says that this stone is found on Cyprus, and resembles alum, that may be cleft (στυπτηρία σχιστῇ). [2] Being flexible, it is made by traders into tissues for the theatre. Thrown into the fire, they flame up, but come out more resplendent without having been attacked by the fire. [3]

APOLLONIUS DYSCOLUS, who lived in the first half of the second century A.D., has the following interesting notice on asbestos: [4] "Sotacus, in his treatise on stones, [5] says in regard to the stone called Carystius [6] that it has woolly and downy excrescences, and that napkins are spun and woven from this mineral. It is twisted also into lamp-wicks which emit a bright light and are inexhaustible. [7] When these napkins are soiled, their cleaning is performed not by means of washing in water, but brush-wood is burnt, the napkin

---

the quarries of which are still preserved (see LENZ, *Mineralogie der alten Griechen und Römer*, p. 59).

[1] Greek ἀμίαντος ("undefilable"), from μιαίνω ("to soil, defile").

[2] Regarding alum see M. BERTHELOT, *Collection des anciens alchimistes grecs*, Vol. I, p. 237.

[3] F. DE MÉLY, *l. c.*, p. 24. The Arabic version (L. LECLERC, *Traité des simples*, Vol. II, p. 414) says that it resembles the alum of Yemen, and speaks of tissues without reference to theatrical use. J. YATES (*Textrinum Antiquorum*, p. 359) remarks that the epithet ἱμαντώδους may have referred to that variety of asbestos which is now called mountain-leather and commonly found with the fibrous asbestos.

[4] *Historiae mirabiles*, XXXVI (*Rerum naturalium scriptores Graeci minores*, ed. KELLER, Vol. I, p. 52).

[5] A work which is lost now. Sotacus lived in the third or perhaps even toward the close of the fourth century B.C. He is chiefly known to us from quotations in Pliny who cites him on seven occasions. Judging from the exact definitions of localities which he gave in order to determine stones and jewels according to their origin, he appears to have travelled a good deal in Hellas and on the Greek islands. The then known world from India to Britannia and Aethiopia supplied him with material for observations; and his definitions, as we see from Pliny, were accepted as models by subsequent scholars. He dealt also with the employment of the single stones, particularly in medicine and magic (compare F. SUSEMIHL, *Geschichte der griechischen Litteratur in der Alexandrinerzeit*, Vol. I, pp. 860—861).

[6] That is, stone from Carystus (see the above citation from Strabo).

[7] Hence arose the name asbestos (ἄσβεστος) which means "inextinguishable."

in question is placed over this fire, and the squalor flows off; [1] while the cloth itself comes forth from the fire brilliant and pure, and is again utilized for the same purposes. The wicks remain burning with oil continually without being consumed. The odor of such a wick, when burnt, tests and detects the presence of epilepsy in persons. [2] This stone is produced in Carystus, from which place it received its name; in great abundance, however, on Cyprus, as you go from Gerandrus to Soli, [3] under rocks to the left of Elmaeum. At the time of the full moon the stone increases, and again it decreases with the waning of the moon." [4]

Pausanias (i, 26) narrates that the golden lamp made by Calli-machus for the temple of Athene Polias in the Acropolis of Athens, which was kept burning day and night, had a wick of Carpasian flax (λίνου Καρπασίου), the only kind of flax that is indestructible by fire. [5] Plutarch (circa A.D. 46—120), in his *De oraculorum defectu*,

---

[1] This is a correct estimation of the process. The throwing into the fire of asbestine cloth, narrated in so many texts, Western and Eastern, is of course not to be taken literally; the cloth was simply put over a charcoal fire. There is no reason to accede to the opinion of J. T. Donald (*Some Misconceptions concerning Asbestos, Engineering and Mining Journal*, Vol. LV, 1899, p. 250) that these stories "are to a large extent mythical; certainly, if true, the articles in question were not made of asbestos."

[2] Pliny (xxviii, 63, § 226) says the same about the smell arising from burnt goat's horns or deer's antlers (morbum ipsum deprehendit caprini cornus vel cervini usti nidor).

[3] A city on the north coast of Cyprus.

[4] A similar observation is referred by Pliny (xxxvii, 67, § 181) to the selenitis ("moon-stone"), which contains an image of the moon, and reflects day by day the form of this luminary while waxing and waning, if this is true (selenitis...imaginem lunae continens, redditque ea in dies singulos crescentis minuentisque sideris speciem, si verum est). According to Dioscorides (v, 159), the selenitis is found at night at the time of the waxing moon, and, pulverized, the stone is administered to epileptics. It thus seems that the last clause of Apollonius, as well as his reference to epilepsy, were inspired by traditions pertaining properly to selenitis. The latter, in my opinion, denotes a variety of mica, and it will be seen that the Chinese also know of a stone in which notions of mica and asbestos are blended. Ibn al-Baitār, in his Arabic rendering of Dioscorides' Materia Medica, translated the Greek *amiantos* by *al-ṭalk* (that is mica, not our talc).

[5] Asbestos from the vicinity of Carpasus, a town in the north-east corner of Cyprus, now called Carpas.

mentions napkins, nets, and kerchiefs of this material, but adds that it was no longer found in his time, only thin veins of it, like hairs, being discoverable in the rock. [1] There was further asbestine cloth for enveloping the ashes of cremated bodies, as stated by Pliny. As in other matters, so likewise on asbestos we owe to Pliny the most detailed notes.

PLINY knew asbestos of two localities,—Arcadia and India. That found in the mountains of Arcadia is of an iron color. [2] He has the following notice regarding asbestine cloth: "An invention has been made of a kind of material which cannot be consumed by flames. It is styled 'live,' and I have seen at banquets table-cloths made from it and burning over a fire. When the dirt was thus removed, they came forth from the fire brighter than water would have cleaned them. Funeral garments are made of this stuff for the kings to separate the ashes of the body from those of the pyre. This substance is found in the deserts of India scorched by the sun, where no rains fall, in the midst of deadly serpents, and thus becomes accustomed to live [3] in the blaze. It is but rarely found, and difficult to weave owing to the shortness of its fibres. Its color is red by nature, and becomes white only through the action of fire. When found in its crude state, it equals the price of excellent pearls. In consequence of its natural properties it is called by the Greeks *asbestinon.* [4] Anaxilaus [5] is responsible for the statement that a tree enveloped by this linen is felled without the

---

[1] Among the Greek alchemists the word "asbestos" assumed the significance "lime;" thus Zosimus wrote a treatise on the latter under the title "Asbestos" (M. BERTHELOT, *Origines de l'alchimie*, p. 185).

[2] Asbestos in Arcadiae montibus nascitur coloris ferrei (XXXVII, 54, § 146).

[3] *Vivere*; this description accounts for the above attribute "live" (*vivum*).

[4] That is, inextinguishable, inconsumable.

[5] A physician and Pythagorean philosopher who was banished by the Emperor Augustus in 28 B.C. on a charge of practising magic.

blows of the axe being audible. Hence this linen occupies the foremost rank the world over." [1]

In another passage Pliny mentions amiantus as resembling alum (*alumen*) in appearance, [2] and losing nothing from the agency of fire. It resists all practices of sorcery, particularly those of the Magi. [3]

The notes of the ancients are very plain, but deficient in facts. They give us the localities where asbestos was found, state the kind of products made from it, and point out its power of resistance to fire. We hear nothing, however, about the mode of mining the mineral, or preparing, spinning, and weaving its fibres. [4] Above all, it should be borne in mind that no theory regarding the origin and nature of asbestos is handed down to us from classical antiquity. Pliny's idea that its fire-resisting quality is bred by the tropical sun of India, can hardly be regarded as such, and is no more than an expression of his personal opinion. Several authors, it is true, have ascribed to Pliny a belief in the vegetal origin of asbestos, but this is an unfounded assumption. DANA [5] peremptorily says that Pliny supposed asbestos to be a vegetable product. BOSTOCK and RILEY, [6] pointing to the word *mappa*, as boldly assert that "he

---

[1] Inventum iam est etiam quod ignibus non absumeretur. Vivum id vocant, ardentesque in focis conviviorum ex eo vidimus mappas sordibus exustis splendescentes igni magis quam possent aquis. Regum inde funebres tunicae corporis favillam ab reliquo separant cinere. Nascitur in desertis adustisque sole Indiae, ubi non cadunt imbres, inter diras serpentes, adsuescitque vivere ardendo, rarum inventu, difficile textu propter brevitatem. Rufus de cetero colos splendescit igni. Cum inventum est, aequat pretia excellentium margaritarum. Vocatur autem a Graecis ἀσβέστινον ex argumento naturae. Anaxilaus auctor est linteo eo circumdatam arborem surdis ictibus et qui non exaudiantur cacdi. Ergo huic lino principatus in toto orbe (XIX, 4).

[2] AULUS GELLIUS (*Noctes atticae* [*circa* A.D. 175] XV, 1) mentions a wooden tower for the defence of the Piraeus, which could not be set on fire by Sulla, because it was coated with alum.

[3] Amiantus alumini similis nihil igni deperdit. Hic veneficiis resistit omnibus, privatim Magorum (XXXVI, 31, § 139).

[4] BLÜMNER, *Technologie*, Vol. I, 2d ed., p. 205.

[5] *System of Mineralogy*, p. 389.

[6] *Natural History of Pliny*, Vol. III, p. 136.

evidently considers asbestos to be a vegetable, and not a mineral production." [1] Pliny indeed makes no statement whatever to the effect that asbestos is a plant or the product of a tree, as we hear, for instance, in China; neither is there any such testimony in any other classical source. On the contrary, all Greek authors distinctly speak of asbestos as a mineral. Moreover, Pliny most positively regarded both asbestos and amiantus as minerals; otherwise he would not have listed them, as we have seen, in his books XXXVI and XXXVII, which are devoted to mineralogy. For this reason I am convinced that throughout classical antiquity asbestos was considered as nothing but a mineral substance. This is most strongly corroborated by the fact that the ancients were familiar with at least three mines in their own dominion,—Carystus, Cyprus, and Arcadia; and the people who mine asbestos are assuredly familiar with its true nature, and cannot possibly believe in its vegetal provenience. Pliny has inserted his principal notice of asbestos in his book on textiles, because it was as a textile that the substance was chiefly utilized and known. Certainly this textile deserved the name "linen;" in fact, it could not have been termed anything else. We ourselves still speak of asbestos-cloth, and entertain no thought of a vegetable product in this connection. There are vegetal, animal, and mineral fibres, and any material woven from these may be called cloth. The verb *nascitur* ("it is born, it grows"), used by Pliny, does not allow of inferences, any more than the word *linum*. This term does not necessarily refer to plant-life; on the contrary, Pliny employs it also with reference to minerals. Thus the Indian *adamas* does not "grow" (that is, occur) in a stratum of gold. [2]

---

[1] Even so cautious a worker as E. O. VON LIPPMANN (*Abhandlungen*, Vol. I, p. 17) wrongly makes Pliny say that asbestos is an incombustible flax. Pliny does not express himself in this manner.

[2] Indici non in auro nascentia (xxxvii, 15, § 56); or the selenitis is said to grow in

The notion of the vegetal character of asbestos, indeed, did not exist in classical antiquity, but it is Hellenistic and seems to have sprung up somewhere in the anterior Orient. The earliest source to which I can trace it is the Greek Alexander Romance (*Pseudo-Callisthenes*, III, 22) in which is described a dining-room of imperishable wood in the palace of Queen Candace,—not exposed to putrefaction, and inconsumable by fire. Other manuscripts, however, read ἀμιάντων and "stones" instead of "wood;" so that the passage is now rendered, "There was there also a dining-room of incombustible amiantus." [1] A Syriac work on natural history of uncertain date, wrongly ascribed to Aristotle, in which Syriac translations of the Homilies of Basilius the Great and the Physiologus and several other unknown books have been utilized, makes a distinct allusion to an "asbestos tree:" "This tree is styled 'The Constant One.' When a man takes a piece of it and flings it into a very hot bath, the latter becomes tepid, as though it had never experienced fire. Also a fire-stove which is set in flames is extinguished and cools off; likewise a baking oven and chimney is extinguished as soon as a piece of that tree is thrown into them." [2] This notice is followed in the same work by the description of the salamander, which, as will be noticed farther on, plays such a signal part in the mediæval legends of asbestos. The tree-asbestos was adopted also by the Arabic writer Abū Dulaf (below, p. 329). It turns up also in China.

The scarcity of information which the ancients have left to us on the subject of asbestos is to some extent made good by three relics of asbestos tissues still preserved in Italy. One found at Puzzuolo in 1633 belonged to the Gallery Barberini. Another, in

---

Arabia (nasci putatur in Arabia [67, § 181]). In a similar manner *croître* was employed in French: R. DE BERQUEN (*Les merveilles des Indes orientales*, p. 15, Paris, 1669), for instance, has, "Cette precieuse pierre croist en plusieurs endroits du monde."

[1] A. AUSFELD, *Der griechische Alexanderroman*, p. 99.

[2] K. AHRENS, *Buch der Naturgegenstände*, p. 80.

the Library of the Vatican, was discovered in 1702 a mile outside of the Gate of Rome, called Porta Maior; it was a corpse-cloth, five feet wide and six feet and a half long, coarsely spun, but as soft and pliant as silk, enclosing the skull and calcined bones of a human body,—discovered in a marble sarcophagus, thus furnishing a remarkable confirmation of Pliny's statement. The deceased, judging from the sculptured marble, was a man of rank who is supposed to have lived not earlier than the time of Constantine. A third piece of asbestine cloth, of considerable dimensions, is shown in the Museum Borbonico at Naples; it was found at Vasto in the Abruzzi. [1]

The early Chinese notices of asbestos bear the same sober character as those of the classical authors.

EARLY IMPORTATION OF ASBESTOS INTO CHINA.—The Chinese first became acquainted with asbestos through their trade with the Roman Orient. Indeed, the first authentic notices of a product from this mineral in the Annals refer to the territory of western Asia. The *Wei lio* 魏略, written by Yü Huan between 239 and 265, [2] enumerates asbestos-cloth among the products frequently found in Ta Ts'in (the Roman Orient). [3] The same statement is made in the Annals of the Later Han Dynasty; [4] likewise in those of the Tsin and Liu Sung Dynasties. [5] The fact that Ta Ts'in produces asbestine cloth is mentioned also in the famous Nestorian inscription of Si-ngan fu. The term used in the Annals is *huo huan pu* 火浣布 (literally, "cloth which can be cleansed by fire"), evidently suggested by the stories of the ancients. After the example of WYLIE, [6] I use

---

[1] J. YATES, *Textrinum Antiquorum*, pp. 359, 360.

[2] See CHAVANNES, *T'oung Pao*, 1905, pp. 519—520; and PELLIOT, *Bull. de l'Ecole française*, Vol. VI, 1906, p. 364.

[3] HIRTH, *China and the Roman Orient*, p. 74.

[4] *Hou Han shu*, Ch. 118, p. 4 b.

[5] HIRTH, *l. c.*, pp. 40, 45, 46, 61; CHAVANNES, *T'oung Pao*, 1907, p. 183.

[6] *Chinese Researches*, section III, p. 141.

the term "fire-proof cloth" as a convenient synonyme, though this meaning is not directly conveyed by the Chinese expression.

The alleged philosopher Lie-tse [1] mentions a tribute of asbestos-cloth to King Mu of the Chou dynasty (1001—946 B.C.) on the part of the Western Jung. Asbestos is characterized there as follows: "The fire-proof cloth, in order to be cleansed, was thrown into the fire. The cloth then assumed the color of fire, and the dirt assumed the color of the cloth. When taken out of the fire and shaken, it was brilliantly white like snow." This text is not authentic, but retrospective, and cannot be older than the Han period. In the same manner as the diamond was a product hailing from the Roman Orient, so also was asbestos. [2]

In like manner the text of the *Chou shu* 周書, [3] alluding to the same event as that of Lie-tse, is of a purely retrospective character, and devoid of chronological value. [4] The matter, indeed, is not connected with King Mu or the Chou dynasty; but the fact is borne out by these two texts that under the Han (206 B.C.—A.D. 220), asbestos-cloth, together with diamond-points, was imported into China over a land-route leading from the Roman Orient by way of Central Asia. [5]

---

[1] Ch. 5, *T'ang wén.*

[2] WYLIE (*l. c.*, p. 142) seems to regard Lie-tse's text as historical; and HIRTH (*China and the Roman Orient*, p. 250) even goes so far as to say, that if the philosopher Lie-tse, whose writings are said to date from the fourth century B.C. (A.D. in Hirth's book is a misprint), can be trusted, asbestos-cloth was known in China as early as a thousand years B.C. E. FABER (*Naturalismus bei den alten Chinesen*, p. 132), in his translation of Lie-tse, justly wondered that things like asbestos were already known in times of such hoary antiquity; but certainly they were not. The alleged *kun-wu* 錕 鋙 sword mentioned in Lie-tse is not, as hitherto believed, a sword, but a diamond-point.

[3] See CHAVANNES, *Mémoires historiques de Se-ma Ts'ien*, Vol. V, p. 457.

[4] The text of the *Chou shu* has passed into the *Po wu chi* (Ch. 2, p. 4 b, ed. printed at Wu-ch'ang).

[5] King Mu was the chosen favorite and hero of Taoist legend-makers, to whose name all marvellous objects of distant trade were attached (in the same manner as King Solomon and Alexander in the West). The introduction of the Western Jung is emblematic of the intermediary rôle played by Turkish tribes in the transmission of goods from western Asia to China.

Wylie was inclined to believe that the earliest allusion to asbestos occurs in the *Shi i ki* 拾遺記, where it is said that the people of Yü-shan 羽山 brought yellow cloth for presentation to the Emperor Shun. This, according to him, is not very distinct; but as we learn from the same authority that the same nation, on two later occasions, brought an offering of fire-proof cloth, it seems not unfair to infer that the former offering was of a similar character. That work, however, as stated by WYLIE elsewhere, [1] has little historical value. It was written by Wang Kia 王嘉 of the fourth century; but this work is not preserved, having been afterwards disarranged and partially destroyed. Even if the passage in question were traceable to Wang Kia, our belief in it would not be strengthened; for no authentic work of the pre-Christian era contains any allusion to this matter. Asbestos was found on Chinese soil only in post-Christian times; and Chinese notions regarding asbestos being, as will be seen, to a large extent based on Western folk-lore, it is reasonable to conclude that the Chinese were not acquainted with asbestos before their contact with the Roman Orient. The various accounts of the *Shi i ki* about tributes of asbestos, however, point to the fact that this material came from Western regions.

General Liang-ki 梁冀, who lived under the Emperor Huan 桓 (147—157) of the Han dynasty, [2] had a costume made from asbestos-cloth, which he used to wear on the occasion of great banquets. He would insist on declining the wine-cup till it was spilled on his suit; and then with feigned anger he would take it off, ordering it to be thrown into the fire. It blazed up as if it were reduced to ashes; but the stains being removed, and the fire extinguished, the cloth appeared bright and clean, as if it had been purified with lees. [3]

---

[1] *Notes on Chinese Literature*, p. 192.

[2] He died in 159 (GILES, *Biographical Dictionary*, p. 478).

[3] Compare WYLIE, *Asbestos*, p. 143. This text is handed down under the name of

A report of incontestable authenticity concerning asbestos-cloth being sent as tribute to China refers to the second month of the year 239, in the time of the Three Kingdoms, when envoys from an unnamed country of the Western Region (*Si Yü*), introduced at the Court by means of double interpreters, offered fire-proof cloth to Ts'i Wang Fang 齊王芳 (240—253) of the Wei dynasty. The Emperor directed his military staff to test it, and to proclaim the result to the officers. [1] The intention was perhaps implied to make use of this material for army purposes. Under the Wei, also, the tradition was upheld that early under the Han, gifts of such cloth had been presented by Western countries. Two sovereigns of the Wei lent expression to an ill-founded scepticism as to the actual existence of this substance,—a belief which was not shared by the Taoist Ko Hung 葛洪 of the fourth century. [2] Ko Hung inaugurates a new period in the study of this subject on the part of the Chinese. Under the Han and throughout the third century, the Chinese accepted asbestos products as a fact, without inquiring into the nature of the mineral or the causes of its wonderful properties. They were satisfied to state merely the effects of its properties. Ko Hung is the first Chinese author to render an account of the origin of asbestos in the romantic spirit appropriate to the Taoist school. The ideas which he expounded, however, are closely inter-

---

Fu-tse 傅子, who lived in the latter part of the fourth century, and appears in P'ei Sung-chi's commentary to *San kuo chi* (*Wei chi*, Ch. 4, p. 1). HIRTH (*China and the Roman Orient*, p. 251) wrongly ascribes it to the text of *Wei chi* itself; he aptly reminds us of the jest practised by the Emperor Charles the Fifth, who astonished his guests after dinner by exposing an asbestos table-cloth to a fire. In the *Encyclopædia Britannica* (Vol. II, p. 714) this anecdote is connected with Charlemagne; R. H. JONES (*Asbestos*, p. 3) allows both Charleses to pass. The one attribution is as true as the other.

[1] *San kuo chi* (*Wei chi*), Ch. 4, p. 1.

[2] These texts have been translated by WYLIE (*l. c.*, pp. 150—151): they are therefore not reproduced here, especially as they bear no immediate relation to our subject, which is to trace the development of Chinese notions of asbestos in their dependence on Western beliefs. Compare also the analogous text in the *Yü kien* 寓簡 (WYLIE, *Notes*, p 165), Ch. 3, p. 2 b (ed. of *Chi pu tsu chai ts'ung shu*).

twined with those which the further development of the matter at the end of the classical period in the Occident brought to life. The sober and prosaic notices of the Han and Wei periods thoroughly coincide with those of the classical authors, while Ko Hung's thoughts are on the same level as those of the post-classical writers. In their efforts to find a plausible explanation for the origin of asbestos, the Taoist nature-philosophers directed their thoughts toward the animal and vegetable kingdoms, now explaining it as the hair of a beast, now as the fibre of a plant, and also, through the introduction of the activity of a volcano, welding these two theories into one. Nobody as yet has unravelled the mystery of how these strange speculations arose. [1] As regards the supposed animal origin of asbestos, the gist of the Chinese accounts in general is that there is a fiery mountain (volcano) on which lives an animal lustrous with fire, about the size of a rodent, covered with hair of unusual length and as fine as silk. Ordinarily it dwells in the midst of the fire, when its hair is of a deep-red color; but sometimes it comes out, and its hair is then white. On a dark night the forest is visible from the reflection of the animal's lustre. It is put to death by being sprinkled with water, whereupon its hair is spun and woven into cloth, which makes what is called fire-proof cloth. If the cloth becomes soiled, it is purified by fire. The solution of this riddle may be betrayed in advance: the Chinese animal yielding asbestos

---

[1] CHAVANNES (*Bulletin de l'École française*, Vol. III, p. 438) indicates an interesting text in *Pei shi* (Ch. 97, p. 2), according to which the Emperor Yang (605—616) of the Sui dynasty despatched Wei Tsie and Tu Hing-man on a mission to the countries of the west; in the kingdom of Shi (Kesh, at present Shāhr-i-sabz) they took ten dancers, lion-skins, and hair of the rat which enters the fire (*huo shu mao*). Chavannes cites the definition given of this animal in the *Ku kin chu*, "The fire-rat enters the fire without burning; its hairs are over ten feet long; they can be made into a textile known as 'cloth washable in the fire.'" "Ce sont des fibres d'amiante ou asbeste qu'on présentait aux Chinois comme étant les poils d'un animal merveilleux," is the comment added by M. Chavannes.

is a disguise of the classical salamander, whose hair or wool was believed by the Arabs and mediæval Europe to furnish the material for asbestos textiles. The history of this subject must be studied in detail to arrive at a correct appreciation of the Chinese traditions, which, on their part, are of sufficient extent and importance to throw light back on the development of the matter in the West.

THE SALAMANDER IN GREEK AND ROMAN LORE.—An animal by the name of salamander is first mentioned by ARISTOTLE (384—322 B.C.): "On the Island of Cyprus, where copper-ore is smelted and accumulates for many days, animals are developed in the fire, somewhat larger than the big flies with short wings that go hopping and running through the fire. They die when removed from the fire. The possibility, however, that the bodily substance of some animals is not destroyed by fire, is proved by the salamander; for this creature, as it is said, will extinguish the fire while passing through it." [1]

AUBERT and WIMMER, in their edition of Aristotle's work, [2] reject this passage as unauthentic, and presumably with good reason. Aristotle does not mention this animal in any other passage, and it is not clear from his text what kind of animal he understands by *salamandra*; it is also difficult to credit a scholar of the intellectual calibre of Aristotle with the belief in animals crossing fire unhurt, which belong, not to natural history, but to the realm of fable.

THEOPHRASTUS (372—287 B.C.), Aristotle's great disciple, mentions the salamander in two of his writings as an animal which he apparently knew from personal experience. He enumerates "the lizard,

---

[1] ῞Οτι δ'ἐνδέχεται μὴ κάεσθαι συστάσεις τινὰς ζῴων, ἡ σαλαμάνδρα ποιεῖ φανερόν· ἅυτη γάρ, ὡς φασί, διὰ πυρὸς βαδίζουσα κατασβέννυσι τὸ πῦρ (*Historia animalium*, v, 19, § 106).

[2] *Aristoteles Tierkunde*, Vol. I, pp. 119, 515.

which is called the salamander," together with birds and the green frog, among the animals whose appearance prognosticates rain. [1] In his treatise on fire he discusses means of counteracting the force of conflagrations; for instance, vinegar, and vinegar mixed with the white of an egg. "If the power of cold is added to such a fluid," he continues, "this co-operates toward the extinction of fire, and this property is said to be found in the salamander; for this creature is cold by its nature, and the fluid flowing out of its body is sticky, and at the same time contains such a juice that it penetrates forward. This is shown by water and fruits which, when touched by it, become injurious, and usually have a deadly effect. The animal's slowness of motion is also of assistance; for the longer it tarries in the fire, the more it will contribute toward its extinction. However, it cannot extinguish a fire of any dimensions, but only one commensurate with its nature and physical ability; and a fire in which it did not dwell long enough will soon light up again." [2] Also Theophrastus, in the same manner as his master, reproduced a popular opinion of his time, as seen by his addition "it is said" ($\varphi\alpha\sigma i$); but compared with Aelian and Pliny, he is rational and reasonable to a high degree. [3]

AELIAN [4] tells the following story of the salamander: "The salamander is not a product of fire, nor does it rise from the latter like the so-called pyrigoni; [5] yet it does not fear fire, but, going against the flame, the animal tries to combat it like an adversary. The witnesses to this fact are the artisans and workmen dealing

---

[1] Καὶ ἡ σαύρα φαινομένη ἣν καλοῦσι σαλαμάνδραν, ἔτι δὲ καὶ χλωρὸς βάτραχος ἐπὶ δένδρου ἔδων ὕδωρ σημαίνει (*De signis tempestatum*, 15; opera, ed. WIMMER, p. 391).

[2] *De igne*, 60 (opera, ed. WIMMER, p. 361).

[3] The important text of Antigonus of Carystus will be discussed in another connection (see below).

[4] *De natura animalium*, II, 31.

[5] The insects mentioned in the text of Aristotle quoted above.

22

with fire. As long as their fires flame up brightly and further their
labor, they pay no attention to this creature; but when the fires
go down and become extinguished, and the bellows blow in vain,
they become aware of the counteraction of the animal. Then they
trace it out and visit their vengeance upon it; thereupon the fire
rises again, and assists their work." In another passage of the
same work (IX, 28) Aelian asserts that the hog, when swallowing
a salamander, is not hurt, while men partaking of its flesh are killed.
The same is expressed by Pliny: "Those in Pamphylia and in the
mountainous parts of Cilicia who eat a boar after it has devoured
a salamander will die, for the danger of poison is by no means
indicated in the odor or taste of the meat; water and wine in
which a salamander has perished, even if it has only drunk of the
beverage, will also have a mortal effect." [1] In the zoölogical portion
of his great work, Pliny describes the animal thus: "The salamander
is an animal of the shape of a lizard, with a star-like design. It
never comes out except during heavy rains, and disappears when
the sky becomes serene. Such intense cold inheres in this animal,
that by its mere contact, fire will be extinguished, not otherwise
than by the action of ice. The milky mucus flowing from its mouth,
whatever part of the human body it may touch, causes all hair to
fall off; and the spot thus touched assumes the appearance of tetter." [2]

In Book XXIX, where he treats the remedies derived from the
animal kingdom, Pliny has devoted another chapter to the salamander.

---

[1] Apros in Pamphylia et Ciliciae montuosis salamandra ab iis devorata qui edere,
moriuntur: neque enim est intellectus ullus in odore vel sapore; et aqua vinumque
interemit salamandra ibi inmortua vel si omnino biberit unde potetur (XI, 53, § 116). In
XXIX, 23, he dilates still further on the subject.

[2] Sicut salamandrae, animal lacertae figura, stellatum, numquam nisi magnis imbribus
proveniens et serenitate desinens. Huic tantus rigor, ut ignem tactu restinguat non alio
modo quam glacies. Eiusdem sanie, quae lactea ore vomitur, quacumque parte corporis
humani contacta toti defluunt pili, idque, quod contactum est, colorem in vitiliginem mutat
(X, 67, § 188).

The most interesting point that he makes there is this: "If the assertion of the Magi were true, that the animal is helpful in conflagrations, since it is the only creature able to extinguish fire, this experience would long ago have been made in Rome; Sextius also rejects this statement as incorrect." [1] This passage shows that there were men who disavowed this popular belief; and they are headed by Dioscorides, who affirms that it has been said, and wrongly, that the salamander remained immune on entering fire. [2] Further, Pliny imputes the superstition to the Persian Magi; and it may, indeed, have spread into the antique world with the diffusion of the Mithraic cult into Rome.

O. KELLER [3] also holds that the fables about the salamander betray Oriental origin, but he has not succeeded in tracing their sources. [4] Pliny's and Aelian's stories doubtless go back to the Alexandrian Physiologus, whether they may have drawn upon this work directly, or received them by way of oral tradition flowing from Alexandria. The *Physiologus* (Ch. 31) states that the salamander entering a fire-stove extinguishes the fire; [5] and the same is found

---

[1] Ex ipsa quae Magi tradunt contra incendia, quoniam ignes sola animalium extinguat, si forent vera, iam esset experta Roma. Sextius...negatque rostingui ignem ab iis.

[2] L. LECLERC, *Traité des simples*, Vol. II, p. 235.

[3] *Antike Tierwelt*, Vol. II, p. 321. Keller neglected the fundamental passage of Theophrastus regarding the salamander.

[4] The evidence produced by Keller in favor of the Oriental origin is rather perplexing. The name "salamander," which cannot be explained from Greek, indubitably comes from Asia. Arabic and Persian offer the name by omitting the syllable *al*, and the word thus abbreviated is said to mean "poison within." It is of course impossible to derive the Greek name from Persian or Arabic; on the contrary, Arabic *samandal* سَمَنْدَل, *samandar, samaidar, semendel, semendul, samand, sandal,* and Persian also *sālāmandirā* سالامندرا, are derived from Greek *salamandra*, as admitted by all competent philologists (F. HOMMEL, *Namen der Säugetiere bei den südsemitischen Völkern*, p. 33; the Ethiopic Physiologus still offers the form *salmandar*; STEINGASS, *Persian-English Dictionary*, p. 642; YULE, in his *Marco Polo*, Vol. I, p. 216). The derivation from Persian *sām* سلم ("fire", not "poison," which is *samm* سم, an Arabic word) and *andar* ("within") certainly rests on mere playful popular etymology.

[5] F. LAUCHERT, *Geschichte des Physiologus*, pp. 27, 261.

in the *Hieroglyphica* of the Egyptian priest Horapollon of the
fourth century A.D. [1] The tradition, accordingly, must have been
current in Egypt as early as the first or second century. Let us
note right here that the *Physiologus* (Ch. 7) tells also the legend
of the phœnix which cremates itself in the Temple of the Sun at
Heliopolis, how on the ensuing day arises from the ashes a worm,
which develops on the second day into a young bird, till on the
third the phœnix itself comes out therefrom in its previous shape;
for this notion has likewise been associated with the attempts to
account for the origin of asbestos,—asbestos, salamander, and phœnix,
all representing or yielding matters going through fire unscathed.
The Physiologus contains no reference to asbestos; and it must be
emphasized that the assimilation of the three has not taken place
in classical antiquity, during which they were clearly separated.
A wondrous and fabulous book of the type of the *Cyranides*, a late
Greek work written between 227 and 400, would not have missed
this opportunity, had such an assimilation then existed among the
Greeks; but it does not mention a fire-proof textile spun from the
animal's hair. [2]

THE SALAMANDER AND PHŒNIX AMONG THE ARABS.—Old D'HERBELOT,[3]
even, knew that the Arabic word *samandar* designates the animal
styled by us "salamander," and that Oriental authors are not in
accord as to its species,—the one taking it for a kind of marten,

---

[1] F. HOMMEL, *Aethiopische Ueb. d. Physiologus*, p. XXXII.

[2] F. DE MÉLY, *Lapidaires grecs*, p. 91. This work defines the salamander as a
quadruped bigger than the green lizard, and Pliny and Dioscorides also take it for a
lizard. O. KELLER's (*l. c.*, p. 318) identification with *Salamandra maculata* — that is,
the animal now called by us salamander (or eft, newt) — seems to me arbitrary. The
amplifications of the Cyranides are interesting: the animal's heart renders him who carries
it with him fearless of fire, intrepid in a conflagration, and incombustible; and when its
heart is worn as an amulet by people burnt with fever, the fever will at once abate, etc.

[3] *Bibliothèque orientale*, Vol. III, p. 192.

its hair being made into a strong stuff, which can be thrown into fire to be cleansed, when it is soiled, without being in the least damaged; others taking it for a kind of bird generated and consumed in the fire, and found only in places where a perpetual fire is entertained; others, again, describing it as an insect or reptile like a lizard,—but neither D'HERBELOT nor YULE[1] noticed that the salamander as a bird (his product "salamander's plumage" being the equivalent of "asbestos") is no other than the masqueraded phœnix of the ancients.[2] The climax of these curious adjustments is reached by Damīrī (1344—1405), in his *Hayāt al-hayawān*, who notes the phœnix under the title "salamander," describes it as an animal like a fox or marten, and attributes to it the yielding of asbestos: "*Samandal* سمندل is a certain bird that eats *al-bīš* البيش (aconite), which is a plant found in the land of China, where it is edible. It is green in that country; and when it is dry, it becomes a kind of food for the people of China without any injurious effect on them. But if it be taken away from China, even to a distance of a hundred cubits, and is then eaten, the eater of it dies instantaneously.[3]

---

[1] *The Book of Ser Marco Polo*, Vol. I, p. 216.

[2] JULIUS CAESAR SCALIGER (*De subtilitate ad Cardanum*, fol. 305 b, Lutetiae, 1557), however, identified it with the phœnix, "which is not entirely fabulous, but, as we read in the navigators, occurs in the interior of India, and is called by the natives *semenda*."

[3] This story is found in (and is probably copied from) Ibn al-Baiṭār (1197—1248), who quotes Ibn Semdjun as follows: "Some physicians report that the plant *bīš* بيش, grows in China toward the frontier of India, in a country called Halāhil, where alone it occurs. It is eaten as a vegetable in the country of Halāhil, toward the frontier of India. In a dried state it is an article of food for the people of the country, who experience no harm from it. When taken out of that country, if only to a distance of a hundred paces, it acts as a poison, instantly killing him who eats of it" (L. LECLERC, *Traité des simples*, Vol. I, p. 298). This text is important, inasmuch as it shows that the consumption of edible aconite did not take place in China, as Damīrī wrongly asserts, but in a border state of the Himalayan region of northern India. Damīrī's allegation appears embarrassing, as "the Chinese do not seem to have considered any of the aconites as edible" (G. A. STUART, *Chinese Materia Medica*, p. 11); neither does BRETSCHNEIDER (*Bot. Sin.*, pt. 3, pp. 252—257) know anything about such a practice. The statement of the *Pie lu* regarding one variety of aconite, that it is of a sweetish taste, only shows that there is a non-

A wonderful thing in connection with the phœnix is that it takes pleasure in fire and in remaining in it. When its skin becomes dirty, it cannot be cleansed except by means of fire. It is found largely in India. [1] It is an animal smaller in size than the fox, piebald in color, with red eyes and a long tail. Sashes are woven of its soft hair; and when they become dirty, they are thrown into fire, upon which they become clean without being burnt. Other authorities assert that the phœnix is a bird found in India, that

---

poisonous aconite in China. On the other hand, we know that in India only two varieties of *Napellus* are poisonous, — *Napellus* proper and *Aconitum rigidum* — while the two others, *Aconitum multifidum* and *A. rotundifolium*, are harmless and are eaten in Bhutan (HOOKER, *Flora of British India*, Vol. I, p. 29). According to FLÜCKIGER and HANBURY (*Pharmacographia*, p. 15), the tubers of *Napellus* are taken in Kunawar as aphrodisiac. Arabic *bīš* is derived from Hindi *bīš*, the latter from Sanskrit *vishā* (*visha*, "poison"), *Aconitum ferox* (*ativishā, Aconitum heterophyllum*; HOERNLE, *Bower Manuscript*, p. 186). The word appears in al-Bērūnī (SACHAU, *Alberuni's India*, Vol. II, p. 159) and in Qazwīnī, who describes how the fabulous poisonous girls of India are reared on it (SILVESTRE DE SACY, *Chrestomathie arabe*, Vol. III, p. 398). Regarding aconite in India, see WATT, *Dictionary of Economic Products of India*, Vol. I, pp. 84—99 (also published as a separate pamphlet in the series *Agricultural Ledger*, No. 3, 1902); in Tibet, H. LAUFER, *Beitr. zur Kenntnis der tib. Med.*, p. 57. Much valuable and interesting material on Western and Eastern beliefs in aconite poison and its effects has been gathered by W. HERTZ, *Sage vom Giftmädchen* (*Abh. bayer. Akad.*, Vol. XX, 1893, pp. 48—52). Of course, it is not the phœnix which feeds on aconite, but the salamander as a venomous animal. Its poisonous character, inherited from the classical authors, is explained by the Arabs through this process of nutrition.

[1] PLINY (XIX, 4) attributed asbestos to the deserts of India, where, under the scorching rays of the tropical sun and among numerous deadly serpents, it acquires the property of resisting fire. Hierocles, a Greek writer of the sixth century A.D., says of the Brahmans of India that their garments are made of the soft and skin-like fibres of stones, which they weave into a stuff that no fire burns or water cleanses; when their clothes get soiled, they are thrown into a blazing fire, and come out quite white and bright (McCRINDLE, *Ancient India as descr. in Class. Lit.*, p. 186). G. WATT (*Dictionary of the Economic Products of India*, Vol. I, p. 338) mentions two localities, — the Gokāk Taluka, in the Belgaum district in the southern Maratha country, where asbestos is used as an external application in ulcers, made into a paste, after rubbing it down with water; and the country to the south and west of the Kurum River, Afghanistan, where it is medicinally employed and made into brooms and rough ropes, and padding for saddles. Watt imparts a vernacular name for asbestos, *shankha* [*çankha*]*-palita*, which he translates "wick made of shells." On Ceylon, asbestos is found, but is not mined commercially (J. C. WILLIS, *Ceylon*, p. 3, Colombo, 1907).

lays its eggs and produces its young in fire. It possesses the property of being unaffected by fire. Sashes are made of its feathers and taken to Syria. If one of them becomes dirty, it is thrown into fire, which consumes the dirt over it, but the sash itself is not burnt. Ibn-Khallikān states, 'I have seen a thick piece of it woven in the shape of a belt for a riding beast throughout its length and breadth. It was put into fire, but the fire had no effect on it whatever. One end of it was then dipped in oil and left over the burning wick of a lamp, upon which it lighted up and remained so for a long time, after which the flame was extinguished; and it was found to be in the same condition as before, unaltered in any way.' He further states, 'I have read in the writing of our shaikh, the very learned Abd-al-Laṭīf, that a piece of *samandal* a cubit in breadth and two cubits in length was presented to the sovereign of Aleppo. They kept on dipping it in oil and lighting it up, until the oil was exhausted, but yet it remained as white as it was.'" Farther on, Damīrī mentions the salamander under the name *samandar* سمندر and *samaidar* سميدر as "a certain animal well known to the people of India and China, according to Ibn-Sīdah."[1]

Damīrī has compiled his information from the writings of his predecessors. The earliest Arabic notice of the *samandal*-phœnix, as far as I know, occurs in the *Adjaib al-Hind* عجايب الهند ("The Wonders of India"), written in the tenth century, where the bird is localized on one of the Islands of Wāqwāq الوقواق: "It can enter fire without burning itself, and remain there long without eating anything but earth."[2] This work, however, while naming the phœnix

---

[1] A. S. G. JAYAKAR, *Ad-Damīrī's Zoological Lexicon*, Vol. II, pt. 1, pp. 79—81 (Bombay, 1908). G. FERRAND (*Textes relatifs à l'Extrême-Orient*, Vol. I, p. 248) objects to Jayakar's translation of *samandal* by "phœnix;" but Jayakar is certainly right. The three ideas of asbestos, salamander, and phœnix are assimilated in this notion.

[2] LITH and DEVIC, *Livre des merveilles de l'Inde*, p. 172. L. M. DEVIC, in his separate translation of this work (p. 204, Paris, 1878), has this comment: "Semendel ou

for the salamander, makes no reference to a fire-proof textile obtained from the animal. As shown below (p. 328), the geographer Yāqūt (1179—1229) mentions the popular belief that asbestos is the plumage of a bird. In regard to the Caliph Māmun, it is told that the Indian King Dehim presented him with a skin of the bird *samandal* which no fire was able to consume. [1]

If the Chinese, as will be seen, made the salamander a rodent, this zoölogical feat meets a parallel among the Arabs. Qazwīnī enumerates the *samandalun* or *sandalun* as his fifth kind of rat, and describes it as a species of rat that enters fire, recording the same as Damīrī relates about the phœnix (above, p. 319); adding

---

Semendoul est le nom arabe et persan de la salamandre, animal fantastique sur la nature duquel les Orientaux ne s'accordent guère; les uns en font un quadrupède, d'autres un oiseau, d'autres enfin un reptile, tous lui attribuant d'ailleurs la faculté de vivre dans le feu sans se brûler. Marco Polo désigne par ce nom l'amiante." No Arabist as yet seems to have conceived the notion that this tradition becomes intelligible only if we combine the three classical traditions concerning asbestos, salamander, and phœnix associated in post-classical time by the common idea of their incombustibility; hence we meet in Arabic literature accounts of asbestos termed "salamander" which is an animal interpreted as a reptile, phœnix, and finally also as a mammal. — G. FERRAND (*Journal asiatique*, 1904, Mai-Juin, pp. 489—509) has advanced the theory that the one of the two Wāqwāq spoken of by the Arabic writers should be identified with Madagascar (the other is Japan, *Wa-kuok* 倭國; compare also the notice of CHAVANNES, *T'oung Pao*, 1904, pp. 484—487). In an additional notice (*Journal asiatique*, 1910, Mars-Avril, pp. 321—327) FERRAND admits that Wāqwāq may be identified also with Java-Sumatra. In his admirable work *Textes relatifs à l'Extrême-Orient* (Vol. I, p. IV), he adds to these possibilities also East Africa. While not contesting the ingenuity of Ferrand's theory, it is not convincing in all parts (it is chiefly based on the supposed etymology of Wāqwāq being derived from the native names for Madagascar, *Vahuaka*, and for the tree *vakua*). The authority of al-Bērūnī, however, is not to be disparaged, according to whom Wāqwāq belongs to the Qumair Islands; the latter, according to his statement, belong to the Dīva Islands (Malediva and Laccadiva); further, as assured by the same author, Qumair is not, as believed by the common people, the name of a tree, but of a people whose color is whitish, and who practise the religion of the Hindu (SACHAU, *Alberuni's India*, Vol. I, p. 210). Wāqwāq is here clearly indicated as an island or insular group in the Indian Ocean with a populace of Hindu culture. The phœnix, as shown by the above extract from Damīrī, is naturalized by the Arabs in India; and it is difficult to believe that the Adjaib should place the bird on Madagascar, in Indonesia, or in East Africa.

[1] G. WEIL, *Geschichte der Chalifen*, Vol. II, p. 253.

at the end, however, that the animal merely looks like a rat, but in reality is none, and that it occurs in the country of Gūr (east of Herāt in Khovaresm). [1] A gloss to the Talmud, which repeatedly alludes to the legends of the salamander, remarks that the animal has the shape of a mouse, and arises when the wood of the myrtle is burnt in a stove during seven consecutive years. [2] It is the same when other Oriental authors make the salamander an animal resembling a marten, except that it differs from it in color; for the salamander is always red, yellow, or green. [3]

THE SALAMANDER AND PHŒNIX IN MEDIÆVAL EUROPE. —In the poetry of the European middle ages the salamander appears first of all in the love-songs of the Provençal Troubadours. Pierre de Cols d'Aorlac regards the erotic fire burning in his heart as so pleasing that it is the more desirable to him, the more it burns him, like the salamander, which is happy in fire and blaze. [4] In the contemporaneous lyrics of Italy we meet the allegories of the salamander and phœnix woven together: the amorous fire (*il foco amoroso*) is likened to that tenanted by the salamander; the poet is consumed by it, but at the same time rejuvenated like the phœnix; or he dies from the effect of the amorous fire like the phœnix, not being endowed with the salamander's property of being able to live in fire; or he rises again to a new life, like the phœnix, and life in fire becomes his second nature, as is the case with the salamander.

---

[1] F. HOMMEL, *Namen der Säugetiere bei den südsemitischen Völkern*, p. 338; JAYAKAR, *Damīrī's Zoological Lexion*, Vol. II, pt. I, p. 80. In another place Qazwīnī mentions also the mineral asbestos (G. JACOB, *Waren beim arabisch-nordischen Verkehr*, p. 18).

[2] L. LEWYSOHN, *Zoologie des Talmuds*, p. 228.

[3] D'HERBELOT, *Bibliothèque orientale*, Vol. III, p. 192.

[4] The idea that the salamander is happiest in fire first occurs in Saint Augustin (*De civitate Dei*, XXI). It is notable how the exaggerations grow. Classical authors stated nothing to that effect, but merely that the salamander coming in contact with fire can extinguish it.

Also the German poetry of the thirteenth century not infrequently mentions the salamander, and incombustible materials spun from its hair. The latter, for instance, occurs in Wolfram von Eschenbach's Parsifal. The earliest mediæval allusion to this pseudo-salamander asbestos seems to be made in a Provençal treatise on birds and animals ("Naturas d'alcus auzels e d'alcunas bestias"), where it is said, "The salamander subsists on pure fire, and from its skin is made a cloth which fire cannot burn." [1] Again the salamander, through the metamorphosis of the phœnix, appears as a bird. Richard de Fournival, who died about 1260, regards the salamander as a white bird subsisting on fire, and from whose plumage are made cloths that can be purified only by fire. [2] According to the Old-French romance of Bauduin de Sebourc, the salamander lives in the terrestrial paradise as a bird with white woolly down made into tissues; and in Partonopeus de Blois a nuptial coat is lined with salamander's down. [3] ALBERTUS MAGNUS (*circa* 1193—1280) [4] seems to be the only mediæval author who knew that salamander's plume was asbestos. [5] KONRAD VON MEGENBERG (1309—74), who in his *Book of Nature* devoted a chapter to the salamander, [6] tells that Pope Alexander possessed a garment of salamander-wool which was washed in fire instead of water.

---

[1] Salamandra vieu de pur foc, e de son pel fa hom un drap que foc nol pot cremar. Compare F. LAUCHERT, *Geschichte des Physiologus*, pp. 186, 188, 189, 202.

[2] HIPPEAU, *Bestiaire d'amour*, p. 20 (Paris, 1860).

[3] W. HERTZ, *Sage vom Giftmädchen*, p. 66 (*Abh. bayer. Akad.*, Vol. XX, 1893). He refers also to the Byzantine poet Manuel Philes (thirteenth century), who, in his didactic poem on the Properties of Animals, classifies the salamander among the birds.

[4] *De secretis mulierum item de virtutibus herbarum lapidum et animalium*, p. 134 (Amstelodami, 1669).

[5] Si vis ignem perpetuum inextinguibilem facere. Accipe lapidem qui Abaston dicitur, et est coloris ferrei et quam plurimum in Arabia reperitur. Si enim lapis ille accendatur nunquam poterit extingui, eo quod habet naturam lanuginis, quae pluma salamandris vocatur, cum modico humidi unctuosi pinguis, inseparabilis est ab ipso, et id fovet ignem accensum in eo.—Albertus' form *abaston* may be compared with the Middle-English forms *asbeston, abeston, abiston, albeston*.

[6] Ed. of F. PFEIFFER, pp. 276—279.

F. LAUCHERT [1] has shown that the mediæval notions of salamander and phœnix are traceable to the Greek Physiologus; [2] but he omitted to point out that the conception of the salamander-asbestos is novel, and peculiar to mediæval times. YULE [3] admits that he cannot tell when the fable arose that asbestos was a substance derived from the salamander. Certain it is, that it did not exist among the classical peoples; certain it is, also, that the early mediæval writers, with the exception of Albertus Magnus, were not aware of the fact that the alleged product of the salamander was nothing but asbestos, and that asbestos as a mineral was unknown to them, [4] while it was known to the Arabs. There can be no doubt that the Arabs (say, roughly, in the tenth and eleventh centuries) spread the legend to Europe [5] by way of Byzance and Spain. The lacune indicated by Yule remains, and it will be seen in the further discussion that this gap in our knowledge is aptly filled by the records of the Chinese.

Marco Polo, with his keen power of observation and his large share of common sense, was the first to shatter the European superstition. It is interesting that he uses the word "salamander" in the sense of asbestos.

"In a mountain of the province of Chingintalas there is a vein of the substance from which salamander is made. For the real truth is that the salamander is no beast, as they allege in our part

---

[1] *Geschichte des Physiologus, l. c.*

[2] Plinian influence is visible in the venomous properties of the "snake salamander, which, when touching even the foot of a tree, poisons all its branches" (LAUCHERT, p. 194; PLINY, XXIX, 23).

[3] In his edition of *Marco Polo*, Vol. I, p. 216.

[4] MEGENBERG (*l. c.*, p. 434) noted asbestos after Isidorus, but did not see its identity with salamander-wool.

[5] It is interesting to note that our own historians of the middle ages did not always grasp the facts in the case; while our Orientalists, owing to the knowledge of Arabic sources, were able to unravel the mystery. Thus A. SCHULTZ (*Das höfische Leben zur Zeit der Minnesänger*, Vol. I, p. 338) mentions without explanation "the textures produced from salamanders and burnt by no fire;" and G. JACOB (*Waren beim arabisch-nordischen Verkehr im Mittelalter*, p. 18), with reference to Qazwīnī, lays bare the fact.

of the world, but is a substance found in the earth; and I will tell you about it.

"Everybody must be aware that it can be no animal's nature to live in fire, seeing that every animal is composed of all the four elements. Now I, Marco Polo, had a Turkish acquaintance of the name of Zurficar, and he was a very clever fellow. And this Turk related to Messer Marco Polo how he had lived three years in that region on behalf of the Great Kaan, in order to procure those Salamanders for him. He said that the way they got them was by digging in that mountain till they found a certain vein. The substance of this vein was then taken and crushed, and when so treated it divides as it were into fibres of wool, which they set forth to dry. When dry, these fibres were pounded in a great copper mortar, and then washed, so as to remove all the earth and to leave only the fibres like fibres of wool. These were then spun, and made into napkins. When first made these napkins are not very white, but by putting them into the fire for a while they come out as white as snow. And so again whenever they become dirty they are bleached by being put in the fire.

"Now this, and nought else, is the truth about the Salamander, and the people of the country all say the same. Any other account of the matter is fabulous nonsense. And I may add that they have at Rome a napkin of this stuff, which the Grand Kaan sent to the Pope to make a wrapper for the Holy Sudarium of Jesus Christ." [1]

This sober account based on information received in China has left a lasting impression upon European science, and has taught how to discriminate between asbestos as a mineral and the salamander as an animal. A. BOETIUS DE BOOT [2] rejected Polo's designation of

---

[1] Ed. of YULE and CORDIER, Vol. I, p. 213. It will be seen farther on that Marco Polo's account is confirmed by the contemporaneous Annals of the Yüan Dynasty.

[2] *Gemmarum et lapidum historia*, p. 383 (Lugduni Batavorum, 1636).

the mineral as salamander, restoring the ancient names "amiantus" and "asbestinus," and ridiculed the belief in any animal living in fire. Relying on Marco Polo, A. KIRCHER [1] has fully discussed the subject from a scientific point of view; and his contemporary, the zoölogist JOHN RAY, [2] was able to state, "Quod Salamandra sine ullo incommodo in igne vivere possit a vulgo creditum, verum a doctioribus dudum abunde refutatum est."

ASBESTOS IN THE NEAR EAST.—Asbestos was well known to the Arabs and Persians, and was much employed by them. [3] A number of valuable notes concerning this matter we owe to the erudition of E. WIEDEMANN. [4] Evliya Effenda narrates that the wonderful carpet presented by Khosru I Nūrshirvān to the monastery which he built near Ütch Kilise was made of asbestos, and that asbestos textiles were manufactured on Cyprus. [5] The Arabic soldiers who hurled naphtha at beleaguered towns were equipped with asbestos garments in order to guard them from accidents which might have happened from handling this inflammable substance. [6] Dimashqī, Abul Fēdā (1273—1331), and Yāqūt (1179—1229) point to Badakshān

---

[1] *La Chine illustrée*, pp. 278—280 (Amsterdam, 1670). Kircher says that he could receive no information as to the stuff sent by the Great Khan to the Pope (see also CORDIER's note in Yule's *Marco Polo*, Vol. I, p. 216; and compare the above quotation from K. von Megenberg).

[2] JOANNES RAIUS, *Synopsis animalium quadrupedum*, p. 273 (Londini, 1693).

[3] A Syriac allusion occurs in the *Historia Monastica* of the Bishop of Margā (A.D. 840): "Prayer made the martyrs like asbestos before the fire" (E. A. W. BUDGE, *The Book of Governors*, Vol. II, p. 499).

[4] *Zur Mechanik und Technik bei den Arabern* (SB. P. M. S. Erlg., Vol. 38, 1906, pp. 39, 40).

[5] The latter notice goes back to Dioscorides (L. LECLERC, *Traité des simples*, Vol. II, p. 414).

[6] The Italian chevalier Aldini, about 1825, conducted a series of experiments in using asbestos garments for the protection of firemen. His idea was revived in Paris, the firemen there having been furnished with such clothes, and after conclusive proof of their practical utility, was followed in London (R. H. JONES, *Asbestos*, pp. 31, 159).

as the place where the mineral was found; the former making special mention of lamp-wicks made from it, into which fire penetrates, while they remain unharmed. Yāqūt has the following report: "In the mines near Badakshān is found the stone *fatīla* (that is, 'stone of the wick'), which resembles papyrus (*bardī*). The people believe that it is the plumage of a bird. [1] It is styled also *al-ṭalq*. It is not consumed by fire. It is placed in oil and kindled with fire, in which case it burns like a lamp-wick. [2] When the oil burns, the stone remains as before, and none of its properties changes. This always takes place whenever it is dipped in oil and burns. When thrown into a blazing fire, it is not hurt by it. Coarse table-cloths are woven from it. These, being soiled, are put into fire to be purified, and whatever dirt is on them is consumed by the flames. They are cleansed, and come out as pure as though they had never been affected by dirt." The erroneous designation *al-ṭalq* is traceable to Ibn al-Baiṭār (1197—1248), who groups around Dioscorides' notice of asbestos Arabic accounts of the mineral *ṭalq* corresponding to our mica. [3]

A very interesting description of asbestos is given by Abū Ubaid al-Bekrī (1040—94) of Cordova in Spain, in his *Geography of Northern Africa*, as follows: [4]—

"Among the singular products of the country of the Negroes is noticeable a tree with long and slender stem, called *turzi*. It grows in the sand, and bears a big and swollen fruit containing within it a white wool which is made into stuffs and garments. These stuffs are capable of remaining in a vehement fire forever without

---

[1] That is, the phœnix. For explanation see above, pp. 318—323.

[2] Compare the statement of Theophrastus (p. 302).

[3] L. Leclerc, *Traité des simples*, Vol. II, pp. 414, 415. Pseudo-Aristotle (Ruska, *Steinbuch des Aristoteles*, p. 174) also describes mica under the same name.

[4] MacGuckin de Slanf, *Description de l'Afrique septentrionale par El-Bekri*, p. 336 (Alger, 1913).

being damaged. The jurist Abd al-Melek affirms that the inhabitants of Al-Lames, a town of that region, wear only clothing of this kind. Near the river Derā is found a substance similar to it. This is a sort of stone, called, in the language of the Berber, *tamatghost*. When rubbed between the hands, it softens to such a degree that it assumes the consistency of linen. It serves for the making of cordage and halters, which are absolutely incombustible. A costume was made from this substance for one of the Zenatian princes who ruled at Sidjilmessa. A man of proved veracity told me that a trader had sent for a napkin made from this mineral for Ferdilend, King of Galicia, in Spain (Ferdinand I of Leon). He offered it to the prince, explaining that it had belonged to one of the disciples of Jesus, and that fire could produce no impression upon it. He furnished the proof under the eyes of the King, who, struck by such a marvel, expended all his wealth to purchase this relic. He sent it to the sovereign of Constantinople, that it might be deposited in the principal church, and in return received a royal crown with the authorization to wear it. Several persons tell of having seen in the house of Abul Fadl of Bagdad the fringe of a napkin made of this substance, which, when put into fire, became whiter than previously. In order to clean such a napkin, which had the appearance of linen, it was sufficient to place it on a fire."

The employment of asbestos for the purpose of a *pia fraus* is related also by an Arabic traveller. Abū Dulaf who wrote the diary of his journey to China about 941 tells of an incombustible tree, growing in the territory of the tribe Bajā (east of Transoxania), from the wood of which the natives make idols; Christian travellers are in the habit of taking this wood along, asserting that it comes from the cross of Christ. Again he relates about the tribe Kharlok that their houses are of incombustible wood. [1] Both Marquart and

---

[1] G. FERRAND, *Relations de voyages arabes, persans et turks rel. à l'Extrême-Orient,*

Ferrand who translated and discussed this text have been unable to cope with this problem. Certainly it is not here the question of a tree, as wrongly supposed by these scholars; still less do we meet here, as suggested by Marquart, the conception that the wood of the cross had miraculously been shooting forth again. What we meet here, in fact is asbestos; and this matter has clearly been expounded as early as 1843 by J. YATES in his classical work *Textrinum Anti- quorum: An Account of the Art of Weaving among the Ancients* (pp. 362—365). Yates sets forth that ignorance of the true nature of asbestos caused it to be employed in the dark ages for purposes of superstition and religious fraud, and cites several important documents to this effect. One of these is taken from the *Chronicon Casinense* ("Chronicle of the Abbey of Monte Casino") of Leo Ostiensis who narrates a story that some monks returning from a pilgrimage to Jerusalem brought home a particle of the cloth with which Jesus wiped the feet of his disciples (particulam lintei, cum quo pedes discipulorum Salvator extersit); and when the genuineness of this relic was doubted, they put it in fire from which it came forth in its previous shape. Thus the authenticity of the relic was convincingly established. Tilingius, in 1684, directly says that impostors exhibit to simple women-folks the stone amiantus, and frequently sell it as

---

Vol. I, pp. 210, 215. Ferrand has misunderstood Marquart, for he ascribes to the latter the supposition that the question is here of teak-wood. On the contrary, MARQUART (*Osteurop. und ostasiat. Streifzüge*, p. 76) has decidedly rejected this idea, and strangely enough proposed to regard the incombustible tree as the birch. Why the birch should be called incombustible I am unable to see. Abū Dulaf is not to be taken too seriously in matters of natural history; and his assigning to certain tribes of certain products, as partially seen also by Marquart, is purely arbitrary or fictitious. The list of his stones presents curious reminiscences of the fabulous stones of the Alexander Romance and the Arabic *lapidaires* based thereon. The most striking of these reminiscences is the stone luminous at night and serving as lamp (*Pseudo-Callisthenes*, II, 42). This stone, accord- ing to the Arabic scribe, is found in the country of the Kirgiz! For this reason I am inclined to think that also his incombustible tree is a purely literary invention from the same source. The Chinese have several accounts of unconsumable trees, partly leaning toward asbestos (see WYLIE, *l. c.*, p. 148).

the wood from the cross of the Savior; they easily take faith therein, since it is not consumed by fire and is veined in the manner of wood. It is equally manifest that Abū Dulaf's incombustible tree which supplied Christians with sacred souvenirs of the cross was nothing but asbestos, and the report of al-Bekri previously mentioned affords additional evidence to this effect. The alleged products ascribed by Abū Dulaf to Central-Asiatic regions are fancifully construed from the legends told in the Alexander Romance, and there, as mentioned above (p. 308), we encounter also the asbestine wood.

Under the Sung dynasty asbestine stuffs were imported into China by the Arabs over the maritime route; they were seven inches wide, differing in length. In the period Chêng-ho 政和 (1111—18), under the Emperor Hui-tsung, asbestine stuffs of half this width were sent as tribute by the Arabs, and at a later date were followed by dishes and baskets of the same material, which on the whole looked like the cloth then made from the product of the cotton-tree, but somewhat darker and almost black in color. When flung into the fire, they came forth brilliant white. [1] Mosul produced asbestine cloth during the middle ages. [2]

THE SALAMANDER-ASBESTOS IN CHINA.—After this review of the development of the relevant beliefs in the West, we are prepared to understand the asbestos traditions of the Chinese. In these, three stages of development are clearly set off. The first, already described, ranging approximately from the Han to the third century, I am tempted to term the "historical or classical" set of beliefs,

---

[1] We shall revert once more to this text, not utilized by Wylie and inserted in the *Tie wei shan ts'ung t'an* 鐵圍山叢談 (Ch. 5, p. 20; edition of *Chi pu tsu chai ts'ung shu*) of Ts'ai T'iao 蔡絛, who lived in the first half of the twelfth century. WYLIE (*Notes*, p. 196) states regarding this author that he treats mostly of events which occurred in his own time, and that the work shows a good deal of research, and may be relied upon as an authority in investigations regarding that period.

[2] HIRTH and ROCKHILL, *Chau Ju-kua*, p. 140.

23

agreeing, as they do, with Greek and Roman lore; the second, from the beginning of the fourth century down to the end of the Sung, denotes the "romantic" period of beliefs, coinciding with those of mediæval Europe and the Arabs; the third, inaugurated by the Yüan or Mongol dynasty, is the "realistic," or, if the word be allowed, "scientific," period, based on the actual discovery of asbestos on Chinese soil. We have to deal here first with the mediæval romanticism inaugurated by the speculations of the adepts of Taoism.

The earliest attempts to explain the origin and composition of asbestos were made by the celebrated alchemist Ko Hung 葛洪 (249—330), in his work *Pao-pʻu-tse*. [1] This author reports on three kinds of asbestos (*huo huan pu* 火浣布) as follows: "As regards the first kind of fire-proof cloth, it is said that there is in the ocean a majestic mound [2] harboring a fire that burns of itself. [3] This fire rises in the spring, and becomes extinguished in the autumn. On this island grows a tree, the wood of which is able to resist the action of fire, and is but slightly scorched by it, assuming a yellow color. The inhabitants make fuel of it in the usual way, but this fuel is not transformed into ashes. When their food has been cooked, they extinguish the firewood by means of water. In the same manner it is put to use again and again, and indeed represents an inexhaustible supply. The barbarians gather the flowers

---

[1] My rendering is based on the text as quoted in the *Wei lio* 緯畧 (Ch. 4, p. 3; ed. of *Shou shan ko tsʻung shu*, Vol. 74). This fundamental source on the subject has been overlooked by Wylie.

[2] *Su kʻiu* 蕭邱. I should be inclined to regard this as the proper name of Volcano Island, if this term were traceable in the Liang Annals, which, as will be seen below, contain the source for this account of Ko Hung; but it does not occur there. Again, the notice of the Annals goes back to the lost reports of Kʻang Tʻai 康泰, on his mission to Fu-nan in the first part of the third century. If Kʻang Tʻai's report had contained the name *Su kʻiu*, we might reasonably conclude that it would have found its way into the Annals; for this reason it may be solely an invention of Ko Hung.

[3] That is, an active volcano.

of these trees, and weave cloth from them. This is the first kind of fire-proof cloth. Further, they also peel the bark of these trees, boil it by means of lime, and work it into cloth, which is coarse and does not come up to the quality of the material prepared from the flowers. This is the second kind of fire-proof cloth. Moreover, there are white rodents (*pai shu* 白鼠) covered with hair, each three inches long, and living in hollow trees. They may enter fire without being burnt, and their hair can be woven into cloth, which is the third kind of fire-proof cloth." [1]

The first two sorts of asbestos established by Ko Hung, and alleged to be of vegetal origin, are certainly imaginary; and how this matter came about will be fully discussed hereafter. Here the fact that concerns us is that Ko Hung is the first Chinese writer in whom the idea of the animal origin of asbestos has crystallized. Certainly, his "white rodent" is nothing but the salamander of the Western legend, whose wool furnishes asbestos. At first sight it is striking, of course, that Ko Hung's notice far precedes in time any Western version of the legend; yet this can rationally be explained. Two conjectures which might be made to get easily over this state of affairs would not prove before the facts. We cannot assume that the legend is spontaneously Chinese in origin and migrated from China to Western Asia: in China it has no basic facts, whereas

---

[1] 抱朴子曰。火浣布有三種。其一曰海中蕭邱有自生火。春起秋滅。洲上生木。木為火焚不糜但小 (gloss: 一無小字) 焦黃。人或得薪俱如常。薪但不成灰。炊熟則以水滅之。使復更用如此不窮。夷人取此木華績以為布一也。又其木皮赤剝之。以灰煮治以為布籠不及華俱可火浣二也。又有白鼠毛長三寸居空木中。入火不灼。其毛可績為布三也 (*Wei lio*, Ch. 4, p. 3).

we have traced its logical development in the West from the combination of salamander and asbestos. Nor would it be possible to regard the account of Ko Hung as unauthentic or as an anachronism, as we have a number of texts, ranging from the fourth to the sixth century, all relating to the same legend. The *Wu lu* 吳錄[1] is credited with the statement that in Ji-nan (Tonking) is captured a fire-rodent whose hair is made into cloth, being styled "fire-proof cloth."[2] According to BRETSCHNEIDER,[3] this book was written in the third century, during the period of the Three Kingdoms (221—280); but it is hard to believe that at that early date the legend of the salamander-asbestos was known in China. The localization in Ji-nan, foreign to Ko Hung, also seems somewhat suspicious. We have noticed above (p. 312) that asbestos was known in the China of that period, and that in the coeval Annals a tribute gift of it from the Western Regions (*Si Yü*) is on record for the year 239, no reference, however, being made to the salamander story. The earliest date that we may assume for the coming into existence of the latter on Chinese soil is the end of the third or the beginning of the fourth century.

It is more interesting that Kuo P'o 郭璞 (276—324), a contemporary of Ko Hung, likewise alludes to the salamander-asbestos; for Kuo P'o, in his commentary to the *Shan hai king*, is made to say the following, as translated by WYLIE:[4] "Ten thousand *li* to the east of Fu-nan is the kingdom of Ké-po. More than five thousand *li* farther east is the burning mountain kingdom, where, although there may be long-continued rain on the mountain, the

---

[1] Records of the Kingdom of Wu, by Chang Pu 張勃 of the third century.
[2] *Wei lio*, Ch. 4, p. 3. WYLIE (*l. c.*, p. 149) quotes this passage from *T'ai p'ing yü lan* (Ch. 820, p. 8), where the locality is defined as Pei-king 北景 in Ji-nan.
[3] *Bot. Sin.*, pt. 1, p. 209, No. 1043.
[4] *L. c.*, p. 146.

fire constantly burns. There is a white rat in the fire, which some-times comes out to the side of the mountain, in order to seek food, when the people catch it and make cloth from the hair, which is what is now called fire-proof cloth." What Wylie transcribes Ké-po is properly Ch'i-po 耆薄; and this is nothing but a variant for the well-known Shê-p'o 闍婆, the old Chinese designation for the island of Java. The fact that in this connection the question really is of Java becomes evident from other parallel texts alluding to the same matter. [1] The name "Shê-p'o" for Java, however, does not appear in Chinese records earlier than the first half of the fifth century, the first embassy coming from there being listed in the year 433: consequently Kuo P'o of the Tsin dynasty cannot have possessed any knowledge of Shê-p'o, which name must be a later interpolation in his text. Aside from this point, however, the story is entirely creditable to him, because the geographical portion of it, as will be seen, is based on the narrative of K'ang T'ai of the third century, and is even more exactly reproduced by him than by Ko Hung. Kuo P'o, however, shuns the account of vegetable asbestos, as related by K'ang T'ai and repeated after him by Ko Hung, and focusses the notion of asbestos exclusively on the white rodent (that is, the salamander) inhabiting an active volcano. K'ang T'ai knew nothing at all about this animal. Ko Hung does not naturalize it anywhere. It is Kuo P'o who took up this legend and placed its home on the Volcano Island first reported by K'ang T'ai: consequently Kuo P'o's story is a compromise reached between the salamander story coming from the West and the tree-asbestos story of Fu-nan, but it is valueless for tracing the region from which the salamander legend hailed. It did not hail from Volcano

---

[1] Compare PELLIOT, *Bull. de l'Ecole française*, Vol. III, p. 264; Vol. IV, p. 270; these texts will be discussed farther on.

Island in the Malay Archipelago, as Kᶜang Tᶜai located there only the alleged tree-asbestos, which in fact is bark-cloth, that has nothing to do with mineral asbestos. Kᶜuo Pᶜo, further, shows his familiarity with the salamander in his edition of the dictionary *Erh ya*. [1] This enumerates ten kinds of tortoise, the tenth of which is termed "fire tortoise" (*huo kuei* 火龜); and Kᶜuo Pᶜo annotates that it is like the "fire rodent" (*huo shu*). [2] The latter animal is not included among those enumerated in the text of the *Erh ya;* that is to say, it is entirely foreign to the ideas of ancient national Chinese culture, but is a borrowed type, which first dawned upon the horizon of the Chinese in the very age of Kᶜuo Pᶜo himself.

Another contemporaneous allusion to the same matter is found in the *Ku kin chu* 古今注, written toward the middle of the fourth century by Tsᶜuei Pao 崔豹, who says that the fire-rodent remains immune when going into fire, and that what is termed "fire-proof cloth" is made from the animal's hair, which is ten feet long. [3] Tsᶜuei Pao, in his succinct and sober statement, thoroughly agrees with Ko Hung, differing from him only in somewhat exaggerating the length of the hair. Yet the same author, in the same work, presents a more fantastic account of the matter, which he traces to the *Book of Marvels* [4] ascribed to the Taoist adept Tung-fang So (born in 160 B.C.). This attribution, as is well known, certainly is fictitious; and the following text bears out this fact again, because it is based on the account of Kᶜang Tᶜai, and must therefore be later than the third century. Tung-fang So, according

---

[1] Ch. B, p. 10 b.

[2] This name has been adopted by the Polyglot Dictionary of Kᶜien-lung (Ch. 31, p. 24) with the literal renderings into Manchu *tuwai singgeri*, Tibetan *me byi*, and Mongol *galχi khulugana*. The explanations given in the Manchu dictionaries show that the salamander-asbestos is understood (see SACHAROV, *Manchu-Russian Dictionary*, p. 765).

[3] *Pᶜei wén yün fu*, Ch. 36, p. 59.

[4] Entitled by him *Shen i chuan* 神異傳, otherwise *Shen i king* 神異經.

to Ts⁏uei Pao, is made to say, [1] "In the southern regions there is a volcano forty *li* in length, and from four to five *li* in width. In the midst of this volcanic fire grow trees unconsumable by fire, and day and night exposed to a scorching heat, over which neither wind nor rain has any power. In the fire lives also a rodent, a hundred catties in weight, and covered with hair over two feet in length, as fine as silk, and white in color. [2] Sometimes it comes out; and by sprinkling water over it, it is put to death. Its hair is then removed and woven into cloth, which is known under the name 'fire-proof cloth.' " Another text, likewise wrongly connected with the name of Tung-fang So, expatiates on the animal with still greater vagaries of fancy, and will be discussed below. We notice that in this Taoist narrative the salamander is made a denizen of Volcano Island, in the same manner as by Kuo P⁏o. We accordingly have two versions of the legend current during the fourth century,— a simple and sober one, accounting for the origin of asbestos from an animal identical with the Western salamander; and an elaborate and fantastic one, aggrandized by Taoist lore under the influence of K⁏ang T⁏ai's report of a Volcano Island in the Malay Archipelago.

The salamander turns up again in that interesting book *Liang se kung tse ki*, relating to the beginning of the sixth century, and written by Chang Yüe (667—730), [3] "Merchants from the Southern

---

[1] *Pien tse lei pien*, Ch. 21, p. 6. The text is quoted also in the commentary to *San kuo chi*, *Wei chi* (Ch. 4, p 1 b), in the *Wei lio* (Ch. 4, p. 3), and in the *Ts⁏i tung ye yü* by Chou Mi.

[2] It must certainly be white, because asbestos coming out of a fire has this color. WYLIE (*l. c.*, p. 145), who translates from a modern edition of *Shen i king*, has the addition, "It ordinarily lives in the fire, and is of a deep-red color; but sometimes it comes out, and its hair is then white."

[3] See this volume, p. 198. The text in question is preserved in the *Wei lio* 緯 畧, Ch. 4, p. 3 b; and in the *Ko chi king yüan*, Ch. 27, p. 13. WYLIE (*l. c.*, p. 143) seems to have translated from another book. His addition, "which the emperor had deposited among the miscellaneous cloths," is not in the text before me.

Sea brought as presents three pieces (*tuan*) [1] of fire-proof cloth. [2] Duke Kie, recognizing it from afar, exclaimed, 'This is fire-proof cloth, indeed: Two pieces are made from twisted bark, [3] and one is made from the hair of a rodent.' On making inquiry of the merchants, their statement exactly agreed with that of the duke. [4] On asking him the difference between the cloth of vegetal and that of animal origin, the duke replied, 'That manufactured from trees is stiff, that from rodents' hair is pliable; this is the point by which to discriminate between them. Take a burning-mirror and ignite the *tsê* trees [5] on the northern side of a hill, and the bark of the trees will soon become changed.' The experiment was made, and it turned out in accordance with his affirmation." [6] The witty duke, accordingly, exploded the old tale of K'ang T'ai, that bark cloth was incombustible and a sort of asbestos. He himself, on former occasions, had doubtless applied the experiment which he recommended in the course of the story, and was possessed of that truly scientific

---

[1] A cloth measure of 18 feet.

[2] According to the text of the *Wei lio*, "Duke Kie, passing a market, noticed traders offering three *tuan* of fire-proof cloth" (杰公至市見商人齎火浣三端).

[3] It is notable that he speaks of twisted, not of woven bark, as K'ang T'ai and his followers did (see p. 347).

[4] This sentence is omitted in the text of the *Wei lio*.

[5] 柘木 *Cudrania tribola*, Hance. Wylie takes this for 斫, or he may have found this reading in his text; for he translates, "Take some wood cut down on the north side of the hill and set a light to it by means of a solar speculum." Duke Kie, of course, did not mean to say this. He wanted to prove by experiment that tree-bark is not incombustible, like asbestos; and with this end in view, it was not necessary to chop the trees.

[6] 南海商人齎火浣布三端。杰公遙識曰。此火浣布也。二是緝木皮所作。一是績鼠皮所作。以詰商人具如杰公之說。因問木鼠之異公曰。木堅毛柔是可別也。以陽燧火山陰柘木爇之木皮改。常試之果驗 (*Ko chi king yüan*, Ch. 27, p. 13).

spirit which does not halt at received traditions, but tries by experiment to get at the root of things. To him true asbestos was only the kind attributed to the salamander,[1] and the duke's wisdom demonstrates that the rodents' hair of the Chinese was really mineral asbestos.

The texts thus arrayed bear out sufficiently the fact that the legend of the salamander-asbestos was popularly current in China from the fourth to the sixth century; and the records of the Chinese very aptly fill the gap which, as we noticed (p. 325), exists in the West between the close of classical antiquity and the traditions of the Arabs and mediæval Europe. The Chinese texts are all prior to those of the Arabs, and it is therefore necessary to conclude that the Chinese and the Arabs must have borrowed the legend from a common source extant in Western Asia at least during the third century. This source is as yet unknown to us, but the conviction of its existence is a postulate without which we cannot intelligently understand the case. There are also indications in Western sources which allow the inference that this prototype resulting in the Chinese and Arabic notions must have lingered in the anterior Orient in the beginning of our era. We have referred to the probable Oriental origin of the salamander legend, and to Pliny's association of it with the Persian Magi; we have pointed out also that it was current in Egypt during the first century A.D., and that Pliny's and Aelian's stories are dependent on the Alexandrian Physiologus. There is accordingly good reason to believe that the

---

[1] This is confirmed by another passage in the same work *Liang se kung tse ki*, in which Volcano Island (火 洲) is mentioned. Here it is said that from the bark of the fiery tree growing there only cloth is made, while fire-proof cloth is produced from the hair of the fire-rodent living on a blazing mound. This text will be found in *T'u shu tsi ch'éng, Pien i tien* 41, Woman Kingdom (*Nü kuo*), *hui k'ao*, p. 2. It is said to have been translated in its entirety by D'HERVEY-ST.-DENYS in his *Mémoire sur le Fou-sang*, which unfortunately is not accessible to me.

salamander legend was known in the Orient on a line stretching from Egypt to Persia, and that the numerous translations of the Physiologus, if nothing else, supported its wide diffusion. At the same time, however, as we know from the Chinese records, asbestos-cloth was in evidence in western Asia, and was traded from there over the routes of Central Asia to China. Salamander and asbestos being familiar to the nations of the Roman Orient, they were in possession of the elements with which to form that legend which proceeded from them to China and at a later date loomed up among the Arabs. It may be supposed that this primeval version, as yet unknown, will turn up some day in an early Syriac source (or possibly in a Greek papyrus): and if a Syriac work should tell us of an asbestos-tree, and immediately join to this a notice of the salamander, [1] we may imagine that the temptation was strong to link those two accounts together.

The germ of this lost Oriental version possibly is traceable to a Greek text, from which it can be shown how the identification of asbestos with the salamander may have been effected. Antigonus of Carystus, who was born between 295 B.C. and 290 B.C., and lived at Athens and Pergamum, [2] has left a small collection of "Wonderful Stories," among which is the following: [3] "There are worm-shaped hairy creatures living in the snow. In Cyprus, where copper-ore is smelted, an animal is engendered a little larger than a fly. The same occurs also in the smelting-furnaces of Carystus. Part of them die when separated from the snow; others, when separated from the fire. The salamander, however, quenches the fire." This text is based on that of Aristotle, given above (p. 314), where

---

[1] See above, p. 308.

[2] Compare U. VON WILAMOWITZ-MÖLLENDORFF, *Ueber Antigonos von Karystos.*

[3] *Historiae mirabiles,* 90, 91 (*Rerum naturalium scriptores Graeci minores,* ed. KELLER, Vol. I, p. 22).

are also mentioned worms found in long-lying snow. [1] Antigonus, however, has here an essential addition, not met with in Aristotle or any other author; and this is that this fire animal occurs also in the furnaces of Carystus. [2] Now, we have seen that, according to Apollonius and Strabo, Carystus on Euboea was one of the principal asbestos-producing regions, and that from this locality the mineral was even named Carystius. Antigonus hailed from Carystus, and this fact may entitle us to the opinion that he was acquainted with the asbestos mined near his home town. True it is, he does not mention asbestos in the few fragments of his writings which are preserved; and there is nothing to indicate that in the above passage he means to include asbestos in the "smelting-furnaces of Carystus." The point which I wish to make, however, is that it was easy to read this interpretation into his text. An Oriental Greek, Syrian, or Arab, for instance, who knew that "Carystius" was a synonyme for "asbestos," could well have been reminded thereof while reading this passage, and the immediate mention of the salamander might then have led him to link the two notions together. [3] In this manner we gain a satisfactory clew as to the probable origin of the salamander-asbestos assimilation, which certainly must have been brought about on the soil of Hellenism,

---

[1] Aristotle does not name the animal living in fire, but, judging from his description, it appears to be an insect. PLINY (XI, 36, § 119), who speaks of the same creature after Aristotle, calls it *pyrallis* or *pyrotocon* (others read *pyrausta*), and describes it as a winged quadruped (*pinnatum quadrupes*) of the size of a larger fly. AELIAN (*Hist. anim.*, II, 2) styles it *pyrigonos* ("fire-born").

[2] Pliny, in harmony with Aristotle, places it only on Cyprus (*in Cypri aerariis fornacibus*), while Aelian gives no locality.

[3] It is possible also that the μυιῶν of the Greek text (from μυῖα, "fly") led to a confusion with μῦς ("mouse"), and gave rise to the conception of the salamander as a rat (Qazwīnī), mouse (Talmud), or rodent (Chinese). On the other hand, it must be admitted that this metamorphosis is capable also of a logical explanation: the salamander-lizard is smooth and hairless; when the salamander was made to yield asbestos, it naturally had to be transformed into an animal with hair-growth.

during the second or in the beginning of the third century A.D.

Besides the salamander of the character of a rodent, we receive another intimation as to the nature of this animal, which answers the classical notions. A work *Sung chi* 宋志 ("Memoirs of the Sung Period"), by Shên Yo 沈約, [1] contains the following notice: "Blazing Island (Yen chou 炎洲) is situated in the southern ocean, and harbors the animal *ki* (or *kie*)-*ku* 猲猸. When it is caught by people, it cannot be wounded by chopping or piercing. They gather fuel, build a fire, bind the animal and throw it into the fire, and yet it will remain unscorched." [2]

The name for this animal, which is clearly differentiated from the rodent that follows, seems to be connected with some Malayan form underlying our word "gecko," described thus by YULE and BURNELL: [3] "A kind of house-lizard. The word is not now in Anglo-Indian use; it is a naturalist's word; and also is French. It was no doubt originally an onomatopœia from the creature's reiterated utterance. Marcel Devic says the word is adopted from Malay *gekok* [*gēkoq*]. This we do not find in Crawfurd, who has *tăké*, *tăkék*, and *goké*, all evidently attempts to represent the utterance. In Burma, the same, or a kindred lizard, is called *tokté*, in like imitation." [4]

---

[1] Quoted in *Ye k'o ts'ung shu* 野客叢書 by Wang Mou 王懋 of the Sung period (*Ko chi king yüan*, Ch. 27, p. 13). Regarding this work see WYLIE, *Notes on Chinese Literature*, p. 161. It was published in 1201.

[2] Then follows the story of the rodent-salamander mingled with the alleged bark-cloth asbestos: "There is, further, the Volcanic Country, constantly enveloped by fire which is not quenched by rain. In this fire there is a white rodent. When the trees in the forests on this burning island have been wetted by rain, their bark becomes scorched; and when exposed to fire, it becomes white. The islanders gather this bark during several months, and weave it into cloth, which makes fire-proof cloth. Either the bark of the trees or the hair of the rodents may yield it."

[3] *Hobson-Jobson*, p. 367.

[4] "Some of the Borneo reptiles produce singular sounds. The commonest among them is a gecko, the *chichak*, which name imitates perfectly the cry which it produces. A much louder and more characteristic cry is that of *Goniocephalus borneensis*, a large

The characters *ki-ku*, in this case, are chosen by the Chinese author only to imitate the sounds of a word like "gecko." As a rule, the animal *ki-ku* is regarded as a mammal. The word first appears under the T'ang in the *Yu yang tsa tsu*, and is synonymous with *fêng li* 風狸, *fêng mu* 風母 ("wind mother"), or *fêng shêng shou* 風生獸 ("wind-born beast"). [1] On the other hand, the Chinese know a saurian, *ko-kiai* 蛤蚧, being a word-formation analogous to the Malayan names of the lizard, and, according to Chinese authors, imitative of the call of the animal. [2]

It thus appears that the rodent-salamander of the Chinese, after all, was a lizard like the salamander of the ancients; and the lizard character of the animal leaks out in the earliest account of the subject by Ko Hung, when he says that the animal lives in hollow trees; for it is the lizard who has acquired this habit. A. R. WALLACE, [3] in describing the lizards of the Aru Islands, observed, "Every shrub and herbaceous plant was alive with them; every rotten trunk or dead branch served as a station for some of these active little insect-hunters."

The fact that it was not the Arabs from whom the Chinese received the salamander-asbestos tale is illustrated, from a negative

---

lizard which lives on trees and has a high and serrated crest down its back. The Malays call this lizard *kog-go*, an imitation of its call-note, which is frequently repeated" (O. BECCARI, *Wanderings in the Great Forests of Borneo*, p. 35). In the *Encyclopædie van Nederlandsch-Indië* (Vol. IV, p. 400) the word is given as *toke*, which is peculiar to Sundanese; it passed also into the language of the Batak on Sumatra; in Malayan it is *tekek* and *tokek*; in Javanese, *tekek*. Compare Moro *tagatak* or *tukatuk*, "lizard" (R. S. PORTER, *Primer of the Moro Dialect*, p. 45). In the same encyclopædia (Vol. I, p. 551) will be found a description of the genus and of the beliefs in its venomous property, which are very similar to those entertained by the ancients in regard to the salamander.

[1] See the texts of *Pên ts'ao kang mu*, Ch. 51 A, p. 20 b; and *Wu li siao shi*, Ch. 10, p. 12.

[2] *Pên ts'ao kang mu*, Ch. 43, p. 6. The oldest text referring to it is the *Ling piao lu i* of the T'ang (compare PFIZMAIER, *Denkwürdigkeiten aus dem Tierreiche Chinas*, *SBAk. Wien*, Vol. 80, 1875, p. 14).

[3] *The Malay Archipelago*, p. 331.

viewpoint, by the absence in China of any specific reference to the phœnix, of which the Arabs make a great case (p. 319). Some Chinese works have a general reference to birds, but the coincidence is not perfect. Thus the apocryphal *Sou shên ki* 搜神記[1] has a volcano in the region of the Kᶜun-lun, inhabited by herbs, trees, birds, and mammals, all existing in blazing fire and yielding fire-proof cloth. [2]

---

[1] WYLIE, *Notes*, p. 192. The passage is in Ch. 13, p. 3 (of the Wu-chᶜang print).

[2] A case of a different character may be mentioned in this place, as it reveals a very curious coincidence between a Chinese and an Arabic text. The interesting work *Tu yang tsa pien* 杜陽雜編, written by Su Ngo 蘇鶚 in the latter part of the ninth century, contains the following story (Ch. B, p. 1; edition of *Pai hai*): "During the year of the reign of the Emperor Shun-tsung 順宗 (A.D. 805) the country Kiü-mi 拘弭 [otherwise 拘彌, the territory of Keria; see CHAVANNES, *Documents sur les Tou-kiue occidentaux*, p. 128] sent as tribute a pair of birds insensible of fire (刼火雀一雄一雌). These birds were uniformly black and of the size of a swallow. Their voice was clear, but did not quite resemble that of ordinary birds. When placed on a fire, the fire was spontaneously extinguished. The Emperor, admiring this wonder, had the birds put in a cage of rock-crystal [rock-crystal being believed to be a transformation of ice and to have a cooling effect], which was hung in the sleeping-apartments of the palace. At night the inmates of the palace tried to set fire to the birds by means of burning wax candles, but entirely failed in damaging their plumage." Abū Ubaid al-Bekrī (1040—94) of Cordova (MAC GUCKIN DE SLANE, *Description de l'Afrique septentrionale par El-Bekri*, p. 43) has the following account: "Nous donnons le récit suivant sur l'autorité d'Abou-'l-Fadl Djāfer ibn Yousof, Arabe de la tribu de Kelb, qui avait rempli les fonctions de secrétaire auprès de Mounis, seigneur de l'Ifrīkiya: 'Nous assistions à un repas donné par Ibn-Ouanemmou le Sanhadjien, seigneur de la ville de Cabes, quand plusieurs campagnards vinrent lui présenter un oiseau de la taille d'un pigeon, mais d'une couleur et d'une forme très singulières. Ils déclarèrent n'avoir jamais vu un oiseau semblable. Le plumage de cet animal offrait les couleurs les plus belles; son bec était long et rouge. Ibn-Ouanemmou demanda aux Arabes, aux Berbers et aux autres personnes présentes s'ils avaient jamais vu un oiseau de cette espèce, et sur leur réponse qu'ils ne le connaissaient pas même de nom, il donna l'ordre de lui couper les ailes et de le lâcher dans le palais. A l'entrée de la nuit, on plaça dans la salle un brasier-fanal allumé, et voilà que l'oiseau se dirigea vers ce meuble et tâcha d'y monter. Les domestiques eurent beau le repousser, il ne cessa d'y revenir. Ibn-Ouanemmou, en ayant été averti, se leva, ainsi que toute la compagnie, afin d'aller voir ce phénomène. Moi-même, dit Djāfer, j'étais un de ceux qui s'y rendirent. Alors, sur l'ordre d'Ibn-Ouanemmou, on laissa agir l'oiseau, qui monta jusqu'au brasier ardent, et se mit à becqueter ses plumes, ainsi que font tous les oiseaux quand ils se chauffent au soleil. On jeta alors dans le brasier des chiffons imprégnés de goudron et une quantité d'autres

While the Chinese, in a somewhat masqueraded form, received the legend of the salamander, they never adopted this word, as did the Arabs and Persians. It was reserved for the Jesuit Father Ferdinand Verbiest (1623—88) to introduce the Chinese, in his *Kun yü t'u shuo*, to an illustration of a European salamander under the title *sa-la-man-ta-la* 撒辣漫大辣, which he says occurs in the country Germania (*Je-êrh-ma-ni-ya*) in Europe: "Its habitat is in cold and moist places, its temper is very cold, its skin is thick, and its strength is such as to extinguish fire; its hair is of mixed color, black and yellow; a black and spotted crest runs along its back down to its tail." The figure by which his note is illustrated shows a cat or fox-like mammal. [1]

THEORY OF THE VEGETAL ORIGIN OF ASBESTOS.—In order to arrive at a correct appreciation of the complex notions developed by Ko Hung and Kuo P'o regarding asbestos, we shall now turn our attention to another matter. In the first half of the third century A.D., K'ang T'ai 康泰 and Chu Ying 朱應 were engaged in a mission to Fu-nan 扶南 (Cambodja), and on their return to China published two works in which were laid down their experiences during this memorable journey. Their record furnished to the compilers of the Chinese Annals a great deal of information on the ancient history

---

objets inflammables, afin d'augmenter l'intensité du feu, mais l'animal n'y fit aucune attention et ne se dérangea même pas. Enfin il sauta hors du brasier et se mit à marcher, ne paraissant avoir éprouvé aucun mal.' Quelques habitants de l'Ifrikiya assurent que, dans la ville de Cabes, ils avaient entendu raconter l'histoire de cet oiseau. Dieu seul sait si elle est vraie." In examining each for itself, we should certainly take both the Chinese and the Arabic story for an abstruse fable. Such a fire-proof bird most assuredly does not exist. On either side we are treated to the report of eye-witnesses. The two stories apparently are independent, although the subject is identical. After all, might this mysterious bird be an offshoot of the salamander-phœnix, restored to life by an overstrained imagination?

[1] *T'u shu tsi ch'éng*, XIX, chapter "Strange Animals," *hui k'ao* 3, p. 9.

of that country. [1]　In the article on Fu-nan, inserted in the Annals of the Liang Dynasty (502—556), [2] we meet a curious notice on asbestos with reference to a Malayan region, as follows: "It is reported that Fu-nan is bounded on the east by the ocean known as Ta-chang 大漲 ('Great Expanse'). [3]　In this ocean is a great island on which the kingdom of Chu-po 諸薄 (Java) is situated. East from this kingdom is the island of Ma-wu 馬五洲. [4] Going again over a thousand *li* in an easterly direction across the Ta-chang Ocean, one reaches Volcano Island. [5]　On this island there

---

[1] PELLIOT, *Bull. de l'Ecole française*, Vol. III, p. 275.

[2] *Liang shu*, Ch. 54, p. 3; likewise in *Nan shi*, Ch. 78, p. 3.

[3] Corresponding to our Chinese Sea, extending from Hai-nan to the Straits of Malacca.

[4] PELLIOT (*Bull.*, Vol. IV, p. 270) is inclined to identify this island with Bali by assuming a clerical error ("Ma-li" for "Ma-wu").

[5] *Tse jan huo chou* 自然火洲 (literally, "the island of fire which burns of itself"). PELLIOT (*Bull.*, Vol. III, p. 265) has justly recognized that the reading "great island" 大洲 in *Liang shu* and *Nan shi* is an error for "fire island." Indeed, the text of *Nan shi* is quoted with the correct reading in the *Wei lio* (Ch. 4, p. 3) of the Sung period, in an essay entitled "Asbestos." WYLIE, in his study *Asbestos in China* (p. 149), not consulted by Pelliot, translated the name by "spontaneous combustion great island." He accordingly accepted the wrong reading, and took the word *jan* in the sense of "to burn." The latter point of view is justified, as, for instance, the *Hüan lan* 玄覽 (*Ko chi king yüan*, Ch. 27, p. 13) writes 燃火之洲. Which of the numerous volcanic islands of the Archipelago, one of the chief volcanic belts on the globe, should be understood by K'ang T'ai's "Volcano Island," certainly is difficult to guess. In my opinion, Timor stands a fair chance of claiming this honor. A. R. WALLACE (*The Malay Archipelago*, p. 5) observes, "To the eastward, the long string of islands from Java, passing by the north of Timor and away to Banda, are probably all due to volcanic action. Timor itself consists of ancient stratified rocks, but is said to have one volcano near its centre." Again on p. 7, "In Timor the most common trees are Eucalypti of several species, so characteristic of Australia, with sandal-wood, acacia, and other sorts in less abundance. These are scattered over the country more or less thickly, but never so as to deserve the name of a forest. Coarse and scanty grasses grow beneath them on the more barren hills, and a luxuriant herbage in the moister localities. In the islands between Timor and Java there is often a more thickly wooded country, abounding in thorny and prickly trees. These seldom reach any great height, and during the force of the dry season they almost completely lose their leaves, allowing the ground beneath them to be parched up, and contrasting strongly with the damp gloomy, ever-verdant forests of the other islands. This peculiar character, which extends in a less degree to the southern peninsula of Celebes and the east end of Java, is most probably owing to the proximity of Australia. The

are trees which grow in the fire. The people in the vicinity of the island peel off the bark, and spin and weave it into cloth hardly a few feet in length. This they work into kerchiefs, which do not differ in appearance from textiles made of palm and hemp fibres, [1] and are of a slightly bluish-black color. When these are in the least soiled, they are thrown into fire and thoroughly purified. This substance is made also into lamp-wicks which never become

south-east monsoon, which lasts for about two-thirds of the year (from March to November), blowing over the northern parts of that country, produces a degree of heat and dryness which assimilates the vegetation and physical aspect of the adjacent islands to its own. A little further eastward in Timor-laut and the Ké Islands, a moister climate prevails, the south-east winds blowing from the Pacific through Torres Straits and over the damp forests of New Guinea, and as a consequence every rocky islet is clothed with verdure, to its very summit. Further west again, as the same dry winds blow over a wider and wider extent of ocean, they have time to absorb fresh moisture, and we accordingly find the island of Java possessing a less and less arid climate, till in the extreme west near Batavia rain occurs more or less all the year round, and the mountains are everywhere clothed with forests of unexampled luxuriance." "The land mammals of Timor are only six in number, one of which is a shrew mouse (*Sorex tenuis*), supposed to be peculiar to the island" (*ibid.*, p. 160).

[1] *Tsiao ma* 蕉麻. Pelliot renders this by "scorched hemp" (*du chanvre roussi*), as if the reading were 焦. Wylie translates the term "raw hemp;" but the word *tsiao* denotes a particular group of plants, the fibre-furnishing palms, and is co-ordinated with the word *ma* ("hemp"). Clothing of palm-fibres was particularly made by the aboriginal tribes of southern China, and known as *hung tsiao pu* 紅蕉布 (*hung tsiao* being a variety of the genus *Musa*; see the *Ch'i ya* 赤雅 by Kuang Lu, Ch. A, p. 5, ed. of *Chi pu tsu chai ts'ung shu*). The so-called Manila hemp of commerce is obtained from the Abaca (*Musa textilis*), the staple material for Filipino weavings (see C. R. Dodge, *Descriptive Catalogue of Useful Fibre Plants of the World*, pp. 248—249, Washington, 1897; and the recent interesting article of C. Elata, *Philippine Fiber Plants*, in the *Philippine Craftsman*, Manila, 1914, pp. 442—456). Marco Polo (ed. of Yule and Cordier, Vol. II, p. 124) mentions that the people of the province of Kuei-chou manufacture stuffs of the bark of certain trees which form very fine summer clothing. I do not believe with Yule (p. 127) that Polo here refers to the so-called grass-cloth, but he indeed means literally cloth woven from the bark-fibres of trees. The Miao in the prefecture of Li-p'ing, province of Kuei-chou, indeed make textiles from tree-bark, called bark-cloth (*p'i pu* 皮布; see *Ta Ts'ing i t'ung chi*, Ch. 400, p. 4). According to Megasthenes (Strabo, xv, 60) the Sarmanes (Sanskrit *çramana*, "ascetic") of India used to wear garments made from the bark of trees. The various kinds of hemp grown in China are briefly enumerated in *Chinese Jute, published by Order of the Inspector General of Customs* (Shanghai, 1891).

24

exhausted." This text presents a somewhat amazing effort at associating heterogeneous ideas. The real affair described is the well-known bast-cloth, common to the Malayan and Polynesian tribes, and peculiar to many other culture-areas, which assuredly is not incombustible; and this product is passed off as asbestos. The reference to the purification in fire and to the making of wicks doubtless proves that asbestos is intended. On the other hand, the resemblance of asbestos-fibres to hemp or flax is well-known. [1]

The term "bark-cloth" is equivocal: it denotes principally two types,—one known under the Polynesian name *tapa*, in which the bast is flayed and pounded or macerated in water till it becomes soft and pliable; [2] and another, in which the bast-fibre shreds into filaments that may be spun and woven. As K῾ang T῾ai refers to the latter process, he must have had textiles of bast-fibre in mind. Ko Hung, as already stated, based his account of asbestos on K῾ang T῾ai's report, and was familiar with both beaten and woven bark-cloth; for he has established two vegetable varieties of asbestos,— one woven from the flowers of trees, the other prepared from bark.

---

[1] Hence our name "earth-flax" (Dutch *steenvlas*, that is, "stone flax;" German *Flachs-stein*).

[2] This method is practised not only by the Malayo-Polynesian stock, but also by the negroes of Africa and the aboriginal tribes of America. Only a few instances from literature may be given, whose number might certainly be augmented by many others. W. MARSDEN (*History of Sumatra*, p. 49, London, 1811) says on this subject, "The original clothing of the Sumatrans is the same with that found by navigators among the inhabitants of the South Sea Islands, and now generally called by the name of Otaheitean cloth. It is still used among the Rejangs for their working dress, and I have one in my possession, procured from these people, consisting of a jacket, short drawers, and a cap for the head. This is the inner bark of a certain species of tree, beaten out to the degree of fineness required; approaching the more to perfection, as it resembles the softer kind of leather, some being nearly equal to the most delicate kid skin; in which character it somewhat differs from the South Sea cloth, as that bears a resemblance rather to paper, or to the manufacture of the loom." In central Celebes the art of weaving is still unknown, and the tribes use only beaten bark cloth derived from a large variety of trees (P. and F. SARASIN, *Reisen auf Celebes*, Vol. I, p. 259, where the process is described). See also DODGE, *l. c.*, pp. 98—101.

Is K^{c}ang T^{c}ai himself responsible for this fanciful combination, or did he merely reproduce a tradition overheard by him in Fu-nan? We know that K^{c}ang T^{c}ai, during his residence in that country in the first part of the third century, encountered a Hindu named Ch^{c}ên-sung 陳宋, who had been despatched there by the King of Central India in response to the mission intrusted to Su-wu 蘇物 by Fan Chan 范旃, King of Fu-nan. Thus K^{c}ang T^{c}ai availed himself of the opportunity of interviewing Ch^{c}ên-sung on all matters concerning India, and on his return to China published a work on the hundred and odd kingdoms of which he had heard. This valuable source of information has unfortunately perished. [1] India and Fu-nan entertained close commercial relations: diamonds, sandal-wood, and saffron being expressly mentioned in the T^{c}ang Annals as products that were exchanged by India with Ta Ts^{c}in, Fu-nan, and Kiao-chi (Tonking). [2] True it is, asbestos is not specified in the list of these products; but K^{c}ang T^{c}ai's story allows us a peep behind the scenes, for it incontrovertibly shows that asbestos was known in Fu-nan during the time of his sojourn. Certainly it could not have come from any Malayan region, where asbestos, as far as I know, is not found or utilized by the native population: it evidently arrived in Fu-nan from India. In A.D. 380 India presented to the Court of China an offering of fire-proof cloth; [3] and this same event is alluded to in the Annals of the Tsin Dynasty, in the life of Fu Kien 苻健 (337—384), [4] in the statement that India offered fire-proof cloth. [5] We remember that Pliny naturalizes asbestos in India, that Hierocles equips the Indian Brahmans with

---

[1] PELLIOT, *Bull.*, Vol. III, p. 276.

[2] *T'ang shu*, Ch. 221 A, p. 10.

[3] *Shi leu kuo ch'un ts'iu*, Ch. 37, p. 11 (compare WYLIE, *l. c.*, p. 143).

[4] GILES, *Biographical Dictionary*, p. 230.

[5] *Tsin shu*, Ch. 112 (compare *Pien tse lei pien*, Ch. 21, p. 6).

asbestos garments, and that the Arabs derived the mineral from Badakshān (pp. 320, 327): hence we are entitled to presume that asbestos was sometimes shipped also from India to Fu-nan in the beginning of the third century. This postulate is necessary to account for the fact that Kᶜang Tᶜai struck correct notions in Fu-nan regarding asbestos,—notions which agree with those of the classical authors. Asbestos products, however, were rare in Fu-nan, as in Hellas and Rome (PLINY, rarum inventu) and everywhere else, and the supply presumably could not keep pace with the demand; therefore the "malign and astute" people of Fu-nan [1] conceived the ruse to trade off Malayan bast-cloth under the name of "asbestos." This at least seems to me the best possible theory explaining Kᶜang Tᶜai's account, as far as the theory of vegetal origin is concerned. A specific example of what the Fu-nan asbestos was is offered by the interesting story of Duke Kie, discussed above, from which it appears that bast-cloth was really shipped to China under the label "asbestos." The merchants who offered this ware hailed from the Southern Sea, and this product must have been identical with what was shown Kᶜang Tᶜai on his visit in Fu-nan. Duke Kie's clever experiment also demonstrates that Kᶜang Tᶜai had merely fallen victim to a mystification.

The influence of the asbestos text in the Liang Annals is apparent not only in the Taoist school of the fourth century, as shown above, but also in several later works. Thus the *Hüan lan* or *Yüan lan* 玄(元)覽, a work of the Tᶜang period (618—906), [2] says, "In Pᶜi-kᶜien 毘騫 there is the Island of Blazing Fire, producing a tree the substance of which can be woven, and which furnishes what is called fire-proof cloth." The geographical term "Pᶜi-kᶜien"

---

[1] Thus they are characterized in the Annals of the Southern Ts'i (PELLIOT, *Bull.*, Vol. III, p. 261).

[2] Cited in *Ko chi king yüan*, Ch. 27, p. 13.

occurs in the Fu-nan account of the Liang Annals as the name of
a great island of the ocean, situated 8000 *li* from Fu-nan, and,
according to PELLIOT, [1] seems to have been along the Irrawaddy
and the Indian Ocean. The information of the *Hüan lan*, of course,
is deficient, as in the Liang Annals Volcano Island has nothing to
do with Pʿi-kʿien, but is located far eastward, in the Malay Archipelago.

In the above translation of the passage of the Liang Annals, the
kingdom of Chu-po has been identified with Java, the name being
a variant of Shê-pʿo, by which Java became known from the first
half of the fifth century. This conclusion is confirmed by a text
ascribed to the *I wu chi* 異 物 志 and contained in the *Tʿai pʿing
yü lan*, [2] in which the Island of Blazing Fire is located in the
kingdom of Se-tiao 斯 調, which is doubtless a misprint for Ye-tiao
葉 調. Now, we owe to the ingenuity of PELLIOT the identification
of this name with the old Sanskrit designation Yavadvīpa, [3] and
this solution of the problem seems to me a well-assured result.
Since the *I wu chi*, in its account of Volcano Island, depends upon
the text of the Liang Annals, it seems equally certain that the
Chu-po country mentioned in the latter is the island of Java. The
passage of the *I wu chi* is worded as follows: "In the kingdom of
Ye-tiao (Java) there is the Island of Blazing Fire, covered with a
fiery plain, which lights up spontaneously in the spring and summer,
and dies away during the autumn and winter. Trees grow there
which do not waste, the branches and bark renewing their fresh
appearance; in the autumn and winter, however, when the fire dies
out, they all wither and droop. It is customary to gather the bark

---

[1] *Bull.*, Vol. III, p. 264.

[2] Ch. 820, p. 9 (edition of Juan Yüan, 1812). The text is quoted also in the commentary to *San kuo chi, Wei chi*, Ch. 4, p. 1.

[3] *Bull.*, Vol. IV, p. 268; and *Tʿoung Pao*, 1912, p. 457.

in the winter for the purpose of making cloth. It is of a slightly bluish-black color. When it is soiled, it is thrown into fire again, and comes out fresh and bright." [1] The interesting point here is that the trees alleged to yield asbestos are set in causal relation with the fire of the volcano, which transmits to the bark its fire-proof quality.

Two other texts may likewise be traced to the Fu-nan account in the Liang Annals. The *Hüan chung ki* 玄中記, written by Kuo 郭 [2] of the fifth century, observes that "there is a volcano in the south, producing a tree which is used for fuel without being consumed; the bark, when woven, makes fire-proof cloth, of which there are two kinds." [3] The *Shu i ki* 述異記 ("Record of Wonderful Matters"), by Jên Fang 任昉, who lived in the beginning of the sixth century, annotates that "the fire of this active volcano in the south is extinguished in the twelfth month whereupon all trees push forth branches; while, when the fire rises again, the leaves drop, the same as in winter in China, When the wood is used for fuel, it is not consumed by the fire; and the bark, when woven, makes fire-proof cloth." This version must be connected with one handed down in the *Wên hien t'ung k'ao* of Ma Tuan-lin, who erroneously says that the Volcano country (*Huo shan*) became known only at the time of the Sui (589—618), and then quotes the following from the "Customs of Fu-nan" (*Fu-nan t'u su* 扶 南土俗), by K'ang T'ai: [4] "Volcano Island is situated somewhat over a thousand *li* east of Ma-wu Island. In the spring the rains set in; and when the rainy season is over, the fire of the volcano

---

[1] Compare WYLIE, *l. c.*, p. 146.

[2] His personal name is unknown.

[3] In agreement with *Pao-p'u-tse* (p. 332).

[4] Compare PELLIOT, *Bull.*, Vol. III, pp. 275 and 276, note 2. My rendering is based on the text in *Yüan kien lei han*, Ch. 233, p. 19.

breaks forth. The trees in the forests of the island, when wetted by the rain, have a black bark, but, when affected by the fire, the bark assumes a white color. [1] The inhabitants of the adjoining isles gather this tree-bark during the spring, and weave it into cloth; they make it also into lamp-wicks. When but a bit soiled, they fling the cloth into fire, and this means purify it. There is, further, a mountain, north of the country Ko-ying (written Kia-ying 加營)[2] and west of Chu-po (Java), 300 *li* in circumference. The active eruption of fire opens from the fourth month, and ceases in the first month. During the period of volcanic activity the trees drop their leaves, as in China during the cold season. In the third month the people betake themselves to this mountain to peel the tree-bark, which is then woven into fire-proof cloth."

The *Lo-yang kia lan ki* 洛陽伽藍記[3] states that the country Kü-se 車斯 produces fire-proof cloth which is made from the bark of trees, and that these trees are not consumed by fire. [4] The number of texts insisting on the vegetal origin of asbestos could doubtless be much increased; but those here assembled are sufficient to show that this doctrine, first traceable to Kᶜang Tᶜai, had obtained a permanent hold on the Chinese mind, despite the contradictory explanation based on the salamander. While the Chinese salamander versions unquestionably go back to Western traditions, I am not convinced that this is the case also with the vegetal theory. As set forth above (p. 306), I do not share the opinion of those who impute to Pliny a belief in a plant origin of asbestos.

---

[1] This observation, of course, relates in reality to asbestos.

[2] See PELLIOT, *Bull.*, Vol. IV, p. 278, note.

[3] Records of the Buddhist Establishments in the Capital Lo-yang, written by Yang Hüan-chi 楊衒之 in 547 or shortly afterwards (BRETSCHNEIDER, *Bot. Sin.*, pt. 1, No. 483; and CHAVANNES, *Bull.*, Vol. III, p. 383).

[4] *T'u shu tsi ch'éng*, chapter on fire (*tsa lu*), p. 11 b. Kü-se is perhaps identical with Kü-shi 車師, designating "Turfan-Dsimsa."

The tree-asbestos of the Alexander Romance and a Syriac work
(p. 308) represents rather isolated instances which show lack of
cohesion, and cannot be unduly emphasized. Asbestos filaments
bear such a striking resemblance to hemp or flax fibres, that it
becomes intelligible that the theory of their identity could have
spontaneously been advanced in various parts of the world. Our
own nomenclature of asbestos varieties is witness thereof. [1] In the
following section I shall try to explain how this theory originated
in Fu-nan. [2]

The Arabs and mediæval Europe, as already observed, were too
much absorbed by the identification of asbestos with the salamander
and phœnix to pay much attention to the idea of vegetal provenience. This view, curiously enough, loomed up in Europe in
Martini's *Atlas Sinensis*. It is told there that there is a kingdom

---

[1] The mountain-tree asbestos of the Chinese meets its parallel in our "mountain wood"
or ligniform asbestos (*xylotil*),—a variety of asbestos which is hard and close grained, generally of a brownish color, and often bearing an exact resemblance to petrified wood. At
first sight it might easily be mistaken for the latter, especially when sufficient iron is
present to give it the ruddy tinge of decayed wood or bark. Under the microscope,
however, the crystal fibre is easily detected, as is also the absence of the vegetable cells
which are always to be found in petrified wood (R. H. Jones, *Asbestos*, p. 14). Also the
Chinese seem to have taken petrified wood for asbestos (see Wylie, *l. c.*, p. 152; and the
writer's *Notes on Turquois*, p. 24).

[2] An analogous example in which the ancients were deluded in regard to a Chinese
product, is presented by Chinese silk taken by several classical authors for thin fleeces
obtained from trees (Yates, *Textrinum Antiquorum*, p. 182). Virgil (*Georgica*, II, 121)
has the verse, "And Seres comb their fleece from silken leaves" (Velleraque ut foliis
depectant tenuia Seres). Strabo (xv, 20) supposed the raw silk material to be a sort of
byssos fibres scraped from the bark of trees. According to Dionysius Periegetes, the
Seres comb the variously colored flowers of the desert land to make precious figured garments, resembling in color the flowers of the meadow (*ibid.*, p. 181). Pliny (vi, 20) speaks
of the Seres famed for the wool found in their forests; they comb off a white down adhering
to the leaves, and steep it in water. The use of water to detach silk from the trees is
insisted on also by Solinus and Ammianus Marcellinus, both of whom propound the vegetal
theory of the origin of silk. Pausanias of the second century denied that the threads
from which the Seres make webs are the produce of bark, and described the silkworm with
fair correctness.

in Tartary styled Taniu, which produces stones; and above these, an herb which fire can never consume. When it is surrounded by flames, it reddens as though it would be entirely burned up; but as soon as the fire is out, it re-assumes its former gray or ash color. It is never very large or high; but it grows like human hair, and has almost the shape of the latter. Its consistency is very feeble and delicate; and when placed in water, it is noted that it turns into mud and is entirely dissolved. [1]

THE VOLCANIC THEORY.—After having discussed the opinions of the animal and vegetal origin of asbestos, another question remains to be answered,—How did the idea of a volcano acting upon the formation of asbestos spring into existence and develop? Besides the volcanic theory propounded by K'ang T'ai, there are a few others that call for attention. The *Shi i ki* [2] records an embassy from the country of the Yü-shan bringing a tribute of fire-proof cloth to the Emperor Wu of the Tsin dynasty in the year 280. On this occasion the envoys of Yü-shan stated that "in their country there is a mountain containing veined stones (*wen shi* 文石) sending forth fire, the appearance of smoke being visible at the horizon throughout the four seasons. This fire was known as the 'cleansing fire.' When unclean clothes were thrown on these blazing stones, however big the accumulation of filth, they were purified in this manner, and came out as new." These clothes, of course, must have been of asbestos-fibres. This story is strange, [3] and is hardly reproduced correctly in the Chinese text, as it is now before us. No reason can be discovered why asbestos-cloth should be cleaned in a volcanic

---

[1] A. KIRCHER, *La Chine illustrée*, p 278 (Amsterdam, 1670). Kircher refutes this error; Martini's story is doubtless derived from the Chinese.

[2] Ch. 9, p. 4 (ed. of *Han Wei ts'ung shu*); compare WYLIE, *l. c.*, p. 143.

[3] In all probability it is a mere echo and bad digestion of K'ang T'ai's narrative.

fire, as any other ordinary fire would answer the same purpose. The true story must have been so worded that asbestos itself was produced by the volcano in question, and that the agency of the volcanic fire to which it was exposed was instrumental in rendering it impervious to fire.[1] We have here, then, a reference to an asbestos-producing volcano situated in the west of China. A burning mountain beyond the Kʻun-lun, upon which any object that is thrown is immediately burnt, is mentioned in the *Shan hai king*;[2] and we have seen that the *Sou shên ki* derives asbestos from this volcano in the Kʻun-lun.[3] Chinese tradition, accordingly, is acquainted with two volcanoes producing asbestos,—one on an island in the eastern part of the Malay Archipelago, first reported by Kʻang Tʻai; and another placed in Central Asia. From none of these territories, however, has asbestos ever become known to us: hence we are compelled to conclude that the volcanic theories of the Chinese records have not been prompted by immediate observation, but are the result of a series of speculative thoughts. These thoughts themselves, on the other hand, have a certain foundation in correct observation: it is in the manner of their concatenation that the speculative element comes in.

It may first be noted that from our scientific viewpoint even the direct association of asbestos with volcanoes is quite correct. In the widest sense of the word, we include under "asbestos" both pyroxene and hornblende; the latter most frequently, the former

---

[1] In a manner similar to that in which Pliny invokes the scorching heat of the tropical sun in the deserts of India as the cause of the fire-proof quality of the mineral.

[2] WYLIE, *l. c.*, p. 146.

[3] The Sung History, according to BRETSCHNEIDER (*Mediæval Researches*, Vol. II, p. 190), describes a volcano north of Urumtsi, which contains sal ammoniac: "Inside there is a perpetual fire, and the smoke sent out from it never ceases; clouds or fogs are never seen around this mountain; in the evening the flames issuing from it resemble torch-light; the bats, from this phenomenon, appear also in a red color." Compare W. OUSELEY, *Oriental Geography of Ebn Haukal*, p. 264.

more rarely, assuming an asbestiform character. Pyroxene, a very common mineral, is a constituent in almost all basic eruptive rocks, and is principally confined to crystalline and volcanic rocks. In different localities it is associated with granite, granular limestone, serpentine, greenstone, basalt, or lavas. Likewise hornblende is an essential constituent of igneous rocks. [1] Nevertheless we cannot grant the Chinese the merit of having made such an observation, which is due solely to our modern geological research. There is, moreover, no volcano in Asia which to our knowledge has ever yielded asbestos, nor do the Chinese pretend to have actually imported the material from a volcanic region. To them the volcano is a romantic place of refuge to explain the perplexing properties of asbestos. The introduction of the volcano must not be explained by reading into it the latest achievements of our geology, but from the thoughts evolved by the nature philosophy of the Chinese, nourished by the glowing accounts accruing from foreign countries. The question will be difficult to settle, whether Kʻang Tʻai owes his theory to himself and his Chinese environment, psychological and educational, or whether he borrowed it outright from the people of Fu-nan. I feel positive of the one fact, that the volcanic point in it was conceived in Fu-nan; for China has no volcanoes, and all Chinese accounts of such relate to countries abroad. [2]

---

[1] R. H. JONES, *Asbestos*, p. 21. Asbestos occurs in high altitudes. In Italy, for instance, it is rarely found at a lower level than five thousand feet, ranging from this upwards to twelve thousand; in fact, up to the line of perpetual snow. Hence the addition "mountain" is so prominent in our names for the varieties; as, "mountain wood," "mountain leather," "mountain paper," "mountain cork," "mountain flax."

[2] There is a negative criterion which illustrates that the Fu-nan tradition of the volcanic asbestos is not due to an impetus from outside. The Arabic authors make frequent allusions to the volcanoes of Java and neighboring islands, but never mention asbestos in this connection. Ibn Khordādbeh, in his *Book of the Routes and Kingdoms* (844—848), tells of a small volcano in Jāba (Java), a hundred cubits square, and only of the height of a lance, on the summit of which flames are visible during the night, while it throws up smoke during the day. The merchant Soleiman, who wrote in 851, speaks of a

To K{c}ang T{c}ai, asbestos-fibres were of vegetal origin, the product of the bark of a tree, somewhat on the order of palm or hemp fibre. The ready-made textile was impervious to fire, and the mind eager to account for this wonder of nature settled on the theory that this property should have been brought about through the action of a natural fire. The material in its crude state had already habituated itself to fire, which had hardened it in such a manner that it could successfully resist all attacks of the element,—an idea also alive in Pliny's mind. People of Fu-nan who had occasion to visit certain Malayan islands with their belt of volcanic mountains observed the great luxury of vegetation which there prevailed, and its endurance despite volcanic eruptions. PLINY tells us of an ash-tree overshadowing the fiery spring of a volcano and always remaining green. [1] Chao Ju-kua, describing the action of Mount Etna, observes, "Once in five years fire and stones break out and flow down as far as the shore, and then go back again. The trees in the woods through which this stream flows are not burned, but the stones it meets in its course are turned to ashes." [2] If there were plants to outlive the ravages of volcanic destruction, the primitive mind argued that the absorption of subterranean fire had made them fire-proof. The fibres of asbestos, being fire-proof, were consequently derived from plants growing on volcanic isles, this association being facili-

---

Mountain of Fire near Jāwaga (Java) which it is impossible to approach; at its foot there is a spring of cold and sweet water; the same is reiterated by Ibn al-Faqīh (902). Masūdī (943) reports a tradition regarding the Malayan volcanoes, according to which, during the thunder-like eruptions, a strange and terrifying voice resounded announcing the death of the king or chief, the sounds being louder or lower in accordance with the importance of the person (see G. FERRAND, *Relations de voyages arabes, persans et turks rel. à l'Extrême-Orient*, Vol. I, pp. 28, 41, 59, 99, 110, 145; and CARRA DE VAUX, *Maçoudi, Livre de l'avertissement*, pp. 90—92). Not one of these or any later Arabic writers mentions asbestos among the products of either Java or any other Malayan region.

[1] Viret aeterno hunc fontem igneum contegens fraxinus (ii, 107, § 240).

[2] Translation of HIRTH and ROCKHILL, p. 154.

tated by the fact that their inhabitants manufactured fabrics of bark-fibres. That this hypothesis was formulated in Fu-nan appears plausible to a high degree; for, aside from the inward probability of this supposition, there is no such account in classical antiquity, Western Asia, or India. Pliny neither correlates asbestos with volcanoes, nor does he speak of asbestos in his discourse on the latter.

The report of Kᶜang Tᶜai, duly adopted by his countrymen, was then crossed by the salamander story inflowing from the Roman Orient, and the imaginative Taoists at once set to work to reach a compromise between the salamander-asbestos and the volcanic tree-bark asbestos. If the vegetable kingdom in certain places could survive a volcanic fire, and if, as stated by Western traditions, the salamander could exist in fire, there was in all the world no reason why the hardy creature could not stand a *volcanic* fire as well. This was the act of Kuo Pᶜo, who ejected the trees and replaced them by the salamander, that now made its home in the blazes of Volcano Island in the Malay Archipelago (p. 335). To the author of the *Sou shên ki* [1] this compromise seemed too radical, and he arbitrated by restoring Kᶜang Tᶜai and bringing Kuo Pᶜo to honor. The vegetable as well as the animal kingdom, in his way of reasoning, can live in volcanic fires; and asbestos is either the product of the bark of these plants, or of the plumage of birds or the hair of beasts. Wang Mou of the Sung period accepted this verdict, and acquiesced in the belief that there is foundation for both these statements. [2]

DISCOVERY OF ASBESTOS ON CHINESE SOIL.—The Annals of the Later Han Dynasty, in the interesting chapter dealing with the

---

[1] Ch. 13, p. 3 (of the Wu-chᶜang print).

[2] WYLIE, *l. c.*, p. 147.

southern Man (*Nan Man*) and the barbarous tribes in the south-west
of China (*Si-nan I* 西南夷), have the following report: "Their
contributions of tribute-cloth, fire-down (*huo tsᶜui* 火毳), parrots,
and elephants, were all conveyed to the Treasury." [1] WYLIE [2]
refers this account to the tribe called Jan-mang 冉駹), [3] mentioned
in this chapter of the Annals a couple of pages before; but it
would seem that it relates in fact to the Pai-ma-ti 白馬氐, [4]
a tribe settled in Sze-chᶜuan Province (north-east of Mao chou). [5]

The term "fire-down," employed in the text of the Annals, is
explained by the commentary as being identical with the term
"fire-proof cloth" (*huo huan pu*); that is to say, it is understood
by the Chinese in the sense of asbestos. The word *tsᶜui* is very
ancient, and appears as early as the time of the *Shi king* [6] with the
significance of clothing woven from the down of birds or the fine
undergrowth of hair of mammals. [7] Such textiles woven from bird's
down are ascribed by the Chinese also to the aboriginal tribes
inhabiting southern China. E. H. PARKER [8] has extracted from the
Ling nan i wu chi the information that the chiefs of southern
China select the finest down of the geese and mix it with the

---

[1] *Hou Han shu*, Ch. 116, p. 11 b.

[2] *L. c.*, p. 150.

[3] He wrongly transcribes the first character *Tan* (compare HIRTH, *China and the Roman Orient*, p. 36). The tribal name *Mang* is doubtless identical with the Mang 莽 studied by G. DEVÉRIA (*Frontière sino-annamite*, p. 159); see also CHAVANNES, *T'oung Pao*, 1906, p. 689.

[4] *Ibid.*, p. 11 a.

[5] Compare the interesting study of J. H. PLATH, *Fremde barbarische Stämme im alten China*, p. 515 (*SB. bayer. Akad.*, 1874). The Pai-ma-ti seem to have extended from Sze-chᶜuan as far as into Kan-su (CHAVANNES, *T'oung Pao*, 1905, p. 528).

[6] LEGGE, *Chinese Classics*, Vol. IV, p. 121.

[7] It is only the soft down of wild birds and wild beasts. The translation "habillement fait en laine," given by BIOT (*Le Tcheou-li*, Vol. II, p. 6), is erroneous, as already pointed out by J. H. PLATH (*Nahrung, Kleidung und Wohnung der alten Chinesen*, p. 37); also COUVREUR has the wrong rendering, "vêtement de laine."

[8] *China Review*, Vol. XIX, p. 191.

threads of white cloth to make coverlets, the warmth and softness of which are not inferior to those of soft floss cushions. In other words, Mr. Parker adds, eider-down quilts were known in China very long ago. D. I. MACGOWAN, in his highly interesting essay *Chinese and Aztec Plumagery*, [1] makes this contribution to the subject: "A work styled 'New Conversations on things seen and heard at Canton,' was written by a native of Su-chou who spent many years in that city in a mercantile capacity in the latter part of the last century. In a short section devoted to bird clothes, he says, 'There are several kinds of birds, the feathers of which are woven into a peculiar cloth by the Southern Barbarians. Among them is the celestial goose velvet, [2] the foundation of the fabric being of silk, into which the feathers were ingeniously and skilfully interwoven, on a common loom, those of a crimson hue being the most expensive. Of these wild goose feathers, two kinds of cloth were made, one for winter, the other for summer wear. Rain could not moisten them; they were called 'rain satin,' and 'rain gauze,' respectively. Canton men imitated the manufacture, employing feathers of the common goose, blending them with cloth. This fabric, though inferior in quality, was much cheaper.'" The tribe Nung 儂 in Kuang-si made a special industry of fabricating a tissue of cotton and goose-down. [3] Kuang Lu 鄺露, who spent several years among the Miao tribes in the service of one of the female chiefs, [4]

---

[1] *American Journal of Science and Arch.*, 2d ser., Vol. XVIII, 1854, p. 59. This important study has been unduly forgotten by the present, and I apprehend also by the preceding, generation. Neither Bretschneider nor Hirth, in their references to *so-fu*, has ever appealed to it, and acquaintance with this treatise would doubtless have led them to better results.

[2] Apparently a literal translation of *t'ien ngo jung* 天鵝絨 ("silk-floss of the wild swan"). I find this term mentioned in the *T'ien kung k'ai wu* (Ch. 2, p. 46) as the name of a fur garment woven from down and feathers of hawks and wild geese.

[3] G. DEVÉRIA, *Frontière sino-annamite*, p. 112.

[4] WYLIE, *Notes on Chinese Literature*, p. 59.

and wrote an interesting account of them in his book *Chʻi yu* 赤雅,[1] mentions the bird-feather textiles under the name *niao chang* 鳥章 and discriminates between fine feather weavings styled *so-fu* 鎖袱[2] and coarse feather textiles termed "goose fishing-nets" (*ngo ki* 鵝罽).

This evidence permits us to infer that the term *huo tsʻui*, as applied to asbestos coming from the South-western Barbarians,[3] signifies "bird-down able to resist fire," and accordingly echoes a tradition current among these barbarians themselves. If nothing else, the peculiar choice of this term, which occurs in no other text, would amply support this opinion. The conclusion that the barbarians themselves worked this fibrous asbestos into a textile would of course not be forcible; at least, it is not imperative, and it is sufficient to assume that they had gotten hold of the raw material. When we further consider that parrots[4] and elephants named in the Annals are local products, the conclusion may be hazarded that also asbestos was found in the same region. This impression is confirmed by a statement of Yang Shên 楊慎 (1488—1559) to the effect that "fire-proof cloth is produced in Kien-chʻang 建昌 in Shu (Sze-chʻuan). This substance is as white as snow, and is obtained from crevices in the stones, being identical with what the Annals of the Yüan Dynasty term 'stone silk-floss' (*shi jung* 石絨)."[5] An asbestos-producing locality in

---

[1] The preface is dated 1635. The passage is in Ch. A, p. 5 b of the reprint, in *Chi pu tsu chai tsʻung shu.*

[2] The Arabic word *ṣūf* صوف (T. WATTERS, *Essays on the Chinese Language*, p. 355).

[3] The occurrence of the term in the Han Annals is an isolated instance.

[4] In the text "trained birds," interpreted as parrots. Parrots are first mentioned in *Tsʻien Han shu* (Ch. 6, p. 6) under the name "birds able to speak" (*nêng yen niao* 能言鳥). They are frequently referred to in the Annals as tribute gifts (for instance, *Kiu Tʻang shu*, Ch. 198, p. 9 b; *Tʻoung Pao*, 1904, p. 40).

[5] *Ko chi king yüan*, Ch. 27, p. 13 (compare WYLIE, *l. c.*, p. 153). Regarding the asbestos of the Yüan see below.

Sze-ch°uan is here clearly pointed out; and this agrees with the statement of F. P. Smith [1] that asbestos is met with in Mao chou, Sze-ch°uan; and, as the Pai-ma-ti were settled near this region, they were very well within reach of asbestos.

It is not surprising that these "barbarians" had come into possession of asbestos; for this mineral is found on the surface in numerous places of this globe, and there are instances on record that it has accidentally been discovered even by primitive tribes. In 1770 P. S. Pallas [2] reported that the Bashkir, a Turkish tribe in the region of Yekaterinburg, had discovered on a mountain a coarse kind of asbestos of yellowish-gray hue, being exposed to the air in large pieces split lengthwise, with brittle fibres which could be pulverized into a hard white wool. In the same area he visited also the Asbestos or Silken Mountain, [3] giving a circumstantial account of the occurrence and mining there of the mineral, and mentioning also that an old woman had possessed the knowledge of weaving it into incombustible linen and gloves and making it into paper. [4]

The most remarkable utilization of asbestos on the part of a primitive tribe is made by the Eskimo. D. Crantz [5] has the

---

[1] *Contributions toward the Mat. Med. of China*, p. 26.

[2] *Reise durch verschiedene Provinzen des russischen Reichs*, Vol. II, p. 134.

[3] In Russian *Sholkovaya Gora* (*ibid.*, p. 184).

[4] R. H. Jones (*Asbestos*, p. 37), not familiar with the interesting account of Pallas, represents the matter as though this site had been discovered only shortly before 1890, and even asserts that the Silken Mountain is said to be entirely composed of asbestos. It seems well out of the question that the Technical Society of Moscow, on whose report Jones falls back, could have made such an absurd statement, for Pallas had already said that the mountain consists principally of slate. His investigation is apt to refute also Jones's preposterous allegation that up to the present time little use has been made of asbestos in Russia and Siberia, "on account of the prevailing ignorance respecting its peculiar properties." As early as 1729 news was spread in Russia of an incombustible linen from Siberia. This referred to an asbestos-quarry discovered there about 1720 (P. J. von Strahlenberg, *Nord- und östliche Teil von Europa und Asia*, p. 311, Stockholm, 1730).

[5] *The History of Greenland*, Vol. I, p. 56, London, 1767.

25

following observation on the occurrence and utilization of asbestos in Greenland: "The amiantus and asbestos or stone-flax are found in plenty in many hills of this country. Even in the Weichstein are found some coarse, soft, ash-gray veins, with greenish, crystalline, transparent *radii* shooting across them. The proper asbestos or stone-flax looks like rotten wood, either of a white-gray, a green, or a red cast. It has in its grain long filaments or threads, and about every finger's length a sort of joint, and the broken end is hard and fine like a hone. But if it is pounded or rubbed, it develops itself to fine white flaxen threads. When this stone is beaten, mollified and washed several times in warm water from its limy part that cemented the threads into a stone, then dried upon a sieve, and afterwards combed with thick combs which the clothiers use, like wool or flax, you may spin yarn out of it and weave it like linen. It has this quality, that it will not burn, but the fire cleanses it instead of lye or suds. The ancients shrouded their dead, and burnt or buried them, in such incombustible linen. They still make purses or such kind of things of it for a curiosity in Tartary and the Pyrenean mountains. Paper might be made of this linen. The purified filaments may also be used as we use cotton in a lamp. But we must not imagine that the Greenlanders have so much invention: They use it dipped in train (for as long as the stone is oily, it burns without consuming) only instead of a match or chip, to light their lamps and keep them in order." In the *Encyclopædia Britannica* [1] it is stated that "by the Eskimo of Labrador asbestos has been used as a lamp-wick." I do not know from what source or authority this statement comes; but, in view of the data of Crantz, it does not sound very probable.

Marco Polo's account has shown us that in the time of the

---

[1] Vol. II, p. 714.

Mongols asbestos was dug, that its preparation and weaving were perfectly understood, and that asbestos products were utilized in China. From this time onward we no longer hear of imported "fire-proof cloth," while the accounts of native asbestos increase. As early as the Sung period an attempt had been made in the Imperial Atelier to spin and weave asbestine fibres imported by the Arabs into cloth, but not with brilliant success. [1]

A positive allusion to a locality where asbestos was found during the Mongol period is made in the biography of the treacherous Uigur minister Ahmed (A-ho-ma), [2] who, in a memorial to the Emperor Kubilai, stated that "Mount Pu-ko-ts'i 布格齊 produces asbestos, which is woven into cloth unconsumable by fire; an officer should be despatched to gather it." In the main section of the Annals [3] the date of this memorial is fixed in the year 1267, and it is added that the Emperor indorsed it and issued an order in compliance with the request. The term for "asbestos" used in this text is *shi jung* 石絨 (literally, "stone silk floss"). We have already seen that Yang Shên (1488—1559) pronounced this term identical with what is generally known as "fire-proof cloth," that is, asbestos; and this identification is certain beyond doubt. [4]

---

[1] *T'ie wei shan ts'ung t'an* (already quoted above, Ch. 5, p. 20 b).

[2] *Yüan shi*, Ch. 205, p. 2 a. He figures among the "Villainous Ministers." Marco Polo has told his story (ed. of YULE and CORDIER, Vol. I, p. 415).

[3] *Yüan shi*, Ch. 6, p. 12.

[4] Giles, Schlegel, and the English and Chinese Standard Dictionary, have adopted it in this sense. The term with the same meaning is used in Japan (GEERTS, *Produits*, p. 450). Also Chang Ning 張寧 of the Ming, author of the *Fang chou tsa yen* 方洲雜言, combines the "stone silk floss" of the Mongols with the ancient tributes of fire-proof cloth (*Pien tse lei pien*, Ch. 21, p. 6; WYLIE, *l. c.*, p. 153). An analogous expression occurs in the form *shi ma* 石麻 ("stone hemp") in the *Tung ming ki* (*P'ei wên yün fu*, Ch. 21, p. 4 b). This text would possess a veritable value if any dependence could be placed on this spurious work (see CHAVANNES and PELLIOT, *Traité manichéen*, p. 145), which may reach back to the middle of the sixth century. The passage in question, however, cannot be exactly dated, nor can the mysterious country Pu-tung be identified

In regard to the location of Monnt Pu-ko-ts⁣i, Wylie, who has already called attention to this passage, [1] observed that it is difficult to identify it; but, "as asbestos is said to be found in Tartary, it is not unreasonable to suppose a coincidence in this also." G. Schlegel [2] writes the name of the mountain 別怯赤山, [3] translating this by "red mountains of Pie-kieh," which he places in Sze-ch⁣uan at 27° 12′ latitude and 102° 53′ longitude. [4]

A. Williamson [5] seems to be the first European author to record the occurrence of the mineral in Shan-tung. Under the title "asbestos" he has the following: "This strange fossil mineral is found at King-kwo-shan, and also at Law-sze-shan. The natives use it for making fire-stoves, crucibles, and other fire-proof purposes. The fibre is good and very feathery, and by the admixture of cotton or hemp could be woven into articles of clothing. Such articles being exposed to fire and having all the alloy consumed, would

---

(it appears only in this passage, as shown by *Pien i tien*, Ch. 42, where Pu-tung is ranked among the unidentified countries of the East, solely with reference to this text). The allusion to asbestos is obvious. The text runs thus: "In the lake Ying-ngo 影娥 池, there are ships fastened by means of 'stone veins' (*shi mo* 石脈) worked into ropes. These 'stone veins' come from the country Pu-tung 晡東, and are as fine as silk floss. They are extracted from the stone, and reeled like hempen cordage. The material is styled 'mineral hemp,' and is also made into cloth." The passage, at any rate, demonstrates that the mineral character of asbestos was known to the Chinese prior to the age of the Yüan, and possibly during the sixth century. The following text from the Persian geography of Ahmed Rāzī of the sixteenth century and relating to Egypt might eventually be enlisted for the explanation of the Chinese story. It is thus translated by C. Huart (*Publ. de l'Ecole des Langues Orientales*, 5th ser., Vol. V, 1905, p. 121): "Dans certaines localités croît une herbe dont on fait les cordages des gros navires; elle donne une lumière à la façon d'une chandelle; quand elle s'éteint, on la fait tourner plusieurs fois et elle redevient lumineuse."

[1] *L. c*, p. 152.

[2] *Nederlandsch-chineesch Woordenboek*, Vol. III, p. 1066.

[3] This is the reading of the *Fang chou tsa yen*.

[4] It would be interesting to settle this question. Thus far, I have failed to find any indications in the *Yüan shi* regarding the site of this mountain.

[5] *Notes on the Productions of Shan-tung* (*Journal China Branch R. As. Soc.*, Vol IV, 1868, p. 70).

afterwards form fire-proof garments, such as ancient history speaks of, and such as are used in legerdemain. But the mineral would make most excellent fire-brick, which would be cheaper and more durable than any others. This is worthy of the consideration of the masters of the steamers on the coast." Unfortunately Williamson did not supply the technical name by which the substance is known to the Chinese. This defect was made good by F. P. SMITH,[1] who furnished the name *pu huei mu* 不灰木 (literally, "wood without ashes;" incombustible wood), and pointed out three localities where it is obtained,—Lu-ngan fu in Shan-si, district of Yü-t'ien in Tsun-hua chou in Chi-li, and Mao chou in Sze-ch'uan. The occurrence in Shan-tung was confirmed by A. FAUVEL,[2] who stated that "asbestos is common in Shan-tung; pounded and mixed with soapstone it is made into crucibles, and very pretty white Chinese furnaces; they are as light as cardboard, and stand any heat; these articles are extensively made in the capital of the provinces." In this account I have full confidence, because Fauvel was a good naturalist and observer, and because I saw and collected such stoves myself. These specimens, six in number,[3] were obtained at Peking in 1903; and from the description given me by Chinese, there could be no doubt that they were really made of asbestos. This impression is corroborated by Professor L. P. Gratacap, Curator of the Department of Mineralogy in the American Museum of Natural History of New York, who states that these stoves "consist of a very finely triturated asbestos, with which (purposely or adventitiously I cannot say) there is an admixture of particles of

---

[1] *Contributions toward the Mat. Med. of China,* p. 26.

[2] *China Review,* Vol. III, 1875, p. 376.

[3] In the American Museum, New York (Cat. Nos. 12427, 12652—12656). A specimen is figured in the *Catalogue of the Chinese Collection for the International Health Exhibition, London, 1884,* p. 82, and is defined there as "lime stove."

limestone; there is evidently also a smearing of clay, which to a slight extent pervades also the asbestiferous mass." As this substance is designated by the Chinese in Peking *pu huei mu*, it is conclusively proved that at present this term relates to a variety of asbestos, though this does not imply that it might not refer also to other lime-like minerals which in our opinion do not come under that category. These asbestos stoves, white in color, enclosed in frames of wood or brass and heated with coal-briquettes, are much utilized in Peking and manufactured about 80 *li* in the hills toward the west of the metropolis. I could not learn the name of the village or locality. [1]

GEERTS [2] pointed out that *pu huei mu* denotes in Japan incrustations of carbonate of lime, which settle around branches of trees immersed in a current of mineral water. This may be; in China this term refers also to petrified wood.

In reading the notes of Li Shi-chên [3] on the subject of *pu huei mu*, we are struck by the fact that he does not make any allusion

---

[1] The *Port Catalogues of the Chinese Customs' Collection at the Austro-Hungarian Universal Exhibition, Vienna, 1873* (p. 56) contain the following entry in the Chefu collection (repeated also in later Exhibition Catalogues of the Customs): "Asbestos, *lung-ku-ni* 龍骨泥; place of production, Shan-tung; used for making fire-stoves, crucibles, etc.; the fibre woven with cotton or hemp is made into fire-proof materials." This information is spurious, and based on a misunderstanding of Williamson, who said that the fibre is good and very feathery, and by the admixture of cotton or hemp *could* be woven into articles of clothing; in fact, of course, it is not so woven by the Chinese, nor is it woven by them at all; at least, there is not the slightest evidence of this. Moreover, the term *lung-ku-ni* has nothing to do with asbestos, but denotes a medical preparation made from powdered dragon-bones, that is, bones of fossil animals.—How badly China is treated by our mineralogists, and even in otherwise complete monographs, is illustrated by the book of R. H. JONES on Asbestos. All that is said there in regard to China amounts to the one sentence (p. 39), "In China also asbestos occurs; but, apart from the manufacture of a coarse kind of cloth, we know little of any purpose to which it is there applied." I have never seen or heard of any asbestos-cloth now manufactured in China.

[2] *Produits*, p. 450 (see also p. 344).

[3] *Pén ts'ao kang mu*, Ch. 9, p. 14 b. The translation given by F. DE MÉLY (*Lapidaires chinois*, p. 85) is an incomplete abstract from the *Pén ts'ao*.

to the "fire-proof cloth;" he does not tell us that it is identical with what anciently was called *huo huan pu*. In fact, the traditions regarding the two products are entirely distinct. Certainly *pu huei mu* refers to the mineral, and *huo huan pu* to the finished textile product.

There is another term, *yang kʻi shi* 陽起石, which likewise refers to a variety of asbestos. It is difficult to see why SMITH[1] and GEERTS[2] were so much exercised about this identification, the one saying that "this variety of hornblende, or greenstone, is scarcely to be called an asbestos, as it is by some writers;" the other even going so far as to impeach some foreign authors on a charge of confusion. Both Smith and Geerts were insufficiently informed on the subject; for what they describe is certainly styled by us "asbestos," whether the Chinese specimens commercially be of good or bad quality. D. HANBURY[3] identified *yang kʻi shi* with "asbestos tremolite,[4] silicate of lime and magnesia;" and this is what we still include under "asbestos." It appears that this stone is used only medicinally.[5] The *English and Chinese Standard Dictionary*[6] lists both *pu huei mu* and *yang kʻi shi* under "asbestos."[7]

---

[1] *L. c.*, p. 27.

[2] *L. c.*, p. 448.

[3] *Notes on Chin. Mat. Med.*, p. 111 (*Pharmaceutical Journal*, 1861); or in his *Science Papers*, p. 218.

[4] This word is derived from Tremola, Mount St. Gotthard, where this variety was first found.

[5] F. DE MÉLY, *Lapidaires chinois*, p. 105; BIOT in Bazin, *Chine moderne*, p. 556.

[6] Vol. I, p. 112.

[7] It should be pointed out, however, that this meaning of *yang kʻi shi* is of comparatively recent origin, the exact date of which remains to be ascertained. In the older texts cited by Li Shi-chên on the subject, nothing can be found to remind us of asbestos; and the early sources are so brief and obscure that they hardly allow of any positive conclusions. Thus the *Pie lu* merely refers to Shan-tung as the place of provenience by saying that *yang kʻi shi* occurs in the hills and valleys of Mount Tsʻi and in Lang-ye, adding that it is the root of mica (*yün mu*, "cloud mother") in the Cloud Mountains (*Yün shan*). Tʻao Hung-king states that this mineral, which is dug together with mica,

Marco Polo proved that he was possessed of a scientific mind when he exploded the salamander legend at the very moment that his Turkish acquaintance told him of how asbestos was dug and spun. The same case might be applied as a test for the scientific ability of the Chinese. True it is, the scholars of the Ming period clearly recognized the identity of the asbestos discovered under the Yüan with the imported fire-proof cloth of old. In vain, however, do we look in the literature of the Chinese for an awakening on their part, and a critical attitude toward the ancient legends, when the mining and working of the material within their boundaries has offered the opportunity ever since the days of the Mongols. The minds of Chinese scholars, at least those of the last centuries, were not trained to observation, and still less to logical conclusions based thereon, especially when these were apt radically to antagonize venerable traditions. The discovery of asbestos in China did not lead to studies by her scholars and to an overthrow of popular errors. On the contrary, the old book-knowledge persisted and triumphed. Wylie quotes the following from Chou Liang-kung 周亮工, an author who lived under the Manchu dynasty and had occasion to see a strip of asbestos cloth: "The ancients said that it was woven from the bark of a tree that grew on a burning mountain; while some say that it is from the hair of a rodent. The statement that it is from the bark of a tree, is the most

---

is very similar to mica, only of greater density; and that *yang k'i shi*, dug in Yi-chou together with alum (*fan shi*), is a bit yellow and black in color, but that it is only the root of alum or mica, and that the true state of affairs is not yet assured. T'ao Hung-king, accordingly, was not positive about the true nature of the substance; it may originally have been a variety of mica or alum. At any rate, it has no practical importance for the historian of asbestos, as the Chinese never made any use of it in the manner of asbestos, but only took it internally as a medicine. It should be remembered that Apollonius has allusions to mica in his account of asbestos (p. 304), and that Dioscorides and Pliny liken asbestos to alum (pp. 303, 308).

probable, as its color is more like hempen than woollen fabrics."
To the credit of the Chinese, however, it must be said that Tsᶜai
Tᶜiao 蔡條 of the Sung period plainly rejected the legend of
the animal origin of asbestos, though he failed to grasp the real
nature of the substance. It will be remembered that this author,
in his work *Tᶜie wei shan tsᶜung tᶜan*, reports the importation on
the part of the Arabs of asbestine cloth and asbestos raw material,
and that the latter was woven into textiles in the Imperial Atelier
of the house of Sung. These facts impressed the Sung scholars
and set them to thinking. Tsᶜai Tᶜiao makes the positive statement
that asbestos is not the hair of a rodent (非鼠毛也), and
that the Chinese manufactures of his time testify to the fact that
the old stories are wrong.

ADDENDA.—In the letter purported to have been addressed by Prester John
to the Byzantine Emperor Manuel, and written about 1165, we read the
following about the salamander yielding the material for asbestine garments
(F. ZARNCKE, *Der Priester Johannes* I, p. 89): "In alia quadam provincia [of
India, the territory of the alleged Royal Presbyter] iuxta torridam zonam sunt
vermes, qui lingua nostra dicuntur salamandrae. Isti vermes non possunt vivere
nisi in igne, et faciunt pelliculam quandam circa se, sicut alii vermes, qui
faciunt sericum. Haec pellicula a dominabus palatii nostri studiose operatur,
et inde habemus vestes et pannos ad omnem usum excellentiae nostrae. Isti
panni non nisi in igne fortiter accenso lavantur." In this description the
salamander is associated with the silkworm working itself an envelope that is
reeled off and spun like silk, the material being incombustible and washed in
fire. In view of the popularity of the stories about Prester John in the
thirteenth century, the "salamander-silk," so frequently mentioned in the texts
of that period, may well be traceable to the passage in question. In one of
the mediæval manuscripts edited by Zarncke (pp. 167, 170), twelve men appear
before King Manuel as ambassadors of the Presbyter, and impress him by
cleaning their robes of salamander-silk in flaming fire. The Presbyter's letter
is instructive for another reason; for it shows, as pointed out on p. 325, that
the identity of the salamander's product with asbestos was not recognized in
the early middle ages. The bread, it is told there, is baked in a vessel made
from asbestos; the pavement is of green topaz, which by nature is cold, to
moderate the heat of asbestos (A pistoribus panis efficitur et in clibano facto

26

ex asbesto ponitur et coquitur. Pavimentum clibani est de topazio viridi, qui naturaliter est frigidus, ut caliditas asbesti temperetur. Alioquin panis non coqueretur sed conbureretur. Tantus est calor asbesti). The walls of a furnace in the bakery (pistrinum) were likewise of asbestos (Est enim furnus factus exterius de lapidibus preciosis et auro, interius caelum et parietes sunt de albesto lapide, cuius natura talis est, quod, semel calefactus sit, deinde inremissibiliter sine igne semper erit calidus). These passages concerning asbestos are wanting in the original text of the letter, and are interpolations occurring in manuscripts of the thirteenth century.

Falstaff, after many uncomplimentary remarks on Bardolph's personal appearance, exclaims, "I have maintained that salamander of yours with fire any time this two and thirty years; God reward me for it!" (SHAKESPEARE, 1 *Henry IV*, III 3, 52). A lizard in the midst of flames was adopted by Francis I as his badge, with the legend, *Nutrisco et extinguo*, "I nourish and extinguish" (E. PHIPSON, *Animal Lore of Shakespeare's Time*, p. 320).

P. 339, note 1. The French translation of the text in question by d'Hervey-St.-Denys has been rendered into English by S. W. WILLIAMS in his article *Notices of Fu-sang* (*J. A. O. S.*, Vol. XI, 1882, p. 98). It appears from this translation as though in the opinion of Duke Kie Volcano Island were situated in the land of the Amazons, about ten thousand *li* north-west of Fu-sang; nor is the cloth from the bark of the fiery tree mentioned in it. In the translation of Williams it runs thus: "In the middle of the kingdom is an island of fire with a burning mountain, whose inhabitants eat hairy snakes to preserve themselves from the heat; rats live on the mountain, from whose fur an incombustible tissue is woven, which is cleaned by putting it into the fire instead of washing it." In fact, the text, as reprinted in *T͟ʻu shu tsi chʻêng*, is worded as follows: "Southward [from the country of Women or Amazons], arriving at the southern shore of Volcano Island, the inhabitants on Mount Yen-kun there subsist on .... crabs and bearded snakes in order to ward off the poisonous vapors of the volcanic heat. In this island there are fiery trees, the bark of which can be wrought into cloth. In the blazing mound live fiery rodents, whose hair can be made into stuffs. These are incombustible, and when soiled, are cleaned by means of fire" (南 至 火 洲 之 南 炎 崐 山 之 上 其 土 人 食 蝑 蟹 髥 蛇 以 辟 熱 毒 洲 中 有 火 木 其 皮 可 以 爲 布 炎 丘 有 火 鼠 其 毛 可 以 爲 褐 皆 焚 之 不 灼 汚 以 火 浣). Yen-kun is an artificially coined term, which does not appear in other texts; it is apparently intended for "blazing (*yen*) Kun-lun." The exact meaning of *sü* 蝑 is not known to me; according to Kʻang-hi it is identical with 蚣 蝑. The interesting feature of the above text is that the asbestos and salamander story is linked together with fabulous accounts of Fu-sang and the Amazons, and it will be remembered that the report of a specular lens coming from Fu-sang is embodied in the same text (this volume, p. 198). If I expressed the view that this lens appears to have been of Western origin, and that Chang Yüe was familiar with traditions relating to Fu-nan, India, and Fu-lin (p. 204), this opinion is confirmed by the present case in which Chang Yüe adapts to his purpose the Fu-nan version of asbestos in combination with the salamander story.

P. 351. The country Se-tiao appears in another text of the *I wu chi*, cited in the *Chêng lei pên ts͏ᶜao* (Ch. 23, fol. 49). There, a plant is briefly described under the name *mo-ch͏ᶜu* 摩 廚 (according to G. A. STUART, *Chinese Materia Medica*, p. 499, unidentified), which grows in Se-tiao; the latter, it is added, is the name of a country. If it could be proved that *mo-ch͏ᶜu* is the transcription of a Javanese name (and this is probable), the case would make an interesting contribution to the identification of Se-tiao with Ye-tiao.

120

藏文姓名的汉文转写

MÉLANGES.

CHINESE TRANSCRIPTIONS OF TIBETAN NAMES.

I have read with keen interest M. Pelliot's study *Quelques transcriptions chinoises de noms tibétains* (this volume, pp. 1—26), which is as instructive and illuminating as his recent, very important contribution *Les noms propres dans les traductions chinoises du Milindapañha* (J. A., 1914, Sept.-Oct., pp. 379—419). M. Pelliot is an excellent phonetician, and commands an admirable knowledge of ancient Chinese phonology, such as is possessed by no other contemporary. It is only to be hoped that he will publish some day, for the benefit of all of us, an *œuvre d'ensemble* on this complex subject, which is still so much obscured. M. Pelliot's criticism is most assuredly welcome, always founded, as it is, on serious and solid information, and inspired by no other motive than the ideal desire to serve the common cause. It is a privilege and a stimulus to co-operate with such a sympathetic and highly intelligent worker, for whom I have an unbounded admiration, and to be guided by his friendly advice and effectual support. Indeed, without committing an indiscretion, I may say that in the present case this criticism was voluntarily solicited on my part, as I have never flattered myself for a moment that all difficulties presented by the Sino-Tibetan transcriptions have been solved by me; on the contrary, I am wide awake to the fact that my feeble attempt in this direction was merely a tentative beginning, which should be continued and improved by an abler hand. I am very happy that M. Pelliot has taken up this problem with such minute care and unquestionable success, and I need hardly assure him of my keen sense of obligation for his untiring efforts and the inspiring instruction which I have derived from his comments.

What M. Pelliot observes under 1—5 on the transcription of the Tibetan prefixes visible in the Chinese final consonants meets with my heartiest approval: indeed, this is the logical amplification of what I myself had noted on the transcription of *t'am-t'uñ* (p. 86, T'oung Pao, 1914).

Pelliot No. 5: The Tibetan reading *K'od-ne brtsan* is correct. No. 6: The Tibetan reading *goñ* is justified, and plainly appears as such in Bushell's plate.

In my first draught of the monument, made in Tibetan letters, it is indeed written *gcǹ*; I do not know now how it happened that it was printed *kuǹ*. No. 7: The last Tibetan word is so indistinct in Bushell's plate that the matter can hardly be decided merely on this basis; but I admit that *čab* or even *tsab* could be read into it, and accordingly that M. Pelliot's conjecture is justifiable. No. 8: I gladly adopt the Tibetan reading *bla ɑbal blon-kru-bzɑǹ myes-rma*. No. 10: The last Tibetan word may well be *ken*, not *yen*. I am unable to recognize *b* after *ha* in Bushell's plate, but it may be traceable in the original stone or in a rubbing. No. 14: The stone is here in such a hopeless condition that certainty of reading is out of the question; what appears quite certain to me is the letter-combination *rgy*. I regret having had the misfortune of over-looking Col. Waddell's study utilized by M. Pelliot: at the time when I wrote, the volume of the *J. R. A. S.*, in which it is contained, was in the hands of my book-binder, and in this way the accident occurred.

My note on Čog-ro, which M. Pelliot (p. 7) does not well comprehend, seems to me quite plain. Indeed, I do not speak of Čog-ro as the name of a man, as insinuated by M. Pelliot, but simply as a name. I never had any other opinion than that Čog-ro is the designation of a locality, which is adopted by the men hailing from there, and is prefixed to their personal names. The "inadvertance" noted by M. Pelliot (p. 9) in regard to my writing *mǹan-pon* and *mǹa-dpon* is only seeming: *mǹan-pon* is the reading of the Tibetan text in the inscription; and *mǹa-dpon*, as explained on p. 76, is the restoration proposed by me. In accordance with the purpose of the passage on p. 86, there was only occasion to cite the latter.

I am not convinced that M. Pelliot's restoration of 鉢掣逋 to *dpal čᶜen-po* (No. 12) is to be preferred to my proposition *ɑba čᶜe-po* (p. 28, *T'oung Pao*, 1914). The character *po* 鉢 was certainly read with a final consonant (*pat*); but there are numerous examples in the transcriptions of Sanskrit where it merely corresponds to *pa* or *ba*, as in *parama*, *utpala*, *pippala*, *pra-* 鉢喇, *udumbara* 優曇鉢羅.[1] In view of Chinese *čᶜòt* 掣 (M. Pelliot wrongly writes 折, not given in the relevant passage of *Sin T'ang shu*, which has 掣), it is not impossible that in ancient Tibetan the word-formation *ᶜčᶜed-po*, as an equivalent of *čᶜe-po*, existed[2] (for analogous cases of this kind see at the end of this notice). The supposition of a pronunciation *čᶜer-po*, proposed by M. Pelliot, is impossible: *čᶜer* (*čᶜe-r*) is a terminative, and cannot be connected with any suffix like *po* or *pa*.[3]

---

[1] See also Baron A. VON STAËL-HOLSTEIN, *Kiea-Ch'ui-Fan-Tsan*, p. 177, No. 151 (*Bibl. Buddhica*, Vol. xv).

[2] My restoration was *čᶜe-po*, not, as M. Pelliot makes me say, *čᶜen-po*.

[3] Compare examples in Jäschke's *Tibetan Dictionary*, p. 161 a, and in the *Dict. tibétain-français*, p. 329 b.

With reference to *tu* 度, M. Pelliot states that it is attested in transcriptions only as *ʼdu*, not as *ʼdag*. The former is doubtless the rule, but instances of *ʼdag* nevertheless occur. JULIEN[1] says that in the *Fa yüan chu lin* this character is used in rendering Sanskrit *dakshiṇa*, and Baron A. VON STAËL-HOLSTEIN[2] quotes an example where it has the value *da*.

In some cases it had seemed to me advisable, even at the sacrifice of rigid adherence to the Chinese transcriptions, to fall back on realities alive in the Tibetan language or in Tibetan records, rather than to resort to conjectural forms for which there is as yet no evidence. M. Pelliot is certainly right in maintaining that the transcription *fu-lu* 拂廬, in theory, would presuppose a Tibetan form *ʼpʻru*. I myself had noted on my index-card that it should lead to a dialectic form *ʼsbru*,[3] but did not express this opinion, because such a word is not known at present. All we can say now is that *fu-lu* represents a word of the general or normal type *sbra*, whatever the possible dialectic variations may be. For the aforementioned reason I adopted the reading *pʻo* in the name Sroṅ-lde-btsan, because *pʻo* ("the male") is a title actually found in connection with royal names. In adopting the reading Sa-sroṅ lde-btsan, proposed by M. Pelliot (No. 23), we face the difficulty that we cannot correlate this with Tibetan historical tradition. Again, if we try to make sense of this, we shall have to change *sroṅ* into *sruṅ* (*sa-sruṅ*, rendering of Sanskrit *bhūmipāla*); but it is not known to me that the Tibetan kings ever assumed such a title.

M. Pelliot's observations on the name Tʻu-fan are very ingenious, and will no doubt contribute toward a definite solution of this problem in the near future. In regard to his etymology of the word *la-pa*, proof seems to be required that *la-pa* is really evolved from the Chinese loan-word in Uigur-Mongol, *labai*, and that the supposed change of meaning really took place. At the outset, this theory is not very probable. A conch-trumpet (*labai*) and a copper or brass bass-tuba (*la-pa*) are entirely distinct and co-existing types of musical instruments, each of which has had its individual history. We know that the conch-trumpet came from India as a sequel of Buddhism. As to *la-pa*, J. A. VAN AALST (*Chinese Music*, p. 59) has aptly compared it with the *chatzozerah* of the Hebrews and the *tuba* of the Romans. Certainly there is no direct interrelation, but transmission through the medium of Persia and Turkistan seems to me a possibility deserving of consideration. First of all, it would be necessary, of course, to trace the history of the word and the object from Chinese records.

As regards the tones, I have to a certain degree modified my former views, since some time ago I had the opportunity of studying the admirable treatise

---

[1] *Méthode*, No. 2106.

[2] *L. c.*, p. 184, No. 222.

[3] The interchange of *sb* and *pʻ* (as well as of *sk* and *kʻ*, *st* and *tʻ*) is well known.

of H. Maspero, *Etudes sur la phonétique historique de la langue annamite*, in which a lucid exposition of the tone system is embodied. MM. Maspero and Pelliot's opinions on the historical development of the tones are very sensible, but in the present state of our knowledge it would be premature to decide positively in favor of the one or the other theory: a great amount of research will be required before we can formulate well-assured deductions. [1]

It is interesting to learn that in the *Mantra mudropadeça* to be published by M. Hackin the inverted *i* serves for the expression of long *ī* in Sanskrit words. This, however, would not signify at the outset that the same graphic expedient should denote *ī* in indigenous Tibetan words: what holds good for the writing of Sanskrit need not be applicable to Tibetan. I had occasion to hear six different Tibetan dialects, and am unable to hear an *ī* in any of these. The case alluded to by M. Pelliot remains to be seen.

As to *mo-muñ* 末蒙, I can now offer a better equivalent for the first element of this compound. The explanation of the word *bud-med*, which I hazarded on p. 97, note (*T'oung Pao*, 1914), is erroneous. The second element, *med*, has nothing to do with the verb *med* ("not to have"), but is indeed an independent base with the significance "female, woman" This is evidenced by the following facts. In the peculiar Bunan language we have a word *tse-med* ("daughter, girl"), the element *tse* being apparently connected with *tsi-tsi* ("child"), occurring in the same idiom. [2] The stem *tsi*, *tse*, is encountered in Lisu *tsa-me*, *tsa-mei*, *tsa-mu*. and *tsa* ("woman") [3] and A-hi Lo-lo *ma-ča-mo*. [4] As *mu*, *mo*, *me* ("female"), is joined to this stem, the element *med* in Bu-nan *tse-med* is likely to have the same meaning. In Lepcha we have two stems, *mót* and *mit* ("female"), used with or without the prefix *a*, and a word *mo* parallel with *mot*. In his *Lepcha Dictionary*, which is based on materials collected by Gen. G. B. Mainwaring, A. GRUNWEDEL (p. 289) tentatively suggested that Lepcha *mit* be regarded as related to *-med* in Tibetan *bud-med*. A differentiation of meaning has been evolved in Lepcha in this manner: that *mit* or *a-mit* particularly refers to women of superior beings (for example, *rum-mit*, "goddess"); and *mo*, *mót* or *a-mo*, *a-mót*, to the female of animals (for instance, *hik mót*, "hen"), but sometimes also to human beings. [5] The

---

[1] On p. 25 M. Pelliot speaks of an hypothesis of mine regarding the function of certain Tibetan prefixes. This is not my hypothesis, but is merely the reproduction of observations and opinions given by Tibetan grammarians.

[2] JASCHKE, *J. A. S. B.*, Vol. 34, pt. 1, 1865, p. 95.

[3] A. ROSE and J. C. BROWN, *Lisu Tribes of the Burma-China Frontier* (*Mem. As Soc. Beng*, Vol III, p 275).

[4] A. LIÉTARD, *T'oung Pao*, 1912. p 19; and *Bull de l'Ecole française*, Vol. IX, 1909, p 552. Lo-lo-p'o *zo-me* and P'u-p'a *za-ma* appear to be associated with Mo-so *ze-mu kua* and Tibetan *b-za*.

[5] G. B. MAINWARING, *A Grammar of the Róng (Lepcha) Language*, pp. 24, 25.

forms *mi-t* and *mó-t* represent derivations from the bases *mi* and *mó* by means of the formative suffix -*t*. [1] The same relation exists in Tibetan between *mo* and the element -*med* ('*möt*) contained in *bud-med*. The same word *med* or *mot* may be recognized in the Chinese transcription *mot* 末. [2] As to the second element of the Sino-Tibetan compound, *moṅ*, *muṅ*, reference may be made to Miao-tse *maṅ* ("spouse"), [3] and to a word for "woman" in Kanaurī, that is given by Pandit Joshi as *mun-riṅ*, and by Bailey as *mōn-riṅz*. From the standpoint of Tibetan, *moṅ* may very well be *mo-ṅ*; that is, a derivative from the base *mo* ("woman") by means of the suffix -*ṅ*, so that we should obtain two derivatives from the same base,—*mo-t* and *mo-ṅ*. Analogous cases in Tibetan are: *rtsa-ba* ("root"), forming *rtsa-d* and *rtsa-ṅ*; *dro* ("warm"), forming *dro-d* ("heat") and *dro-n*; *lči* ("heavy"), forming *lči-d* ("weight") and *lči-n*; *nu-ma* ("breast"), forming *nu-d-pa* and *s-nu-n-pa* ("to suckle"); *rga-ba* ("aged"), forming *rga-d-pa* and *rga-n-pa*. The word *mot-moṅ* (in Tibetan presumably written *med-moṅ*) preserved in the Tᶜang Annals, accordingly, is a compound consisting of two synonymes, each meaning "woman."

B. LAUFER.

---

[1] For examples of such formations in Tibetan see SCHIEFNER, *Mélanges asiatiques*, Vol I, p 346.

[2] In Kanaurī or Kanāwarī, *med-po* is said to mean "master, owner, proprietor;" and *med-mo*, "mistress," both words being borrowed from Tibetan (T. R. JOSHI, *Grammar and Dictionary of Kanāwari*, p. 103, Calcutta, 1909). In the Kanaurī-English vocabulary published by T. G. BAILEY (*J. R. A. S*, 1911, pp. 315—364), these words are not given. Our Tibetan dictionaries have not recorded the two words with those meanings; but we know that in Tibetan, *med-po* means "a man who owns nothing, a pauper," and *med-mo*, "a penniless woman." The data given by Pandit Joshi, on the contrary, would presuppose a base *med* with the meaning "to own," which it is difficult to credit. There may be a misunderstanding on the part of the Pandit, or the two words may be peculiar to Kanaurī without bearing any relation to Tibetan.

[3] P. VIAL, *Les Lolos*, p. 36.

121

钻石——在中国和希腊民间传说中的研究
附：书评一则

Field Museum of Natural History

Publication 184

Anthropological Series

Vol. XV, No. 1

THE DIAMOND

A STUDY IN CHINESE AND HELLENISTIC FOLK-LORE

BY

Berthold Laufer

Curator of Anthropology

Chicago

1915

CONTENTS

THE DIAMOND

A Study in Chinese and Hellenistic Folk-Lore

INTRODUCTORY.— Of all the wonders and treasures of the Hellenistic-Roman Orient, it was the large variety of beautiful precious stones that created the most profound and lasting impression on the minds of the Chinese. During the time of their early antiquity the number of gems known to them was exceedingly limited, and mainly restricted to certain untransparent, colored stones fit for carving; while the transparent jewel with its qualities of lustre, cut, polished, and set ready for wearing, was a matter wholly unknown to them. Only contact with Hellenistic civilization and with India opened their eyes to this new world, and together with the new commodities a stream of Occidental folk-lore poured into the valleys of China. That a chapter from a series of discussions devoted to Chinese-Hellenistic relations[1] is taken up by a detailed study of the history of the diamond, is chiefly because this very subject affords a most instructive example of the diffusion of classical ideas to the Farthest East. The mind of the Chinese offered a complete blank in this respect, being unacquainted with the diamond, and was therefore easily susceptible to the reception of foreign notions along this line.[2] India was the distributing-centre of diamonds to western Asia, Hellas and Rome, on the one hand, and to south-eastern

---

[1] Two other contributions along this line have thus far been published: The Story of the Pinna and the Syrian Lamb (*Journal of American Folk-Lore*, Vol. XXVIII, 1915, pp. 103–128) and Asbestos and Salamander (*T'oung Pao*, 1915, pp. 297–371).

[2] GEERTS (Les produits de la nature japonaise et chinoise, p. 201) stated in 1878 that the diamond had not yet been found in China or Japan. Diamonds have been discovered in Shan-tung Province only during recent years (compare A. A. FAUVEL, Les diamants chinois, *Comptes-rendus Soc. de l'industrie minière*, 1899, pp. 271–281; Chinese Diamonds, *Mines and Minerals*, Vol. XXIII, 1902–03, p. 552). The late F. H. CHALFANT (in the work Shantung, the Sacred Province of China, ed. by FORSYTH, p. 346) gives this account: "Fifty-five *li* south-east of I-chou-fu lie the diamond fields. The stones are found on the low watershed between two streams, distributed through a very shallow soil over a reddish sandstone conglomerate. A determined effort was made by the same German company that operated the gold mine near I-chou, to develop the diamond field, but the enterprise was not a commercial success. It is the opinion of the German experts that the stones were deposited in their present position by the action of water at the time when, according to the theory, there was a connection between the two rivers. It is supposed that the source of the supply is somewhere in the mountains of Mêng-yin. Meanwhile, diamonds, some of them of very good quality, are constantly picked up at the locality described and occasionally at other points." The mines were abandoned by the

5

Asia and China on the other hand.   Nevertheless the ideas conceived by the Chinese regarding the diamond do not coincide with those entertained in India, but harmonize with those which we find expounded in classical literature.   This fact is due to the direct importation of diamonds from the Hellenistic Orient to China; but it has been entirely unknown heretofore, and this is another reason which will justify this investigation now made for the first time.   Its significance lies not only in the field of Chinese research, but in that of classical archæology as well.   The copious and reliable accounts of Chinese authors advance our knowledge of the subject to a considerable degree beyond the point where the classical writers leave us, and elucidate several problems as yet unsettled.   It will be seen on the pages to follow that the use of the diamond-point in the ancient world, doubted or disowned by many scholars, now becomes a securely-established fact, and also that the acquaintance of the ancients with the true diamond rises from the sphere of sceptical speculation into a certain and permanent fact. Likewise the much-ventilated question as to whether the ancients employed diamond-dust, and cut and polished the diamond, will be presented in a new light.

LEGEND OF THE DIAMOND VALLEY.— The *Liang se kung ki*,[1] one of the most curious books of Chinese literature, contains the following account: "In the period T'ien-kien (502–520) of the Liang dynasty,

---

Germans in 1907, as the diamonds proved to be of little value for gems, while answering well for industrial purposes (*Engineering and Mining Journal*, Vol. LXXXIV, 1907, p. 1159).   An anonymous writer in *Mines and Minerals* (Vol. XXIII, 1903, p. 552) reports as follows on Chinese diamond-digging: "The Chinese procure the diamonds by the following method: After the summer rains which, according to them, produce diamonds on the surface of the soil, whence the uselessness of digging to find them, they walk back and forth over the sand of the torrents.   The fragments of diamonds, on account of their sharp points and edges, penetrate the rye straw of their sabots to the exclusion of other gravel.   When they think there is a sufficient quantity they make a pile of the sabots and burn them.   The ashes are afterwards passed through a sieve to separate the diamonds.   Those which we saw were small, varying from the size of a grain of millet to that of a hemp seed.   They are generally of a light-yellow color like those of the Cape, though there are some perfectly white. When they find them of sufficient size they break them, as they told us, in order to make drill points, for, not knowing how to cut them, the Chinese in general do not consider them as precious stones.   They prefer the jade, the amethyst, the carnelian, and the agate.   Only the rich Chinese of the ports and of Peking have bought cut diamonds, imported from India or Europe, to ornament their hats or their rings, since the Dutch first brought them into China in the sixteenth century.   The Shan-tung collectors sell them throughout China, and their trade is of considerable importance."   The exact date of this modern diamond-digging is not known to me, but it seems not to be earlier than the latter part of the nineteenth century.   I can find no reference to it in Chinese literature.

[1] Or *Liang se kung tse ki* (see BRETSCHNEIDER, Bot. Sin., pt. 1, No. 451), that is, Memoirs of the Four Worthies or Lords of the Liang Dynasty (502–556), who were

Prince Kie of Shu (Sze-ch'uan) paid a visit to the Emperor Wu,[1] and, in the course of conversations which he held with the Emperor's scholars on distant lands, told this story: 'In the west, arriving at the Mediterranean,[2] there is in the sea an island of two hundred square miles (*li*). On this island is a large forest abundant in trees with precious stones, and inhabited by over ten thousand families. These men show great ability in cleverly working gems,[3] which are named for the country Fu-lin 拂林. In a northwesterly direction from the island is a ravine hollowed out like a bowl, more than a thousand feet deep. They throw flesh into this valley. Birds take it up in their beaks, whereupon they drop the precious stones. The biggest of these have a weight of five catties.' There is a saying that this is the treasury of the Devarāja of the Rūpadhātu 色界天王."[4]

From several points of view this text is of fundamental importance. First of all, it contains the earliest mention in Chinese records of the country Fu-lin, antedating our previous knowledge of it by a century.

Huei-ch'uang, Wan-kie, Wei-t'uan, and Chang-ki; the work was written by Chang Yüe (667–730), a statesman, poet, and painter of the T'ang period. The text translated above is given in *T'u shu tsi ch'ĕng*, section on National Economy 321, chapter on Precious Commodities (*pao huo*); it is reprinted in the writer's Optical Lenses (*T'oung Pao*, 1915, p. 204).

[1] He was the first emperor of the Liang dynasty and bore the name Siao Yen; he lived from 464 to 549.

[2] Literally, "the Western Sea" (*Si hai*). Compare HIRTH, The Mystery of Fu-lin II (*Journal Am. Or. Soc.*, Vol. XXXIII, 1913, p. 195).

[3] Literally, "implements or vessels of precious stones" (*pao k'i*), among which also antique intaglios are presumably included.

[4] A Sanskrit-Buddhist term meaning "the Celestial King of the Region of Forms." Region of Forms is the second of the three Brahmanic worlds (*trailokya*). The detailed discussion of this subject on the part of O. FRANKE (Chinesische Tempelinschrift, *Abhandl. preuss. Akad.*, 1907, pp. 47–50) is especially worth reading. There are four Celestial or Great Kings guarding the four quarters of the world, each posted on a side of the world-mountain Sumeru. The one here in question is Kubera or Vaiçravaṇa, the regent of the north and God of Wealth, the ruler of the aerial demons, called Yaksha. In earlier Buddhist art he is represented as standing on a Yaksha (see the writer's Chinese Clay Figures, pp. 297 *et seq.*); in later art he is figured holding in his right hand a standard and in his left an ichneumon (*nakula*) spitting jewels (compare A. FOUCHER, *Bull. de l'Ecole française*, Vol. III, p. 655). This animal is known as the inveterate enemy of snakes; and snakes, in Indian belief, are the guardians of precious stones and other treasures. By devouring the snakes, the ichneumon (or, to use its Anglo-Indian name, mangoose) appropriates their jewels, and has hence developed into the attribute of Kubera. The reference to the Indian God of Wealth in the above text is, of course, not an element inherent in the story, as it was transmitted from Fu-lin, but an interpolation of the Chinese author prompted by a reflection regarding a tradition hailing from India. This Indian story has been recorded by him in another passage of the same work, and will be discussed farther on (p. 18).

Professor HIRTH, a lifetime student of the complex Fu-lin problem,[1] encountered the first notices of Fu-lin in the Annals of the T'ang Dynasty, and an incidental reference to it in the Annals of the Sui Dynasty, written between 629 and 636, thus tracing the first appearance of the name to the first half of the seventh century. CHAVANNES[2] called attention to a text written in 607, in which Fu-lin is mentioned, with reference to a passage translated by him from the *Ts'e fu yüan kuei*, where the name is written in the same manner as in our text above.[3] The latter distinctly relates to the period T'ien-kien (502-520), and, further, is chronologically determined through the mention of the Liang Emperor Wu. Accordingly we are here confronted with the earliest allusion to the country Fu-lin in the beginning of the sixth century. The fact that the well-known Fu-lin discussed by Hirth and Chavannes, and no other, is involved in this passage, is evidenced by the very contents of the text, which, as will be demonstrated presently, harbors a tradition emanating from the Hellenistic Orient. It is notable that our text writes the second element of the name 林 instead of 菻, as the later documents do; it is obvious that a popular interpretation is intended here, the "forest" (*lin*) of the jewels being read into *Fu-lin:* as if it were "forest of Fu." This is not the place to revive the much-ventilated question of the etymology of this name, or to take sides with the interpretations proposed by HIRTH and CHAVANNES;[4] but brief reference should be made to the recent theory of PELLIOT,[5] according to whom the word *Fu-lin* is the product of the name *Rōm*, prompted by a supposed intermediary form *Frōm*, which issued from Armenian *Hrom* or *Horom* and Pahlavī *Hrōm*. Pelliot thinks also that the name *Fu-lin* appears in China with certainty around 550, and that it is possibly still older, which perfectly harmonizes with the result obtained from the above text.

The story about the capture of the precious stones is almost enigmatical in its terse brevity, but it at once becomes intelligible if we recognize it as an abridged form of a well-known Western legend. The oldest hitherto accessible version of it is contained in the writings of

---

[1] In his book China and the Roman Orient, and in his studies The Mystery of Fu-lin (*Journal Am. Or. Soc.*, Vol. XXX, 1909, pp. 1-31; Vol. XXXIII, 1913, pp. 195-208).

[2] *T'oung Pao*, 1904, p. 38.

[3] The same mode of writing occurs in *Yu yang tsa tsu* and in a poem of the T'ang Emperor T'ai-tsung (see *P'ei wen yün fu*, Ch. 27, p. 25).

[4] The latter has developed the conflicting views of both sides in *T'oung Pao*, 1913, p. 798.

[5] *Journal asiatique* (Mars-Avril, 1914), p. 498.

EPIPHANIUS, Bishop of Constantia in Cyprus (*circa* 315–403).[1] In his discourse on the twelve jewels forming the breastplate of the High Priest of Jerusalem, the following tale is narrated of the hyacinth. The theatre of action is a deep valley in a desert of great Scythia, entirely surrounded by rocky mountains rising straight like walls; so that from their summits the bottom of the valley is not visible, but only a sullen mist like chaos. The men despatched there in search of those stones by the kings, who reside in the neighborhood, slay sheep, strip them of their skins, and fling them from the rocks into the immense chaos of the valley. The stones then adhere to the flesh of the sheep. The eagles that loiter on the cliffs above scent the flesh, pounce down upon it in the valley, carry the carcasses off to devour them, and thus the stones remain on the top of the mountains. The convicts condemned to gather the stones go to the spots where the flesh of the sheep has been carried away by the eagles, find and take the stones. All these stones, whatever the diversity of their color, are of value as precious stones, but have this effect: that, when placed over a violent charcoal fire, they themselves are but slightly hurt, while the coal is instantly extinguished. This stone is reputed to be useful to women in aiding parturition; it is said also to dispel phantoms in a similar manner.[2]

---

[1] Epiphanii opera, ed. DINDORF, Vol. IV, p. 190 (Leipzig, 1862). The text in question is reproduced also by J. RUSKA (Steinbuch des Aristoteles, p. 15).

[2] The notion that the stones gathered by eagles aid in parturition rests on the belief of the ancients that the so-called *aëtites* or "eagle-stone," found in the nests of eagles, possesses remarkable properties having this effect. According to PLINY (x, 3, § 12; and XXXVI, 21, § 151), who distinguishes four varieties, this stone, so to speak, has the quality of being pregnant; for when shaken, another stone is heard to rattle within, as though it were enclosed in its womb. A male and a female stone are always found together; and without them, the eagles would be unable to propagate. Hence the young of the eagle are never more than two in number. PHILOSTRATUS, in his Life of Apollonius from Tyana, notes that the eagles never build their nests without first placing there an eagle-stone (F. DE MÉLY, Lapidaires grecs, p. 27). This stone is regarded as ferruginous geodes, a globular mass of clay iron-stone, which sometimes is hollow, sometimes encloses another stone or a little water. According to the Physiologus (XIX), the parturition-stone is found in India, whither the female vulture repairs to obtain it. From the Physiologus the story passed into the Arabic writers (J. RUSKA, Steinbuch des Aristoteles, p. 165; Steinbuch des Qazwīnī, pp. 18, 38; L. LECLERC, Traité des simples, Vol. I, pp. 121–123). O. KELLER (Tiere des classischen Altertums, p. 269) regards the legend of the eagle-stone as Egyptian, because it is mentioned by Horapollo (II, 49); but his work Hieroglyphica belongs to the fourth century A.D., while even THEOPHRASTUS (De lapidibus, 5) speaks of parturient stones. It seems more plausible that, as intimated by the Physiologus, the story hails from India. The physician Razi, who died in 923 or 932, observes (LECLERC, *l. c.*) that he encountered in some books of India the statement that a woman is easily delivered when the stone is placed on her abdomen. Regarding similar notions in China compare F. DE MÉLY, L'alchimie chez les Chinois (*Journal asiatique*, 1895, Sept.-Oct., p. 336) and Lapidaires chinois, p. LXIII.

The coincidence of this tale with our Chinese text is striking, the chief points — the deep valley, the flesh thrown down as bait, the birds bringing up the stones with it — being identical. The coincidence is the more remarkable, as the subsequent additional features with which the legend has been embellished in the West are lacking in the Chinese version. For this reason the conclusion is justified that the latter, directly traceable to a version of the type of Epiphanius, was transmitted straightway to China, as revealed by the very words of the Chinese account, from Fu-lin, a part of the Roman Empire.

In the second oldest Western version we encounter two new elements,— Alexander the Great and snakes guarding the stones. The oldest Arabic work on mineralogy, wrongly connected with the name of Aristotle and composed before the middle of the ninth century, has the following under the "diamond:"[1] "Nobody but my disciple Alexander reached the valley in which diamonds are found. It lies in the east along the extreme frontier of Khorasan, and its bottom cannot be penetrated by human eyes.[2] Alexander, after having advanced thus far, was prevented from proceeding by a host of snakes. In this valley are found snakes which by gazing at a man cause his death. He therefore caused mirrors to be made for them; and when they thus beheld themselves, they perished, while Alexander's men could look at them.[3] Thereupon Alexander contrived another ruse: he had sheep slaughtered, skinned, and flung on the bottom of the valley. The diamonds adhered to the flesh. The birds of prey seized them and brought part of them up. The soldiers pursued the birds and took whatever of their spoils they dropped." This account might lead us to suspect that the legend may have formed part of the Romance of Alexander, the archetype of which is preserved in the book known as that of Pseudo-Callisthenes, and produced at Alexandria in Egypt in the second century A.D.[4] In fact, however, it does not appear there, nor in any of the other early Western or Oriental cycles of the Alexander legends. The first Alexander legend in which it was incorporated is

---

[1] J. RUSKA, Steinbuch des Aristoteles, p. 150.

[2] Almost identical with the phraseology of Epiphanius: "Ita ut signis desuper, a summitatibus montium tanquam de muris aspiciat solum convallis, pervidere non possit."

[3] A reminiscence of the basilisk, that hideous serpent-like monster described by Pliny (VIII, 33). The mediæval poets have the basilisk die when it beholds itself in a mirror (F. LAUCHERT, Geschichte des Physiologus, p. 186).

[4] According to current opinion. A. AUSFELD (Der griechische Alexanderroman, p. 242, Leipzig, 1907), however, in his fundamental investigation of the Greek work, dates the oldest recension of Pseudo-Callisthenes with great probability in the second century B.C.

the *Iskander-nāmeh* of the Persian poet Nizāmī (1141–1203);[1] here we likewise meet the snakes, and it is now clear that Aristotle's *lapidarium* was the source of Nizāmī's episode.[2] It is well known that in the Arabic stories of Sindbad the Sailor, Sindbad, deposited by the Rokh in the Diamond Valley, observes how merchants throw down flesh, which is carried upward by vultures (also Nizāmī speaks of vultures) together with the diamonds sticking to it; enveloped by this flesh, he is lifted in the same manner.[3] The gradual growth of the legend from the simple form in which Epiphanius had clothed it is interesting to follow. In the celebrated Arabic "Book of the Wonders of India,"[4] written about A.D. 960, our legend is told by a traveller who had penetrated into the countries of India, and who localized it in Kashmir. He introduces a new element,— a fire constantly burning in the valley day and night,

---

[1] J. RUSKA, Steinbuch des Aristoteles, p. 14.

[2] Qazwīnī (1203–83) has the same story somewhat more amplified (J. RUSKA, Steinbuch aus der Kosmographie des al-Qazwīnī, p. 35); but it is interesting that he communicates two versions of it,— one being a close adaptation of Aristotle's account, the other staged on Serendīb (Ceylon) [where diamonds are not found] and not connected with the name of Alexander. It is obvious that the Arabic polyhistor, in his notice of the diamond, is reproducing two different sources,— the first being introduced by the words "Aristotle says;" the second, by the words "Another says." It is clear also that in this anonymous version the snakes are a purely incidental accessory which was lacking in the original text. "The mines are located in the mountains of Serendīb, in a valley of great depth, in which there are deadly snakes." The snakes, however, are put out of commission in the capture of the diamonds, which is due to the action of the vultures; and in order to justify the introduction of the reptiles, it is added at the end that large stones have to remain in the valley, as it cannot be reached for fear of the snakes. This observation is not without value for tracing the origin and growth of the legend. It shows that the feature of the snakes, however tempting this suggestion of its Indian origin may be to a superficial judgment, was not conceived in India, but in the Arabic-Persian sphere of the Alexander legends, with the evident object of aggrandizing the exploits of the conqueror. Qazwīnī's duplicity of versions is mirrored by MARCO POLO (ed. of YULE and CORDIER, Vol. II, pp. 360–361), who likewise offers two variants,— one with serpents, and another without them. The dependence of Qazwīnī's story on that in Aristotle's *lapidarium* has already been recognized by E. ROHDE (Der griechische Roman, p. 193, note, 3d ed., Leipzig, 1914). Ruska is right in his conclusion that the traditions concerning stones are relatively independent, and particularly so from the Alexander cycle; many a story in its origin had no connection with Alexander, but was subsequently associated with him in the same manner as King Solomon became the centre of numerous legendary fabrics. This follows in particular from the thorough investigation of A. AUSFELD (Der griechische Alexanderroman), who devoted a lifetime of study to the Greek romance of Alexander, and in whose purified text, representing the oldest accessible version, these mineralogical fables do not appear.

[3] Compare also BENJAMIN OF TUDELA, p. 82 (ed. of GRÜNHUT and ADLER, Jerusalem, 1903).

[4] P. A. VAN DER LITH and L. M. DEVIC, Livre des merveilles de l'Inde, p. 128 (Leiden, 1883–86); or L. M. DEVIC, Les merveilles de l'Inde, p. 109 (Paris, 1878).

summer and winter. The serpents are distributed around the fire; sheep's flesh, eagles, and capture of the stones, are the same features as previously mentioned, but the dangers of the work are magnified: the flesh may be devoured by the flames; the eagle, drawing too near the fire, may likewise be burnt; and the captors may perish from the peril of the fire and the serpents.[1]

In the Sung period (960–1278) the story was vaguely known to Chou Mi.[2] In his work *Ts'i tung ye yü*, as quoted by Li Shi-chên, he says that, according to oral accounts, diamonds come from the Western Countries (*Si yü*) and the Uigurs; that the stones stick to the food taken by eagles on the summits of high mountains, thus enter their bowels, and appear in their droppings, which are searched by men for the stones in the desert of Gobi, north of the Yellow River. The honest author adds, "I do not know whether it is so or not." Fang I-chi, the author of the *Wu li siao shi*,[3] who wrote in the first half of the seventeenth century, criticises Chou Mi's story as erroneous and not

---

[1] An echo of a certain motive of the legend of the Diamond Valley seems to reverberate in the Shamir legend of the Semitic peoples. The most interesting form of this legend is found in Qazwīnī (RUSKA, Steinbuch aus der Kosmographie, p. 16), who calls the stone *sāmūr* and characterizes it as the stone cutting all other stones. Solomon endeavors to obtain it that the stones required for the temple might be cut noiselessly. Only the eagle knows the place to find it, but the secret must be elicited from the bird through a ruse. The eggs are removed from its nest, enclosed in a glass bottle, and restored to their place. The returning eagle cannot break the glass with its pinions, and seeks for a piece of the stone in question, which he throws toward the vessel, breaking it into halves without noise. The eagle replies to Solomon's query that the stone is brought from a mountain in the west, termed Mount Sāmūr, whither Solomon sends the Djinns, who get a goodly supply for him. In this legend the stone *sāmūr* doubtless is intended for the diamond, and the motive of the eagle knowing its whereabouts is the same as in the legend of the Diamond Valley. The Talmud has strangely disfigured this story which is very sensibly told by Qazwīnī, and has transformed the stone *shamir* into a worm of the size of a barley-grain, capable of splitting and engraving the hardest objects, so that the *shamir* figures among the fabulous animals of the Talmud (L. LEWYSOHN, Zoologie des Talmud, p. 351). The worm (and simultaneously) diamond *shamir* has been entrusted to the wood-cock who took it to the summit of an uninhabited mountain; this is analogous to the birds or eagles bringing the diamonds up from the snake valley, and it is very tempting to assume that the snakes may have given rise to the curious Talmudic conception of the diamond as a worm. Lewysohn is of the opinion that the word *shamir* conveys the notion of hardness, and, for example, denotes iron, which is harder than stone, and also the diamond.— The Hebrew word *shamir* appears in Jeremiah (XVII, 1), Ezekiel (III, 9), and Zechariah (VII, 12), and is supposed to refer to the diamond ("adamant stone" in the English Bible); more probably it is the emery. In the opinion of some scholars, Greek σμύρις ("emery") is derived from the Hebrew word. For further bibliographical data on the Shamir legend see T. ZACHARIAE, Zeitschr. Vereins für Volkskunde, Vol. XXIV, 1914, p. 423.

[2] A celebrated and fertile author, who was born about 1230, and died before 1320 (see PELLIOT, T'oung Pao, 1913, pp. 367, 368).

[3] Ch. 8, p. 22 (edition of *Ning tsing t'ang*, 1884).

clear. Both authors were evidently not acquainted with the older version of the *Liang se kung ki*.

A new impetus to the legend was given during the Mongol period in the thirteenth century, when it was revived among the Arabs, in China, and in Europe. Reference has already been made to Qazwīnī (1203–83), who attributes it to the Valley of the Moon among the mountains of Serendīb (Ceylon); and the geographer Edrīsī localizes it in the land of the Kīrkhīr (probably Kirghiz) in Upper Asia. The Arabic mineralogist Ahmed Tīfāshī, who died in 1253, even gives two versions,— one referring to the hyacinth (in agreement with Epiphanius) of Ceylon, the other to the diamonds of India.[1] The former is vividly told, and the serpents "able to swallow an entire man" have duly been introduced; the latter is briefly jotted down, with a reference to the former chapter.

Ch'ang Tê, the Chinese envoy who was sent in 1259 to Hulagu, King of Persia, mentions in his diary, among the wonders of the Western countries, the diamond, of which he correctly says that it comes from India. "The people take flesh," his story goes, "and throw it into the great valley. Then birds come and eat this flesh, after which diamonds are found in their excrement."[2] It is obvious that Ch'ang Tê recorded the legend as

---

[1] A. RAINERI BISCIA, Fior di pensieri sulle pietre preziose di Ahmed Teifascite, pp. 21, 54 (2d ed., Bologna, 1906). As this work may not be in the hands of every reader, the text of the longer version may here be given: "Narra Ahmed Teifascite, a cui il sommo Iddio usi misericordia, che in alcuni anni non piovendo punto in quel montuoso territorio de Rahun, ed i suoi torrenti non trasportando per conseguenza verun lapillo di giacinto, coloro i quali bramano nulladimeno di farne acquisto, ricorrono al seguente compenso. Siccome sulla cima del prefato monte trovansi, ed annidano molte aquile, stante la total mancanza di abitatori, così prendono quelli un grosso animale, lo scannano, lo scorticano, e dopo averlo tagliato e diviso in larghi pezzi li lasciano alle falde dello stesso monte, e se n'allontanano. Osservando quelle aquile siffatti pezzi di carne corrono tosto per rapirli, e li trasportano verso dei loro nidi; ma giacchè cammin facendo sono costrette di posarli qualche volta in terra, n'accade perciò che attacansi a cotesti pezzi di carne diverse pietruzze o lapilli di giacinto. In seguito ripigliando le aquile stesse il volo coi rispettivi pezzi di carne, e venendo tra loro a contesa per rapporto ai medesimi, si dà la combinazione che nella mischia ne cadono alcuni fuori dal predetto monte; lo che veduto dalle persone ivi a bella posta concorse vanno subito a raccogliere da tali pezzi tutta quella copia di giacinto, che vi è rimasta attaccata. La parte inferiore dell'indicato monte è ingombrata da folti boschi, da larghi e profondi fossi, e burroni, non che da alberi d'alto fusto, ove trovansi vari serpenti che inghiottiscono un uomo intero. Per tal cagione niuno può salir su quel monte e vedere le maraviglie che in esso contengonsi."

[2] BRETSCHNEIDER, Mediæval Researches, Vol. I, p. 152. Bretschneider states that the legend is very ancient, but refers only to Sindbad the Sailor from a second-hand source, and to Marco Polo. The text of the passage will be found in G. SCHLEGEL (Nederlandsch-chineesch Woordenboek, Vol. I, p. 860). Compare MARCO POLO (ed. of YULE and CORDIER, Vol. II, p. 361): "The people go to the nests of those white eagles, of which there are many, and in their droppings they find plenty of diamonds which the birds have swallowed in devouring the meat that was cast into the valleys."

heard by him in the West, and that his version does not depend upon the older one of the *Liang se kung ki*, which evidently was not known to him. This case is interesting, for it shows that the same Western story was handed on to the Chinese at different times and from different sources.

About the same time, MARCO POLO chronicled the diamond story[1] which he learned in India, and its close agreement in the main points with the Arabic authors is amazing. The Venetian was not the first European, however, to record it; as pointed out by Yule, it is one of the many stories in the scrap-book of the Byzantine historian Tzetzes.[2]

Nicolo Conti of the fifteenth century relates it of a mountain called Albenigaras, fifteen days' journey in a northerly direction from Vijayanagar; and it is told again, apparently after Conti, by Julius Cæsar Scaliger. As a popular tale it is found not only in Armenia,[3] as stated by Yule, but also in Russia.[4]

---

[1] YULE and CORDIER, The Book of Ser Marco Polo, Vol. II, p. 360. The bewitching of the serpents by means of mirrors is wanting. The feature of the eagles feeding upon the serpents appears to be a thoroughly Indian notion, absent in the Arabic accounts.

[2] One of the earliest mediæval sources that contains the story is the fantastic description of India and the country of Prester John, written by Elysæus in the latter part of the twelfth century, and edited by F. ZARNCKE (Der Priester Johannes II, pp. 120-127). This text is as follows: "Quomodo autem carbunculi reperiantur audiamus. Ibi est vallis quaedam, in qua carbunculi reperiuntur. Nullus autem hominum accedere potest prae pavore griffonum et profunditate vallis. Et cum habere volunt lapides, occidunt pecora et accipiunt cadavera, et in nocte accedunt ad summitatem vallis et deiciunt ea in vallem, et sic inprimuntur lapides in cadavera, et acuti sunt. Veniunt autem grifones et assumunt cadavera et educunt ea. Eductis ergo cadaveribus perduntur carbunculi, et sic inveniuntur in campis."

[3] Probably due to the fact that it was adopted by the Armenian *lapidarium* of the seventeenth century, translated into Russian by K. P. PATKANOV (p. 3). Of especial interest is the fact that the snakes are dissociated from the two Armenian versions known to us. This is the more curious, as the *lapidarium* fastens the story upon Alexander: consequently some Oriental form of the Romance of Alexander must have pre-existed, in which the snakes did not yet figure. For the benefit of those who may not have access to VON HAXTHAUSEN's Transcaucasia (London, 1854), the source of the Armenian popular story (p. 360), its text may here follow: "In Hindostan there is a deep and rocky valley, in which all kinds of precious stones, of incalculable value, lie scattered upon the ground; when the sun shines upon them, they glisten like a sea of glowing, many-colored fire. The people see this from the summits of the surrounding hills, but no one can enter the valley, partly because there is no path to it and they could only be let down the steep rocks, and partly because the heat is so great that no one could endure it for a minute. Merchants come hither from foreign countries; they take an ox and hew it in pieces, which they fix upon long poles, and cast into the valley of gems. Then huge birds of prey hover around, descend into the valley, and carry off the pieces of flesh. But the merchants observe closely the direction in which the birds fly, and the places where they alight to feed, and there they frequently find the most valuable gems."

[4] AZBUKOVNIK, Tales of the Russian People (in Russian), Vol. II, p. 161. As the story is here told in regard to the hyacinth, it appears to go back directly to the account of Epiphanius.

Under the Ming (1368–1643) the story was repeated by Ts'ao Chao in his work *Ko ku yao lun*, which he published in 1387. His version is as follows: "Diamond-sand comes from Tibet (*Si-fan*). On the high summits of mountains with deep valleys, unapproachable to men, they make perches for the eagles, on which they set out food. The birds eat the flesh on the mountains and drop their ordure into desert places. This is gathered, and the stones are found in it."[1]

As regards the origin of our legend, two distinct opinions have been voiced. YULE[2] and ROHDE[3] point to its great resemblance to what Herodotus (III, 111) tells of the manner in which cinnamon was obtained by the Arabs; and a certain amount of affinity between the two cannot be denied. Great birds, says Herodotus, make use of cinnamon-sticks to build their nests, fastened with mud to high rocks, up which no foot of man is able to climb. So the Arabians resort to the artifice of cutting up the carcasses of beasts of burden and placing the pieces near the nests, whereupon they withdraw to a distance; and the old birds, swooping down, seize the flesh and bring it up into their nests. As the pieces are large, they break through the nest and fall to the ground, when the Arabians return and collect the cinnamon. The interval between Herodotus and Epiphanius is too great to be spanned or to allow us to link their stories in close historical bonds. There must be many intermediary links unknown to us. They evidently belong, as two individual variations, to the same type of legend, and seem to point to the fact that the latter existed in the near Orient for a long time.[4] The Chinese text recorded in the beginning of the sixth century, from which we started, furnishes additional testimony to this effect.

V. BALL[5] is inclined to think that the story "appears to be founded on the very common practice in India, on the opening of a mine, of offering up cattle to propitiate the evil spirits who are supposed to guard treasures — these being represented by the serpents in the myth. At such sacrifices in India, birds of prey invariably assemble to pick up

---

[1] *Ko chi king yüan*, Ch. 33, p. 3 b.

[2] *L. c.*, p. 363.

[3] Der griechische Roman, p. 193.

[4] Certain elements of the story may be found also in PLINY's (XXXVII, 33) curious legend of the stone *callaina*, which has wrongly been identified with the turquois: Some say that these stones are found in Arabia in the nests of the birds called "blackheads" (Sunt qui in Arabia inveniri eas dicant in nidis avium, quas melancoryphos vocant). Pliny then reports the occurrence of the stones on inaccessible rocks which people cannot climb, and mentions the danger connected with the venture of seeking them. Capturing them with slings certainly is a different feature, characteristic of another cycle of legends.

[5] Translation of Tavernier's Travels in India, Vol. II, p. 461.

what they can, and in that fact we probably have the remainder of the foundation of the story. It is probable also that the story by Pliny and other early writers, of the diamond being softened by the blood of a he-goat, had its origin in such sacrifices."[1] This subjective explana-

---

[1] This tradition, which, as will be seen below, has a curious parallel in China, is entirely independent of the Diamond-Valley story, and bears no relation to it. It is regrettable that Ball does not betray who the "other early writers" are. Pliny, in fact, is the earliest and only ancient writer to have it on record; Augustinus (fifth century), Isidorus (who died in 636) and Marbod (1035-1123) have merely reiterated it after Pliny, and Pliny's story certainly is not borrowed from India. W. CROOKE (Things Indian, p. 135) is inclined to think that if Ball's explanation be correct, the early diamond-diggers must have been non-Aryans, who did not regard the cow as sacred. The "early diamond-diggers" are a bit of exaggeration: in no Indian record of very early date does any mention of the diamond occur. Crooke's information on this point lacks somewhat the necessary precision. According to him, "diamonds were from very early times valued in India. The Purāṇas speak of them as divided into castes, and Marco Polo describes them as found in the kingdom of Mutfili." The Purāṇa were at the best composed in the first centuries A.D., and more probably much later. The knowledge of the diamond, certainly, does not go back in India into that unfathomable antiquity, as pretended by some mineralogical and other authors (for instance, G. WATT, Dictionary of Economic Products of India, Vol. III, p. 93). It was wholly unknown in the Vedic period, from which no specific names of precious stones are handed down at all. The word *maṇi*, which has sometimes been taken to mean the diamond (MACDONELL and KEITH, Vedic Index of Names and Subjects, Vol. II, p. 119), simply denotes a bead used for personal ornamentation and as an amulet, and the arbitrary notion that it might refer to the diamond is disproved by the fact that it could be strung on a thread. The word *vajra*, which at a subsequent period became an attribute of the diamond, originally served for the designation of a club-shaped weapon and of Indra's thunderbolt in particular (MACDONELL, Vedic Mythology, p. 55). Philological considerations show us that the diamond had no place in times of Indian antiquity, for no plain and specific word has been appropriated for it in any ancient Indian language. Either, as in the case of *vajra*, a word long familiar with another meaning was transferred to it, or epithets briefly indicating some characteristic feature of the stone were created. S. K. AIYANGAR (Note upon Diamonds in South India, *Quarterly Journal of the Mythic Society*, Vol. III, p. 129, Madras, 1914) calls attention to the fact that the first systematic reference to diamonds is made in the Arthaçāstra of Kauṭilya (see V. A. SMITH, Early History of India, 3d ed., pp. 151-153). He mentions six kinds of diamonds classified according to their mines, and described as differing in lustre and degree of hardness. He points out those of regular crystalline form and those of irregular shape. The best diamond should be large, heavy, capable of bearing blows, regular in shape, able to scratch the surface of metal vessels, refractive and brilliant. Aiyangar dates the work in question "probably at the commencement of the third century B.C." This date, however, is a mooted point (compare L. FINOT, *Bull. de l'Ecole française*, Vol. XII, 1912, pp. 1-4), which it would be out of place to discuss here. More probably, it is in the early Pāli scriptures of Buddhism that we can trace the first unmistakable references to the diamond. In the Questions of King Milinda (*Milindapañha*, translation of RHYS DAVIDS, p. 128) we read that the diamond ought to have three qualities: it should be pure throughout; it cannot be alloyed with another substance; and it is mounted together with the most costly gems. The first alludes metaphorically to the monk's purity in his means of livelihood; the second, to his keeping aloof from the company of the wicked; the third, to his association with men of highest excellence, with men who have entered the first or second or third stage of

tion is hardly convincing. It presupposes that the legend originated in India, but this postulate is not proved. That the later Arabic authors and Marco Polo place the locality in India, means nothing. Epiphanius lays the plot in Scythia; the Chinese version is laid in Fu-lin, and that

---

the Noble Path, with the jewel treasures of the Arhats. The Milindapañha may be dated with a fair degree of certainty: Milinda, who holds conversations with a Buddhist sage, is the Greek King Menandros, who ruled approximately between 125 and 95 B.C. in the north-west of India; and the dialogues attributed to him may have been composed in the beginning of our era (M. WINTERNITZ, Geschichte der indischen Litteratur, Vol. II, p. 140; V. A. SMITH, Early History of India, p. 225). It is therefore quite sufficient to believe that the diamond became known in India during the Buddhist epoch in the first centuries B.C., say, roughly, from the sixth to the fourth century. The precious stones mentioned in Milindapañha are enumerated by L. FINOT (Lapidaires indiens, p. XIX). The earliest descriptions of the diamond on the part of the Indians are by Varāhamihira (A.D. 505–587; see H. KERN, Verspreide Geschriften, Vol. II, p. 97) and by Buddhabhaṭṭa, who wrote prior to the sixth century A.D. Since the word *vajra* designates both Indra's thunderbolt and the diamond, it is in many cases difficult to decide which of the two is meant (A. FOUCHER, Etudes sur l'iconographie bouddhique de l'Inde, Vol. II, p. 15, left the point undecided, rendering *vajrāsana* by "siège de diamant ou du foudre"); and the same obstacle turns up again in Chinese-Buddhist literature, where the term *kin-kang* as the translation of Sanskrit *vajra* covers the two notions; so that, for instance, PELLIOT (*Bull. de l'Ecole française*, Vol. II, p. 146) raises the question, "Quel est le sens précis de *kin-kang*?" Whether the title of the Sūtra *Vajracchedikā*, for instance, is correctly translated by "diamond-cutter," as has been done, is much open to doubt. If it should mean "sharply cutting, like a diamond" (WINTERNITZ, *l. c.*, p. 249), why could it not mean as well "sharply cutting, like a thunderbolt"? The thunderbolt, generally described as metallic, is also sharp; and Indra whets it like a knife, or as a bull its horns. Though a Chinese commentator of that work observes that, as the diamond excels all other precious gems in brilliance and indestructibility, so also the wisdom of this work transcends and shall outlive all other knowledge known to philosophy (W. GEMMELL, The Diamond Sutra, p. 47), it is but a late afterthought, and proves nothing as to the original Indian concept. The most curious misconceptions have arisen about the so-called "Diamond-Seat" (*Vajrāsana*). This is the name of the throne or seat on which Çākyamuni, the founder of Buddhism, reached perfect enlightenment under the sacred fig-tree at Gayā. The Chinese pilgrim Hüan Tsang, who visited the place during his memorable journey in India, remarks that it was made from diamond (*Ta T'ang si yü ki*, Ch. 8, p. 14, ed. of *Shou shan ko ts'ung shu;* JULIEN, Mémoires sur les contrées occidentales, Vol. I, p. 460; WATTERS, On Yuan Chwang's Travels, Vol. II, p. 114); but this is incredible, if for no other reason, because he proceeds to say that this throne measured over a hundred paces in circuit. While this may be solely the outcome of a popular tradition growing out of an interpretation of the name, Hüan Tsang himself explains well how this name arose. It is derived, according to him, from the circumstance that here the thousand Buddhas of this eon (*kalpa*) enter the *vajrasamādhi* ("diamond ecstasy"), the designation for a certain degree of contemplative ecstasy. Moreover, in the Biography of Hüan Tsang (JULIEN, Histoire de la vie de Hiouen-Thsang, p. 139) it is more explicitly stated that the employment of the word "diamond" in the term "Diamond-Seat" signifies that this throne is firm, solid, indestructible, and capable of resisting all shocks of the world. In other words, it is used metaphorically; Buddha's own firmness and determination in the long struggle for obtaining enlightenment and salvation, his fortitude in overcoming the hostile forces of Māra, the Evil One, being transferred to the seat which he occupied immovably during

of Pseudo-Aristotle in Khorasan, etc. No ancient Sanskrit or Pāli version of the story has as yet become known; and the weight of evidence is in favor of the Arabs having propagated it farther eastward in the ninth and tenth centuries, while it was known in China long before that time. The snakes and eagles, of course, could be translated into Indian thought as Nāga and Garuḍa;[1] but, again, the Indians do not tell us of such a tradition in connection with these two mythical creatures. Even granted that the addition of the snakes in Pseudo-Aristotle might be due to a secondary influence or to some latent undercurrent of Indian conception which possibly penetrated into Syria, the Indian origin of the legend would not be proved, either: for Epiphanius has no snakes; and the old Chinese version lacks them too, and has "birds" instead of eagles. We remember, however, that the Chinese text winds up with an allusion to a Buddhist notion, the Devarāja of the Rūpadhātu; but neither is this evidence of an Indian provenience of the legend, which, as unambiguously stated in the text of Chang Yüe, hailed from Fu-lin. This additional annotation, certainly not devised in Fu-lin, was derived by the author from another tradition, which we now propose to examine, and which will shed unexpected light on the position held by India in the diffusion of this tale.

A contribution to the question whether the legend of the Diamond

---

that interval. The counterpart of this sacred site may be viewed in China on the Island of P'u-t'o, in the so-called "P'an-t'o Rock," which is styled "Diamond Precious Stone," on which, according to local legend, the Bodhisatva Avalokiteçvara (Kuan-yin) sat enthroned; this Diamond-Seat, however, is nothing but a rocky bowlder, the top of which is reached by means of a ladder, where contemplative monks may often be seen absorbed by the religious practice of meditation (*dhyāna*; compare R. F. JOHNSTON, Buddhist China, p. 313, London, 1913). The Vajrāsana of Buddha, accordingly, has as much to do with the diamond in its quality of stone as, for instance, Dante's diamond throne on which the angel of God is seated (L'angel di Dio, sedendo in su la soglia, Che mi sembiava pietra di diamante.— *Purgatorio*, IX, 104–105). Here also it is a metaphor, referring, according to the one, to the firmness and constancy of the confessor, or, according to others, to the symbol of the solid fundament of the Church (Divina Commedia, ed. SCARTAZZINI, p. 371). In a text of the Japanese Shin sect, the question is of a "heart strong as the diamond" in the sense of a diamond-hard faith (H. HAAS, Amida Buddha, p. 122). Also the heart of the hardened sinner is compared with the diamond in Buddhist literature (H. WENZEL, Nāgārjuna's Friendly Epistle, p. 24, stanza 83; S. BEAL, The Suhrillekha or Friendly Letter, p. 31, stanza 85, London, 1892). The Manicheans used the word in a similar manner by way of illustration, when it is said in one of their writings that the Messenger of Light is the precious diamond pillar supporting the multitude of beings (CHAVANNES and PELLIOT, Traité manichéen, p. 90).

[1] MARCO POLO (*l. c.*) explains the presence of the serpents in a natural manner: "Moreover in those mountains great serpents are rife to a marvellous degree, besides other vermin, and this owing to the great heat. The serpents are also the most venomous in existence, insomuch that any one going to that region runs fearful peril; for many have been destroyed by these evil reptiles."

Valley was known in ancient India is furnished by the same work, *Liang se kung tse ki*, as supplied to us with the Fu-lin version of the legend. Here we read this story: "A large junk of Fu-nan (Cambodja) which had come from western India arrived (in China) and offered for sale a mirror of a peculiar variety of rock-crystal,[1] one foot and four inches across its surface, and forty catties in weight. On the surface and in the interior it was pure white and transparent, and displayed many-colored objects on its obverse. When held against the light and examined, its substance was not discernible. On inquiry for the price, it was given at a million strings of copper coins. The Emperor ordered the officials to raise this sum, but the treasury did not hold enough. Those traders said, 'This mirror is due to the action of the Devarāja of the Rūpadhātu.[2] On felicitous and joyful occasions he causes the trees of the gods[3] to pour down a shower of precious stones, and the mountains receive them. The mountains conceal and seize the stones, so that they are difficult to obtain. The flesh of big animals is cast into the mountains; and when the flesh in these hiding-places becomes so putrefied that it phosphoresces, it resembles a precious stone. Birds carry it off in their beaks, and this is the jewel from which this mirror is made.' Nobody in the empire understood this and dared pay that price."[4] This account gives us a clew as to how it happened that the Devarāja of the Rūpadhātu was linked with the aforesaid legend hailing from Fu-lin. Both legends are on record in the same book, and the author combined the one report with the other. There is no reason to wonder that the story of the Fu-nan traders was not comprehended in China. We ourselves should be completely at sea, did not the Western legends enlighten the mystery. The story-teller from Fu-nan either did not express himself very clearly or was not perfectly understood by his interpreter, or the text of the *Liang se kung tse ki* has come down to us in corrupt shape. It is indubitable, however, that the story here on record is an echo of the legend of the Diamond Valley. All its essential features clearly stand out,— the inaccessible mountains hoarding the stones, the casting of flesh on them, and birds securing the stones. The narrative is only obscure in omitting to state that the jewels ad-

---

[1] Compare the writer's note on this subject in *T'oung Pao*, 1915, p. 200.

[2] See above, p. 7.

[3] This term corresponds to Sanskrit *devataru* ("tree of the gods"), a designation for the five miraculous trees to be found in Indra's Heaven,— *kalpavṛiksha, pārijāta, mandāra, saṁtāna,* and *haricandana* (compare HOPKINS, *Journal Am. Or. Soc.,* Vol. XXX, 1910, pp. 352, 353).

[4] *T'ai p'ing yü lan*, Ch. 808, p. 6 (the Chinese text will be found in *T'oung Pao*, 1915, p. 202).

here to the flesh which is devoured by the birds, while the puerile intimation that the putrefaction of the flesh transforms it into stone is
interpolated.   The Fu-nan merchants had come to China from the
shores of western India, and brought from there the expensive crystal
mirror.   With it came the story, and thus some form of the legend of
the Diamond Valley must have existed in the western part of India at
least in the beginning of the sixth century A.D.   Certainly it was a
much fuller and more intelligent version than that presented to us
through the medium of the Fu-nan seafarers.   Be this as it may, also
India took its place in this universal concert of Asiatic nations; and
our Chinese text has fortunately preserved the only Indian version
thus far known, and now first revealed and explained.   It is most interesting that the Indian tradition belongs to the type of the plain
dramatic version, in which the by-play of the serpents is wanting; so
is the Garuḍa; and the only specific Indian traits are the tree of the
gods and the Devarāja Kubera.   Aside from these incidents, which
are inconclusive in stamping the legend as Indian in its origin, it
thoroughly tallies with that of Epiphanius.   For this and also chronological reasons it follows that Fu-lin was the centre from which the
legend spread simultaneously to India and China.   G. HUET[1] has recently given another interesting example of a story originating in
western Asia, a weak echo of which was carried into India.

It is therefore my opinion that the legend of the Valley of Diamonds
or Precious Stones in its two early variations, as represented by Epiphanius and Pseudo-Aristotle, whatever its antecedents and its possible
associations with earlier stories of the Herodotian type may have been,
originated in the Hellenistic Orient, and was propagated from this centre
to China, to India, to the Arabs, and to Persia.   The Chinese tradition
of the *Liang se kung tse ki*, being an exact parallel to that of Epiphanius
and approaching it more closely in time than any of the Arabic and
other versions, being earlier and purer than that of Pseudo-Aristotle,
presents an important contribution to the question, and shows that
traditions of Fu-lin flowed into China long before its name was recorded
in her official annals.   The Chinese and Indian versions bear out still
another significant point that may enable us to reconstruct the original
form in which the subject was propagated in the Hellenistic world.   It
is manifest that Epiphanius, while by a lucky chance our earliest source
on the matter, does not preserve the story in its primeval or pure form;
he pursues a theological tendency by lining it up in his discourse on the

---

[1] Le conte du "mort reconnaissant" et le livre de Tobie (*Revue de l'histoire des
religions*, Vol. LXXI, 1915, pp. 1–29).

stones in the breastplate of the Jewish High Priest, and focuses it on the hyacinth, which makes for too narrow a specialization to be creditable to the original. Certainly Epiphanius is not the author of the story, but merely its propagandist; it was folk-lore of his time which he imbibed and employed for his specific purpose. This point of view is upheld by our Chinese text, which records the story as a tradition coming from the Hellenistic Orient, and which clearly indicates also its object. The precious stones of anterior Asia had always wrought an unbounded fascination on the minds of the Chinese, and the scope of this tradition is to account for the enormous wealth in jewels possessed by the country Fu-lin. Here we have a bit of humorous wit, as offered by the inhabitants of Fu-lin in explanation of numerous queries addressed to them by foreign traders: it was a story freely circulating in Fu-lin, not centring around the hyacinth, but relating to precious stones in the widest sense. Such appears to have been the original story, and thus it is preserved to us by the Chinese. That Pseudo-Aristotle and his successors (except Tīfāshī with his relapse into the hyacinth) chose the diamond, is easily intelligible, the diamond being always deemed the foremost and most valuable of all precious stones.[1]

INDESTRUCTIBILITY OF THE DIAMOND.— The Taoist adept Ko Hung (fourth century A.D.) has the following notice on the diamond: "The kingdom of Fu-nan (Cambodja) produces diamonds (*kin kang* 金剛) which are capable of cutting jade. In their appearance they resemble fluor-spar.[2] They grow on stones like stalactites,[3] on the bottom of the sea to the depth of a thousand feet. Men dive in search for the stones, and ascend at the close of a day. The diamond when struck by an iron hammer is not damaged; the latter, on the contrary, will be

---

[1] J. H. KRAUSE, Pyrgoteles, p. 29. The diamond is forestalled in the text of Epiphanius by the reference to the incombustible property of the stones.

[2] *Ts'e shi ying* 紫石英, thus identified by D. HANBURY, Notes on Chinese Materia Medica (*Pharmaceutical Journal*, 1861, p. 110), or Science Papers, p. 218. E. BIOT identified it with rock-crystal and smoky quartz (PAUTHIER and BAZIN, Chine moderne, Vol. II, p. 556).

[3] *Chung ju shi* 鐘乳石, identified by D. HANBURY (*l. c.*), with carbonate of lime in stalactitic masses, obtained from caves. The Chinese name, however, does not signify, as stated by Hanbury, "hanging- (like a bell) milk-stone," but the term *chung ju* refers to the mammillary protuberances or knobs on the ancient Chinese bells (see HIRTH, *Boas Anniversary Volume*, pp. 251, 257). GILES (No. 5691) has the name in the form *shi chung ju*, "stone-bell teats,— stalactites." Reduced to a powder the stone is used as a tonic. Compare F. PORTER SMITH, Contributions toward the Materia Medica of China, p. 204; GEERTS, Produits de la nature japonaise et chinoise, p. 342; F. DE MÉLY, Lapidaires chinois, pp. 92, 254. Important Chinese notes on this mineral are contained in the *Yün lin shi p'u* of Tu Wan (Ch. c, p. 8), *Ling-wai tai ta* of 1178 by Chou K'ü-fei (Ch. 7, p. 13), and *Pên ts'ao kang mu* (Ch. 9, p. 17 b).

spoiled. If, however, a blow is dealt at the diamond by means of a ram's horn,[1] it will at once be dissolved, and break like ice."[2]

The motive, diamonds being fished from the ocean, is an old Indian fable. We meet it in the *Suppāraka-jātaka*, No. 463 in the famous Pāli collection of Buddha's birth-stories. According to this legend, the diamonds are to be found in the Khuramāla Sea. The Bodhisatva was on board ship, acting as skipper for a party of merchants. He reflected that if he told them this was a diamond sea, they would sink the ship in their greed by collecting the diamonds. So he told them nothing; but having brought the ship to, he got a rope, and lowered a net as if to catch fish. With this he brought in a haul of diamonds, and stored them in the ship; then he caused the wares of little value to be cast overboard.[3] Of course, the Indian mineralogists knew better than that, and even enumerate eight sites where the diamond was found.[4]

---

[1] According to another reading, "antelope, or chamois horn" (*ling yang kio*). The latter is said to be solid and to occur only in the High-Rock Mountains (*Kao shi shan*) of Annam (*Wu li siao shi*, Ch. 8, p. 21 b; and *T'u shu tsi ch'êng, Pien i tien*, Annam, *hui k'ao* 6, p. 8 b).

[2] *Pên ts'ao kang mu*, Ch. 10, p. 12. Compare P. PELLIOT, Le Fou-nan (*Bull. de l'Ecole française*, Vol. III, 1903, p. 281). The same notice has been embodied in the account of the country of Fu-nan contained in the New Annals of the T'ang Dynasty (*T'ang shu*, Ch. 222 B, p. 2; and PELLIOT, *l. c.*, p. 274). Fu-nan, of course, did not produce diamonds, as said by the T'ang Annals in this passage, but imported them from India, as attested by a statement in the same Annals (*T'ang shu*, Ch. 221 A, p. 10 b) to the effect that India trades diamonds with Ta Ts'in (the Roman Orient), Fu-nan, and Kiao-chi. As both Indian diamonds and legends concerning them were encountered by the Chinese in Fu-nan, it was pardonable for them to believe that diamonds were a product of that country. Chao Ju-kua (translation of HIRTH and ROCKHILL, p. 111) says that the diamond of India will not melt, though exposed to the fire a hundred times.

[3] E. B. COWELL, The Jātaka, Vol. IV, p. 88. Compare also the Tibetan Dsang-lun, Ch. 30 (I. J. SCHMIDT, Der Weise und der Thor, pp. 227 *et seq.*); and SCHIEFNER, Tāranātha, p. 43. The Hindu mineralogists entertain also the notion that the diamond floats on the water (L. FINOT, Lapidaires indiens, p. XLVIII); and there is a fabulous account of a diamond of marine origin in the *Tsa pao tsang king* (BUNYIU NANJIO, Catalogue, No. 1329; CHAVANNES, Cinq cents contes et apologues, Vol. III, p. 1), translated from Sanskrit into Chinese in A.D. 472. A merchant from southern India who had an expert knowledge of pearls traversed several kingdoms, showing everywhere a pearl, the specific qualities of which nobody could recognize till he met Buddha, who said, "This wishing-jewel (*cintāmaṇi*) originates from the huge fish *makara*, whose body is two hundred and eighty thousand *li* (Chinese leagues) long. The name of this gem is 'hard like the diamond' (*kin-kang kien*, Chinese rendering of Sanskrit *vajrasāra*, an attribute of the diamond). It has the property of producing at once precious objects, clothing, and food, and securing everything according to one's wish. He who obtains this gem cannot be hurt by poison, or be burnt by fire." My translation is based on the text, as quoted in *Yüan kien lei han* (Ch. 364, p. 15 b), the wording of which to some extent dissents from that translated by M. CHAVANNES (*l. c.*, p. 77).

[4] L. FINOT, Lapidaires indiens, p. XXV.

In the Jātaka, the notion of the pearl being born from the ocean[1] has been transferred to the diamond. Q. Curtius Rufus echoes this native tradition when, in his description of India, he says that the sea casts upon the shores precious stones and pearls, these offscourings of the boiling sea being valued at the price which fashion sets on coveted luxuries.[2]

The Chinese tradition transmitted from Fu-nan — that iron does not break the diamond, but that the latter breaks iron — is reflected in the same manner by PLINY, who says that the stones are tested upon the anvil, and resist the blows with the result that the iron rebounds, and the anvil splits asunder.[3] This certainly is pure fiction and merely a popular illustration of the hardness of the stone.[4] This notion has accordingly migrated, and the Physiologus presents the missing link between East and West by asserting that the diamond cannot be damaged by iron, fire, or smoke.[5] In India we meet the same test, inasmuch as a diamond is regarded as genuine if it is struck with other stones or iron hammers without bursting.[6] The fact that the Arabic treatises on mineralogy reiterate the same story need not be discussed here; for the account of Ko Hung is far older than these, and proves that long before the advent of the Arabs it passed from India to Fu-nan and from Fu-nan to China.

Discussing the phenomena of sympathy and apathy ruling in nature, PLINY sets forth that this indomitable power which contemns the two most violent agents of nature, iron and fire,[7] is broken by the blood of

---

[1] *Ibid.*, p. XXXII. A Sanskrit epithet of the pearl is *samudraja* ("sea-born").

[2] J. W. McCRINDLE, Invasion of India by Alexander, p. 187.

[3] Incudibus hi deprehenduntur ita respuentes ictus ut ferrum utrimque dissultet, incudes ipsae etiam exiliant (XXXVII, 15, § 57). Compare BLÜMNER, Technologie, Vol. III, p. 230.

[4] The diamond is hard, but not tough, and can easily be broken with the blow of a hammer. It is as brittle as at least the average of crystallized minerals (FAR-RINGTON, Gems and Gem Minerals, p. 70). The fabulous notion of the ancients was first refuted by GARCIA DA ORTA (or, ab Horto), in his work on the Drugs of India, which appeared in Portuguese at Goa in 1563. "It is out of the question," he says, "that the diamond resists the hammer; on the contrary, it can be pulverized by means of a small hammer, and may easily be pounded in a mortar with an iron pestle, the powder being used for the grinding of other diamonds" (compare J. RUSKA, Der Diamant in der Medizin, *Festschrift Baas*, p. 129). In the Italian translation of Garcia (p. 182, Venice, 1582) the passage runs thus: "Non è il vero, che il diamante resista alla botta del martello, percioche con ogni picciolo martello si riduce in polvere, e con grandissima facilità si pesta col pistello di ferro; e in questo modo lo pestano coloro, che con la sua polvere poliscono gli altri diamanti."

[5] F. LAUCHERT, Geschichte des Physiologus, p. 34.

[6] R. GARBE, Die indischen Mineralien, p. 82.

[7] PLINY, accordingly, was of the opinion that the diamond is able to resist fire, and DIOSCORIDES (L. LECLERC, Traité des simples, Vol. III, p. 272) acquiesced in

a ram, which, however, must be fresh and warm. The stone must be well steeped in it, and receive repeated blows, and even then will break anvils and iron hammers unless they be of excellent temper.[1] This fantasy has passed into the writings of ST. AUGUSTIN,[2] and, further, into our mediæval poets, who interpreted the ram's blood as the blood of Christ, likewise into our *lapidaires*.[3]

---

this belief. THEOPHRASTUS (De lapidibus, 19; opera ed. F. WIMMER, p. 343), in a passing manner, alludes to the incombustibility of the diamond by ascribing the same property to the carbuncle (*anthrax*); the lack of humidity in these stones renders them impervious to fire (compare KRAUSE, Pyrgoteles, p. 15 and note 4). APOLLONIUS DYSCOLUS, in the first half of the second century A.D. (Rerum naturalium scriptores Graeci minores, ed. KELLER, Vol. I, p. 50), says that the diamond, when exposed to a fire, is not heated.

[1] Siquidem illa invicta vis, duarum violentissimarum naturae rerum ferri igniumque contemptrix, hircino rumpitur sanguine, neque aliter quam recenti calidoque macerata et sic quoque multis ictibus, tunc etiam praeterquam eximias incudes malleosque ferreos frangens (*ibid.*, § 59); also in the same work, xx, procemium: sanguine hircino rumpente.

[2] Qui lapis nec ferro nec igni nec alia vi ulla perhibetur praeter hircinum sanguinem vinci (De civitate Dei, XXI, 4). Also ISIDORUS, Origines, XII, 1, 14; and MARBODUS, De lapidibus pretiosis, 1.

[3] F. LAUCHERT, Geschichte des Physiologus, p. 179. L. PANNIER (Les Lapidaires français du moyen âge, p. 36):
"Par fer ne par foú n'iert ovréé
S'el sang del buc chiald n'est trempéé."
F. PFEIFFER, Buch der Natur von Konrad von Megenberg, p. 433; ALBERTUS MAGNUS, De virtutibus lapidum, p. 135 (Amstelodami, 1669). The origin of the Plinian story is hard to explain, as there is no other ancient or Oriental source that contains it. C. W. KING (Antique Gems, p. 107) thinks it is a jeweller's story, probably invented to keep up the mystery of the business. BLÜMNER (Technologie, Vol. III, p. 231) supposes either that the ancient lapidaries really used ram's blood in good faith, without examining whether the diamond could also be broken without it, or that they merely pretended such a procedure to the laymen as an alleged artifice of their trade. These rationalistic speculations, unsupported by evidence, are unsatisfactory. More plausible is the view of E. O. VON LIPPMANN (Abhandlungen und Vorträge, Vol. I, p. 83), that the blood of the ram, owing to the sensual lust of this animal, was regarded as particularly hot. As is well known, a ram was the animal sacred to Bacchus (O. KELLER, Antike Tierwelt, Vol. I, p. 305); and ram's blood was a remedy administered in cases of dysentery (F. DE MÉLY, Lapidaires grecs, p. 92). What merits special attention, however, is that Capricorn as asterisk of the zodiac, according to Manilius, belonged to Vesta; and that everything in need of fire, like mines, working of metals, even bakery, was under its influence. Moreover, in ancient astrology, the twelve signs of the zodiac are associated with twelve precious stones, and in this series *adamas* belongs to Capricorn (see the list in F. BOLL, Stoicheia, No. 1, p. 40). The idea of ram's blood acting upon the diamond, therefore, seems to be finally traceable to an astrological origin. A curious custom relating to ram's horn is reported by Strabo (XVI, 4, § 17). When the Troglodytæ of Ethiopia bury their dead, some of them bind the corpse from the neck to the legs with twigs of the buckthorn [*Paliurus;* an infusion of this plant, according to Strabo, forms the drink of these people in general]. They at once throw stones over the body, at the same time laughing and rejoicing, until they have covered its face. Thereupon

That our Chinese text above speaks of a ram's horn may be due to the fact that this modification was caused by the error of a scribe or by some misunderstanding of the Western tradition regarding ram's blood. More probably the people of Fu-nan (Cambodja), or even of India, are responsible for the alteration, which in this form was then picked up by the Chinese. The adequateness of the latter interpretation follows from an interesting passage in the book *Hüan chung ki* of the fifth century, quoted by Li Shi-chên, which concludes a notice of the diamond with the statement that in the countries of the West the nature of Buddha is metaphorically likened to the diamond, and ram's horn to the "impurity of passion" (*fan nao* 煩惱). This compound is a technical Buddhist term, being a translation of Sanskrit *kleça-kashāya*, the third of a series of five *kashāya*, five impurities or spheres of corruption.[1] Taken individually, these two emblematic figures of speech are unobjectionable; but what would it mean, that a ram's horn, symbolic of the impurity of passion, can break the Buddha, who has the nature of the diamond? This, from a Buddhistic angle, is unintelligible; the opposite would be true. The foundation of this symbolism, plainly, cannot be of Buddhistic origin; but the impetus was apparently received from a Christian source, and was re-interpreted in India. The matter

---

they place over it a ram's horn and go away. In this case the ram's horn doubtless figures also as an instrument of extraordinary strength: it overpowers the body and soul of the deceased, keeping his spirit down and preventing it from a return to the former home, where it might do harm to the survivors. Therefore the mourners rejoice in accomplishing their purpose. Ram's heads were extensively employed in Greek art (H. WINNEFELD, Altgriech. Bronzebecken aus Leontini, *Progr. Winckelmannsfest*, No. 59, 1899). Ball's opinion that ram's blood is the outcome of Indian sacrifices held on the opening of a mine, discussed above on p. 15, is untenable, as there is no Indian tradition connecting the diamond with ram's blood. The baselessness of this theory is further demonstrated by the fact that the Chinese have altered the classical "ram's blood" into a "ram's horn;" and the Chinese account hailed from Fu-nan (Cambodja), a country with a strong impact of Indian civilization. The transformation, therefore, seems to have been effected in an Indian region. For this reason it is impossible to seek the origin of this idea in India, where apparently it was not understood and was changed into a "horn," which appears to have been regarded there as stronger than blood. As to the classical idea of heat suggested by ram's blood, it is noteworthy, however, that in late Indian art, Agni, the God of Fire, is represented as riding on a gray goat, flames of fire streaming round about him, his crown also being surrounded by fire (B. ZIEGENBALG, Genealogy of the South-Indian Gods, p. 191, Madras, 1869). Thus the conception of the ram or goat as an animal of fire is brought out,— a fire of such vehemence as to subdue the hardest body of nature.

[1] See EITEL, Handbook of Chinese Buddhism, p. 67; CHAVANNES, Cinq cents contes et apologues, Vol. I, p. 17; and O. FRANKE, Chin. Tempelinschrift, p. 51. F. DE MÉLY (Lapidaires chinois, p. 124) incorrectly understands that "in India the nature of Buddha is compared with the diamond; and his sadness, with the horn of the antelope *ling*."

will only become intelligible if we substitute "ram's blood" for "ram's horn" and interpret "ram's blood" as the blood of the Lamb, the Christian Saviour. This symbolic explanation has indeed been attached in the West to Pliny's ram's blood subduing the diamond. The idea is not found in the Physiologus, which compares the diamond itself with Christ (analogous to Buddha as the diamond), but it turns up in the mediæval poets. Frauenlob explains the destruction of the diamond through buck's blood as the salvation, saying that the *adamas* (diamond) of the hard curse was broken by the blood of Christ.[1]

DIAMOND AND LEAD.— Dioscorides of the first century A.D. observes on the diamond, "It is one of the properties of the diamond to break the stones against which it is brought into contact and pressed. It acts alike on all bodies of the nature of stone, with the exception of lead. Lead attacks and subdues it. While it resists fire and iron, it allows itself to be broken by lead, and this is the expedient employed to pulverize it."[2]

The oldest Arabic book on stones, sailing under the flag of Aristotle, reports in the chapter on the diamond, probably drawing from Dioscorides, that it cannot be overpowered by any other stone save lead, which is capable of pulverizing it.[3]

In a Syriac and Arabic treatise on alchemy of the ninth or tenth century, edited and translated by R. DUVAL, it is said that lead makes the diamond suffer; the translator understands this in the sense that lead serves for the working of the diamond, adding in a note that one worked the diamond and other precious stones, enclosed in sheets of lead, by means of ruby or diamond dust.[4] The action of lead on the diamond certainly is imaginary. This idea conveys the impression of having received its impetus from the circle of the alchemists. Muhammed Ibn Mansūr, who wrote a treatise on mineralogy in Persian during the thirteenth century, says regarding this point, "On the anvil, the diamond is not broken under the hammer, but rather penetrates into the anvil. In order to break the diamond, it is placed between lead, the latter being struck with a mallet, whereupon the stone is broken. Others, instead of using lead, envelop the diamond in resin or

---

[1] Compare F. LAUCHERT, Geschichte des Physiologus, p. 179. In the Cathedral of Troyes there is a sculpture from the end of the thirteenth century, representing the Lamb of God under the unusual form of a ram with large horns and bearing the Cross of the Resurrection. A. N. DIDRON (Christian Iconography, Vol. I, pp. 325, 326) styles this work a "most unaccountable anomaly," but the symbolism set forth above surely accounts for it.

[2] L. LECLERC, Traité des simples, Vol. III, p. 272.

[3] J. RUSKA, Steinbuch des Aristoteles, p. 149 (compare p. 76).

[4] M. BERTHELOT, La chimie au moyen âge, Vol. II, pp. 124, 136.

wax."[1] The Armenian *lapidarium* of the seventeenth century[2] is most explicit on the matter: "The diamond is bruised by means of lead in the following manner: lead is hammered out into a foil, on which the diamond is put; and when completely wrapped up with it, it is placed on an iron anvil, the lead being struck with an iron hammer. The diamond crumbles into pieces from these blows, but remains in the leaden foil, and is not dispersed into various directions, as it is prevented from so doing by the ductility of the lead. Released from the latter, the broken diamond is fit for work. In want of lead, the diamond is covered with wax and wrapped up in twelve layers of paper, whereupon it is smashed by hammer-blows. In order to secure it in pure condition and without loss, the whole mass is flung into boiling water, causing the wax to melt, the paper to float on the surface of the water, and the diamond-splinters to sink to the bottom of the vessel. Then it is pounded in a steel mortar and is at once ready for industrial purposes. With this pounded diamond (diamond-dust) the jewellers polish good and coarse diamonds." The practical object in the use of lead is here clearly indicated; but what appears in this work of recent date as a merely technical process was in its origin a superstitious act, as is explained by Tīfāshī, who wrote toward the middle of the thirteenth century. According to this author, the diamond, as stated by Pliny, is a golden stone; and in the same manner as gold is affected by lead, lead is able to pulverize the diamond.[3]

This Western idea has likewise migrated into China, and turns up in the *Tan fang kien yüan*, an alchemical work by Tu Ku-t'ao of the Sung period, according to whom lead can reduce the diamond to fragments.[4] This author terms the stone "metal-hard awl or drill" (*kin kang tsuan* 金剛鑽); that is, "diamond-point" (*kin kang* being the usual name for the diamond). According to Li Shi-chên, the author of the *Pên*

---

[1] J. VON HAMMER, *Fundgruben des Orients*, Vol. VI, p. 132 (Wien, 1818); M. CLÉMENT-MULLOT, Essai sur la minéralogie arabe, p. 131 (*Journal asiatique*, 6th series, Vol. XI, 1868). Al-Akfānī expresses himself in a similar manner (WIEDEMANN, Zur Mineralogie im Islam, p. 218).

[2] Russian translation of K. P. PATKANOV, p. 1.

[3] A. RAINERI BISCIA, Fior di pensieri, p. 53 (2d ed., Bologna, 1906).

[4] *Pên ts'ao kang mu*, Ch. 10, p. 12. The author speaks of a certain kind of lead styled "lead with purple back" (*tse pei yüan* 紫背鉛), in regard to which the *Pên ts'ao kang mu* only says that it is a variety of lead very pure and hard, able to cut the diamond (compare GEERTS, Les produits de la nature japonaise et chinoise, p. 605). Geerts annotates, "Ceci est une de ces absurdités que l'on trouve si souvent chez les auteurs chinois à côté de renseignements exacts et utiles." Certainly, the Chinese are not responsible for this "absurdity," which comes straight from our classical antiquity.

ts'ao kang mu, this name first occurs in the dictionary *Shi ming*, while the usual mineralogical designation is *kin kang shi* ("metal-hard stone"). Also Pseudo-Aristotle has the diamond "boring" all kinds of stones and pearls, and Qazwīnī styles it a "borer." Li Shi-chên says that "by means of diamond-sand jade can be perforated and porcelain repaired, hence the name awl (*tsuan*)."[1] An interesting analogy to this conception occurs in the Arabic stories of Sindbad the Sailor, dating in the ninth century. Sindbad tells, "Walking along the valley I found that its soil was of diamond, the stone wherewith they pierce jewels and precious stones and porcelain and onyx, for that it is a hard dense stone, whereon neither iron nor steel has effect, neither can we cut off aught therefrom nor break it, save by means of the load-stone." We shall now discuss one of the most interesting problems bearing on the diamond,— the ancient employment of the diamond-point.

THE DIAMOND-POINT.— In the book going under the name of the alleged philosopher Lie-tse, which in the text now before us is hardly earlier than the Han period, we read the following story:[2] "When King Mu of the Chou Dynasty (1001–945 B.C.) was on an expedition against the Western Jung, the latter presented him with a sword of *kun-wu* 錕鋙之劒 and with fire-proof cloth (asbestos). The sword was one foot and eight inches in length, was forged from steel, and had a red blade; when handled, it would cut hard stone (jade) as though it were merely clayish earth." The object of these notes is to discuss the nature of the substance *kun-wu*. Asbestine stuffs were received by the Chinese from the Roman Orient, and likewise the curious tales connected with them. If asbestos came from that direction, our first impression in the matter is that also the substance *kun-wu* appears to have been derived from the same quarter; and this supposition will be proved correct by a study of Chinese traditions.

---

[1] It is interesting that the Chinese, while they worked jade and porcelain, and, as will be seen farther below, also pearls, by means of diamond-points, did not know the fact that the latter can cut glass,— perhaps merely for the reason that they never understood how to make plate-glass. The ancients did not cut glass, either, with the diamond, and this practice does not seem to have originated before the sixteenth century (compare BECKMANN, Beiträge zur Geschichte der Erfindungen, Vol. III, p. 543). In recent times, however, the Chinese applied the diamond also to glass. Archdeacon GRAY, in his interesting book Walks in the City of Canton (p. 238, Hongkong, 1875), tells how the glaziers of Canton cut with a diamond the designs traced with ink upon the surface of glass globes and readily effect this labor by running the diamond along these ink-lines.

[2] Ch. 5, *T'ang wên*, at the end (compare E. FABER, Naturalismus bei den alten Chinesen, p. 132; L. WIEGER, Pères du système taoiste, p. 149; A. WYLIE, Chinese Researches, pt. III, p. 142). The work of Lie-tse is first mentioned as a book in eight chapters in *Ts'ien Han shu* (Ch. 30, p. 12 b).

The *kun-wu* sword of Lie-tse has repeatedly tried the ingenuity of sinologues. HIRTH,[1] who accepted the text at its surface value, regarded this sword as the oldest example in Chinese records of a weapon made from iron or steel; and while the passage could not be regarded as testimony for the antiquity of the sword-industry in China, it seems to him to reflect the legendary views of that epoch and to hint at the fact that the forging of swords in the iron-producing regions of the north-west of China was originally invested in the hands of the Huns. Thus Hirth finally arrived at the conclusion that the *kun-wu* sword may actually mean "sword of the Huns." FABER, the first translator of Lie-tse, regarded it as a Damascus blade; and FORKE[2] accepted this view. F. PORTER SMITH[3] was the first to speak of a *kun-wu* stone, intimating that "extraordinary stories are told of a stone called *kun-wu*, large enough to be made into a knife, very brilliant, and able to cut gems with ease." He also grouped this stone correctly with the diamond, but did not cope with the problem involved.

The *Shi chou ki* ("Records of Ten Insular Realms"), a fantastic description of foreign lands, attributed to the Taoist adept Tung-fang So, who was born in 168 B.C.,[4] has the following story: "On the Floating Island (Liu chou) which is situated in the Western Ocean is gathered a quantity of stones called *kun-wu* 昆吾石. When fused, this stone turns into iron, from which are made cutting-instruments brilliant and reflecting light like crystal, capable of cutting through objects of hard stone (jade) as though they were merely clayish earth."[5]

Li Shi-chên, in his *Pên ts'ao kang mu*,[6] quotes the same story in his notice of the diamond, and winds up with the explanation that the *kun-wu* stone is the largest of diamonds. The text of the *Shi chou ki*, as quoted by him, offers an important variant. According to his reading, *kun-wu* stones occur in the Floating Sand (Liu-sha) of the Western Ocean.[7] The latter term, as already shown, in the Chinese

---

[1] Chinesische Ansichten über Bronzetrommeln, pp. 20, 21.

[2] *Mitteilungen des Seminars*, Vol. VII, 1, p. 162. This opinion was justly criticised by the late E. HUBER (*Bull. de l'Ecole française*, Vol. IV, p. 1129).

[3] Contributions toward the Materia Medica of China, p. 75.

[4] The work is adopted in the Taoist Canon (L. WIEGER, Taoisme, Vol. I, No. 593). The authorship of Tung-fang So is purely legendary, and the book is doubtless centuries later. Exactly the same text is given also in the *Lung yü ho t'u* (quoted in *Yüan kien lei han*, Ch. 323, p. 1; and in the commentary to *Shi ki*, Ch. 117, p. 2 b), a work which appears to have existed in the fourth or fifth century (see BRETSCHNEIDER, Bot. Sin., pt. 1, No. 500).

[5] *P'ei wên yün fu*, Ch. 100 A, p. 16; or *Yüan kien lei han*, Ch. 26, p. 32 b.

[6] Ch. 10, p. 12.

[7] Also the *Wu li siao shi* (Ch. 8, p. 22) has this reading.

records relative to the Hellenistic Orient, refers to the Mediterranean; and Liu-sha is well known as a geographical term of somewhat vague definition, first used in the Annals of the Later Han Dynasty, and said to be in the west of Ta Ts'in, the Chinese designation of the Roman Orient.[1] Liu-sha, in my opinion, is the model of Liu chou, the Floating Island being distilled from Floating Sand in favor of the Ten Islands mechanically constructed in that fabulous book. Accordingly, we have here a distinct tradition relegating the *kun-wu* stone to the Anterior Orient; and Li Shi-chên's identification with the diamond appears plausible to a high degree. His opinion is strongly corroborated by another text cited by him. This is the *Hüan chung ki* by Kuo[2] of the fifth century, who reports as follows: "The country of Ta Ts'in produces diamonds (*kin-kang*), termed also 'jade-cutting swords or knives.' The largest reach a length of over a foot, the smallest are of the size of a rice or millet grain.[3] Hard stone can be cut by means of it all round, and on examination it turns out that it is the largest of diamonds. This is what the Buddhist priests substitute for the tooth of Buddha."[4] Chou Mi, quoted above regarding the legend of the Dia-

---

[1] HIRTH, China and the Roman Orient, pp. 42, 292. F. DE MÉLY (Lapidaires chinois, p. 124) translates "River Liu sha," and omits the "Western Ocean." The term Liu-sha existed in early antiquity and occurs for the first time in the *Shu king*, chap. *Yü kung* (LEGGE, Chinese Classics, Vol. III, pp. 132, 133, 150), denoting the then known farthest west of the country, the desert extending west of the district of Tun-huang in Kan-su. It is cited also in the elegy *Li sao* by Kū Yüan (XIII, 89; LEGGE, Journal R. As. Soc., 1895, pp. 595, 863), in the records of the Buddhist pilgrims (CHAVANNES, Religieux éminents, p. 12), and in the memoirs of the mediæval travellers (BRETSCHNEIDER, Mediæval Researches, Vol. I, p. 27; Vol. II, p. 144). See also PELLIOT, *Journal asiatique*, 1914 (Mai-Juin), p. 505.

[2] His personal name is unknown.

[3] PLINY (XXXVII, 15, § 57) speaks of a kind of diamond as large as a grain of millet (milii magnitudine) and called *cenchros;* that is, the Greek word for "millet."

[4] F. DE MÉLY (Lapidaires chinois, p. 124) incorrectly understands by this passage that the bonzes of India adorn with diamonds the tooth of Buddha. In fact, a diamond itself was passed off as Buddha's-tooth relic. A specific case to this effect is on record: "In the period Chêng-kuan (627–650) there was a Brahmanic priest who asserted that he had obtained a tooth of Buddha which when struck resisted any blow with unheard-of strength. Fu Yi heard of it, and said to his son, 'It is not a tooth of Buddha; I have heard that the diamond (*kin-kang shi*) is the strongest of all objects, that nothing can resist it, and that only an antelope-horn can break it; you may proceed to make the experiment by knocking it, and it will crash and break'" (*P'ei wên yün fu*, Ch. 100 A, p. 40 b). Fu Yi, who was a resolute opponent of Buddhism and was raised to the office of grand historiographer by the founder of the T'ang dynasty (he died in 639; see *Mémoires concernant les Chinois*, Vol. V, pp. 122, 159; LEGGE, Journal Roy. As. Soc., 1893, p. 800), was certainly right. Compare H. DORÉ, Recherches sur les superstitions en Chine, Vol. VIII, p. 310. Also PALLADIUS (Chinese-Russian Dictionary, Vol. II, p. 203 a) is inexact in saying that the Buddhists passed off the diamond as Buddha's tooth in China, where the diamond was unknown. Regarding Buddha's-tooth relic, besides the various

mond Valley, states, "The workers in jade polish jade by the persevering application of river-gravel, and carve it by means of a diamond-point. Its shape is like that of the ordure of rodents;[1] it is of very black color, and is at once like stone and like iron." Chou Mi apparently speaks of the impure, black form of the diamond, which is still used by us for industrial purposes, the tipping of drills and similar boring-instruments.[2] These texts render it sufficiently clear that the *kun-wu* stone of the *Shi chou ki*, which is found in the Hellenistic Orient, is the diamond,[3] and that the cutting-instrument made from it is a diamond-point. The alleged transmutation of the stone into iron is further elucidated by the much-discussed passage of Pliny, "When by a lucky chance the diamond happens to be broken, it is triturated into such minute splinters that they can hardly be sighted. These are much demanded by gem-engravers and are enclosed in iron. There is no hard substance that they could not easily cut by means of this instrument." [4]

---

accounts of Hüan Tsang, see Fa Hien, Ch. 38 (LEGGE, Record of Buddhistic King-doms, pp. 105–107); CHAVANNES, Mémoire sur les religieux éminents, p. 55; DE GROOT, *Album Kern*, p. 134; YULE and CORDIER, Book of Ser Marco Polo, Vol. II, pp. 319, 329–330, etc. The Pāli Chronicle of Ceylon describes a statue of Buddha, in which the body and members were made of jewels of different colors; the commentary adds that the teeth were made of diamonds (W. GEIGER, Mahāvaṁsa, p. 204). It accordingly was an Indian idea (not an artifice conceived in China) that the diamond could be substituted for Buddha's tooth. It is curious that Pseudo-Aristotle warns against taking the diamond in the mouth, because it destroys the teeth (RUSKA, Steinbuch des Aristoteles, p. 150). The poet Su Shi (1036–1101), in his work *Wu lei siang kan chi* (WYLIE, Notes, p. 165), remarks that antelope-horn is able to break Buddha's tooth to pieces; in this case, Buddha's tooth is a synonyme for the diamond, and we have an echo of Ko Hung's legend above referred to (p. 21).

[1] *Shu shi* 鼠矢, incorrectly rendered by F. DE MÉLY (Lapidaires chinois, p. 124) by "arrow-point." The word *shi* is here not "arrow," but "ordure, dung" (*shi* in the third tone); the text of the *Wu li siao shi* indeed writes *shi* 屎, which is the proper character; and *Ko chi king yüan* (Ch. 33, p. 3 b), in quoting the same text of Chou Mi, offers the variant *shu fên* 鼠糞, which has the same meaning.

[2] Known in the trade as "bort,"— defective diamonds or fragments of diamonds which are useless as gems.

[3] The reflective and refractive power of the diamond is well illustrated in the definition of that book, "brilliant and reflecting light like crystal." The coincidence with PLINY'S (XXXVII, 15, § 56) description of the Indian *adamas* is remarkable, "which occurs not in gold, but in a substance somewhat cognate to crystal, not differing from the latter in its transparent coloration" (Indici non in auro nascentis et quadam crystalli cognatione, siquidem et colore tralucido non differt). The opinion that diamond, according to its composition, was a glass-like stone of the nature of rock-crystal, prevailed in Europe till the end of the eighteenth century, when it was refuted by Bergmann in 1777, and experiments demonstrated that the diamond is a combustible body (F. VON KOBELL, Geschichte der Mineralogie, p. 388).

[4] Cum feliciter contigit rumpere, in tam parvas friatur crustas, ut cerni vix possint. Expetuntur hae scalptoribus ferroque includuntur, nullam non duritiam

Dioscorides of the first century A.D. distinguishes four kinds of diamonds, the third of which is called "ferruginous" because it resembles iron, but iron is heavier; it is found in Yemen. According to him, the adamantine fragments are stuck into iron handles, being thus ready to perforate stones, rubies, and pearls.[1] The concept of a mysterious association of the diamond with iron survived till our middle ages. KONRAD VON MEGENBERG, in his Book of Nature, written in 1349–50,[2] observes that, according to the treatises on stones, the virtue of the diamond is much greater if its foundation be made of iron, in case it is to be set in a ring; but the ring should be of gold to be in keeping with the dignity of the stone.

If we now glance back at the text of Lie-tse, from which we started, we shall easily recognize that the *kun-wu* sword mentioned in it is in fact only a mask for the diamond-point; for Lie-tse, with reference to this sword, avails himself of exactly the same definition as the *Shi chou ki*, expressed in the identical words,— "cutting hard stone (jade) as though it were merely clayish earth,"— and the jade-cutting knife (*tao*) is unequivocally identified with the diamond in the *Hüan chung ki*. The passage in Lie-tse, therefore, rests on a misunderstanding or a too liberal interpretation of the word *tao* 刀 , which means a cutting-instrument in the widest sense, used for carving, chopping, trimming, paring, scraping, etc. It may certainly mean a dagger or sword with a single edge; and Lie-tse, or whoever fabricated the book inscribed with his name, exaggerated it into the double-edged sword *kien*.[3] Then he was certainly obliged to permit himself the further change of making this sword of tempered steel;[4] and by prefixing the classifier *kin* ('metal') to the words *kun* and *wu*, the masquerade was complete for eluding the most perspicacious sinologues.[5] Lie-tse's *kun-wu* sword is a romantic

---

ex facili cavantes (XXXVII, 15, § 60). It is not necessary, as proposed by F. DE MÉLY (Lapidaires chinois, p. 257), to make a distinction between *kin kang shi* ("diamond") and *kin kang ts'uan* ("emery"). It plainly follows from the Chinese texts that the latter is the diamond-point (see below, p. 34).

[1] Compare L. LECLERC, Traité des simples, Vol. III, p. 272.

[2] Ed. of F. PFEIFFER, p. 433.

[3] The conception of the diamond as a sword had perhaps been conveyed to China from an outside quarter. In the language of the Kirgiz, the word *almas*, designating the "diamond" (from Arabic *almās*), has also the significance "steel" (in the same manner as the Greek *adamas*, from which the Arabic word is derived), and *ak almas* ("white diamond") is a poetical term for a "sword" (W. RADLOFF, Wörterbuch der Türk-Dialecte, Vol. I, col. 438).

[4] This metamorphosis was possibly somehow connected with the original meaning "steel" inherent in the Greek word *adamas*.

[5] The missing link is found in another passage of the *Shi chou ki*, where the same event is described as in Lie-tse. It runs as follows: "At the time of King Mu of the

fiction evolved from the *kun-wu* diamond-points heard of and imported from the Hellenistic Orient. It has nothing to do with the sword industry of the Huns or Chinese, as speculated by Hirth; nor is it a Damascus blade, as suggested by Faber and Forke. Such books as Lie-tse and many others of like calibre cannot be utilized as historical sources for archæological argumentation; their stories must first be analyzed, critically dissected, scrutinized, and correlated with other texts, Chinese as well as Western, to receive that stamp of valuation which is properly due them. It is now clear also why Lie-tse links the *kun-wu* sword with asbestos, inasmuch as the two are products of the Hellenistic Orient. The circumstance that both are credited to King Mu is a meaningless fable. King Mu was the chosen favorite and hero of Taoist legend-makers, to whose name all marvellous objects of distant trade were attached (in the same manner as King Solomon and Alexander in the West). The introduction of the Western Jung on this occasion possibly is emblematic of the intermediary rôle which was played by Turkish tribes in the transmission of goods from the Anterior Orient and Persia to China.[1]

As regards the history of the diamond, we learn that the Chinese, before they became acquainted with the stone as a gem, received the first intimation of it in the shape of diamond-points for mechanical work, sent from the Hellenistic Orient,— known first (at the time of the Han) under the name *kun-wu;* in the third century (under the Tsin), as will be shown below, under the name *kin-kang;* and later on, as *kin-kang tsuan.* It seems that the Chinese made little or no

---

Chou dynasty the Western Hu presented a jade-cutting knife of *kun-wu,* one foot long, capable of cutting jade as though it were merely clayish earth." In this text (quoted in *P'ei wĕn yün fu,* Ch. 19, p. 13) the word *tao* is used, and *kun-wu* is plainly written without the classifiers *kin.* Here we have the model after which Lie-tse worked. The term *kun-wu tao,* written in the same style as in *Shi chou ki,* appears once more in the biography of the painter Li Kung-lin (*Sung shi,* Ch. 444, p. 7), who died in 1106. The Emperor had obtained a seal of nephrite, which his scholars, despite long deliberations, could not decipher till Li Kung-lin diagnosed it as the famous seal of Ts'in Shi Huang-ti made by Li Se in the third century B.C. (compare CHAVANNES, *T'oung Pao,* 1904, p. 496). On this occasion the painter said that the substance nephrite is hard, but not quite so hard as a diamond-point (*kun-wu tao*).

[1] It is interesting that the diamond appears also in the cycle of Si-wang-mu, the legendary motives of which, in my opinion, to a large extent go back to the Hellenistic Orient. In the *Han Wu-ti nei chuan* (p. 2 b; ed. of *Shou shan ko ts'ung shu*), the goddess appears wearing in her girdle a magic seal of diamond (*kin-kang ling si*). The work in question, carried by an unfounded tradition into the Han period, is a production of much later times, but seems to have existed in the second half of the sixth century (PELLIOT, *Bulletin de l'Ecole française,* Vol. IX, p. 243; and *Journal asiatique,* 1912, Juillet-Août, p. 149).

use of the diamond for ornamental purposes, and did not understand how to work it.[1]

Not only have the Chinese stories about the diamond-point, but there is also proof for the fact that this implement was among them a living reality turned to practical use. Li Sün, the author of the *Hai yao pen ts'ao*,— an account of the drugs of southern countries, written in the second half of the eighth century,[2]—discusses the genuine pearl found in the southern ocean, and observes that it can be perforated only by the diamond-point (*kin-kang tsuan*).[3] The poet Yüan Chên (779–831), his contemporary, says in a stanza, "The diamond-point bores jade, the sword of finely tempered steel[4] severs the floating down."

The preceding accounts have conveyed the impression that the diamond-points employed by the Chinese were plain implements of the shape of an awl tipped with a diamond. A different instrument is described in the *Hüan chung ki*, a work of the fifth century, which has already been quoted from the *Pên ts'ao kang mu*. In the great cyclopædia *T'ai p'ing yü lan*[5] the passage of this book concerning the diamond is handed down as follows: "The diamond comes from India and the country of Ta Ts'in (the Roman Orient). It is styled also 'jade-cutting knife,' as it cuts jade like an iron knife. The largest reach a

---

[1] The *Nan chou i wu chi* (Account of Remarkable Objects in the Southern Provinces, by Wan Chen of the third century) states that the diamond is a stone, in appearance resembling a pearl, hard, sharp, and matchless; and that *foreigners* are fond of setting it in rings, which they wear in order to ward off evil influences and poison (*T'ai p'ing yü lan*, Ch. 813, p. 10).— The Polyglot Dictionary of K'ien-lung (Ch. 22, p. 65) discriminates between *kin-kang tsuan* ("diamond-point") and *kin-kang shi* ("diamond stone"). The former corresponds to Manchu *paltari*, Tibetan *p'a-lam*, and Mongol *ocir alama;* the latter, to Manchu *palta wehe* (*wehe*, "stone"), Tibetan *rdo p'a-lam* (*rdo*, "stone"), and Mongol *alama cilagu* (the latter likewise means "stone"). The Manchu words are artificial formations based on the Tibetan word. Mongol *alama* apparently goes back to Arabic *almās* (Russian *almaz*), Uigur and other Turkish dialects *almas* (Osmanli *elmas*), ultimately traceable to Greek-Latin *adamas*. Al-Akfānī writes the word *al-mās*, the initials of the stem being mistaken by him for the native article *al* (WIEDEMANN, Zur Mineralogie im Islam, p. 218).

[2] BRETSCHNEIDER, Bot. Sin., pt. I, p. 45.

[3] *Pên ts'ao kang mu*, Ch. 46, p. 3 b; *Chêng lei pên ts'ao*, Ch. 20, fol. 12 b (edition of 1523). Al-Akfānī says in the same manner that the pearl is perforated only by means of the diamond (E. WIEDEMANN, Zur Mineralogie im Islam, p. 221).

[4] *Pin t'ie.* Julien's opinion that the diamond is understood by this term is erroneous, and was justly antagonized by MAYERS (*China Review*, Vol. IV, 1875, p. 175). Regarding this steel imported into China by Persians and Arabs, see BRETSCHNEIDER, Mediæval Researches, Vol. I, p. 146; WATTERS, Essays on the Chinese Language, p. 434; HIRTH and ROCKHILL, Chau Ju-kua, p. 19.

[5] Ch. 813, p. 10 (edition of Juan Yüan, 1812).

length of over a foot, the smallest are of the size of a rice-grain. In order to cut jade, it is necessary to make a large gold ring, which is held between the fingers; this ring is inserted into the jade-cutting knife, which thus becomes fit for work." This description is not very clear, but I am under the impression that an instrument on the order of our roller-cutter is understood.

This investigation may be regarded also as a definite solution of a problem of classical archæology, which for a long time was the subject of an extended and heated controversy.[1] The Chinese, though receiving the diamond-point from the Occident, have preserved to us more copious notes and clearer and fuller texts regarding this subject than the classical authors; and if hitherto it was possible to cast doubts on Pliny's description of diamond-splinters (above, p. 31), which have been taken by some authors for diamond-dust, this scepticism is no longer justified in the light of Chinese information. What Pliny describes is indeed the diamond-point, and the accurate descriptions of the Chinese fully bear out this fact.

DIAMOND AND GOLD.— The earliest passage of fundamental historical value in which the diamond is clearly indicated occurs in the *Tsin k'i kü chu* 晉起居注,[2] and is handed down to us in two different versions. One of these runs as follows:[3] "In the third year of the period Hien-ning (A.D. 277), Tun-huang[4] presented to the Emperor diamonds (*kin-kang*). Diamonds are the rulers in the midst of gold (or preside in the proximity of gold 主金中). They are neither washed,[5] nor can they be melted. They can cut jade, and come from (or are produced in) India." The other version of this text, ascribed to

---

[1] The chief arguments are discussed below on pp. 42–46.

[2] The term *k'i kü chu* 起居注 designates a peculiar class of historical records dealing with the acts of prominent persons and sovereigns. The first in existence related to the Han Emperor Wu. The well-known *Mu t'ien-tse chuan* (Life of the Emperor Mu) agreed in style and make-up with the *k'i kü chu* which were extant under the Sui dynasty (see *Sui shu*, Ch. 33, p. 7). Under the Tsin quite a number of books of this class were written, which are enumerated in the chapter on Sui literature quoted. Judging from the titles there given, each must have embraced a fixed year-period; hence the passage quoted above must have been contained in the *Tsin Hien-ning k'i kü chu*, that is, Annotations on the Conditions of the Period Hienning (275–280) of the Tsin Dynasty, a work in ten chapters, written by Li Kuei 李軌. Nineteen other titles of works of this type referring to the Tsin period, and apparently all contemporary records, are preserved in the *Sui shu* and were utilized at that time; thus the *Tsin k'i kü chu* is quoted in the biography of Yü-wên K'ai 宇文愷 in the Sui Annals.

[3] *T'ai p'ing yü lan*, Ch. 813, p. 10.

[4] In the north-western corner of Kan-su, near the border of Turkistan.

[5] As is the case with gold-sand.

the same work, is recorded thus:[1] "In the thirteenth year of the reign of the Emperor Wu (A.D. 277) there was a man in Tun-huang, who presented the Court with diamond jewels (*kin-kang pao*). These are produced in the midst of gold (生金中). Their color is like that of fluor-spar,[2] and in their appearance they resemble a grain of buckwheat. Though many times fused, they do not melt. They can cut jade as though it were merely clayish earth." It is manifest that these two texts, from their coincidence chronologically, are but variants referring to one and the same event, under the Tsin dynasty (265–419); and it is likewise apparent that the text as preserved in the *T'ai p'ing yü lan*, the great cyclopædia published by Li Fang in 983, bears the stamp of true originality, while that in the *P'ien tse lei pien* is made up of scraps borrowed from the *Pao p'u tse* of Ko Hung (p. 21) and Lie-tse's notice of *kun-wu* (p. 28).[3] From this memorable passage we may gather several interesting facts: diamonds were traded in the second part of the third century from India by way of Turkistan to Tun-huang for further transmission inland into China proper; and the chief characteristics of the stone were then perfectly grasped by the Chinese, particularly its property of cutting other hard stones. The most important gain, however, for our specific purpose, is the observation that a bit of Plinian folk-lore is mingled with the Chinese account. We are at once reminded of Pliny's statement that *adamas* was the name given to a nodosity of gold, sometimes, though but rarely, found in the mines in company with gold, and that it seemed to occur only in gold.[4] Pseudo-

---

[1] *P'ien tse lei pien*, Ch. 71, p. 11 b.

[2] See above, p. 21.

[3] A third variant occurs in *Yüan kien lei han* (Ch. 361, p. 18 b), where the term "diamond" is, strangely enough, suppressed. This text runs thus: "The Books of the Tsin by Wang Yin say that in the third year of the period Hien-ning (A.D. 277), according to the *K'i kü chu*, from the district of Tun-huang were brought to the Court objects found in gold caves, which originate in gold, are infusible, and can cut jade."

[4] Ita appellatur auri nodus in metallis repertus perquam raro [comes auri] nec nisi in auro nasci videbatur (XXXVII, 15, § 55). Also Plato is credited with having entertained a similar notion (KRAUSE, Pyrgoteles, p. 10; H. O. LENZ, Mineralogie der alten Griechen und Römer, p. 16; BLÜMNER, Technologie, Vol. III, p. 230; and in Pauly's Realenzyklopädie, Vol. IX, col. 322); although others, like E. O. VON LIPPMANN (Abhandlungen und Vorträge, Vol. II, p. 39), are not convinced that Plato's *adamas* means the diamond. The note in BOSTOCK and RILEY'S translation of Pliny (Vol. VI, p. 406) — that "this statement cannot apply to the diamond as known to us, though occasionally grains of gold have been found in the vicinity of the diamond" — is not to the point. On the contrary, it is a well-established fact that the diamond does occur in connection with gold; and this experience even led to the discovery of diamond-mines in the Ural. Owing to the similarity between the Brazilian and Uralic gold and platina sites, Alexander von Humboldt, in 1823,

Aristotle, in the introduction to his work, philosophizes on the forces of nature attracting or avoiding one another. To these belongs gold that comes as gold-dust from the mine. When the diamond encounters a grain of it, it pounces on the gold, wherever it may be in its mine, till the union is accomplished.[1] Qazwīnī speaks of an amicable relationship between gold and the diamond, for if the diamond comes near gold, it clings to the latter; also it is said that the diamond is found only in gold-mines.[2] A commentary to the *Shan hai king*[3] has the following: "The diamond which is produced abroad belongs to the class of stones, but resembles gold (or metal) and has a brilliant splendor. It can cut jade. The foreigners wear it in the belief that it wards off evil influences." It is therefore highly probable that the first element (*kin*) in the Chinese compound *kin-kang* was really intended to convey the meaning "gold" (not "metal" in general), and that the term was framed in consequence of that tradition reaching Tun-huang, and ultimately traceable to classical antiquity. A further intimation as to the significance of the newly-coined term we receive in the same period, that of the Tsin dynasty, when the stone and its nature were perfectly known in China. Indeed, it is several times alluded to in the official Annals of the Tsin Dynasty (265–419). At that time "a saying was current among the people of Liang,[4] that the principle of the diamond of the Western countries is strength, and that for this reason the name *kin-kang* was conferred upon it in Liang."[5] In combining this information with the previous text of the *Tsin k'i kü chu*, we arrive at the conclusion that the term *kin-kang* reflects two traditions,— the word *kin* referring to the origin of the diamond in gold, the word *kang* alluding to its

---

expressed the idea that the diamond accompanying these two metals in Brazil should be discovered also in the Ural; under the guidance of this prognostic, the first diamonds were really found there in 1829 (BAUER, Edelsteinkunde, 2d ed., p. 292). The diamonds of California have been found in association with gold-bearing gravels, while washing for gold (FARRINGTON, Gems and Gem Minerals, p. 87). The statement of Pliny proves that he indeed speaks of the diamond.

[1] J. RUSKA, Steinbuch des Aristoteles, p. 129.

[2] RUSKA, Steinbuch aus der Kosmographie des al-Qazwīnī, p. 6.

[3] Quoted in *Yüan kien lei han*, Ch. 26, p. 46.

[4] Liang is the name of one of the nine provinces (*chou*) into which China was anciently divided by the culture-hero and semi-historical Emperor Yü, comprising what is at present Sze-ch'uan and parts of Shen-si, Kan-su, and Hu-pei (regarding the boundaries of Liang-chou, see particularly LEGGE, Chinese Classics, Vol. III, pp. 119–120). Liang-chou was one of the nineteen provinces into which China was divided under the Tsin dynasty, with Wu-wei (in Kan-su) as capital (compare PITON, *China Review*, Vol. XI, p. 299).

[5] *Tsin shu*, Ch. 14, p. 16. The Annals of the Tsin Dynasty were compiled by Fang Hüan-ling (578–648).

extreme hardness, likewise emphasized by Pliny; *kin-kang*, accordingly, means "the hard stone originating in gold."[1]

In our middle ages we meet the notion of adamantine gold which is credited with the same properties as the diamond.   In the famous letter, purported to have been addressed by Prester John to the Byzantine Emperor Manuel, and written about 1165, a floor in the bakery of the alleged palace of the Royal Presbyter in India is described as being of adamantine gold, the strength of which can be destroyed neither by iron, nor fire, nor any other remedy, save buck's blood.[2]

THE TERM "KUN-WU."— It is difficult to decide the origin of the word *kun-wu*.   It would be tempting to regard it as a transcription of the Greek or West-Asiatic word denoting the diamond-point; unfortunately, however, the Greek designation for this implement is not known.   More probably the Chinese term may be derived from an idiom spoken in Central Asia; at any rate, the word itself was employed in China before the introduction of diamond-points from the West.   In a poem of Se-ma Siang-ju, who died in 117 B.C., we meet a precious stone named *kun-wu* 琨珸 , as occurring in Sze-ch'uan, on the nature of which the opinions of the commentators dissent.[3]   The *Han shu yin i* explains it as the name of a mountain which produces excellent gold. Shi-tse or Shi Kiao (about 280 B.C.) explains it as "gold" or "metal of Kun-wu" 昆吾之金, which may mean that he takes the latter as

---

[1] In the study of Chinese texts some precaution is necessary in the handling of the term *kin kang*, which does not always refer to the diamond, but sometimes presents a complete sentence with the meaning "gold is hard."   Three examples of this kind are known to me.   One occurs in *Nan shi* (biography of Chang T'ung; see *Pien tse lei pien*, Ch. 71, p. 11 b): "Gold is hard, water is soft: this is the difference in their natural properties."   In *Tsin shu* (Ch. 95, p. 13 b; biography of Wang Kia) we meet the sentence 金剛火彊.   This, of course, could mean "the diamond is conquered by fire,"— a sentence which, from the standpoint of our scientific experience, would be perfectly correct; from a Chinese viewpoint, however, it would be sheer nonsense, the Chinese as well as the ancients entertaining the belief that fire does not affect the diamond (p. 23).   The passage really signifies, "Gold is hard, yet is overcome (melted) by fire."   The correctness of this translation is confirmed by a passage in a work *Yi shi fêng kio* (quoted in *Pien tse lei pien, l. c.*), where the same saying occurs in parallelism with two preceding sentences: "Branches of trees fall and return to their roots; water flows from the roots and returns to the branches; gold is hard, yet is overcome by fire; every one returns to his native place."

[2] Pavimentum vero est de auro adamantino, fortitudo cuius neque ferro neque igne neque alio medicamine potest confringi sine yrcino [hircino] sanguine (F. ZARNCKE, Der Priester Johannes I, p. 93).   Compare the analogous passage in the same document, "Infra domum sunt duae magnae molae, optime ad molendum dispositae, factae de adamante lapide, quem namque lapidem neque lapis neque ignis neque ferrum potest confringere."   Both these passages are not contained in the original draught of the letter, but are interpolations from manuscripts of the thirteenth century.

[3] *Shi ki*, Ch. 117, p. 2 b.

the name of the locality whence the ore came. Se-ma Piao (240–305) interprets it as a stone ranking next to jade. Then follows in his text the story of *kun-wu* in Liu-sha, quoted from the *Lung yü ho t'u*, which has been discussed above. I do not know whether this is a separate editorial comment, or was included in the commentary of Se-ma Piao. At all events, the fact is borne out that the word *kun-wu* in the *Shi ki*, and that referring to the West, are considered by the Chinese as identical, and that the mode of writing (with or without the classifier 'jade') is immaterial.[1] We know that in times of old numerous characters were written without the classifiers, which were but subsequently added. The writing *kun-wu* in Lie-tse with the classifier 'metal' plainly manifests itself as a secondary move,[2] and the simple *kun-wu* without any determinative classifier doubtless represents the primary stage. This is shown also by the existence of a character 琨, where the element *kun* is combined with the classifier 'stone.'[3] If in the *Shi ki* the word *kun-wu* is linked with the classifier 'jade;' and if, further, this term appears coupled with nine other designations of stones, the whole series of ten being introduced by the words "following are the stones,"—the interpretation "gold" is absurd, and that of Se-ma Piao has only a chance. It would therefore be possible that *kun-wu* originally served for naming some hard stone indigenous to Sze-ch'uan, and was subsequently transferred to the imported diamond-point. The name for the stone may have been inspired by that of the mountain Kun-wu, stones being frequently named in China for the mountains or localities from which they are derived. On the other hand, there is a text in which the name *Kun-wu* in this connection is conceived as that of a clan or family by the addition of the word *shi* 氏. This is the *Chou shu*,[4] which relates the tradition that the Western Countries offered fire-proof cloth (asbestos), and the Kun-wu Clan presented jade-cutting knives. It seems certain that this version has no basis in reality, but presents a makeshift to account for the troublesome word *kun-wu*. How it sprang into existence may be explained from the fact that there was in ancient times, under the Hia dynasty, a rebel by the name Kun-wu, mentioned in the *Shi king* and *Shi ki*;[5] but it is obvious that this family name bears

---

[1] In *Ts'ien Han shu*, where the same text is reproduced, *kun-wu* is written without the classifiers.

[2] In all likelihood this is merely a device of later editors of Lie-tse's text. There are editions in which the plain *kun-wu* without the classifier is written (see *P'ei wên yün fu*, Ch. 91, p. 16b).

[3] *P'ei wên yün fu*, Ch. 100 A, p. 25.

[4] Regarding this work see CHAVANNES, Mémoires historiques de Se-ma Ts'ien, Vol. V, p. 457. The passage is quoted in *Po wu chi*, Ch. 2, p. 4b (Wu-ch'ang edition).

[5] LEGGE, Chinese Classics, Vol. III, p. 642; CHAVANNES, *l. c.*, Vol. I, p. 180.

no relation to the name of the mountain in Sze-ch'uan, the stone hailing from it, and the diamond-point coming from the West.[1]

Ko Hung informs us that "the Emperor Wên of the Wei dynasty (220–226), who professed to be well informed with regard to every object in nature, declared that there were no such things in the world as a knife that would cut jade, and fire-proof cloth; which opinion he recorded in an essay on the subject. Afterwards it happened that both these articles were brought to court within a year; the Emperor was surprised, and caused the essay to be destroyed; this course being un-avoidable when he found the statements to be without foundation."[2] General Liang-ki, who lived at the time of the Emperor Huan (147–167), is said to have possessed asbestos and "jade-cutting knives."[3] The book handed down under the name of K'ung-ts'ung-tse[4] contains the tradition that the Prince of Ts'in obtained from the Western Jung a sharp knife capable of cutting jade as though it were wood. The poet Kiang Yen (443–504) wrote a poem on a bronze sword, in the preface of which he observes that there are also red knives of cast copper capable of cutting jade like clayish earth,— apparently a reminiscence of the passage of Lie-tse, only the latter's "iron" is replaced by "copper." In the preceding texts the term kun-wu is avoided, and only the phrase "jade-cutter" (ko yü tao) has survived.

TOXICOLOGY OF THE DIAMOND.— Contrary to his common practice, Li Shi-chên does not state whether the diamond is poisonous or not. As to the curative powers of the stone, he asserts that when set into hair-spangles, finger-rings, or girdle-ornaments, it wards off uncanny influences, evil, and poisonous vapors.[5] On this point the Chinese agree with PLINY, according to whom adamas overcomes and neutralizes

---

[1] Also HIRTH (Chinesische Ansichten über Bronzetrommeln, p. 20) persuaded himself that this proper name is not connected with what he believed to be the "kun-wu sword." It is difficult, however, to credit the theory that the name kun-wu, as tentatively proposed by Hirth, could be a transcription on an equal footing with Hiung-nu (Huns). Aside from phonetic obstacles, the fact remains that the Chinese notices of kun-wu do not point in the direction of the Huns, but refer to Liu-sha in Ta Ts'in (the Roman Orient).

[2] A. WYLIE, Chinese Researches, pt. III, p. 151.

[3] Yüan kien lei han, Ch. 225, p. 2; and WYLIE, l. c., p. 143.

[4] The son of K'ung Fu, a descendant of Confucius in the ninth degree, who died in 210 B.C. (GILES, Biographical Dictionary, p. 401). It is doubtful whether the book which we nowadays possess under the title K'ung-ts'ung-tse (incorporated in the Han Wei ts'ung shu) is the one which he wrote (compare CHAVANNES, Mémoires historiques de Se-ma Ts'ien, Vol. V, p. 432). The passage referred to is quoted in P'ei wên yün fu, Ch. 91, p. 21.

[5] The source for this statement doubtless is the Nan chou i wu chi, quoted on p. 34, which ascribes this notion to foreigners.

poisons, dispels insanity, and drives away groundless apprehensions from the mind.[1] The coincidence would not be so remarkable were it not for the fact that in mediæval Mohammedanism the theory of diamonds being poisonous had been developed. This idea first looms up in Pseudo-Aristotle, who is also the first to stage the snakes in the Diamond Valley, and cautions his readers against taking the diamond in their mouths, because the saliva of the snakes adheres to it so that it deals out death.[2] According to al-Bērūnī, the people of Khorasan and Iraq employ the diamond only for purposes of boring and poisoning.[3] This superstition was carried by the Mohammedans into India, where the belief had prevailed that the diamond wards off from its wearer the danger of poison.[4] The people of India now adhere to the superstition that diamond-dust is at once the least painful, the most active, and most infallible of all poisons. In our own time, when Mulhar Ráo of Baroda attempted to poison Col. Phayre, diamond-dust mixed with arsenic was used.[5] A. BOETIUS DE BOOT (1550–1632)[6] was the first modern mineralogical writer who refuted the old misconception, demonstrating that the diamond has no poisonous properties whatever.

IMITATION DIAMONDS.— While all the principal motives of the lore garnered by the Chinese around the diamond come from classical regions, I can discover but a single notion traceable to India. PLINY has written a short chapter on the method of testing precious stones,[7] but he does not tell us how to discriminate between real and counterfeit diamonds. According to the Hindu mineralogists, iron, topaz, hyacinth, rock-crystal, cat's-eye, and glass served for the imitation of the diamond; and the forgery was disclosed by means of acids, scratching,

---

[1] Adamas et venena vincit atque inrita facit et lymphationes abigit metusque vanos expellit a mente (XXXVII, 15, § 61).

[2] J. RUSKA, Steinbuch des Aristoteles, p. 150; and Diamant in der Medizin (*Festschrift Baas*, pp. 121–125); likewise al-Akfānī (E. WIEDEMANN, Zur Mineralogie im Islam, p. 219). Qazwīnī (J. RUSKA, Steinbuch aus der Kosmographie des al-Kazwīnī, p. 35) quotes Ibn Sīnā as saying that the venomous property imputed by Aristotle to the diamond is a hollow pretence, and that Aristotle is ignorant of the fact that snake-poison, after flowing out, loses its baleful effect, especially when some time has elapsed. This sensible remark does not prevent Qazwīnī, in copying his second anonymous source relating to the diamond, from alleging that "it is an extremely mortal poison."

[3] E. WIEDEMANN, *Der Islam*, Vol. II, p. 352.

[4] L. FINOT, Lapidaires indiens, p. 10. Varāhamihira (A.D. 505–587) states that a good diamond dispels foes, danger from thunder-strokes or poison, and promises many enjoyments (H. KERN, Verspreide Geschriften, Vol. II, p. 98).

[5] W. CROOKE, Things Indian, p. 379.

[6] Gemmarum et lapidum historia, p. 124 (ed. of A. Toll, Lugduni Batavorum, 1636); compare also J. RUSKA, *Festschrift Baas*, pp. 125–127.

[7] XXXVII, 76.

and the touchstone. The *Agastimata* is specific on this point by anathematizing forgers and recommending the following recipe: "The vile man who fabricates false diamonds will sink into an awful hell, charged with a sin equal to murder. When a connoisseur believes that he recognizes an artificial diamond, he should test it by means of acids or vinegar, or through application of heat: if false, it will lose color; if true, it will double its lustre. It may also be washed and brought in contact with rice: thus it will at once be reduced to a powder."[1] The *Ts'i tung ye yü* of Chou Mi, previously quoted, imparts this advice: "In order to distinguish genuine from counterfeit diamonds, expose the stone to red-heat and steep it in vinegar: if it retains its former appearance and does not split, it is real. When the diamond-point happens to become blunt, it should be heated till it reddens; and on cooling off, it will again have a sharp point."[2] The first experiment is identical with that proposed in the Sanskrit text. As to the second, we again encounter a striking parallel in Pliny: "There is such great difference in stones, that some cannot be engraved by means of iron, others may be cut only with a blunt graver, all, however, by means of the diamond; heating of the graver considerably intensifies the effect."[3]

ACQUAINTANCE OF THE ANCIENTS WITH THE DIAMOND.— The previous notes have been based on the supposition that the stone termed *adamas* by the ancients, and that called *kun-wu* (or subsequently *kin-kang*) by the Chinese, are identical with what we understand by "diamond." This identification, however, has been called into doubt by students of classical antiquity as well as by sinologues. It is therefore necessary to scrutinize their arguments. Our investigation has clearly brought out two points,— first, that the Chinese notices of the diamond-point (*kun-wu*) agree with Pliny's account of the same implement; and, second, that Chinese traditions regarding the stone *kin-kang* perfectly coincide with those of the ancients and the Arabs concerning *adamas* and *almās*, the latter word being derived from the former. If,

---

[1] L. FINOT, Lapidaires indiens, p. xxx.

[2] F. DE MÉLY (Lapidaires chinois, p. 124) has misunderstood this passage by referring it to the stone in lieu of the diamond-point. "S'il a des facettes émoussées, on le chauffe au rouge, on le laisse refroidir, et ses facettes redeviennent aiguës." This point of view is untenable. First, the facets of a diamond are neither blunt nor sharp; second, a faceted diamond, as will be shown in detail farther on, was always unknown to the Chinese, who for the first time noticed cut diamonds in the possession of the Macao Portuguese; and, third, the parallelism with Pliny proves my conception of the Chinese text to be correct.

[3] Iam tanta differentia est, ut aliae ferro scalpi non possint, aliae non nisi retuso, omnes autem adamante. Plurimum vero in iis terebrarum proficit fervor (XXXVII, 76, § 200). Compare KRAUSE, Pyrgoteles, p. 231.

accordingly, the *adamas* of the Greeks and Romans be the diamond, the continuity of Western and Eastern traditions renders it plain that the Chinese stone *kin-kang* must be exactly the same; if, however, *adamas* should denote another stone, the claim for *kin-kang* as the diamond must lose its force. Eminent archæologists like Lessing, Krause, Blümner, and Babelon, have championed the view that Pliny's *adamas* is our diamond.[1] The opposition chiefly came from the camp of mineralogists. E. S. DANA[2] remarked upon the word *adamas*, "This name was applied by the ancients to several minerals differing much in their physical properties. A few of these are quartz, specular iron ore, emery, and other substances of rather high degrees of hardness, which cannot now be identified. It is doubtful whether Pliny had any acquaintance with the real diamond." This rather sweeping statement does not testify to a sound interpretation of Pliny's text. A recent author asserts,[3] "It is more than doubtful if the true diamond was known to the ancients. The consensus of the best opinions is that the adamas was a variety of corundum, probably our white sapphire." Let us now examine what the foundation of these "best opinions" is.

The very first sentence with which PLINY opens his discussion of *adamas* is apt to refute these peremptory assertions: "The greatest value among the objects of human property, not merely among precious stones, is due to the adamas, for a long time known only to kings, and even to very few of these."[4] The most highly prized and valued of all antique gems, the "joy of opulence,"[5] should be quartz, specular iron ore, emery, and other substances which cannot now be identified! The ancients were not so narrow-minded that almost any stone picked up anywhere in nature could have been regarded as their precious stone foremost in the scale of valuation. If the peoples of India likewise regarded the diamond as the first of the jewels, if their treatises on mineralogy assign to it the first place,[6] and if Pliny is familiar with the

---

[1] Also so eminent an historian of natural sciences as E. O. VON LIPPMANN (Abhandlungen und Vorträge, Vol. I, p. 9) grants to Pliny a knowledge of the diamond.

[2] System of Mineralogy, p. 3, 1850. In the new edition of 1893 this passage has been omitted; the first distinct mention of the diamond is ascribed to Manilius (!), and Pliny's *adamas* is allowed to be the diamond in part.

[3] D. OSBORNE, Engraved Gems, p. 271 (New York, 1912).

[4] Maximum in rebus humanis, non solum inter gemmas, pretium habet adamas, diu non nisi regibus et iis admodum paucis cognitus (XXXVII, 15, § 55; again 78, § 204).

[5] Opum gaudium (PLINY, procemium of Lib. XX).

[6] L. FINOT, Lapidaires indiens, p. XXIV. Buddhabhaṭṭa (*ibid.*, p. 6) says, "Owing to the great virtue attributed by the sages to the diamond, it must be studied in the

adamas of India, it is fairly certain that also the *adamas* is the diamond; it is, at any rate, infinitely more certain than that the jewel
first known only to kings should have been quartz, specular iron ore,
emery, or some other unidentified substance.   That emery is not meant
by Pliny becomes evident from the fact that emery was well known
to the ancients under the name *naxium*.[1]   The Indian diamond is perfectly well described by Pliny as an hexangular crystal resembling
two pyramids placed base to base; that is, the octahedral form in
which the diamond commonly crystallizes.[2]   Whether the five other
varieties spoken of by Pliny are real diamonds or not is of no consequence in this connection; two of these he himself brands as degenerate stones.   The name very probably served in this case as a bare
trademark.   Diamonds at that time were scarce, and the demand was
satisfied by inferior stones.   That such were sold under the name of
"diamond" does not prove that the ancients were not acquainted with
the true diamond.   The diamond of India was known to them,[3] and

---

first place."   P. S. IYENGAR (The Diamonds of South India, *Quarterly Journal of
the Mythic Society*, Vol. III, 1914, p. 118) observes, "Among the Hindu, both ancient
and modern, the diamond is always regarded as the first of the nine precious gems
(*navaratna*)."

[1] BLÜMNER, Technologie, Vol. III, pp. 198, 286.   In Greek it is styled σμύρις.
"Emery is the stone employed by the engravers for the cutting of gems" (DIOS-
CORIDES, CLXVI).

[2] This passage has embarrassed some interpreters of Pliny (H. O. LENZ, Mineralogie der alten Griechen und Römer, p. 163; A. NIES, Zur Mineralogie des Plinius,
p. 5), because they did not grasp the fact that it is the octahedron which has six
points or corners (sexangulus); and thus such inadequate translations were matured
as "its highly polished hexangular and hexahedral form" (BOSTOCK and RILEY,
Natural History of Pliny, Vol. VI, p. 406).   No body, of course, can simultaneously
be hexangular and hexahedral, the hexahedron being a cube with six sides and four
points.   Pliny's wording is plain and concise, and his description tallies with the
Sanskrit definition of the diamond as "six-cornered" (*shaṭkona, shaṭkoṭi,* or *shaḍāra;*
see R. GARBE [Die indischen Mineralien, p. 80], who had wit enough to see that this
term hints at the octahedron and correctly answers to the diamond; likewise L.
FINOT, Lapidaires indiens, p. XXVII).   It is not impossible that the Plinian definition
is an echo of a tradition hailing, with the diamond, directly from India.

[3] The Indian diamond is mentioned also by PTOLEMY, according to whom the
greatest bulk of diamonds was found with the Savara tribe (PAULY, Realenzyklopädie, Vol. I, col. 344), by the Periplus Maris Erythraei (56, ed. FABRICIUS, p. 98),
and by DIONYSIUS PERIEGETES (second century A.D.) in his poem describing the
habitable earth (Orbis descriptio, Verse 1119).   The diamond is doubtless included
also among the precious stones cast by the sea upon the shores of India, mentioned
by CURTIUS RUFUS, and among STRABO's precious stones, some of which the Indians
collect from among the pebbles of the river, and others of which they dig out of the
earth (McCRINDLE, Invasion of India by Alexander, pp. 187–188).   Alexander's
expedition made the Greeks familiar with the diamond, hence it is mentioned by
THEOPHRASTUS (De lapidibus, 19), who compares the carbuncle with the adamas.   I
do not agree with the objections raised by some authors against Theophrastus'

the Periplus[1] expressly relates of the exportation from India of diamonds and hyacinths.   Further, the Annals of the T'ang Dynasty[2] come to our aid with the statement that India has diamonds, sandal-wood, and saffron, and barters these articles with Ta Ts'in (the Roman Orient), Fu-nan, and Kiao-chi.   The fact therefore remains, as attested by the Chinese, that India shipped diamonds to the West.[3]

There is, moreover, in the chapter of Pliny, positive evidence voicing the cause of the diamond.   He is familiar with the hardness of the stone, which is beyond expression (quippe duritia est inenarrabilis); and, owing to its indomitable powers, the Greeks bestowed on it the name adamas ("unconquerable").[4]   He is acquainted, as set forth on p. 31, with the technical use of diamond splinters, which cut the very hardest substances known.   If one of the apocryphal varieties of the diamond, styled siderites (from Greek sideros, "iron"), a stone which shines like iron, is reported to differ in its main properties from the true diamond, inasmuch as it will break when struck by the hammer, and admit of being perforated by other kinds of adamas, this observation

---

acquaintance with the diamond.   H. BRETZL (Botanische Forschungen des Alexanderzuges) has well established the fact that he commanded an admirable knowledge of the vegetation of India; thus he may well have heard also of the Indian diamond from his same informants.   It is not necessary to assume, however, that he knew the diamond from autopsy, as he does not describe it, but mentions it only passingly in the single passage referred to; also H. O. LENZ (Mineralogie der alten Griechen und Römer, p. 19) holds the same opinion.   It is difficult to see that Theophrastus could have compared with the carbuncle any other stone than the diamond.

[1] Ch. 56 (ed. of FABRICIUS, p. 98).   G. F. KUNZ (Curious Lore of Precious Stones, p. 72) observes, "The writer is disinclined to believe that the ancients knew the diamond."   The same author, however, believes in the existence of diamonds in ancient India; but Rome then coveted all the precious stones of India, and he who accepts the Indian diamond as a fact must be consistent in granting it to the ancients, too.

[2] T'ang shu, Ch. 221A, p. 10b.

[3] Indian diamonds were apparently traded also to Ethiopia, for Pliny records the opinion of the ancients that the adamas was only to be discovered in the mines of Ethiopia between the temple of Mercury and the island of Meroë (veteres eum in Aethiopum metallis tantum inveniri existimavere inter delubrum Mercuri et insulam Meroën).   Ajasson's comment that the Ethiopia here mentioned is in reality India, and that the "Temple of Mercury" means the Brahmaloka, or "Temple of Brahma" (it does not mean "temple," but "world" of Brahma) is of course wrong. The reference to Meroë, the capital of Ethiopia, at once renders this opinion impossible; besides, Pliny's geographical terminology is always distinct as to the use of India and Ethiopia.   The tradition of Ethiopic diamonds is confirmed by the Greek Romance of Alexander (III, 23), in which Queen Candace in the palace of Meroë presents Alexander with a crown of diamonds (adamas; see A. AUSFELD, Der griechische Alexanderroman, pp. 101, 192).

[4] Invictum is given by Pliny himself (procemium of lib. xx) as if it were a translation of the Greek word.   The Physiologus says that the stone is called adamas because it overpowers everything, but itself cannot be overpowered.

plainly bears out the fact that Pliny and his contemporaries knew very well the properties of the real diamond, and, moreover, that diamond affects diamond. In short, due allowance being made for inaccuracies of the tradition of the Plinian text and the imperfect state of mineralogical knowledge of that period, no fair criticism can escape from the conclusion that Pliny's *adamas* is nothing but the diamond. The fact that also other stones superficially resembling diamonds were at that time taken for or passed off as diamonds, cannot change a jot of this conclusion. Such substitutes have been in vogue everywhere and at all times, and they are not even spared our own age.[1] Pliny's condemnation of these as not belonging to the genus (degeneres) and only enjoying the authority of the name (nominis tantum auctoritatem habent) reveals his discriminative critical faculty and his ability to distinguish the real thing from the frame-up. The perpetuity of the Plinian observations in regard to the *adamas* among the Arabs, Persians, Armenians, Hindu, and Chinese, who all have focussed on the diamond this classical lore inherited by him, throws additional evidence of most weighty and substantial character into the balance of the ancients' thorough acquaintance with the real diamond. The Arabs, assuredly, were not feeble-minded idiots when they coined their word *almās* from the classical *adamas* for the designation of the diamond, and this test of the language persists to the present day. The Arab traders and jewellers certainly were sufficiently wide awake to know what a diamond is, and their Hindu and Chinese colleagues were just as keen in recognizing diamonds, long before any science of mineralogy was established in Europe. The world-wide propagation of the same notions, the same lore, the same valuation connected with the stone, is iron-hard proof for the fact that in the West and East alike this stone was the diamond. This uniformity, coherence, perpetuity, and universality of tradition form a still mightier stronghold than the interpretation of the Plinian text. For this double reason there can be no doubt also that the *kin-kang* of Chinese tradition is the diamond.

CUT DIAMONDS.— Another question is whether the ancients were cognizant of the diamond in its rough natural state only, or whether they understood how to cut and polish it. This problem has caused

---

[1] There were rock-crystals found in northern Europe in the seventeenth century and passed under the name of diamond. JOHANNES SCHEFFER (Lappland, p. 416, Frankfurt, 1675) tells that the lapidaries sometimes used to polish these crystals or diamonds of Lapland and to sell them as good diamonds, even frequently deceive experts with them, because they are not inferior in lustre to the Oriental stones. In the eighteenth century crystal was still called "false diamond" (J. KUNCKELL, Ars Vitraria, p. 451, Nürnberg, 1743).

an endless controversy. LESSING, in his "Briefe antiquarischen Inhalts" (No. 32), which it is still as enjoyable as profitable seriously to study, has shown with a great amount of acumen that the ancients possessed no knowledge whatever of diamond-dust, and therefore did not know how to polish the diamond. This opinion, however, did not remain uncontradicted. The opposite view is heralded by BLÜMNER,[1] who argues, "Despite the lack of positive testimony, we cannot forbear assuming that the ancients understood, though possibly imperfectly, how to polish the diamond. Since only in this state is the stone capable of displaying its marvellous lustre, play of colors, and translucency, its extraordinary valuation among the ancients would not be very intelligible had they known it merely as an uncut gem." This argument is rather sentimental and intuitive than well founded. As far as the plain facts are concerned, Lessing is right; and, what is even more remarkable, has remained right from 1768, the date at which he wrote, up to the present. No cut diamond of classical antiquity has as yet come to light; and in order to pass audaciously over the body of Pliny, and have us believe what he does not say, such a palpable piece of evidence would be indispensable. As a matter of fact, neither Pliny nor any other ancient writer loses a word about diamond-dust; nor does he mention that the diamond can be cut and polished, or that it was so treated; nor does he express himself on the adamantine lustre.[2] This silence is sufficiently ominous to guard ourselves, I should think, against the rash assumption that the ancients might have cut the diamond. Its high appreciation is quite conceivable without the application of this process, for even the uncut diamond possesses brilliancy and lustre enough to allure a human soul. The possibility would remain that the ancients may have received worked diamonds, ready made, straight from India.[3]

---

[1] Technologie, Vol. III, p. 233.

[2] BECKMANN (Beiträge zur Geschichte der Erfindungen, Vol. III, p. 541) held that the ancients employed diamond-dust for the cutting of stones other than the diamond, but he denied that they polished the diamond with its own dust. This is certainly a contradiction in itself: if the ancients knew the utility of diamond-dust, there is no reason why they should not have applied it to the diamond; and if they did not facet diamonds, it is very plain that they lacked the knowledge of diamond-dust. BAUER (Edelsteinkunde, p. 302, 2d ed.) observes, "In how far the ancients understood how to polish diamonds, or at least to improve existing crystal surfaces by polishing, is not known with certainty. From the traditions handed down, however, it becomes evident that this art was not wholly unknown to the ancients." The latter statement is without basis.

[3] This hypothesis was formulated by H. O. LENZ (Mineralogie der alten Griechen und Römer, pp. 39, 164, Gotha, 1861), who concluded from what the ancients said regarding the brilliancy of the stone that diamonds cut and polished in the country of their origin were traded to Europe.

Here, again, it is unfortunate that our knowledge fails us: the ancient Indian sources exhibit the same lack of information on the identical points as does Pliny. S. K. AIYANGAR[1] justly points out that in the description of the diamond, as given in the Arthaçastra (quoted above, p. 16), "there is nothing to warrant the inference that diamonds were artificially cut; but, perhaps, the fact that diamonds were used to bore holes in other substances makes it clear that lapidary work was not unknown." A very late work on gems, the *Agastimata*, in an appendix of still later date, contains a curious passage in which the cutting of diamonds is prohibited: "The stone which is cut with a blade, or which is worn out by repeated friction, becomes useless, and its benevolent virtue disappears; the stone, on the contrary, which is absolutely natural has all its virtue." L. FINOT,[2] to whom we owe the edition and translation of this work, rightly points out that cutting and polishing are clearly understood here; but another passage in the same treatise speaks of it as a normal process, without forbidding what precedes the setting of diamonds for ornaments, and we regret with Finot that these passages cannot be dated. GARCIA AB HORTO, who wrote in 1563, informs us that by the people of India natural diamonds were preferred to the cut ones, in opposition to the Portuguese.[3] TAVERNIER (1605–89) describes the diamond-polishing in the Indian mines by means of diamond-dust.[4] In the face of the *Agastimata* and Garcia's statements, suspicion is ripe that diamond-cutting was introduced into India only by the Portuguese,[5] and that the employment of uncut stones was the really national fashion of India. The passage in the additional chapter of the *Agastimata*, as stated, cannot be dated with certainty, but it seems more probable that it falls within the time of the Portuguese era of India than that it

---

[1] *Quarterly Journal of the Mythic Society*, Vol. III, p. 130.

[2] Lapidaires indiens, p. xxx.

[3] Si come una vergine si preferisce ad una donna corrotta, cosi il diamante dalla natura polito, e acconcio s'ha da preferire à quello, che dall'arte è stato lavorato. Al contrario fanno i Portughesi, stimando più quelli, che sono dall'artificio dell' huomo acconci, e lavorati (Italian edition, p. 180).

[4] "There are at this mine numerous diamond-cutters, and each has only a steel wheel of about the size of our plates. They place but one stone on each wheel, and pour water incessantly on the wheel until they have found the 'grain' of the stone. The 'grain' being found, they pour on oil and do not spare diamond-dust, although it is expensive, in order to make the stone run faster, and they weight it much more heavily than we do. . . . The Indians are unable to give the stones so lively a polish as we give them in Europe; and this, I believe, is due to the fact that their wheel does not run so smoothly as ours" (ed. of V. BALL, Vol. II, pp. 57, 58).

[5] Also BAUER (Edelsteinkunde, p. 302, 2d ed.) is of the opinion that the diamond-cutting of Europe, which was developed from the end of the middle ages, has not remained without influence upon India, and that perhaps the process was introduced from Europe into India, or was at least resuscitated there.

should be much earlier. It is safer to adopt this point of view, as the *Ratnaparīkshā* of Buddhabhaṭṭa, who presumably wrote somewhat earlier than the sixth century, does not mention the cutting of diamonds,[1] nor does the mineralogical treatise of Narahari from the fifteenth century.[2] At all events, we have as yet no ancient source of Indian literature in which the cutting of diamonds is distinctly set forth. The discovery of such a passage, or, what is still more preferable, archæological evidence in the shape of ancient cut diamonds, may possibly correct our knowledge in the future. For the present it seems best to adhere to the view that the polishing of diamonds was foreign to ancient India, and a process but recently taught by European instructors. Certainly, we should not base our present conclusions on hoped-for future discoveries, which may even never be made, nor should we shift evidence appropriate to the last centuries into times of antiquity, nor is there reason to persuade ourselves that the knowledge of the diamond on the part of the Indians goes back to the period of a boundless antiquity (see p. 16). The Chinese contribute nothing to the elucidation of this problem; and certain it is that they merely kept the diamonds in the condition in which they received them from the Roman Orient, Fu-nan, India, and the Arabs, without attempting to improve the appearance of the stones. The European tradition that Ludwig van Berquen of Brügge in 1476 was the "inventor" of the process of polishing diamonds by means of diamond-dust, is, of course, nothing more than a conventional story (*une fable convenue*). As shown by BAUER,[3] diamonds were roughly or superficially polished as early as the middle ages; and Berquen improved the process and arranged the facets with stricter regularity, whereby the color effect was essentially enhanced.[4] The early history of the technique in Europe is not yet exactly ascertained.[5]

---

[1] L. FINOT (*l. c.*, p. xxx), it is true, alludes to a passage of this work where, in his opinion, it is apparently the question of diamond-polishing. The text, however, runs thus: "The sages must not employ for ornament a diamond with a visible flaw; it can serve only for the polishing of gems, and its value is slight." This only means that deficient diamonds were used for the working of stones other than the diamond.

[2] R. GARBE, Die indischen Mineralien, pp. 80–83.

[3] *L. c.*, p. 303.

[4] The Berquen legend was firmly established in the seventeenth century, under the influence of one of his descendants. ROBERT DE BERQUEN (in his book Les merveilles des Indes orientales et occidentales, p. 13, Paris, 1669), after disdainfully talking about the rough diamonds obtained from India, soars into this panegyric of his ancestor: "Le Ciel doua ce Louis de Berquen qui estoit natif de Bruges, comme un autre Bezellée, de cet esprit singulier ou genie, pour en trouver de luy mesme l'invention et en venir heureusement à bout." Then follows the story of the "invention."

[5] H. SÖKELAND (*Zeitschrift für Ethnologie*, Vol. XXIII, 1891, Verhandlungen, p. 621) took up this question again, and thought that definite proof had not been

On the other hand, we have two testimonies in witness of the fact that, even though a certain crude method of treating diamonds may have lingered in the Orient, the superior European achievements along this line were received by Oriental nations as a surprising novelty. The Armenian *lapidarium* of the seventeenth century states,[1] "No one besides the Franks (Europeans) understands how to polish and to bore the diamond. The polished stone of four carats is sold at ten thousand *otmani*. The Franks at Aleppo say that the diamond, though it is the king of all precious stones, is of no utility without polishing, because in its raw state admixtures will remain, which may often not be noticeable in the cut stone." The Chinese made their first acquaintance with polished diamonds among the Portuguese of Macao, who, they say, base their valuation on this quality.[2]

ACQUAINTANCE OF THE CHINESE WITH THE DIAMOND.— Let us now examine the objections which have been raised by sinologues to the identification of the term *kin-kang* with the diamond. F. PORTER SMITH,[3] who made rather inexact statements on the subject, in 1871 contested that *kin-kang* denotes the real diamond, and treated it under the title "corundum," which arbitrarily he takes for "a kind of adamantine spar." Corundum, he states, crystallizes in six-sided prisms, but the Chinese siliceous stone is said to be octahedral in form. If this be really said by the Chinese, it is evidence that the stone in question is the diamond, not corundum; and the latter, in its main varieties of ruby and sapphire, is well known to the Chinese under a number of terms. Blackish emery, containing iron, it is thought by Smith, is also described

---

brought forward for the assertion that the ancients did not employ diamond-dust; but he recruited no new facts for the discussion, and merely referred to the old fable that the Bishop Marbodus (1035–1123) should have been familiar with diamond-dust. MARBODUS, however, in his famous treatise De lapidibus pretiosis, most obviously speaks only of diamond-splinters (huius fragmentis gemmae sculptuntur acutis; in the earliest French translation, dés piecéttes |Ki en esclatent aguéttes| Les altres gemmes sunt talliées| E gentement aparelliéés.— L. PANNIER, Lapidaires français du moyen âge, p. 36), as translated correctly also by KING (Antique Gems, p. 392); and he does so, not because he was possibly acquainted with them, but because he copied this matter, as most of his data, from Pliny. Likewise KONRAD VON MEGENBERG, in his Book of Nature written 1349–50 (ed. of F. PFEIFFER, p. 433), states only that other hard precious stones are graved with pointed diamond-pieces. It means little, as insisted upon by Sökeland, that A. Hirth and Mariette second the cause of the ancients in the use of diamond-dust, as their opinion is not based on any text to this effect (such does not exist), but merely on the impression received from certain engraved gems. The conclusion, however, that these could not have been worked otherwise than by means of diamond-dust, is unwarranted, and plainly contradicted by Pliny's data regarding the treatment of precious stones.

[1] Russian translation of PATKANOV, p. 4.

[2] *Wu li siao shi*, Ch. 8, p. 22.

[3] Contributions toward the Materia Medica of China, pp. 74, 85.

under this heading in the *Pên ts'ao*. We have seen that what is described in this work, owing to the strict conformity with classical traditions, refers to nothing but the diamond; and it was the black diamonds which were chosen as graving-implements. According to Smith, Cambodja, India, Asia Minor, the country of the Hui-k'i (Uigur), and other countries of Asia, are said to possess this stone. Cambodja is intended for Fu-nan; and the country of the Uigur, as has been shown, is merely the theatre of action for the legend of the Diamond Valley in the version of Chou Mi (this statement is devoid of any geographical value). If the prefecture of Shun-ning in Yün-nan, as stated by Smith, yields the present supply of corundum used in cutting gems, this is an entirely different question. If the name *kin-kang* is bestowed on corundum-points, it is a commercial term, which does not disprove that the *kin-kang* of ancient tradition was the diamond, or prove that it was a kind of corundum. The diamond-points formerly imported were naturally scarce; and the Chinese, recognizing the high usefulness of this implement, were certainly eager to discover a similar material in their country, fit to take the place of the imported article.[1] This is a process which repeated itself in China numerous times: the impetus received from abroad acted as a stimulus to domestic research. If such a stone was ultimately found, it was termed *kin-kang*, not because this stone was confounded with the diamond, but for the natural reason that it was turned to the same use as the diamond-point; in other words, the name in this case does not relate to the stone as a mineralogical species, but to the stone in its function as an implement. Consequently it is inadmissible to draw any scientific inferences from the modern application of the word *kin-kang* as to the character of the stone mentioned in the earlier records of the Chinese.

A. J. C. GEERTS,[2] in his very useful, though occasionally uncritical work, charges the Chinese books with the defect of having constantly confounded the diamond with corundum, adamantine spar, pyrope,

---

[1] This is proved by the Arabs. The Arabic *lapidarium* of the ninth century, attributed by tradition to Aristotle, demonstrates that Chinese emery was known to the Arabs: the localities where it is found are the islands of the Chinese Sea, and it occurs there as a coarse sand in which are also larger and smaller hard stones (RUSKA, Steinbuch des Aristoteles, p. 151). The Arabs certainly did not confound this Chinese emery with the diamond, nor did the Chinese. This is demonstrated also by Ibn Khordādbeh, who wrote his Book of the Routes and Kingdoms between 844 and 848, and according to whom diamond and emery, the latter for polishing metal, were exported from Ceylon (G. FERRAND, Relations de voyages arabes, persans et turks rel. à l'Extrême-Orient, Vol. I, p. 31). Diamond and emery, accordingly, were distinct matters in the eyes of the Arabs, Ceylonese, and Chinese.

[2] Les produits de la nature japonaise et chinoise, pp. 201–202, 356–358 (Yokohama, 1878, 1883).

almandine, zircon, etc. This list is somewhat extended; and whoever deems its length insufficient may stretch it *ad libitum* under screen of the "etc." A charge of confusion is an easy means of overcoming a difficult subject and setting a valve on serious investigation. It is to be apprehended lest in this case the confusion is rather in the mind of Geerts than in that of the Chinese, and results from his failure to read the Chinese texts with critical eyes. The first conspicuous confusion of Geerts is, that on p. 202 he grants Li Shi-chên the privilege of indicating the true diamond,[1] while this license is abrogated on p. 357: "The place of the *kin-kang* between iron pyrite and aluminous schist is contrary to the idea that this author intended to designate under this name the diamond." What neither Geerts, nor his predecessor Smith, nor his successor de Mély, understood, is the plain fact that Li Shi-chên does not speak at all of the diamond as a stone, but of the diamond-point as an implement. For this reason it is embodied in the chapter on stones, and is logically followed by a discussion of stone needles used in acupuncture. The term "kin-kang stone" means to Li Shi-chên nothing but the diamond-point. The fact that, besides, the diamond was known to the Chinese as a precious stone, is evidenced by the text of the *Tsin k'i kü chu* (p. 35), where the diamond is spoken of as a precious stone (*pao*), and by the *Ko chi king yüan*,[2] where the stone is designated as a "diamond jewel" (*kin-kang pao*) and classed with jade and gems in the chapter on precious objects (*chên pao lei*).[3] It is not necessary to push any further this criticism of Geerts, who hazards other eccentric conclusions in this section. The evidence brought together is overwhelming in demonstrating that the *kin-kang* in the texts offered by Li Shi-chên, and in ancient Chinese tradition generally, is the diamond. This uniform interpretation, inspired by an analysis of all traditions in the known ancient world, instead of an appeal to confusion with a choice of fanciful possibilities, seems to be the best guarantor for the exactness of the result.

---

[1] The text referred to is that of Pao-p'u-tse regarding Fu-nan; but it is Li Shi-chên who is made responsible for it by Geerts. This uncritical method of Smith, Geerts, and de Mély, who load everything on to the *Pên ts'ao* or its author Li Shi-chên, without taking the trouble to unravel the various sources quoted by him and to study the traditions with historical criticism, is the principal reason for their failure in reaching positive results.

[2] Ch. 33, p. 3 b.

[3] In the great cyclopædia *T'ai p'ing yü lan* (Ch. 813) the notes on the diamond are arranged in the section on metals, being preceded by those on copper and iron. The cyclopædia *T'u shu tsi ch'êng* has adopted the scheme of Li Shi-chên, placing the diamond in the division "stones." It is content to reiterate simply Li Shi-chên's notes, so that this is one of the poorest chapters of this thesaurus.

The solidity and exactness of Chinese tradition is vividly illustrated also by another fact. The term *kin-kang* for the diamond was coined by the Chinese as a free adaptation of the Sanskrit word *vajra*, and, like the latter, signifies with them both the mythical weapon of Indra and the Indian diamond. We noticed that in the oldest historical account of the diamond relative to the year A.D. 277 this precious stone is stated as coming from India, but that at the same time traditions of classical antiquity are blended with this early narrative. Again, the Chinese fully recognized the stone in the diamond-points furnished to them in the channel of trade with the Hellenistic Orient, and were perfectly aware of the fact that diamonds were utilized in the Roman Empire.[1] In the most diverse parts of the world, wherever commercial, diplomatic, or political enterprise carried them, the Chinese observed the diamond, and in every case applied to it correctly the term *kin-kang*. Thus, according to their Annals, the diamond was found among the precious stones peculiar to the culture of Persia under the Sassanians.[2]

Among the early mentions of diamonds is that of diamond finger-rings sent in A.D. 430 as tribute from the kingdom Ho-lo-tan on the Island of Java.[3] In all periods of their history, the Chinese, indeed,

---

[1] The *Hüan chung ki* of the fifth century expressly states that diamonds come from (or are produced in) India and Ta Ts'in (*T'ai p'ing yü lan*, Ch. 813, p. 10).

[2] *Pei shi*, Ch. 97, p. 7 b; *Wei shu*, Ch. 102, p. 5 b; and *Sui shu*, Ch. 83, p. 7 b. DIONYSIUS PERIEGETES, who lived at the time of the Emperor Hadrian (117–138), in his poem Orbis descriptio (Verse 318), says that the diamond is found in the proximity of the country of the Agathyrsi residing north of the Istros (Danube); and AMMIANUS MARCELLINUS (XXII, 8; ed. NISARD, p. 175) states that the diamond abounds among this people (Agathyrsi, apud quos adamantis est copia lapidis). BLÜMNER (Technologie, Vol. III, p. 232; and in PAULY's Realenzyklopädie, Vol. IX, col. 323) infers from these data that the diamond-mines recently rediscovered in the Ural seem to have been known to the ancients; but this conclusion is not forcible. The mines in the Ural began to be opened only from 1829 (the question is not of a rediscovery), and there is no evidence that diamonds were found there at any earlier time. Aside from this fact, a respectable distance separated the Ural from the habitat of the Agathyrsi, who occupied the territory of what is now Siebenbürgen. Already HERODOTUS (IV, 104) knew them as men given to luxury and very fond of wearing gold ornaments. The interesting point is that the Agathyrsi, as shown by JUSTI (Grundriss der iranischen Philologie, Vol. II, p. 442), judging from the remains of their language, belonged to the Scythian stock of peoples, speaking an Iranian language. The notes of Dionysius and Ammianus, therefore, confirm for a Western tribe of this extended family what the Chinese report about Irān proper, and it may be that the diamond was known to all members of the Iranian group in the first centuries of our era.

[3] PELLIOT (*Bull. de l'Ecole française*, Vol. IV, p. 271), who has indicated this passage, sees some difficulties in the term *kin kang chi huan*. While admitting that *kin-kang* is the diamond, he thinks that this translation does not fit the case, and proposes to understand the term in the sense of "rings of rock-crystal." I see no difficulty in assuming that finger-rings of metal set with a diamond are here in question. This passage, indeed, is not the only one to mention diamond rings. In

were familiar with the diamond. To Chao Ju-kua of the Sung period, India was known as a diamond-producing country, though what he relates about the stone is copied from the text of Pao-p'u-tse, quoted above (p. 21).[1]

Judging from Marco Polo's report,[2] the best diamonds of India found their way to the Court of the Great Khan.

The Annals of the Ming record embassies from Lu-mi (Rum) in 1548 and 1554, presenting diamonds among other objects.[3] In the Ming period eight kinds of precious stones were known from Hormuz, the emporium at the entrance of the Persian Gulf; the fifth of these was the diamond.[4] At the same time diamonds were known on Java.[5]

---

the year A.D. 428 of the Liu Sung dynasty, the King of Kia-p'i-li (Kapila) in India sent diamond rings to the Chinese Court (*Sung shu*, Ch. 97, p. 4). The *Nan fang i wu chi* (Account of Remarkable Products of Southern China, by Fang Ts'ien-li of the fifth century or earlier: BRETSCHNEIDER, Bot. Sin., pt. 1, No. 544) relates that foreigners are fond of adorning rings with diamonds and wearing these (*T'ai p'ing yü lan*, Ch. 813, p. 10); and Li Shi-chên (above, p. 40) is familiar with diamond finger-rings. The Records of Champa (*Lin yi ki*) relate that the King of Lin-yi (Champa), Fan-ming-ta, presented to the Court diamond finger-rings (*T'u shu tsi ch'êng, Pien i tien* 96, hui k'ao 1, p. 11b; or *T'ai p'ing yü lan, l. c.*). Daggers and krisses are set with diamonds in Java, and they are used for inlaying on lance-heads (*Int. Archiv für Ethnographie*, Vol. III, 1890, pp. 94–97, 101). The ancients already employed the diamond as a ring-stone (BLÜMNER, Technologie, Vol. III, p. 232).

[1] HIRTH and ROCKHILL, Chau Ju-kua, p. 111.

[2] Edition of YULE and CORDIER, Vol. II, p. 361.

[3] BRETSCHNEIDER, *China Review*, Vol. V, p. 177.

[4] *Si yang ch'ao kung tien lu*, Ch. C, p. 7 (ed. of *Pie hia chai ts'ung shu*), written in 1520 by Huang Sing-tsêng (regarding this work see Chinese Clay Figures, p. 165, note 3; MAYERS, *China Review*, Vol. III, p. 220; and ROCKHILL, *T'oung Pao*, 1915, p. 76).

[5] *Ibid.*, Ch. A, p. 9.—It is somewhat surprising that the Chinese were not acquainted with the diamonds of Borneo; at least in none of their documents touching their relations with the island is any mention made of the diamonds found there. A good description of the Borneo mines, their sites, working-methods, output, etc., is given by M. E. BOUTAN (Le Diamant, pp. 223–228, with map, Paris, 1886), M. BAUER (Edelsteinkunde, 2d ed., pp. 274–281), and in an article of the Encyclopædie van Nederlandsch-Indië (Vol. I, pp. 445–446). None of these sources, however, bears on the question as to when these mines were opened, or when the first diamonds were discovered, and whether this was done by natives or Europeans. As nearly as I can make out, Borneo diamonds were known in the European market in the latter part of the seventeenth century. In a small anonymous book entitled The History of Jewels, and of the Principal Riches of the East and West, taken from the Relation of Divers of the most Famous Travellers of Our Age (London, 1671, printed by T. N. for Hobart Kemp, at the Sign of the Ship in the Upper Walk of the New Exchange) I find the following: "Let me therefore tell you, that none has been yet able in all the world to discover more than five places, from whence the diamond is brought, viz., two rivers and three mines. The first of the two rivers is in the Isle Borneo, under the equator, on the east of the Chersonesus of Gold, and is called Succadan. The stones fetched from thence are usually clear and of a good water,

STONES OF NOCTURNAL LUMINOSITY.— We noticed that the diamond and the traditions connected with it reached the Chinese chiefly from the Hellenistic Orient. We should therefore be justified in expecting also that the historical texts relative to Ta Ts'in and inserted in the Chinese annals might contain references to this stone; but in Hirth's classical work "China and the Roman Orient," where all these documents are carefully assembled and minutely studied, the diamond is not even mentioned.[1] This, at first sight, is very striking; but it would be permissible to think that the diamond is hidden there under a name not yet recognized as such. In the first principal account of Ta Ts'in embodied in the Annals of the Posterior Han Dynasty,[2] we read that

and almost all bright and brisk, whereof no other reason can be given, but that they are found at the bottom of a river amongst sand which is pure, and has no mixture, or tincture of other earth, as in other places. These stones are not discovered till after the waters which fall like huge torrents from the mountains, are all passed, and men have much to do to attain them, since few persons go to traffic in this isle; and forasmuch as the inhabitants do fall upon strangers who come ashore, unless it be by a particular favor. Besides that, the Queen does rarely permit any to transport them; and so soon as ever any one hath found one of them they are obliged to bring it to her. Yet for all that they pass up and down, and now and then the Hollanders buy them in Batavia. Some few are found there, but the largest do not exceed five carats, although in the year 1648, there was one to be sold in Batavia of 22 carats. I have made mention of the Queen of Borneo, and not of the King, because that the isle is always commanded by a woman, for that people, who will have no prince but what is legitimate, would not be otherwise assured of the birth of males, but can not doubt of those of the females, who are necessarily of the blood royal on their mother's side, she never marrying, yet having always the command."

[1] India's trade in diamonds with Ta Ts'in, already pointed out, is mentioned in the chapter on India, inserted in the T'ang Annals (Ch. 221 A, p. 10 b).

[2] *Hou Han shu*, Ch. 118, p. 4 b. Both the night-shining jewel and the moonlight pearl are mentioned together also in the Nestorian inscription of Si-ngan fu and in the Chinese Manichean treatise (CHAVANNES and PELLIOT, Traité manichéen, p. 68). In the latter it is compassion that is likened to the "gem, bright like the moon, which is the first among all jewels." The *T'ung tien* of Tu Yu (written from 766 to 801) ascribes genuine pearls, night-shining and moon-bright gems, to the country of the Pigmies north-west of Sogdiana (*T'ai p'ing yü lan*, Ch. 796, p. 7 b). In that fabulous work *Tung ming ki*, which seems to go back to the middle of the sixth century (CHAVANNES and PELLIOT, *l. c.*, p. 145), the Emperor Wu of the Han dynasty is said to have obtained in 102 B.C. a white gem. (白珠; the word *chu* means not only "pearl, bead," but also "gems generally"), which the Emperor wrapped up in a piece of brocade. It was as if it reflected the light of the moon, whence it was styled "moon-reflecting gem" (*chao yüe chu*; see *P'ei wĕn yün fu*, Ch. 7A, p. 107). The *San Ts'in ki*, a book of the fifth century, has on record that in the tumulus of the Emperor Ts'in Shi pearls shining at night (*ye kuang chu*) formed a palace of the sun and moon, and that moonlight pearls (*ming yüe chu*) suspended in the grave emitted light by day and night (*T'u shu tsi ch'ĕng*, chapter on pearls, *ki shi*, I, p. 3 b). The word *p'i* used in the term *ye kuang p'i*, at first sight, is striking, as it refers to a perforated circular jade disk, such as occurs in ancient China (see Jade, p. 154), but does not occur in the Hellenistic Orient. It is therefore probable that the term already pre-existed in China, and was merely transferred to a jewel of the Roman Orient

"the country contains much gold, silver, and rare precious stones, particularly the jewel that shines at night (*ye kuang p'i* 夜光璧), or the 'jewel of noctural luminosity,' and the moonlight pearl (or 'pearl as

---

which was reported to the Chinese to shine at night. This holds good also of the term *ming yüe chu*. In *T'oung Pao* (1913, p. 341) and Chinese Clay Figures (p. 151) I pointed out that the two terms are employed as early as the *Shi ki* of Se-ma Ts'ien. The passage occurs in the Biography of Li Se (Ch. 87, p. 2 b), who is ill-famed for the extermination of Confucian literature under the Emperor Ts'in Shi, and who died in 208 B.C. (GILES, Biographical Dictionary, p. 464). In another passage of the same work the two terms "moonlight (or moon-bright) pearl" and "night-shining jade-disk" are coupled together, used in a figurative sense (PÉTILLON, Allusions littéraires, p. 242; LOCKHART, Manual of Chinese Quotations, p. 397). A third passage leaves no doubt of what Se-ma Ts'ien understood by a moonlight pearl. In his chapter treating divination from the tortoise-shell (Ch. 128, p. 2 b), he defines the term thus: "The moonlight pearl is produced in rivers and in the sea, hidden in the oyster-shell, while the water-dragon attacks it. When the sovereign obtains it, he will hold in submission for a long time the foreign tribes residing in the four quarters of the empire." The moonlight pearl, accordingly, was to Se-ma Ts'ien and his contemporaries a river or marine pearl of fine quality, worthy of a king, a foreign origin of it not being necessarily implied. The philosopher Mo Ti or Mo-tse, who seems to have lived after Confucius and before Mêng-tse, mentions the night-shining pearl (*ye kuang chi chu*) in an enumeration of prominent treasures; but I am not convinced of the authenticity of the text published under his name, which was doubtless fabricated by his disciples (compare GRUBE, Geschichte der chinesischen Litteratur, p. 129), and tampered with by subsequent editors. The mention of this pearl in Mo Ti and in other alleged early Taoist writers (compare the questionable text of the *Shi i ki*, quoted by DE GROOT, Religious System of China, Vol. I, p. 278) may be a retrospective interpolation as well. Se-ma Ts'ien must be regarded as the only early author whose references in this case may be relied upon as authentic and contemporaneous. (The uncritical notes of T. DE LACOUPERIE, *Babylonian and Oriental Record*, Vol. VI, 1893, p. 271, with their fantastic comment, are without value.) It seems to me, that, in applying the identical terms to real objects encountered in the Hellenistic Orient, the Chinese named these with reference to that passage of Se-ma Ts'ien by way of a literary allusion, and that for this reason the word *p'i*, in this instance, is not to be accepted literally, as has been done by CHAVANNES (*T'oung Pao*, 1907, p. 181: "l'anneau qui brille pendant la nuit"), but that the term *ye kuang p'i* represents an undivided unit denoting a precious stone. Further, this is corroborated by two facts,— first, that the ancients speak of precious stones, not of rings or disks brilliant at night; and, second, that Yü Huan (220–265), in his *Wei lio*, has altered the term *ye kuang p'i* into *ye kuang chu* ("night-shining pearl or gem") with regard to Ta Ts'in, evidently guided by a correct feeling that this modification would more appropriately conform to the object. Moreover, there are neither in Greek nor in Latin any exact equivalents which might have served as models for the two Chinese expressions; the Chinese, indeed, possessed the latter before coming into contact with the Hellenistic-Roman world; *ye kuang* ("light of the night") is an ancient term to designate the moon, which appears in Huai-nan-tse (SCHLEGEL, Uranographie chinoise, p. 610). This point of terminology, however, must be distinguished from the matter-of-fact problem. Whatever the origin of the Chinese terms may be, from the time of intercourse with Ta Ts'in, they strictly refer to a certain group of gems occupying a conspicuous place in the antique world and deeply impressing the minds of the Chinese. All subsequent Chinese allusions to such gems, even though connected with domestic localities, imply distinct reminiscences of the former indelible experience made in the Hellenistic Orient.

clear as the moon,' *yüe ming chu* 月明珠)." HIRTH[1] and CHAVANNES[2]
have united a certain number of classical texts, in order to show that
the notion of precious stones, and especially carbuncles, shining at
night, was widely propagated in Greek and Roman times; the case,
however, deserves a more critical examination. It seems to me, first
of all, that a distinction must be made between *ye kuang p'i* and *yüe
ming chu.* These two different terms must needs refer to two diverse
groups of stones and correspondingly different traditions. It is not
difficult to identify the latter of the two, if we examine our Pliny.
This is Pliny's *astrion*, of which he says, "Of a like white radiance[3] is
the stone called *astrion*, cognate to crystal, and occurring in India and
on the littoral of Patalene. In its interior, radiating from the centre,
shines a star with the full brilliancy of the moon. Some account for
the name by saying that the stone placed opposite to the stars ab-
sorbs their refulgence and emits it again."[4] Pliny's "fulgore pleno
lunae" appears as the basis for the Chinese term *yüe ming chu* (literally,
"moon shining pearl") with reference to this precious stone, as found
in the anterior Orient.[5] HIRTH (*l. c.*) refers us to Herodotus (II, 44),
who mentions a temple of Hercules at Tyre in Phœnicia with two pillars,— one of pure gold, the other of *smaragdos*, — shining with great
brilliancy at night. Hirth takes this *smaragdos* for "emerald stone;"
it is certain, however, that the word in this passage does not mean
"emerald," but denotes a greenish building-stone of a color similar to
the emerald,[6] perhaps, as BLÜMNER[7] is inclined to think, green porphyry.
This passage, accordingly, affords no evidence that the Chinese "stone

---

[1] China and the Roman Orient, pp. 242–244.

[2] *T'oung Pao*, 1907, p. 181.

[3] With reference to the white stone *asteria*, dealt with in the preceding chapter.

[4] Similiter candida est quae vocatur astrion, crystallo propinqua, in India nascens
et in Patalenes litoribus. Huic intus a centro stella lucet fulgore pleno lunae.
Quidam causam nominis reddunt quod astris opposita fulgorem rapiat et regerat
(XXXVII, 48, § 132).

[5] The much-discussed question as to the stone to be understood by Pliny's
astrion does not concern us here. The opinion that it is identical with what is now
called *asteria* ("star stone") is the most probable one (compare BLÜMNER, Technologie, Vol. III, p. 234). The most detailed study of the subject, not quoted by
Krause or Blümner, is that by J. M. GÜTHE, Über den Astrios-Edelstein des Cajus
Plinius Secundus (München, 1810). Judging from the recent report of D. B. STER-
RETT (Gems and Precious Stones in 1913, p. 704, Washington, 1914), this stone seems
to become fashionable again in jewelry. Possibly also Pliny's *selenitis* (67, § 181),
which has within it a figure of the moon and day by day reflects her various phases,
may be sought in the Chinese "moonlight gem," as already supposed by D'HERBELOT
(Bibliothèque orientale, Vol. IV, p. 398).

[6] KRAUSE, Pyrgoteles, p. 37.

[7] Technologie, Vol. III, p. 240.

luminous at night" might be the emerald; nor can it be invoked as a contribution to the problem, as the Chinese do not speak of pillars, but of a precious stone. Hirth, further, quotes an account from Pliny contained in his notes on the *smaragdus*. It is difficult to see what relation it is supposed to have with the subject under discussion, as Pliny does not say a word about these stones shining at night. The story runs thus: "They say that on this island above the tomb of a petty king, Hermias, near the fisheries, there was the marble statue of a lion, with eyes of smaragdi set in, flashing their light into the sea with such force that the tunnies were frightened away and fled, till the fishermen, long marvelling at this unusual phenomenon, replaced the stones by others."[1] The plot of Pliny's story is certainly laid in the daytime, not during the night; fishes, as is well known, being attracted at night by luminous phenomena spreading over the surface of the water, and even being caught by the glare of torch-light. At any rate, the passage contains nothing about jewels brightening the night. Chavannes, more fortunately, points to Lucian (De dea syria), who describes a statue of the Syrian goddess in Hierapolis bearing a gem on her head called *lychnis:* "From this stone flashes a great light in the night-time, so that the whole temple gleams brightly as by the light of myriads of candles, but in the daytime the brightness grows faint; the gem has the likeness of a bright fire."[2] The name *lychnis* is connected with Greek *lychnos* ("a portable lamp"). According to Pliny, the stone is so called from its lustre being heightened by the light of a lamp, when its tints are particularly pleasing.[3] Pliny does not say that the lychnis shines at night,[4] but his definition indicates well how this tradition arose. Pseudo-Callisthenes (II, 42) makes Alexander the Great spear a fish, in whose bowels was found a white stone so brilliant that every one believed it was a lamp. Alexander set it in gold, and used it as a lamp at night.[5] The origin of this trivial story is perspicuous enough.

---

[1] Ferunt in ea insula tumulo reguli Hermiae iuxta cetarias marmoreo leoni fuisse inditos oculos e smaragdis ita radiantibus etiam in gurgitem, ut territi thynni refugerent, diu mirantibus novitatem piscatoribus, donec mutavere oculis gemmas (XXXVII, 17, § 66). Compare Krause, Pyrgoteles, p. 38.

[2] H. A. Strong, The Syrian Goddess, p. 72 (London, 1913).

[3] Ex eodem genere ardentium est lychnis appellata a lucernarum adsensu, tum praecipuae gratiae (XXXVII, 29, § 103). Dionysius Periegetes compares the lychnis with the flame of fire (Krause, *l. c.*, p. 22). Of the various identifications proposed for this stone, that of tourmaline has the greatest likelihood, as Pliny refers to its magnetic property, inasmuch as, when heated or rubbed between the fingers, it will attract chaff and papyrus-fibres.

[4] He does not say so, in fact, with regard to any stone.

[5] It should be noted, however, that in the oldest accessible form of the Romance of Alexander, as critically restored by A. Ausfeld (Der griechische Alexanderroman,

It is welded from two elements,— a reflex of the ring of Polycrates[1] rediscovered in the stomach of a fish, and the tradition underlying the Plinian explanation of the lychnis.   It is accordingly the lychnis which, through exaggeration of a tradition inspired by the name, gave rise to a fable of stones luminous at night.[2]

A story of AELIAN[3] merits particular attention: Herakleïs, a virtuous widow of Tarent, nursed a young stork that had broken its leg.   The grateful bird, a year after its release, dropped a stone into the woman's lap.   Awakening at night, she noticed that the stone spread light and lustre, illuminating the room as though a torch had been brought in. The author adds that it was a very precious stone, without further determination.[4]   This story meets with a parallel in a curious anecdote of China, told in the *Shi i ki*, that, when Prince Chao of Yen was once seated on a terrace, black birds with white heads flocked there together, holding in their beaks perfectly resplendent pearls (*tung kuang chu* 洞光珠), measuring one foot all round.   These pearls were black as lacquer, and emitted light in the interior of a house to such a degree that even the spirits could not obscure their supernatural essence.[5] Still more striking in its resemblance to Aelian's story is one in the *Sou shên ki:*[6] "The marquis of Sui once encountered a wounded snake, and had it cured by means of drugs.   After the lapse of a year [as in Aelian] the snake appeared with a luminous gem in its mouth to repay his kindness.   This gem was an inch in diameter, perfectly white, and emitted at night a light of the brightness of the moon, so that the room was lighted as by a torch."   The gem was styled "gem of the marquis of

---

p. 84), this incident is not contained; it is contained in the uncritical edition of C. Müller of 1846. If Ausfeld (p. 242) is right in placing the primeval text of Pseudo-Callisthenes in the second century B.C., the episode in question, which indubitably is a later interpolation, is not older than the second or third century A.D.

[1] HERODOTUS, III, 41–42.—The stone in this signet-ring, according to HERODOTUS, was a *smaragdos;* according to PLINY (XXXVII, 1), a sardonyx (compare KRAUSE, Pyrgoteles, p. 135).

[2] As a fabulous stone found in the river Hydaspes, the lychnis is mentioned in the unauthentic treatise De fluviis, wrongly ascribed to Plutarch (F. DE MÉLY, Lapidaires grecs, p. 29).

[3] Hist. animalium, VIII, 22.

[4] A. MARX, in his interesting study Griechische Märchen von dankbaren Tieren (p. 52, Stuttgart, 1889), justly comments that the stone mentioned in this tale is the lychnites or lychnis, because, according to Philostratus (Apollonius from Tyana, II, 14), this was the stone placed by the storks in their nests in order to guard them from snakes, and because the lychnis spreads such marvellous light in the dark and possesses many magical virtues (Orphica, 271).

[5] *P'ei wên yün fu*, Ch. 7A, p. 107.

[6] *T'u shu tsi ch'êng*, chapter on pearls, *ki shi*, I, p. 1 b.

Sui," "gem of the spiritual snake," or "moonlight pearl."[1] The same Chinese work offers another parallel that is still closer to Aelian, inasmuch as the bird in question is a crane, which would naturally take the place of the stork not occurring in China. "K'uai Ts'an nursed his mother in a most filial manner. There nested on his house a crane, which was shot by men practising archery, and in a wretched condition returned to Ts'an's place. Ts'an nursed the bird and healed its wound, and, the cure being effected, released it. Subsequently it happened one night that cranes arrived before the door of his house. Ts'an seized a torch, and, on examination, noted that a couple of cranes, male and female, had come, carrying in their beaks moon-bright pearls (*ming yüe chu*) to recompense his good deed."[2] The coincidences in these three Chinese versions and the story of the Greek author, even in unimportant details, are so striking, that an historical connection between the two is obvious. The dependence of the Chinese upon the Greek story is evidenced by the feature of the moon-bright pearls, whose actual existence is ascribed by the Chinese to the Hellenistic Orient.[3]

HIRTH has conjectured that the Chinese name "jewel that shines at night" possibly is an allusion to the ancient name *carbunculus*, corresponding to Greek *anthrax* (the ruby). Pliny, however, in the chapter devoted to this stone, has no report about its shining at night. He insists, quite naturally, on its "fire," from which it has received its name, *carbunculus* meaning "a red-hot coal."[4] The only blade of straw to which the above hypothesis might cling may be found in the words quoted by Pliny from Archelaus, who affirmed that these stones indoors appear purple in color; in the open air, however, flaming.[5] What I translate by "indoors" means literally, "when the roof overshadows one." This phrase evidently implies no allusion to a dark room, but is used in the sense of "in the shadow of a house," in opposition to the following open-air inspection of the stones. The only ancient text known to me, that mentions a ruby shining at night (and styled "color of marine purple"), is a small Greek alchemical work

---

[1] Compare A. FORKE, Lun-hêng, pt. I, p. 378; and PÉTILLON (Allusions littéraires, p. 243), who quotes this story from Huai-nan-tse.

[2] *L. c.*, *ki shi*, I, p. 6 b.

[3] In a wider sense this typical story belongs to the cycle of the grateful animals, a favorite subject of the Greeks in the Alexandrian epoch (compare A. MARX, Griechische Märchen von dankbaren Tieren; and F. SUSEMIHL, Geschichte der griechischen Litteratur in der Alexandrinerzeit, Vol. I, p. 856).

[4] Compare THEOPHRASTUS, De lapidibus, 18 (opera ed. WIMMER, p. 343).

[5] Eosdem obumbrante tecto purpureos videri, sub caelo flammeos (XXXVII, 25, § 95).

translated by M. BERTHELOT,[1] which cannot lay claim to great antiquity. For the purpose of identification, tourmaline (*lychnis*), and

---

[1] Introduction à l'étude de la chimie, p. 272 (Paris, 1889). Not only HIRTH, but also MAYERS (Chinese Reader's Manual, p. 25), T. DE LACOUPERIE (*Babylonian and Oriental Record*, Vol. VI, 1893, p. 274), and CHAVANNES (*T'oung Pao*, 1907, p. 181), without giving reference to any passage, are unanimous in the belief that the carbuncle is the chief night-shining jewel of the ancients. It would be interesting to learn what alleged passage in an ancient author these scholars had in mind. As far as I know, the carbuncle appears as a night-shining stone only in the mineralogical writings of the middle ages, for the first time presumably in the fundamental work De lapidibus pretiosis of MARBODUS (1035–1123), the famous French Bishop of Rennes. In the earliest French translation of his book (L. PANNIER, Lapidaires français du moyen âge, p. 52) the passage runs thus:

> "Scherbuncles gette de sei ráis.
> Plus ardant piere n'i a máis:
> De sa clarté la noit resplent,
> Mais le júr n'en fera neiént."

In the famous letter, purported to have been addressed by Prester John to the Byzantine Emperor Manuel, and written about the year 1165, we find the carbuncle mentioned in three passages (57, 90, 93; F. ZARNCKE, Der Priester Johannes I, pp. 91, 95, 96), in the fanciful and extravagant description of the palace of the Royal Presbyter in India: "In extremitatibus vero super culmen palacii sunt duo poma aurea, et in unoquoque sunt duo carbunculi, ut aurum splendeat in die et carbunculi luceant in nocte.— Longitudo unius cuiusque columpnae est LX cubitorum, grossitudo est, quantum duo homines suis ulnis circumcingere possunt, et unaquaeque in suo cacumine habet unum carbunculum adeo magnum, ut est magna amphora, quibus illuminatur palatium ut mundus illuminatur a sole.— Nulla fenestra nec aliquod foramen est ibi, ne claritas carbunculorum et aliorum lapidum claritate serenissimi caeli et solis aliquo modo possit obnubilari." KONRAD VON MEGENBERG (1309–78), in his Book of Nature (ed. of F. PFEIFFER, p. 437), extols the carbuncle as the noblest of all stones, combining all their virtues. Its color is fiery, and it is even more brilliant at night than in the daytime; during the day it is dark, but at night it shines so brightly that night almost becomes day. This belief still prevailed in the seventeenth century, as may be gleaned from the following interesting passage of A. BOETIUS DE BOOT (Gemmarum et lapidum historia, p. 140, ed. of A. Toll, Lugduni Batavorum, 1636): "Magna fama est carbunculi. Is vulgo putatur in tenebris carbonis instar lucere; fortassis quia pyropus, seu anthrax appellatus a veteribus fuit. Verum hactenus nemo unquam vere asserere ausus fuit, se gemmam noctu lucentem vidisse. Garcias ab Horto proregis Indiae medicus refert se allocutum fuisse, qui se vidisse affirmarent. Sed iis fidem non habuit. Ludovicus Vartomannus regem Pegæ tantae magnitudinis, et splendoris habere scribit, ut qui regem in tenebris conspicatus fuerit, eum splendere quasi a Sole illustretur existimet, sed nec ille vidit. Si itaque gemmam noctu lucentem natura producat, ea vere carbunculus fuerit, atque hoc modo ab aliis gemmis distinguetur, omnesque alias dignitate superabit. Multi autumant gemmas in tenebris lucentes, a natura gigni non posse; verum falluntur. Nam ut lignis putridis, nicedulis, halecumque squammis, et animalium oculis, natura lucem dare potest; non video cur gemmis idonea suppeditata materia (in tanta rerum creatarum abundantia) tribuere non possit. An itaque habeatur, aut non, incertum adhuc est. Doctissimorum tamen virorum omnium sententia huiusmodi gemmae non inveniuntur. Hinc fit quod rubentes, et transparentes gemmae omnes; ab iis carbunculi, anthraces, pyropi, et carbones nuncupentur. Quia videlicet carbonis instar lucent, ac ignis instar flammeos hinc inde radios iaciunt."

possibly to a certain extent ruby,[1] remain, while emerald must be discarded.[2]

In my opinion, the diamond should be added to the series. The Chinese, at least in modern times, use the epithet *ye kuang* ("brilliant at night") as a synonyme of the diamond.[3] This notion apparently goes back to an ancient tradition; for the *Nan Yüe chi* ("Description of Southern China")[4] relates that the kingdom of Po-lo-ki 波羅基

---

[1] The pilgrim Hüan Tsang (*Ta T'ang si yü ki*, Ch. 11, p. 6; ed. of *Shou shan ko ts'ung shu*) narrates that beside the king's palace was the Buddha's-Tooth Shrine, brightly decorated with jewels. From its roof rose a signal-post, on the top of which was a large ruby (*padmarāga*), which shed a brilliant light, and could be seen shining like a bright star day and night for a great distance (compare WATTERS, On Yuan Chwang's Travels, Vol. II, p. 235; BEAL, Buddhist Records, Vol. II, p. 248; the translation of JULIEN, Mémoires sur les contrées occidentales, Vol. II, p. 32 — "recouvert d'un enduit brillant comme le diamant" — is incorrect, and the whole rendering of the passage is not exact). In view of what is set forth below regarding phosphorescence, it should be remarked right here that any natural phenomenon proceeding from the stone cannot come into question in this case. Moon and star light or artificial illumination of the building must be held responsible for the ruby being visible at night. Thus the causes leading to the conception of stones shining in darkness evidently are different. Also in the case of LUCIAN's lychnis in the temple of Hierapolis, I am not inclined to believe in a natural phenomenon, but rather in a miracle produced by priestly artifice, which supplied the source of light from a hidden corner, and hypnotized the multitude into the belief that it emanated from the stone. With reference to the above passage of Hüan Tsang, it should be added that COSMAS INDICOPLEUSTES (Christian Topography, translated by McCRINDLE, p. 365) mentions a gem in the possession of the King of Ceylon (Taprobane), "as large as a great pine-cone, fiery red, and when seen flashing from a distance, especially if the sun's rays are playing around it, being a matchless sight;"but he does not tell of its shining at night. Friar ODORIC OF PORDENONE of the fourteenth century ascribes a similar gem to the King of the Nicobars (YULE, Cathay, new ed., Vol. II, p. 169): "He carrieth also in his hand a certain precious stone called a ruby, a good span in length and breadth, so that when he hath this stone in his hand it shows like a flame of fire. And this, it is said, is the most noble and valuable gem that existeth at this day in the world, and the great emperor of the Tartars of Cathay hath never been able to get it into his possession either by force or by money or by any device whatever."

[2] BECKMANN (Beiträge zur Geschichte der Erfindungen, Vol. III, p. 553) tentatively included among the luminous stones of the ancients also fluor-spar; but, as admitted by himself, the phosphorescent property of this mineral was not recognized before the seventeenth century. Moreover, whatever may have been said to the contrary (BLÜMNER, Technologie, Vol. III, p. 276; and LENZ, *l. c.*, p. 23), it is extremely doubtful to me whether the ancients were acquainted with fluor-spar. This supposition is not well founded on matter-of-fact evidence, but merely inferred from certain properties of the mineral which became known in our own time, and which were subsequently read into certain accounts of the ancients.— Other stones to which the property of nocturnal luminosity is ascribed are purely fabulous, as, for instance, the "stone attracting other stones," described by Philostratus as sparkling at night like fire (F. DE MÉLY, Lapidaires grecs, pp. 27–28).

[3] J. DOOLITTLE, Vocabulary and Handbook of the Chinese Language, Vol. I, p. 132.

[4] Written by Shên Huai-yüan of the fifth century (BRETSCHNEIDER, Bot. Sin., pt. I, No. 559). The text is cited in *T'ai p'ing yü lan*, Ch. 813, p. 10.

produces diamonds, the lustre of which illuminates the dark night. According to Chao Ju-kua,[1] the King of Ceylon possessed a gem five inches in diameter, which could not be consumed by fire, and at night emitted a brilliancy like a torch. As incombustibility was credited to the diamond, this jewel shining at night, in all probability, was a diamond.[2] Another reason why the diamond should be included in this class will be discussed in the following section.

PHOSPHORESCENCE OF PRECIOUS STONES.— As this subject of stones "luminous at night" has heretofore not been properly comprehended by sinologues and others, it may not be amiss to add some explanatory notes.[3] As a matter of fact, of course, stones cannot shine at night: the lustre of any gem is an optical property, and depends upon the effects of light, solar or artificial, which is reflected back to the human eye.[4] The classical and Chinese reports of stones emitting rays of light in darkness, accordingly, have nothing to do with optical phenomena, or, in particular, with so-called "adamantine lustre." If these stories, partially, should refer to a phenomenon of reality, there is but one that can come into question,— that of phosphorescence. This is a property of some gems, which, after rubbing, heating, exposure to light, or an electrical discharge, radiate a light known as phosphorescence; since the glow, although often of different colors, resembles that of phosphorus. This property is particularly exhibited in the diamond, which, on being rubbed with a cloth or across the fibres of a piece of wood, gives out a light plainly visible in a dark room. It is, however, not a general property of all diamonds, but only efficient in certain stones.[5] Though

---

[1] *Chu fan chi* (ed. ROCKHILL), Ch. A, p. 10; translation of HIRTH and ROCKHILL, p. 73.

[2] An indirect testimony for the diamond being counted among the night-shining stones in the West may be deduced from the passage in the Physiologus, that the diamond is not found in the daytime, but only at night, which may imply, that, in order to be found at night, it must then emit light (compare F. LAUCHERT, Geschichte des Physiologus, p. 28; E. PETERS, Der griechische Physiologus, p. 96; F. HOMMEL, Aethiopische Übersetzung des Physiologus, p. 77; K. AHRENS, Buch der Naturgegenstände, p. 82).— D'HERBELOT (Bibliothèque orientale, Vol. IV, p. 398) already knew that it was a natural property of the diamond to shine in darkness.

[3] The subject in general has been dealt with by G. F. KUNZ (Curious Lore of Precious Stones, pp. 161–175).

[4] The Chinese scholar Sung Lien (1310–81) had a certain idea thereof. In a Dissertation on Sun, Moon, and Stars (*Ji yüe wu sing lun*) he speaks of a "gem like the full moon" (*yüe man ju chu*), whose substance, in principle, has no lustre; but it borrows its lustre from the sun, that half of it turned away from the sun being constantly dark, and the other half turned toward the sun being constantly bright (*P'ei wên yün fu*, Ch. 7A, p. 109).

[5] Compare FARRINGTON, Gems and Gem Minerals, pp. 34, 70. Among all minerals, phosphorescence is best exhibited by fluorite, nearly all specimens of which,

occurring also in other precious stones, the phosphorescent light is most brilliant and intensified in the diamond; and for this reason it would seem plausible that the diamond should have held the foremost rank among the stones luminous at night.

There remains, however, a grave obstacle in the way of this explanation, which must not be overlooked; and this is that the ancient authors who have written on precious stones are entirely reticent on the subject of their phosphorescent quality. It is indeed taught that this phenomenon was observed for the first time only by the physicist Robert Boyle in 1663.[1] This, of course, does not mean that it was entirely unknown before that time, and that it could not have revealed itself to a layman by a chance accident.

M. BERTHELOT,[2] however, has discovered in the collection of Greek alchemists a small treatise propounding the processes "of coloring the artificial precious stones, emeralds, carbuncles, and hyacinths, after the book drawn from the sanctuary of the temple." He believes that artificial coloring of stones is said in this text to impart to them the property of phosphorescence, and that there is no doubt that the ancients made precious stones phosporescent in darkness through the employment of superficial tinctures derived from substances such as bile of marine animals, the analogous properties of which are known to us. I must confess that this conclusion, though emanating from so high and respectable an authority, for whom I have a profound admiration, is not quite convincing to me. First, it seems open to doubt whether the Greek recipe really took the desired effect, as long as this is not experimentally established; second, if it did, it does not furnish proof that the ancients were acquainted with the phenomenon of the phosphorescence of precious stones, as we understand it, which is a physical property inherent in the stone, while in the Greek text the phosphorescence is alleged to result from animal products brought in contact with the stone, not from the stone itself. The text published by Berthelot, while it may tend to prove that certain ancient alchemists knew something about the phosphorescence of certain animal organs, is not at all apt to show that the same tendency in precious stones was familiar to them; on the contrary, it would be much more likely to have

---

when gently heated, will emit a visible light. Its color varies with different varieties, and is usually not the same as the natural color of the mineral. The tints exhibited are usually greenish, bluish, or purplish.

[1] BAUER, Precious Stones, p. 138.

[2] Sur un procédé antique pour rendre les pierres précieuses et les vitrifications phosphorescentes (*Annales de chimie et physique*, 6th series, Vol. XIV, 1888, pp. 429–432); reprinted in his Introduction à l'étude de la chimie, pp. 271–274 (Paris, 1889).

been unknown to them, if that artificial process were ever really applied to stones.

Also from India we receive an intimation as to alleged acquaintance with the fact of phosphorescence before Boyle. The learned Hindu PRAPHULLA CHANDRA RAY,[1] professor of chemistry at the Presidency College, Calcutta, has this to say: "It is sometimes asserted that the phosphorescence of diamond was first observed in 1663 by the celebrated Robert Boyle. Bhoja (eleventh century A.D.), however, mentions this property." Fortunately for us, the Sanskrit text of this passage is added, which reads, "andhakāre ca dīpyate" (translated by Ray, "it phosphoresces in the dark"); but these words simply mean, "it shines in the dark." It is accordingly not the case of Bhoja being familiar with the phosphorescent property of the diamond, but the subjective case of Professor Ray, who knows of Boyle's discovery, and projects this knowledge into his author. It reflects more credit on the well-meant patriotism of the Hindu than on his power of logic. His interpretation being conceded, we could as well infer from the numerous passages of classical and Chinese authors, where precious stones luminous in the dark are spoken of, that also Greeks, Romans, and Chinese possessed an intimate acquaintance with the phenomenon in question.[2] But serious science cannot afford to speed its conclusions up to this rapid tempo; and if the fact remains that no Greek, Roman, Sanskrit, or Chinese text has as yet come to the fore, from which such an inference as to conscious knowledge of the phosphorescence of precious stones can reasonably and without violence be deducted, it is safer to hold judgment in abeyance or to regard the result as negative.[3]

---

[1] A History of Hindu Chemistry, Vol. II, p. 40 (2d ed., Calcutta, 1909).

[2] It is noteworthy that neither the Arabic nor the Indian mineralogists have accounts of precious stones luminous at night. What the Arabs offer of this sort is an entirely different affair. The *lapidarium* of Pseudo-Aristotle mentions a fabulous stone under the name "strange stone," which is found in the dark ocean, has rays in its interior, and is visible at night, its veins being brilliant as though they were laughing faces (a corrupted reading which originally was "brilliant like a mirror;" J. RUSKA, Steinbuch des Aristoteles, pp. 20, 167). The "stone bringing sleep" is red, and large pieces of it radiate at night a glow of fire, and in the daytime smoke emanates from it (*ibid.*, p. 166).

[3] In the passage of the Orphica, "the diamond-like crystal, when placed on an altar, sent forth a flame without the aid of fire," KUNZ (Curious Lore of Precious Stones, p. 163) believes he sees an indication that the phosphorescence of the diamond had already been noted before the second or third century of our era; but the plain text does not bear out this far-fetched interpretation. The Greek author has in mind the well-known burning-lenses of crystal, described also by Pliny (see the writer's article on this subject in T'oung Pao, 1915, pp. 169–228), and compares their reflective power with that of the diamond; he says nothing further than that the lustre of the diamond vies with that of a crystal lens. There is no allusion to the fact that this happens in darkness, and consequently no reference to phosphorescence.

While direct evidence is lacking, an interesting observation may be based on Pliny, which, it seems to me, is conclusive to some degree; and this is the curious circumstance that Pliny is familiar with the magnetic or electrical property of just those gems which have the best claim to being identified with the stones luminous at night of the Chinese,— tourmaline and diamond. In regard to the former (*lychnis*) he states that these stones, when heated by the sun or rubbed by the fingers, will attract chaff and scraps of papyrus.[1] As to the diamond, he remarks that its hostility toward the magnet goes so far, that, when placed near it, it will not allow of its attracting iron; or if the magnet has already seized the iron, it will itself attract the metal and turn it away from the magnet.[2] The fact is correct that diamond becomes strongly electric on friction, so that it will pick up pieces of paper and other light substances, though it is not a conductor of electricity, differing in this respect from graphite.[3] Whether the diamond, as asserted by Pliny, can check the attractive power of the magnet, seems to be a controversial point. GARCIA AB HORTO was the first to antagonize Pliny's allegation, on the ground of many experiments made by him.[4] C. W. KING[5] has the following observation: "This stone is highly electric, attracting light substances when heated by friction, and, as we have already noticed,[6] has the peculiarity of becoming phospho-

---

[1] Has sole excalfactas aut attritu digitorum paleas et chartarum fila ad se rapere (XXXVII, 29, § 103).

[2] Adamas dissidet cum magnete in tantum, ut iuxta positus ferrum non patiatur abstrahi aut, si admotus magnes adprehenderit, rapiat atque auferat (XXXVII, 15, § 61).

[3] "All gems when rubbed upon cloth become, like glass, positively electrified. Gems differ, however, in the length of time during which they will retain an electrical charge. Thus tourmaline and topaz remain electric under favorable conditions for several hours; but diamond loses its electricity within half an hour" (FARRINGTON, Gems and Gem Minerals, pp. 34, 70). The Arabs attribute to the garnet (*bijādī*) the power of attracting wood and straw (J. RUSKA, Steinbuch des Aristoteles, p. 144). I do not believe with Ruska that this statement may be caused by confusing the garnet with amber. Though Vullers and Steingass, in their Persian Dictionaries, assign to the word *bijādī* or *bejād* the meanings "garnet" and "amber," the latter interpretation is evidently suggested by the reference to the attractive power.

[4] Nè meno è il vero che tolga la virtù alla calamita di tirare il ferro; percioche ne ho fatto io molte volte esperienza, e l'ho trovata favola (Italian edition of 1582, p. 182).

[5] Antique Gems, p. 71.

[6] In the passage referred to (p. 27) KING says that "the property of phosphorescence is possessed by no other gem except the diamond, and this only retains it for a few minutes after having been exposed to a hot sun and then immediately carried into a dark room. This singular quality must often have attracted the notice of Orientals on entering their gloomy chambers after exposure to their blazing sun, and thus have afforded sufficient foundation to the wonderful tales built upon the simple

rescent in the dark after long exposure to the sun. The ancients also ascribed magnetic powers to the diamond in even a greater degree than to the loadstone, so much so that they believed the latter was totally deprived of this quality in the presence of the diamond; but this notion is quite ungrounded. Their sole idea of magnetism was the property of attraction; therefore seeing that the diamond possessed this for light objects, the step to ascribing to it a superiority in this as in all other respects over the loadstone was an easy one for their lively imaginations." Ajasson, however, holds that if the diamond is placed in the magnetic line or current of the loadstone, it attracts iron equally with the loadstone, and consequently neutralizes the attractive power of the loadstone in a considerable degree.[1] Be this as it may, Pliny, at any rate, was well informed on the electrical quality of the diamond; and if this experiment in the case of diamond and tourmaline was brought about by rubbing the stones, it is not impossible that in this manner also a phosphorescence was occasionally produced and observed. A few such observations may easily have given rise to fabulous exaggerations of stones illumining the night.

Were phosphorescent phenomena known to the Chinese? First of all, they were known in that subconscious and elementary form in which we find such conceptions in the domain of our own folk-lore. The philosopher Huai-nan-tse of the second century B.C. says that old *huai* trees (*Sophora japonica*) produce fire, and that blood preserved for a long time produces a phenomenon called *lin* 燐 .[2] This word is justly assigned the meaning "flitting light" and "will-o'-the-wisp, as seen over battle-fields." It is defined in the ancient dictionary *Shuo wên* as proceeding from the dead bodies of soldiers and the blood of cattle and horses, popularly styled "fires of the departed souls."[3] The philosopher Wang Ch'ung of the first century A.D. criticised this belief of his contemporaries as follows: "When a man has died on a battle-field, they say that his blood becomes a will-o'-the-wisp. The blood is the vital force of the living. The will-o'-the-wisp seen by people while walking at night has no human form; it is desultory and

---

fact by their luxuriant imaginations." I am somewhat inclined toward the same opinion; but we should not lose sight of the fact that the phenomenon itself, as far as precious stones are concerned, is not described in any ancient record, while we may trust to the future that such will turn up some day in a Greek papyrus. As the matter stands at present, we have at the best a theory founded on circumstantial evidence deduced from the ancients' knowledge of the magnetic property of precious stones.

[1] BOSTOCK and RILEY, Natural History of Pliny, Vol. VI, p. 408.
[2] Quoted under this word in K'ang-hi's Dictionary.
[3] The text is cited in COUVREUR's Dictionnaire chinois-français, p. 496.

concentrated like a light. Though being the blood of a dead man, it does not resemble a human shape in form. How, then, could a man whose vital force is gone, still appear with a human body?"[1] At the present day, when the Chinese in a very creditable manner coined a nomenclature to render our scientific terminology, they chose this word *lin* (ignis fatuus) to express our term "phosphorescence."[2] This shows that they have a feeling that this phenomenon underlies the popular notions conveyed by their word.[3]

The *Po wu chi* by Chang Hua (232–300)[4] has the following interesting text, which shows also that the Chinese had a certain experience of electric phenomena: "On battle-fields the blood of fallen men and horses accumulates and is transformed into will-o'-the-wisps. These adhere to the soil and to plants like dewdrops, and generally are not visible. Wanderers sometimes strike against them, and they cling to their bodies, emitting light. On being wiped off, they are scattered around into numberless particles, which yield a crepitating sound, as though beans were being roasted. They thrive only in quiet places for any length of time, and may soon be extinguished. The people affected by them become perturbed, as though they were mentally unbalanced, and remain for some days in an erratic state of mind. At present when people comb their hair, or are engaged in dressing or undressing, sparks may be noticed along the line of the comb or the folds of the dress, also accompanied by a crepitating sound."[5]

We noticed above that the phosphorescing of certain organs of marine animals was known to Greek alchemists. The counterpart of this observation is found in Chinese accounts of the eyes of whales, especially those of female whales, making "moonlight pearls" (*ming*

---

[1] A. FORKE, Lun-hêng, pt. i, p. 193.

[2] It appears from the *Ku kin chu* of Ts'uei Pao of the fourth century (Ch. B, p. 6 b; ed. of *Han Wei ts'ung shu*) that the phosphorescence of the glow-worm or firefly was styled also *lin* and likewise *ye kuang* ("wild fire," or "fire of the wilderness").

[3] GILES (No. 6717) assigns this significance also to the word *lan* in the compound *yü lan* ("phosphorescence of fishes").

[4] Compare Notes on Turquois, p. 22. The passage is in Ch. 9, p. 2, of the Wu-ch'ang edition.

[5] Also in Japan it was believed that will-o'-the-wisps represent the souls of people (hence called *hito-dama*, "man's soul"), which are floating away over the eaves and roof as a transparent globe of impalpable essence (ASTON, Shinto, p. 50; M. REVON, Le Shintoisme, pp. 111, 302). Interesting information on this subject relative to Japan is given by GEERTS (Les produits de la nature japonaise et chinoise, pp. 186–187). Compare also some notes of M. W. DE VISSER (The Dragon in China and Japan, pp. 213–214); and the same author's detailed study Fire and Ignes Fatui in China and Japan (*Mitteilungen des Seminars für oriental. Sprachen*, Vol. XVII, pt. I, 1914, pp. 97–193).

yüe chu);[1] this was recorded by Ts'uei Pao in the middle of the fourth century.[2] The fact that this was not mere fancy, but that such whale-eye pearls were a product of actual use, is illustrated by the Moho, a Tungusian tribe of the Sungari, who sent these in the year 719 as tribute to the Chinese Court.[3] The fabulous work *Shu i ki* says that in the southern sea there is a pearl which is the pupil from the eye of a whale, and in which one may behold his reflection at night, whence it is called "brilliancy of the night" (*ye kuang*).[4] Varāhamihira (A.D. 505–587), in his Bṛihat-Saṁhitā (Ch. 81, § 23), speaks of a pearl coming from dolphins, resembling the eye of a fish, highly purifying, and of great worth.[5]

Fish-eyes seem to have been enlisted for this purpose in old Japan. The Annals of the Sui Dynasty[6] attribute to Japan a wishing-jewel (*ju i pao chu*, rendering of Sanskrit *cintāmaṇi*) of dark color, as big as a fowl's egg, and radiating at night, said to be the pupil of a fish-eye.[7]

Of other substances of animal origin credited by the Chinese with the property of nocturnal luminosity may be mentioned rhinoceros-horn, discussed by the writer on a former occasion.[8] While at that time I referred the earliest conception of this matter to Ko Hung of the fourth century and to a work of the T'ang period, I am now in a position to trace it to an author of the third century A.D., Wan Chên, who wrote the work *Nan chou i wu chi* ("Account of Remarkable Objects in the Southern Provinces").[9] This writer assumes the existence of a divine or spiritual rhinoceros, whose horn emits a dazzling splendor. The interesting point, however, is that it is just an ordinary horn when examined in the daytime, whereas in the darkness of night the single veins of the horn are effulgent like a torch.[10] In regard to exhibiting luminous properties at night, instances of the real pearl, which is likewise

---

[1] The same term as that ascribed to the Hellenistic Orient and identified above with the *astrion* of Pliny.

[2] The complete text is given by the writer in *T'oung Pao*, 1913, p. 341.

[3] *T'ang shu*, Ch. 219, p. 6.

[4] *P'ei wên yün fu*, Ch. 7A, p. 107; or Ch. 22A, p. 76b. This attribute again is identical with that conferred on the precious stone of the Hellenistic Orient.

[5] H. KERN, Verspreide Geschriften, p. 100 ('s-Gravenhage, 1914).

[6] *Sui shu*, Ch. 81, p. 7.

[7] In all probability this jewel was a Buddhist relic brought over to Japan from India. Reference has been made above (p. 22) to the Buddhist legend, according to which the *cintāmaṇi* originates from the fabulous fish *makara*. The Chinese author Lu Tien (1042–1102), in his *P'i ya*, expresses the view that the *cintāmaṇi* is the pupil of the eye of a fish (*Wu li siao shi*, Ch. 7, p. 13).

[8] Chinese Clay Figures, pp. 138, 151.

[9] BRETSCHNEIDER, Bot. Sin., pt. 1, Nos. 452, 539; and *Sui shu*, Ch. 33, p. 10.

[10] The passage is quoted in the cyclopædia *T'ai p'ing yü lan* (published by Li Fang in 983), Ch. 890, p. 3 (edition of Juan Yüan, 1812).

an animal product, have already been cited (p. 56). A few more cases may here be added. In A.D. 86 moonlight pearls as big as fowl's eggs, 4.8 inches in circumference, were produced in Yü-chang and Hai-hun.[1] In the work *Kuang chi*, by Kuo I-kung of the sixth century,[2] are distinguished three kinds of pearl-like gems, — the gem *mu-nan* 木難 of yellow color,[3] the bright gem (*ming chu* 明珠), and the large gem resplendent at night (*ye kuang ta chu* 夜光大珠), all an inch in diameter, or two inches in circumference, the best qualities coming from Huang-chi;[4] these are perfectly round, and when placed on a plane do not stop rolling for a whole day.[5]

---

[1] Both localities are situated in the prefecture of Nan-ch'ang, Kiang-si Province. This notice is given in the *Ku kin chu* of Ts'uei Pao (fourth century), cited in *T'ai p'ing yü lan*, Ch. 803, p. 6.

[2] BRETSCHNEIDER, Bot. Sin., pt. I, No. 376; and PELLIOT, *Bull. de l'Ecole française*, Vol. IV, p. 172.

[3] In another passage of the same work (cited in *P'ei wên yün fu*, Ch. 7A, p. 107; and *T'ai p'ing yü lan*, Ch. 809, p. 4 b) it is said that this gem of yellow hue originates in the eastern countries. In this case, the name for the gem is *mo-nan* 莫難, which appears to be a phonetic variant of *mu-nan*. The same form is found in the *Ku kin chu* (Ch. c, p. 5 b; ed. of *Han Wei ts'ung shu*), where *shui* 水 *nan* is given as a synonyme, and where it is remarked that the stone is yellow and occurs in the countries of the Eastern Barbarians. Aside from these indications placing the home of the stone vaguely in the East, we have other accounts that attribute it to the Hellenistic Orient. The *Nan Yüe chi* (by Shên Huai-yüan of the fifth century; quoted in *P'ei wên yün fu*, Ch. 7A, p. 102 b) states that *mu-nan* are pearls or beads of greenish color, produced by the saliva of a bird with golden wings, and that they are prized in the country of Ta Ts'in. The *Hüan chung ki* (*T'ai p'ing yü lan, l. c.*) likewise informs us that Ta Ts'in is the place of production. The Annals of the T'ang Dynasty ascribe *mu-nan* to Fu-lin (HIRTH, China and the Roman Orient, p. 59); and Ma Tuan-lin explains them as evolved from the coagulated saliva of a bird (*ibid.*, p. 80),— doubtless the echo of a Western tradition. The *Shi i ki* tells of an auspicious bird living on the fabulous isle Ying-chou, and spitting manifold pearls when singing and moving its wings. An exact description of the stone *mu-nan* is not on record. The *Pên ts'ao kang mu* lists it among the precious stones of yellow color. Yang Shên (1488–1559) identifies it with the emerald (written by him *tsie-ma-lu* instead of *tsie-mu-lu*, see Notes on Turquois, p. 55). Fang I-chi, in his *Wu li siao shi* (Ch. 7, p. 14), proposes to regard it as the yellow *yakut* of the Arabs. These speculations are recent after-thoughts of doubtful value.

[4] Regarding the location of this country see Chinese Clay Figures, p. 80.

[5] *T'u shu tsi ch'êng*, chapter on pearls, *hui k'ao*, I, p. 6 b. The latter statement reminds one of Pigafetta's account regarding the two pearls of the King of Brunei (west coast of Borneo), as large as hen's eggs, and so perfectly round that if placed on a smooth table they cannot be made to stand still (see HIRTH and ROCKHILL, Chau Ju-kua, p. 159).— Li Shi-chên speaks of "thunder-beads" dropping from the jaws of a divine dragon and lighting an entire house at night (see Jade, p. 64). These are certainly not on a par with the other "prehistoric" implements enumerated by him in the same text, as believed by DE VISSER (The Dragon, p. 88), but this matter has crept in here by way of wrong analogy. These alleged thunder-beads are simply a transformation of the snake-pearls of Indian folk-lore.

Also coral has been credited with the same property. The work *Si king tsa ki* ("Miscellaneous Records of the Western Capital," that is, Si-ngan fu) relates: "In the pond Tsi-ts'ui there are coral-trees twelve feet high. Each trunk produces three stems, which send forth 426 branches. These had been presented by Chao T'o, King of Nan Yüe (Annam), and were styled 'beacon-fire trees.' At night they emitted a brilliant light as though they would go up in flames."[1]

Whether in each of the instances cited the case rests on real observation is difficult to decide. Some accounts may be purely fabulous or imaginary, and the luminous property may have freely been transposed from one substance to another. Taken all together, however, we cannot deny that certain phenomena of phosphorescence might to a certain degree have been known to the ancient Chinese in some way or other, although the phenomenon itself was not intelligently understood. A recent author, Sung Ying-sing, who wrote in 1628 (2d ed., 1637) the *T'ien kung k'ai wu*, a treatise on technology, gives an interesting account of the pearl-fishery, and discredits the belief in night-shining pearls. He remarks, "The pearls styled 'moonlight and night-shining' in times of old are those which, when viewed under the eaves in broad daylight on a sunny day, exhibit a fine thread of flashing light; it is uncertain, however, that the night-shining pearls are finest, for it is not true that there are pearls emitting light at the hour of the dusk or night." There is, however, no account on record to show that the Chinese ever understood how to render precious stones phosphorescent; and since this experiment is difficult, there is hardly reason to believe that they should ever have attempted it. Altogether we have to regard the traditions about gems luminous at night, not as the result of scientific effort, but as folk-lore connecting the Orient with the Occident, Chinese society with the Hellenistic world.

---

[1] *T'ai p'ing yü lan*, Ch. 807, p. 5; or *T'u shu tsi ch'êng*, chapter on coral, *ki shi*, p. 1 (see also *Pien i tien* 94, Annam, *hui k'ao* VI, p. 8b, where this event is referred to the beginning of the Han dynasty).

INDEX

Adamantine gold, 38.
Aelian, 59.
Aëtites, 9.
Agastimata, 42, 48.
Agathyrsi, diamond in country of, 53.
Ajasson, 45, 67.
Akfānī, 27, 34, 41.
Albertus Magnus, 24.
Alexander, Romance of, 10, 11, 14, 45, 58.
Almās, Arabic designation of the diamond, 32, 34, 42, 46.
Ammianus, 53.
Apollonius, on diamond, 24.
Armenian version of legend of Diamond Valley, 14.
Arthaçāstra, on diamond, 16, 48.
Asbestos, 28, 33, 39, 40.
Astrion, 57.
Augustinus, 16, 24.
Ausfeld, A., 10, 11, 45, 58.

Ball, V., 15, 48.
Bauer, M., 37, 47, 48, 49, 54, 64.
Beckmann, J., 28, 47, 62.
Benjamin of Tudela, 11.
Berquen, L. van, alleged inventor of diamond-polishing, 49.
Berthelot, M., 26, 61, 64.
al-Bērūnī, 41.
Biot, E., 21.
Biscia, A. R., 13, 27.
Blümner, H., 24, 36, 44, 47, 53, 57, 62.
Boll, F., 24.
Boot de, 41, 61.
Borneo, diamonds of, 54.
Boutan, M. E., monograph on diamond, 54.
Boyle, R., 64, 65.
Buddha, associated with the diamond, 17, 25; diamond passed as his tooth, 30.

California, diamonds of, 37.
Callaina, 15.
Cambodja, see Fu-nan.
Carbuncle, in the legend of Diamond Valley, 14, note 2; 44, 60; luminous at night, 61; 64.
Chalfant, F. H., on diamonds of Shan-tung, 5.
Champa, diamond-rings from, 54.
Chang Hua, 68.
Ch'ang Tê, 13.
Chao Ju-kua, on diamonds of India, 22, 54; 63.

Chavannes, E., 8, 18, 22, 25, 30, 31, 33, 39, 40, 56, 57, 58, 61.
Chou K'ü-fei, 21.
Chou Mi, 12, 42, 51.
Cintāmaṇi, 22, 69.
Conti, N., 14.
Coral, luminous at night, 71.
Cosmas, 62.
Crooke, W., 16, 41.
Curtius, 23, 44.
Cut diamonds, unknown in classical antiquity, India, and China, 46–50; imported into China from India and Europe, 6 note; introduced into India and China by Portuguese, 48, 50.

Dana, E. S., 43.
Dante, 18.
Diamond-point, 27, 28–35.
Diamond-sand, from Tibet, 15; regarded as poisonous in India, 41.
Diamond-Seat, of Buddha, 17, 18.
Diamond throne, in Dante, 18.
Diamonds, of Shan-tung, 5; of India, 16, 44; in Iran, 53; of Java, 54; of Borneo, 54.
Dionysius Periegetes, 44, 53, 58.
Dioscorides, 23, 26, 32, 44.
Duval, R., 26.

Eagle-stone, 9.
Edrīsī, 13.
Electric phenomena, known to Chinese, 68.
Elysæus, legend of Diamond Valley by, 14 note 2.
Emerald, 57, 62, 64, 70.
Emery, 12, 44, 50; of China, mentioned by Arabs, 51.
Epiphanius, 9, 10, 15, 17, 18, 20, 21.
Ethiopia, diamonds in, 45.

Faber, E., 28, 29, 33.
Fang I-chi, 12, 70.
Farrington, O. C., 23, 37, 63.
Fauvel, on Chinese diamonds, 5.
Ferrand, G., 51.
Finot, L., 16, 17, 22, 41, 42, 43, 44, 48, 49.
Fire, does not affect diamond, 23, 38.
Fish-eyes, employed as pearls, 69.
Fluor-spar, known to Chinese, 21, 36; not known to the ancients, 62.
Forke, A., 29, 33, 60, 68.

73

CLASSICAL PHILOLOGY

VOLUME XIII
JANUARY—OCTOBER, 1918

THE UNIVERSITY OF CHICAGO PRESS
CHICAGO, ILLINOIS

to Latinists and the general public. It is unquestionably the best metrical translation of Lucretius into English that has yet appeared.

W. A. MERRILL

UNIVERSITY OF CALIFORNIA

The Diamond. A Study in Chinese and Hellenistic Folk-Lore. By BERTHOLD LAUFER. Field Museum of Natural History, Publication 184, Anthropological Series, Vol. XV, No. 1. Chicago, 1915. Pp. 75.

The first fifteen pages of this learned study are the ones of chief interest to students of the classics. In the *Memoirs of the Four Worthies or Lords of the Liang Dynasty*, written by Chang Yüe (667–730 A.D.), a story is told about an island in the Western Sea (Mediterranean) where there is an inaccessible ravine in which precious stones lie. The inhabitants throw flesh into this ravine. Birds pick up the flesh in their beaks and as they fly, they drop the precious stones. The men of the country are clever workers of gems, which are called Fu-lin after the name of the country. Fu-lin is the Chinese name for some part of the Roman Empire, probably Syria.

A legend similar in all its essential features is found in Epiphanius, bishop of Constantia in Cyprus in the fourth century. Dr. Laufer points out the close likeness of this legend to the story of the Arabians and their curious method of obtaining cinnamon told by Herodotus iii. 111, and to a somewhat similar tale in Pliny *N.H.* xxxviii. 33, but prudently refrains from attempting to link them closely. The source of the legend he finds in the Hellenistic Orient. To one already impressed with the fact that Hellenistic artistic motives influenced early Chinese and even Japanese art in a marked degree, the thesis is in itself reasonable, and Dr. Laufer's proofs are convincing.

There are two further points of interest in this study for the classicist and archaeologist. The author is convinced (pp. 42–46) that the *adamas* of the ancients was actually the diamond, but concludes that ancient gem-workers did not understand the process of cutting and polishing diamonds to add to their luster (pp. 46–50). The study contains other information which will be attractive chiefly to Sinologists and to those interested in the history of the diamond.

In this pamphlet Dr. Laufer has presented another useful link of the broken chain of evidence which connects Hellenistic-Roman civilization with the Far East. Curiously enough the classical archaeologist and the classicist seem to regard the evidence upon this new sphere of Greek influence either with suspicion or with apathy.

W. L. WESTERMANN

UNIVERSITY OF WISCONSIN

122

两件中国皇家玉器

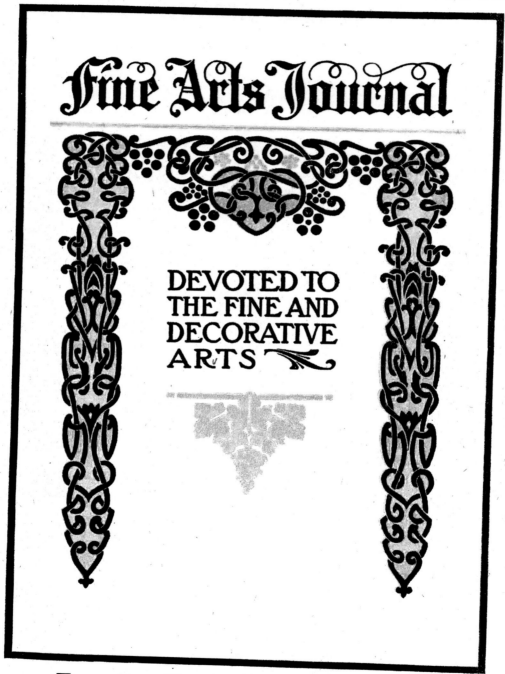

Fine Arts Journal

DEVOTED TO THE FINE AND DECORATIVE ARTS

Twenty Five Cents the Copy
Three Dollars the Year

Vol. XXXII   CHICAGO, JUNE, 1915   Number SIX

The Contents of this issue
are protected by law.
By F. J. CAMPBELL, Publisher
[Published Monthly]

Entered at Chicago Post
Office as SECOND CLASS
MATTER, May 4, 1900,
Under Act, March 3, 1879

CONTENTS

All Art Lovers and Friends of The Chicago Beautiful Plan Should Subscribe Now

Subscription Price ....$3.00 Per Year
Foreign Rate..........................$4.00 Per Year
Canada...............................$3.25 Per Year

Address FRANK J. CAMPBELL, Publisher,
163 W. Washington Street, Chicago

IMPERIAL CHINESE JADE SEAL OF EMPRESS JUI [1796–1797]

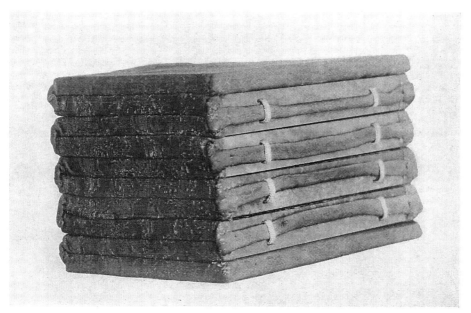

IMPERIAL CHINESE JADE BOOK [CLOSED] OF EMPEROR KANG-HSI [1662-1722

Two Chinese Imperial Jades

By BERTHOLD LAUFER

Associate Curator of Asiatic Ethnology, Field Museum, Chicago

THANKS to the generosity of Mr. Fritz von Frantzius, the well-known maecen and patron of art, the Field Museum of Natural History has recently been made the grateful recipient of a very important gift in the form of two Chinese jades which possess an intrinsic value from an artistic, antiquarian, and historical point of view. Both objects are precious relics of the Manchu dynasty, which was overthrown in 1911. It is well known that since the establishment of the Chinese republic the imperial family retired into seclusion, and being reduced to a limited income, was compelled to coin numerous heirlooms and art treasures on the market abroad. It is believed that in this manner these two rare and unique objects landed in this country and found their final resting-place in Chicago.

One of these is a complete jade-book bound in imperial yellow silk brocade and wrapped up in a silk shawl with dragons woven in gold threads. In its appearance it is a veritable book composed of ten rectangular slabs of uniform dimensions (11x5x½ inches) carved from K h o t a n nephrite of exquisite quality, thirty pounds in weight. It is the most solid book in the world, and certainly one of the finest ever executed, making a strong appeal to our esthetic sense. The binding has been effected by perforating each tablet at four points along the two narrow sides; the borings made by means of a tubular drill run a short distance and turn under a right

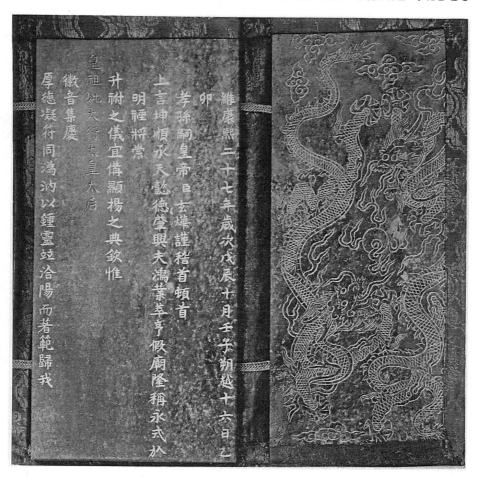

IMPERIAL JADE BOOK OF THE EMPEROR KANG–SHI,
[OPEN] SHOWING TABLETS I AND II

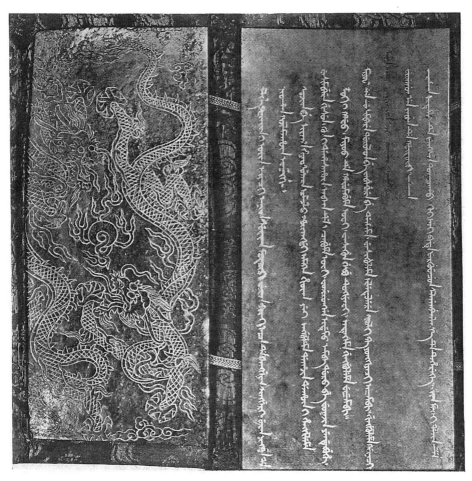

LAST TWO TABLETS [PAGES IX AND X] OF THE
IMPERIAL JADE BOOK OF THE EMPEROR KANG-HSI

angle toward the back—in view of the extreme hardness of the mineral a wonderful feat of technical skill. Yellow silk ribbons pass through these perforations and over the back of the slab, the tablets thus being held together. The first and last inner pages are each engraved with a pair of rampant five-clawed dragons, soaring in clouds and making for the flamed pearl, the grooves being incrusted with gold. The rest is taken up by a composition of the Emperor Kang-hsi (1662-1722), the ablest ruler produced by the Manchu. It was in his era that Chinese art of modern times, patronized by this intelligent sovereign, reached its climax. No porcelains have earned a greater admiration than those produced in the Kang-hsi period. The same high eulogy must be bestowed upon the glyptic works created by the hand of the lapidary, and in this particular field the artisans of China are unrivaled. An Arabic writer has not unjustly said that when Allah distributed mental gifts among mankind he placed them to the Greeks—in their heads; to the Arabs—in their tongues; and to the Chinese—in their hands.

The text is written in Chinese, the Chinese version occupying the tablets 2 to 5 and being accompanied by a translation into Manchu, which is engraved on the tablets 6 to 9. Different colors of jade have been chosen for the Chinese and the Manchu, greenish jade for the former, and a gray clouded jade for the latter. The engraved lines of the characters are filled with indigo-blue, symbolic of the color of the sky, while imperial names are prominently emphasized by being inlaid with gold foil. The document recorded on these tablets, being of a religious and ceremonial character, is intensely interesting; it is an act of canonization: the Emperor, speaking himself, confers upon his grandmother, the Empress Wen, the posthumous title Hiao Chuang, which means "filial and sedate." In the belief of the Chinese the ancestors continue to live, and largely contribute to regulate and control the actions of the living generation. Ancestors may be promoted in rank, and honors may be bestowed upon them; also they may be deprived of rank and title.

The said Empress was the consort of Tien-tsung (1626-1643) and the mother of the first sovereign of the Manchu dynasty, Shun-chi (1644-1661), who was born in 1636. As he was only a boy eight years old at the time of the Manchu conquest of China, his mother was appointed Dowager Empress in 1644. In 1662 she received the title Empress Grandmother. She died on January 27, 1688, and was canonized on November 8 of the same year, the date of our jade book. The interval which elapsed between these two dates gives us a welcome clew to the time consumed on the production of the jade book: the work was accomplished within a period of nine months or less, and this is astounding in view of the difficulty of graving jade which can be attacked only by means of diamond-points, emery, or ruby.

The Emperor performed an act of filial piety growing out of the tenets of ancestral worship by invoking his grandmother in order to bestow upon her the posthumous title by which she should be known to posterity. The purpose of having this state document carved in jade was twofold: first, to keep a permanent record of it, and, secondly, to convey the message directly to the spirit of the Empress in the celestial regions. In order to accomplish this end, they resorted to jade and the azure-blue coating laid over the writing. Jade, ranking as the most valuable of all jewels in the eyes of the Chinese, was believed to embody qualities of solar light and to communicate directly with heavenly powers by means of its transcendental properties. In ancient times the Chinese emperors followed the observance to announce to the great nature-deities heaven and earth their accession to the throne, the announcement being made on five tablets of jade.

Of all precious materials, jade at all times claimed the first rank in the estimation of the Chinese. It has been their most highly appreciated jewel from remote ages, and has been idealized by their poets and philosophers as the quintessence of nature, as the embodiment of all virtues, instilling into the hearts of its wearers the principles of good conduct and all noble and sublime thoughts. It was interred with the dead to preserve their bodies from decay and to promote their resurrection. The last tribute paid to a departed friend was to send his family a piece of carved jade to be placed in his grave.

The other jade object added to the collections of the Field Museum is an imperial seal which was conferred upon the Empress Jui, consort of the Emperor Kia-king, on February 12, 1796, the day when she received her official appointment as Empress of China. The square base of the seal, which has a weight of six pounds and three quarters, is surmounted by a pair of coalescing, four-clawed dragons, serving as handle, the whole being carved from a single block of stone, a very charming transparent plant-green contrasting from white inclusions. The carver has brought out the finest details of the mythical monster, even the teeth, and his harmonious achievement signifies a veritable triumph of spirit and skill over matter. The seal consists of twenty-one Chinese characters written in ancient style, accompanied by a parallel Manchu version, which is composed of eulogistic attributes of the Empress. Only the sovereign and his legitimate consort were entitled to the possession of jade seals. Under the Manchu, the number of official seals at the disposal of the Emperor amounted to twenty-five.

These two memorable palace-treasures, presented by Mr. Fritz von Frantzius, are now on exhibition in the Field Museum of Natural History, in Hall 45 of the East Wing. They are a valuable addition to the collection of jade formerly gathered in China under the auspices of Mrs. T. B. Blackstone, and must be ranked among the finest and most interesting productions of jade in this country. Mr. von Frantzius has manifested his interest in the institution in various other ways: he has recently donated likewise an important collection of Japanese coinage, about eight hundred specimens, which had been brought together at the instigation of the Japanese government.

123

《中国皮影戏》序和导论

Abhandlungen
der Königlich Bayerischen Akademie der Wissenschaften
Philosophisch - philologische und historische Klasse
XXVIII. Band, 1. Abhandlung

Chinesische Schattenspiele

Übersetzt
von
Wilhelm Grube

Auf Grund des Nachlasses durchgesehen und abgeschlossen
von
Emil Krebs

Herausgegeben und eingeleitet
von
Berthold Laufer

Vorgelegt am 8. Juni 1912

München 1915
Verlag der Königlich Bayerischen Akademie der Wissenschaften
in Kommission des G. Franzschen Verlags (J. Roth)

Inhalt.

1*

Vorwort.

Die Grundlage dieser aus Wilhelm Grubes Nachlaß veröffentlichten Arbeit bildet eine handschriftlich aufgezeichnete Sammlung chinesischer Schattenspieltexte (19 Hefte), die der Unterzeichnete im Jahre 1901 von einer Schattenspielertruppe in Peking samt deren aus etwa tausend Figuren bestehenden Apparat für das American Museum in New-York erwarb. Von einer Anzahl der Singspiele wurden phonographische Aufnahmen hergestellt, die zum Teil durch Erich Fischer im Psychologischen Institut der Universität Berlin bearbeitet worden sind. Für das Studium der Texte kam Wilhelm Grube bei seiner ausgedehnten Kenntnis der chinesischen Volkskunde und des volkstümlichen chinesischen Dramas insbesondere als die geeignetste Persönlichkeit in Frage. Im Sommer 1904 wurde die Angelegenheit mit ihm besprochen und der Vorschlag, die Herausgabe und Übersetzung der Texte zu übernehmen, begegnete einer ebenso bereitwilligen als verständnisvollen Aufnahme. Im Herbst desselben Jahres wurde daher das chinesische Manuskript von New-York an Professor Grube in Berlin gesandt, der bereits im März 1905 berichten konnte, daß er dreizehn Stücke übersetzt habe, von denen die Hanswurstpossen bei weitem die interessantesten seien, und daß er sein Hauptaugenmerk auf diese Gattung zu richten gedenke. Am 23. Januar 1906 schrieb Professor Grube:

„Von den Schattenspieltexten habe ich den weitaus größten Teil durchgenommen und übersetzt. Die Hauptschwierigkeit dabei war die Herstellung des Textes, der in einem geradezu schauderhaften Zustande ist. Von 25 Stücken habe ich mir eine korrekte Abschrift angefertigt, die schon 250 Seiten füllt. Nun aber steht mir noch die Durchsicht und Übersetzung eines, wie es scheint, nicht uninteressanten mythologischen Zauberdramas [Nr. III der Sammlung] bevor, welches sehr viel umfangreicher ist als die übrigen und allein vier von den im ganzen neunzehn Heften füllt. Zu diesem Zweck muß ich mir noch etwas Schonzeit ausbitten, da ich außer den Schattenspieltexten gleichzeitig noch andere laufende Arbeiten unter der Feder habe. Die dringendsten davon sind zwei Beiträge für die Kultur der Gegenwart (darunter ein größerer), die bis zum 1. März geliefert werden müssen. Außerdem vier Bogen für das Religionsgeschichtliche Textbuch von Bertholet, bis zum 1. Juli fällig. Trotzdem arbeite ich fast täglich sachte an den Schattenspielen weiter, woraus Sie sehen können, daß es nicht an gutem Willen fehlt; nur ist die Sache schwieriger und auch zeitraubender als ich anfangs dachte.‟

Als erste Probe erschien im Jahre 1906 in der Boas Festschrift (Boas Anniversary Volume) „Die Huldigungsfeier der Acht Genien für den Gott des Langen Lebens. Ein chinesischer Schattenspieltext übersetzt von Wilhelm Grube.‟ Leider sollte der unermüdliche Forscher den Abschluß seiner aufopfernden Arbeit nicht mehr erleben: ein langwieriges Herzleiden setzte am 4. Juli 1908 seiner rastlosen Tätigkeit ein Ziel. Im Februar 1908 hatte der Herausgeber noch die Freude, ihn auf kurze Zeit in seinem Studierzimmer in Berlin zu sehen und neue Arbeitspläne mit ihm zu besprechen. Ende Juli, als er in Peking eintraf, erreichte ihn die Trauerbotschaft von dem allzu frühen Ableben seines verehrten Lehrers. Eine Reise nach Tibet verhinderte den Unterzeichneten, sich des hinterlassenen Manuskripts anzunehmen. Herr Legationsrat Emil Krebs, erster Dolmetscher der Kaiserlich Deutschen Gesandtschaft in Peking, selbst ein Schüler und treuer Verehrer des Dahingegangenen, erbot sich, das unvollendete Werk abzuschließen. Auch an dieser Stelle sei Herrn Krebs für seine in selbstlosester Weise unternommene, überaus mühevolle und gewissenhafte Arbeit wärmster Dank abgestattet. Seine Mitarbeiterschaft ist der Sache in hohem Grade zugute gekommen, da die Texte in der Pekinger Volkssprache abgefaßt sind, die er mit voller Meisterschaft beherrscht. Zahlreiche Anspielungen auf örtliche Verhältnisse haben in ihm den rechten Interpreten gefunden; in schwierigen Fällen konnte er auch den Rat von Pekinger Schattenspielern einholen. Insbesondere erstreckt sich die Arbeit des Herrn Krebs nach drei Seiten hin. Er hat vor allem eine Anzahl (im ganzen 23, dazu die Solos) von Grube nicht übersetzter Stücke selbständig und mit großem Geschick übertragen. Diese Übersetzungen sind in jedem einzelnen Falle als solche kenntlich gemacht. Er hat sodann unvollendete Übersetzungen Grubes abgeschlossen, die bereits fertigen Übersetzungen einer gründlichen Durchsicht unterzogen, die dem Dahingeschiedenen versagt war, dieselben mit der Urschrift verglichen und Verbesserungen sowie Erläuterungen hinzugefügt. Er hat ferner die chinesischen Texte nachgeprüft, die noch nicht abgeschriebenen kritisch bearbeitet und den gesamten Textstoff in druckfertigen Zustand gebracht. Auf Veranlassung von Frau Professor Grube sind die chinesischen Texte in der Druckerei der Katholischen Mission in Yen-chou, Schan-tung, unter Leitung des Herrn Krebs gedruckt worden; dieser Band, im gleichen Format wie diese Abhandlungen, ist von Otto Harrassowitz in Leipzig zu beziehen.

Der Natur der Sache nach kam für die Umschreibung chinesischer Namen nur der Pekinger Dialekt in Frage. Das Studium der chinesischen Volkssprachen ist bisher in recht bescheidenem Maße betrieben worden. In der Mundart von Peking sind nur einige Volkslieder und eine kleine Sammlung von Schnurren veröffentlicht worden. Wir geben uns der Hoffnung hin, daß diejenigen, welche sich zu praktischen oder wissenschaftlichen Zwecken mit der Sprache der Hauptstadt befassen, in diesem Werke reichen und anregenden Stoff finden werden. Es kann nicht genug bedauert werden, daß es Grube nicht mehr vergönnt gewesen ist, uns die Erzeugnisse der chinesischen Schattenbühne zu erklären. Niemand hat ein so feinsinniges und tiefes Verständnis für das Seelenleben des chinesischen Volkes besessen als er.

Der Herausgeber möchte nicht verfehlen, der Direktion des American Museum of Natural History in New-York für die liberale Überlassung der Handschrift seinen Dank auszusprechen, ebenso der Königlich Bayerischen Akademie der Wissenschaften, die hiermit Grubes Vermächtnis der Öffentlichkeit übergibt. Wenn diese seine letzte Arbeit der in Deutschland aufstrebenden Chinaforschung neue Bahnen weisen und zu weiteren Untersuchungen auf dem vielversprechenden Gebiete des chinesischen Dramas und Bühnenwesens anregen sollte, so wird dem Andenken des verdienten Forschers am besten gedient sein. Die Tätigkeit des Herausgebers beschränkt sich auf die Durchsicht und Vorbereitung des Manuskripts zum Druck, Hinzufügung erklärender Anmerkungen, Anordnung des Stoffes und Erledigung der Korrekturen.

Chicago, 24. August 1915.

Berthold Laufer.

Einleitung.

Eine wirkliche Geschichte des Schattenspiels, in dem Sinne wie wir eine Geschichte des literarischen Dramas besitzen, wird wohl niemals geschrieben werden können. Überall, wo das Schattenspiel im Orient gepflegt wurde, war es eine Volksbelustigung, die Unterhaltung der breiten Massen, zu der sich die gelehrte Schriftstellerei nur selten herabließ. Wir finden daher in den orientalischen Literaturen nur abgerissene und fragmentarische Notizen, aus denen sich besten Falls ein allgemeiner Überblick der Geschichte des Spiels gewinnen läßt. Um dieselbe zu rekonstruieren, ist es vor allem erforderlich, auch die Technik des Spiels und der Figuren sowie die von den Schattenspielern gepflegten Stücke selbst zu Rate zu ziehen. Die eine Tatsache tritt klar in der Geschichte des Schattenspiels hervor, daß seine Wiege in Asien gestanden hat und daß wir es dem Orient zu verdanken haben: das klassische Altertum, der Hellenismus, das europäische Mittelalter und die Renaissance kennen das Schattenspiel nicht; es tritt erst gegen die Mitte des siebzehnten Jahrhunderts in Italien und etwas später in Deutschland und Frankreich auf. Dagegen haben die Hellenen das Puppenspiel geübt; und diese Tatsache hätte allein genügen sollen, um vor der übereilten Verknüpfung des Schattenspiels mit dem Puppenspiel zu warnen. Eine mehr als zehnjährige Beschäftigung mit dem Gegenstand hat mich zu der Überzeugung geführt, daß Schattenspiel und Puppenspiel eine durchaus verschiedene Geschichte, einen anders gearteten Ursprung haben und daß der Ausgleich, der zwischen beiden stattgefunden hat, nur die letzte Phase der Entwicklung, das Ergebnis der letzten Jahrhunderte, darstellt. Während ich zu der Ansicht gelangt bin, daß das Schattenspiel in China bodenständig ist, läßt sich an der Hand chinesischer Quellen der exakte Beweis führen, daß das Puppenspiel erst im siebenten Jahrhundert n. Chr. von Turkistan, mit größter Wahrscheinlichkeit von dem sang- und tanzfrohen Kucha, eingeführt worden ist: nicht nur haben damals die Chinesen den mittelgriechischen Namen κοῦκλα, der ihnen bis zum heutigen Tage verblieben ist, zur Bezeichnung der Marionetten eingeführt, sondern diese auch mit derselben eigentümlichen Technik übernommen, wie sie bereits von den Hellenen ausgebildet war. Diese Probleme hoffe ich demnächst in einer ausführlichen Abhandlung unter Vorlage des einschlägigen Quellenmaterials zu erörtern. In diesen Grubes Werk einleitenden Zeilen, deren Raum naturgemäß begrenzt ist, lassen sich die Ergebnisse meiner Studien nur in einem kurzen Umriß darlegen.

Um die Geschichte des Schattenspiels im allgemeinen und das islamische Schattentheater insbesondere hat sich niemand größere Verdienste erworben als Georg Jacob. In langjährig-ausdauernder und erfolgreicher Arbeit hat er Baustein für Baustein gesammelt und uns die große kulturgeschichtliche Bedeutung des Gegenstandes eindringlich vor Augen geführt (vgl. G. Jacob, Das Schattentheater in seiner Wanderung vom Morgenland zum Abendland, Berlin 1901; Bibliographie über das Schattentheater, zweite vermehrte Ausgabe, Erlangen 1902; Türkische Literaturgeschichte in Einzeldarstellungen, Heft I: Das türkische Schattentheater, Berlin 1900; Erwähnungen des Schattentheaters in der Welt-Literatur, dritte vermehrte Ausgabe der Bibliographie über das Schattentheater, Berlin 1906; Erwähnungen des Schattentheaters und der Zauberlaternen bis zum Jahre 1700, erweiterter bibliographischer Nachweis, Berlin 1912;

Geschichte des Schattentheaters, Berlin 1907). Über das arabische Schattenspiel sind wir dank den Forschungen Enno Littmanns gut unterrichtet (E. Littmann, Ein arabisches Karagöz-Spiel. ZDMG, Bd. LIV, 1900, S. 661—680 und Arabische Schattenspiele, Berlin 1901; vgl. auch C. Prüfer, Das Schiffsspiel, ein Schattenspiel aus Kairo, Beitr. zur Kenntnis des Orients, Bd. III. S. 154—169 und Ein ägyptisches Schattenspiel, Erlangen 1906; F. Kern, Das ägyptische Schattentheater, als Anhang zu J. Horovitz, Spuren griechischer Mimen im Orient, S. 98, Berlin 1905). Unter vielen anderen gehört es zu Jacobs Verdiensten, die zuerst von Quatremère herangezogene Stelle aus Raschid-eddīn's Geschichte der Mongolen ins rechte Licht gestellt zu haben (Keleti Szemle, Bd. I, 1900, S. 233—236). Hier haben wir die älteste Erwähnung des chinesischen Schattenspiels in der mohammedanischen Literatur. Unter der Regierung Ogotais (1229—45), des dritten Sohnes und Nachfolgers von Chinggis Khan, erschienen Spielleute aus dem nördlichen China, die hinter einem Vorhang wundersame Spiele aufführten (bei Quatremère: wunderbare Figuren zeigten); die einzelnen Völker waren durch verschiedene Typen vertreten. Unter anderen trat ein weißbärtiger Greis auf, dessen Turban an den Schweif eines Rosses gebunden war, und der mit dem Antlitz auf der Erde dahingeschleift wurde. Er sollte einen gefangenen rebellischen Mohammedaner vorstellen. Der Kaiser wies entrüstet eine solche Verspottung der islamischen Völker zurück. Dieser Bericht ist aus zwei Gründen von Bedeutung. Er zeigt uns einmal die Berührung des chinesischen Schattenspiels mit dem Islam und ich denke, daß Jacob (Geschichte des Schattenspiels, S. 20) im Recht ist, wenn er den Mongolen die Vermittlerrolle zuweist und sich die Wanderung des Spiels nach dem Westen als einen allmählichen, mehrmals wiederholten Prozeß vorstellt. Sodann verbürgt uns Raschid-eddīn die Bekanntschaft der Chinesen mit dem Schattenspiel zur Zeit der mongolischen Yüan-Dynastie und zeigt uns den chinesischen Schattenspieler in seinem eigentlichen Element, dem der politischen Verspottung, die ihm noch gegenwärtig eigen ist. In meinen Sammlungen befinden sich zahlreiche Figuren turbantragender Mohammedaner, welche dem Witz des Schattenspielers zur Zielscheibe dienen. Der Bericht des persischen Chronisten ist daher vollkommen glaubwürdig. F. von Luschan bemerkt in seiner interessanten Abhandlung über das türkische Schattenspiel (Int. Archiv für Ethn., Bd. II, 1889. S. 140): „Daß alle diese Schattenspiele an verschiedenen Orten unabhängig von einander erfunden worden sind, scheint kaum denkbar; wir werden für die verschiedenen Formen desselben vielmehr eine gemeinsame Quelle annehmen dürfen, welche vermutlich in China zu suchen ist." Kunos (Keleti Szemle, Bd. I, S. 141) nimmt an, daß das Schattenspiel chinesischen Ursprungs sei und durch Vermittlung der Perser zu den Türken gelangte. Die chinesische Überlieferung versetzt den Beginn des Schattenspiels in die Tage der Han-Dynastie unter die Regierung des Kaisers Wu (140—87 v. Chr.). Se-ma Ts'ien, der Vater der chinesischen Geschichte, berichtet in seinen Annalen (Schi ki) wie folgt: „Im folgenden Jahre (121 v. Chr.) erschien vor dem Kaiser ein Mann aus dem Lande Ts'i [in Schantung], Schao Wong, um seine Fertigkeiten in Bezug auf die Manen und Geister zu zeigen. Des Kaisers Lieblingsgemahlin Wang war gerade gestorben. Mit Hilfe seiner Kunst ließ Schao Wong des Nachts die Gestalt der Frau Wang und des Herdgottes erscheinen. Der Sohn des Himmels erschaute sie hinter einem Vorhang aus der Ferne. Er ehrte Schao Wong mit dem Titel „Marschall der gelehrten Vollkommenheit", überhäufte ihn mit Geschenken und behandelte ihn mit den für Gäste üblichen Riten" (vgl. E. Chavannes, Mémoires historiques de Se-ma Ts'ien, vol. III, p. 470; für die Geschichte des Schattenspiels ist die Stelle schon verwendet bei G. Schlegel, Chinesische Bräuche und Spiele in Europa, S. 28, Breslau 1869). Diese Erzählung gehört nicht zu den zahlreichen späteren, zurückdatierten Erfindungen der Chinesen, sondern ist ein gut beglaubigter Bericht des zeitgenössischen Annalisten. Bevor wir die Quelle erörtern, wird es gut sein, eine etwas spätere Version derselben Geschichte hier anzufügen, die uns De Groot (Religious System of China, vol. IV, p. 87) erschlossen hat. Yü Pao, der Verfasser einer Legendensammlung mit dem Titel Sou schen ki, die um das Jahr 320 entstanden ist (s. Wylie, Notes on Chinese Literature, p. 192), berichtet folgendermaßen: „Der Kaiser Wu der Han-Dynastie hatte eine Gemahlin namens Li und als sie verschied, mußte er ihrer unaufhörlich gedenken. Da meldete sich Schao Wong (oder Li Schao Wong) aus Ts'i, der in den geheimen Künsten bewandert war, beim Kaiser und sagte, daß er den Geist der verstorbenen Gemahlin zitieren könne. In jener Nacht spannte er einen Vorhang auf, der von Lampen und Fackeln erleuchtet wurde, und gebot dem Kaiser hinter einem anderen Vorhang zu sitzen und aus der Ferne zuzuschauen. Da gewahrte er in dem Vorhang das Bild eines schönen Weibes, dessen Gestalt der verstorbenen Frau Li glich. Der Kaiser näherte sich dem Vorhang, setzte sich und ging wieder herum, ohne jedoch imstande zu sein sie zu sehen. Dieses Ereignis vermehrte seinen Kummer und er machte seinen Gefühlen in einigen Versen Luft." G. Jacob (Geschichte des Schattentheaters, S. 16) hat die Ansicht ausgesprochen, daß die Zauberkünste des Schao Wong eher an die Hexe von Endor, die Helena-Erscheinungen im Volksbuch von Doktor Faust und Verwandtes erinnern und daß sie mit dem Ursprung des Schattenspiels nichts zu schaffen haben. Dieser einseitigen Auffassung vermag ich mich nicht anzuschließen. Gewiß, Schao Wong war kein Schattenspieler und was er im Bilde vorführte, war kein in Gesprächsform gesetztes Schattenspiel; aber wir können nicht erwarten, daß ein solches spontan erwachsen ist, es muß wie jede Erscheinung der menschlichen Kultur eine Entwicklung gehabt haben und der durchaus verbürgte und glaubwürdige chinesische Bericht gibt uns in der Tat ein beachtenswertes Hilfsmittel zur Feststellung dieser Entwicklung an die Hand. Die Forscher, die sich bislang mit der Geschichte des Schattenspiels befaßt haben, vermochten keine Deutung seiner Entstehung zu geben. Eine solche kann auch weder vom indischen noch vom javanischen noch vom islamischen Standpunkt versucht werden. In China liegt dieser Ursprungsgedanke klar vor Augen. Das Schattenspiel ist ein Spiel der Schatten und die Schatten sind die ursprünglich von Beschwörern zitierten Geister oder Manen der Abgeschiedenen. Die Schatten sind die Schattenseelen, durch kunstvoll aus Papier oder Leder geschnittene Figuren in Profil dargestellt und als wirkliche Schatten auf dem Vorhang erscheinend. Das Schattenspiel hat einen religiösen Ursprung und war von Hause aus eine spiritistische Sitzung. Schao Wong war ein Geisterseher: er konnte die Lebenden mit den Geistern der Verstorbenen in Verbindung setzen und ihre Schatten als wirkliche Schattenbilder erscheinen lassen. Sein Ehrgeiz jedoch trieb ihn weiter; er wollte den leichtgläubigen Kaiser mit der gesamten Götterwelt in Beziehung bringen. Vermittelst wolkengeschmückter und phantastisch aufgeputzter Wagen, die sein kaiserlicher Gönner und er selbst bestiegen, sollten die bösen Geister vertrieben werden. Inmitten eines Palastes errichtete er eine bewohnbare Terrasse und malte auf derselben die Götter des Himmels, der Erde und des Gestirns T'ai-i sowie alle Manen und Geister; dann traf er die erforderlichen Vorbereitungen für die Opfer, welche die Götter des Himmels anziehen sollten. Der Erfolg blieb indessen aus; nach Ablauf von mehr als Jahresfrist nahm seine Geschicklichkeit mehr und mehr ab und die Geister erschienen nicht. Um seinem gefährdeten Ruf wieder aufzuhelfen, setzte er ein Schriftstück auf Seide auf und ließ es durch einen Ochsen verschlingen. Mit verstellter Miene verkündete er, daß im Bauche dieses Tieres ein Wunder zu finden sei. Der Ochse wurde geschlachtet und man fand das Dokument, dessen Worte recht seltsam klangen. Der Kaiser erkannte die Handschrift seines Günstlings, forschte die Diener aus und entdeckte den Betrug. Der „Marschall der gelehrten Vollkommenheit" wurde denn auf kaiserlichen Befehl heimlich zu den Geistern befördert, im Jahre 119 v. Chr., nur zwei Jahre nach jener glänzenden Vorstellung, die ihm die verhängnisvolle Beförderung eingebracht hatte. Um diese Vorgänge zu würdigen, muß man sich gegenwärtig halten, daß Schao Wong nur einer von vielen war und daß das China seiner Zeit von Alchimisten, Magiern und religiösen Charlatanen aller Art wimmelte; es handelt sich bei ihm um einen typischen Fall, der dem religiösen Zeitgeist entsprach. De Groot hat die Bedeutung der obigen Erzählung richtig erkannt, wenn er die von dem Zauberer herbeigerufene Seele als den Schatten der Kaiserin bezeichnet. Das betreffende Kapitel in De Groots Werk, das den Titel führt „Animistic Ideas as suggested by Shadows", enthält zugleich die psychologische Begründung meiner Auffassung von dem religiösen, insbesondere spiritistischen Ursprung des Schattenspiels im alten China. Der Schatten war und ist nach chinesischer Volksvorstellung ein wesentlicher Bestandteil der Persönlichkeit und übt einen großen Einfluß auf ihr Schicksal aus; Schatten und Seele werden oft für gleich erachtet; während der Sarg in die Gruft gesenkt wird, treten die Leidtragenden einige Schritte zurück, damit ihr Schatten nicht in die Gruft falle und mitbegraben werde (W. Grube, Religion und Kultus der Chinesen, S. 192). Ein Überlebsel dieser alten Anschauung hat sich noch in einem Bühnenausdruck erhalten: im Hintergrund der Bühne ist an jeder Seite eine Tür angebracht, durch die eine treten die Schauspieler auf, die andere dient ihnen als Ausgang; diese Türen sind unter dem Namen „Türen der Seelen" (kuei mên) bekannt (vgl. M. Bazin, Théatre chinois, p. XLI, der geradezu „Schattentür", la porte des ombres, übersetzt), weil hier die Seelen der Personen der alten Geschichte ein- und ausgehen.

Im weiteren Sinne dürfen wir wohl sagen, daß die uralte Ahnenverehrung den Anstoß zum Schattenspiel, zunächst zur Anfertigung schattenwerfender Figuren gegeben hat, welche die Ahnen vorstellen sollten. Der Wunsch, einen geliebten Verstorbenen wiederzusehen, führte, wie im Falle des Kaisers Wu, zur Nachbildung seiner toten Gemahlin. Dieses Beispiel ist natürlich nur ein Symbol, aus dem sich die Norm abstrahieren läßt; wenn uns gerade dieser besondere Fall überliefert ist, so liegt es daran, daß es sich um eine Begebenheit innerhalb des kaiserlichen Hauses handelt; um die Gepflogenheiten des Volkes kümmerte sich die chinesische Geschichtschreibung wenig oder gar nicht. Wenn das T'an sou, ein Werk des elften Jahrhunderts, berichtet, daß von der Zeit des Kaisers Wu die Nachwelt ihre Schattenspiele erhalten, daß aber in der auf Wu folgenden Zeit man nichts darüber gehört habe (Hirth, Keleti Szemle. Bd. II, 1901, S. 78), so ist die Ursache für diese Erscheinung teils in dem eben genannten Grunde zu suchen, teils in dem tieferen Grunde, daß es ein Bühnendrama nicht vor der Periode T'ang (618—906) gegeben hat. Ein bühnengerechtes Schattenspiel konnte sich naturgemäß erst entwickeln, als die dramatische Gattung überhaupt ins Leben gerufen war (s. W. Grube, Geschichte der chinesischen Literatur, S. 362 u. f.). Leider sind wir über die Geschichte des chinesischen Dramas und Bühnenwesens höchst mangelhaft unterrichtet und tiefer eingreifende Untersuchungen über den Gegenstand liegen noch nicht vor. Daß es zur T'ang-Zeit ein Schattenspiel gegeben hat, ist mir aus mehreren inneren Gründen wahrscheinlich; einwandfreie Textstellen zur Unterstützung dieser Ansicht lassen sich jedoch noch nicht nachweisen. Erst für die folgende Sung-Periode (960—1278) haben wir sichere Belege. Hirth (l. c.) hat auf zwei interessante Texte aufmerksam gemacht, die unser Thema beleuchten. Der eine ist in dem bereits erwähnten T'an sou enthalten und erzählt, daß zur Zeit des Kaisers Jên-tsung (1023—65) unter den Marktleuten öffentliche Erzähler auftraten, welche Episoden aus der Geschichte der Drei Staaten (san kuo) vortrugen und ihren Vortrag durch Schattenfiguren erläuterten, die den Trachtenstil jener Periode (des dritten Jahrhunderts) zeigten. „Das ist der Anfang der bis auf unsere Tage [Ende des elften Jahrhunderts] erhaltenen Bilder der Kämpfe der drei Staaten Schu, Wei und Wu", schließt dieser Bericht. Stück No. VIII (S. 129) der vorliegenden Sammlung behandelt eine Episode aus dieser romanhaften Geschichte der Drei Reiche und ich selbst habe in Han-k'ou eine Serie von Schattenspielen gesehen, deren Stoffe aus derselben Quelle geschöpft sind. Als Verfasser des Romans San kuo yen i gilt ein gewisser Lo Kuan-chung, der im dreizehnten Jahrhundert gelebt haben soll (Wylie, Notes on Chinese Literature, p. 202; Legge. JRAS. 1893, pp. 803—805), über dessen Persönlichkeit aber sonst nichts bekannt ist. Jacob (Geschichte des Schattentheaters, S. 17) hat darin einen Widerspruch zu der Stelle des T'an sou zu erkennen geglaubt und vermutet, daß Lo Kuan-chung entweder nicht der Verfasser des Romans sein oder nicht zur Mongolenzeit gelebt haben kann. Die Dinge liegen freilich anders: nicht der gelehrte Kunstroman, wie er jetzt in zahlreichen Ausgaben verbreitet ist, war das Frühere, sondern die mündlichen Vorträge der Rhapsoden. Die Geschichtenerzähler waren die wirklichen Dichter und ihre Berichte wurden schließlich von einem Verfasser zu einem einheitlichen literarischen Kunsterzeugnis verwoben. Die Epoche der Drei Reiche mit ihren endlosen Fehden war die romantische Periode, die Ritterzeit der chinesischen Geschichte, deren Geist in manchem Spiel unserer Sammlung zu verspüren ist, ein gutes Beispiel für die Beständigkeit und Kontinuität der chinesischen Tradition. Auch jetzt noch bilden die Episoden aus der Geschichte der Drei Reiche den beliebtesten Stoff der öffentlichen Erzähler (vgl. über diese W. Grube, Zur Pekinger Volkskunde, S. 101). Die andere von Hirth angezogene und aus dem Tu ch'êng ki schêng vom Jahre 1235 entlehnte Notiz ist schon von Schlegel (Chinesische Bräuche und Spiele in Europa, S. 29) übersetzt worden. Sie besagt, daß man in der Hauptstadt Hang-chou die Schattenspielfiguren anfangs aus Papier ausgeschnitten, später aber aus bunt bemaltem Leder angefertigt habe; dabei erhielten die

guten und ehrlichen Charaktere regelrechte Menschengesichter, während man die Schurken mit
Teufelsfratzen darstellte; dem Gebrauch der Bühne entsprechend wurden Gute und Böse alle-
gorisch angedeutet. Wenn ich nicht irre, ist dieser Bericht einem etwas älteren Werke ent-
lehnt, dem Mong liang lu (Kap. 20, p. 13b der Ausgabe in der Sammlung Chi pu tsu
tsai ts'ung schu), einer sehr interessanten Beschreibung von Hang-chou, die den Wu Tse-mu,
dort gebürtig, zum Verfasser hat. In seinem Buche findet sich der von Hirth übersetzte Passus
in demselben Wortlaut, aber in etwas erweiterter Form. Vor allem ist beachtenswert, daß nach
Wu Tse-mu das Schattenspiel, bevor es Hang-chou erreichte, in Pien-king (d. i. K'ai-fong-fu,
jetzt Hauptstadt der Provinz Ho-nan) heimisch war und daß dort zuerst die Figuren aus ein-
fachem Papier hergestellt wurden, während später in geschickter Weise Handwerker sie aus
Leder machten und buntfarbig ausschmückten; die Lederfiguren waren unzerstörbar. In Hang-
chou gab es damals vier Unternehmer in diesem Fache, die, um die Figuren zu zeigen, sich
eines ausgespannten Tuches bedienten und sich wenig von den Geschichtenerzählern unterschieden; auch ihre Textbücher waren mit denen der letzteren fast identisch und stellten eine
Mischung von Wahrheit und Dichtung vor. Dann folgt die Stelle betreffs der Unterscheidung
der guten und bösen Charaktere wie oben. In Hang-chou wird noch gegenwärtig das Schattenspiel eifrig gepflegt und die Figuren werden auch jetzt aus fein gegerbtem Schafsleder verfertigt, weshalb der Volksmund das Schattenspiel „Schafslederspiel" (yang p'i hi) nennt.

Zu den von Hirth gegebenen Belegen für das Auftreten des Schattenspiels in der Sung-
Periode kann ich noch einige weitere anfügen. Eine kurze Anspielung auf das Schattenspiel
findet sich in einem Gedicht des Fan Ch'êng-ta (P'ei wên yün fu, Kap. 63), eines bekannten
Dichters und Beamten, der von 1126 bis 1193 lebte und eine sehr interessante Abhandlung
über die Geographie und Naturprodukte des südlichen China schrieb, auch ein Werk über die
Zucht von Chrysanthemen. Statt des gewöhnlichen Ausdrucks ying hi (wörtlich „Schattenspiel") bedient sich dieser Autor der Bezeichnung ying têng hi („Schattenlampenspiel"). Da
beim Schattenspiel der durchscheinende Vorhang durch dahinter aufgestellte Lampen beleuchtet
werden muß, so weist dieser Name deutlich auf unser Schattenspiel hin. Wir haben gesehen,
daß nach dem Mong liang lu das Schattenspiel in K'ai-fong-fu, der Hauptstadt der sogenannten
Nördlichen Sung Dynastie (960—1126), blühte. Diese Angabe wird durch eine interessante
Stelle im Tung king mong hua lu bestätigt. Dieses Werk, das eine Schilderung der Östlichen Hauptstadt (Tung king, d. i. K'ai-fong-fu) enthält, wurde von Mong Yüan-lao, einem
Zeitgenossen der Nördlichen Sung, im Anfang des zwölften Jahrhunderts verfaßt (s. den Kata-
log der Kaiserlichen Bibliothek, Kap. 70, p. 30b; das Werk ist abgedruckt in der Sammlung
Hio tsin t'ao yüan). Hier wird berichtet, daß es vor dem Jahre 1102 in der Hauptstadt
fünfzehn Buden gegeben habe, in denen Vorstellungen volkstümlicher Künste stattfanden; sechs
dieser Anstalten waren der Pflege des Schattenspiels gewidmet und führten die Namen Chao
Ts'i, Ts'ao Pao-i, Chu Po-êrh, Mu K'un-t'o, Fung Sêng und Ko Tsu. Gegen Ende des elften
Jahrhunderts muß also das Schattenspiel schon einen hervorragenden Platz im Volksleben ein-
genommen haben und es gab Schattenspieler von Ruf.

Aus diesen Berichten ergibt sich die interessante Tatsache, daß das chinesische Schattenspiel zunächst episch oder rein erzählend im seinem Charakter war, daß somit das Schattenspiel
in dramatischer Form eine spätere Entwicklungsstufe darstellt, welche die erzählende Form allmählich verdrängt hat. Bei dieser war die Kunst des Vortrags die Hauptsache; der Erzähler
war die Hauptperson, der Mittelpunkt der Anziehung, während die Figuren im Hintergrunde
blieben und der Erläuterung und Versinnbildlichung der Rede, der Demonstration, dienten. Von
der Zeit an, da sich das Schattenspiel dramatisch umgestaltete, verschwand die Persönlichkeit
des Vortragenden von seinem Podium hinter den Vorhang, der zur Schattenbühne wurde, um
die Rolle oder Rollen der auf der Bühne Handelnden zu übernehmen; er wurde unsichtbar für
das Publikum, das nur seine Stimme als Organ der bewegten Figuren vernahm und seine Aufmerksamkeit lediglich auf deren Handeln richtete. Wie die Wandlung von der Schattenerzählung
zum Schattenschauspiel vor sich gegangen ist, entzieht sich vorläufig unserer Kenntnis, aber
den gegenwärtigen Verhältnissen nach zu urteilen muß sich die Umbildung unter dem Einfluß
der Schauspielerbühne vollzogen haben. Denn wie unsere Sammlung lehrt, steht in unseren

II*

Tagen, und wahrscheinlich schon seit einigen Jahrhunderten, dem Schattenspieler der ganze Schauspielplan des großen Theaters offen, wenn er sich auch in seiner Weise die Stücke mundgerecht macht. Da viele Dramen nach Historien, Anekdoten und Novellen gearbeitet sind, so tritt der erzählende Charakter noch deutlich zu Tage. Die Form des alten Schattenspiels der Sung ist jetzt noch in Siam lebendig, insofern dort Sagen aus dem Ramayana erzählend vorgetragen und durch Schattenfiguren erklärt werden (s. F. W. K. Müller, Nang. Siamesische Schattenspielfiguren, mit zwölf Tafeln, Supplement zu Band VII des Int. Archiv f. Ethn., 1894); ein eindringendes Studium des siamesischen Schattenspiels in seiner Heimat wäre allerdings wünschenswert. Nach den uns bis jetzt vorliegenden Quellen läßt sich nicht beurteilen, ob schon unter den Sung ein dramatisch ausgebildetes Schattenspiel bestanden habe. Die positiven Zeugnisse des T'an sou und Mong liang lu lassen nur auf erzählendes Schattenspiel schließen. Überdies gibt uns letzteres Werk im allgemeinen über die Volksbelustigungen von Hang-chou Aufschluß und erwähnt besonders der Puppenspiele, die eine ziemlich große Auswahl von Stoffen zur Verfügung hatten und unter anderem sogar „vermischte dramatische Spiele" (tsa ki) zur Aufführung brachten. Diese Gattung des Schauspiels ist nicht, wie von einigen Sinologen behauptet worden ist, erst im Zeitalter der Yüan-Dynastie entstanden, sondern wurde bereits unter den Nördlichen Sung (960—1126) entwickelt, um unter den Kin und Yüan vervollkommnet zu werden (so schon bei Palladius, Chinesisch-russisches Wörterbuch, Bd. II, S. 186).[1] Die Möglichkeit zu einem dramatischen Schattenspiel war daher in der Epoche der Sung unzweifelhaft vorhanden; ob dasselbe aber wirklich bestanden hat, muß vorläufig, da unsere Quellen nichts darüber aussagen, eine unentschiedene Frage bleiben.

R. Pischel hat in einer geistreichen Abhandlung (Das altindische Schattenspiel, SPAW 1906, S. 482—502) die Hypothese von einem Ursprung des Schattenspiels in Indien aufgestellt und alte Erwähnungen desselben in der Therīgāthā und im Mahabharata zu erweisen gesucht. G. Jacob (Geschichte des Schattentheaters, S. 5—8, und Die Erwähnungen des Schattentheaters und der Zauberlaternen bis zum Jahre 1700, S. 3) hat diese Ergebnisse ohne weiteres angenommen, aber bei aller Verehrung und Bewunderung des leider zu früh dahingeschiedenen genialen Gelehrten muß ich doch bekennen, daß mir Pischels Beweisstücke keineswegs überzeugend sind. Sie stützen sich im wesentlichen auf die Deutung gewisser Termini und wenn deren Auffassung als „Schattenspiel" auch zulässig sein sollte, so ist damit noch nicht erwiesen, daß es sich wirklich um ein dramatisches Schattenspiel handelt. Der Orient kennt Schattenspiele verschiedener Art. Wir alle haben als Knaben das Spiel geübt, vermittelst verschiedener Hand- und Fingerstellungen Schattenbilder von Tieren an die Wand zu werfen. Dieses Spiel ist überall im Osten bekannt, in Indien, China, Korea und Japan (s. besonders S. Culin, Korean Games, p. 9). In Japan, wo es den Namen kage-ye („Schattenbilder") führt, hat man es zu einer Art Kunst ausgebildet; ich sah einst in einem japanischen Theater einen Darsteller, der auf dem Rücken liegend unter gleichzeitiger Bewegung beider Hände auf einen Wandschirm Schatten projizierte, welche lebensvolle Szenen mit Häusern, Bäumen und agierenden Menschen vorstellten; daß dabei noch andere Hilfsmittel, wie in den Händen gehaltene Papierfiguren mitwirkten, ist wohl wahrscheinlich. Wer in China und Japan gelebt hat, ist auch mit den Schatten vertraut, welche die in erleuchteter Stube befindlichen Personen auf die Papierfenster oder Papiertüren werfen und die dem Außenstehenden den Anblick eines eigentümlichen Schattenspiels gewähren. Die volkstümliche Kunst des japanischen Farbendrucks hat von diesem reizvollen Motiv Gebrauch gemacht. Wenn in der Therīgāthā die Nonne Subhā einen aufdringlichen Bewerber mit den Worten zurückweist, daß er sich wie ein Blinder „auf ein Schattenspiel im Menschengedränge stürze", wie Pischel (S. 488) übersetzt, so ist doch die einfachste Erklärung, daß das natürliche Spiel huschender Schatten gemeint ist, die von einer dahinwogenden Menschenmenge geworfen werden. Die Annahme eines Schattenspiels im Sinne einer Theateraufführung scheint mir weder zwingend noch wahrscheinlich zu sein. Der Nachdruck in der betreffenden Stelle liegt in dem Vergleich mit dem Blinden: ein Blinder, der einen

---

[1] Eine Sammlung von dreißig Stücken dieser Art ist unter dem Titel Ku kin tsa ki san schi chung im Jahre 1914 von der Universität Kyōto herausgegeben worden.

Menschen haschen will wie der Bewerber die fromme Nonne, mag sich irrtümlich sehr wohl auf des Menschen Schatten stürzen. Aber welchen Sinn sollte es haben, daß der Blinde sich auf die Schattenbühne stürzen sollte, die er nicht einmal sehen kann? Ein anderer Gedanke, der offenbar Pischel entgangen ist, tritt doch darin hervor, daß sich Subhā selbst mit einem Schatten vergleicht: als buddhistische Nonne führt sie nicht mehr das körperliche Dasein des Weibes, sondern ist nur der Schatten eines Weibes. Was sie dem Antragsteller bedeutet, läßt sich, der dichterischen Form entkleidet, einfach in die Worte fassen: du magst ebenso gut einen Schatten begehren oder umarmen als mich. Von einem wirklichen Schattenspiel ist in dieser Stelle keine Rede. Das einzige tatsächliche Zeugnis, das Pischel für indisches Schattenspiel anführt, ist Nīlakaṇṭha's Glosse zu dem Ausdruck rūpopajīvana im Mahābhārata: „Rūpopajīvana ist bei den Südländern als jalamaṇḍapikā bekannt. Dabei wird, nachdem man ein dünnes Tuch aufgespannt hat, durch Figuren aus Leder das Treiben der Könige, Minister usw. vor Augen geführt." Hier handelt es sich in der Tat um ein Figurenschattenspiel. Ob aber Nīlakaṇṭha's Erklärung, die jedenfalls für seine Zeit zutrifft, auf die betreffende Stelle im Mahābhārata Anwendung finden muß, ist eine andere Frage. Auffällig bleibt jedenfalls, daß die Termini rūpopajīvana und rūpyarūpaka nicht mit der indischen Tradition vom Schattenspiel, wenn es überhaupt eine solche gibt, verknüpft erscheinen und auch Pischel (S. 501) hat diese Lücke in seiner Forschung empfunden. Für verfehlt halte ich auch die Ansicht, daß das chāyānāṭaka eine literarische Weiterbildung des alten, volkstümlichen Schattenspiels sei. Denn einmal ist der Beweis nicht erbracht, daß das Schattenspiel in Indien alt und je volkstümlich gewesen ist: sodann beruht die Auffassung des chāyānāṭaka als eines Schattenspiels doch nur auf der wörtlichen Übersetzung dieses Namens. Eine indische Tradition, daß diese dramatische Gattung ein wirkliches Schattenspiel gewesen sei, gibt es nicht und Pischel selbst bringt keine Spur eines Beweises dafür vor. Der Satz „mag das chāyānāṭaka zur Zeit des Subhaṭa mit Schattenspiel-figuren oder durch Schauspieler selbst vorgeführt worden sein, das scheint sicher zu sein, daß es eine literarische Fortbildung des alten volkstümlichen Schattenspiels ist" scheint mir eine etwas kühne und stark subjektive Interpretation. Aus dem Charakter dieser Stücke selbst (vgl. L. H. Gray, The Dūtāṅgada of Subhaṭa, JAOS vol. XXXII, 1912, pp. 58—77) ist nichts darüber zu entnehmen, daß es sich um Schattenspiele handelt. Die Ansicht von Rājendralāla Mitra, daß diese angeblichen Schattenspiele bühnenmäßige Zwischenstücke seien, scheint mir weit mehr innere Wahrscheinlichkeit für sich zu haben. Pischels Hinweis auf das siamesische Schattenspiel kann für Indien nichts beweisen, denn jenes geht sicher auf das javanische Vorbild zurück und ich sehe vorläufig keinen Grund, warum Java nicht die Heimat für das indische Schattenspiel sein sollte. In Java stand während des Mittelalters das Schattenspiel in hoher Blüte, während Indien keine literarische Überlieferung betreffs des Schattenspiels besitzt: die Sanskrit-Werke über Dramaturgie erwähnen es mit keiner Silbe und das einzige überhaupt vorhandene Zeugnis ist und bleibt die knappe Bemerkung des Nīlakaṇṭha; sein Hinweis auf Süd-indien ist vielleicht nicht zufällig und mag auf eine Berührung des südindischen Schattenspiels mit Java hindeuten. Vorläufig wissen wir vom indischen Schattenspiel nichts.[1]) In Calcutta habe ich trotz eifriger Nachforschungen nichts darüber erfahren können, ebensowenig ist es mir bekannt, daß Berichte in der Literatur über indisches Schattenspiel vorliegen und daß irgend-ein Museum indische Schattenspielfiguren besitzt. Um uns den Glauben an das hohe Alter des Spiels in Indien beizubringen, wäre zunächst doch der Nachweis erforderlich, daß es gegen-wärtig in Indien gepflegt wird. Sollte dies der Fall sein, so könnte das Studium seiner Technik und seiner Stücke möglicherweise zu einem Ergebnis führen. Pischels schwach begründete Hypothese läßt die Frage offen, wie das Spiel auf indischem Boden entstanden ist oder ent-standen sein könnte, und läßt uns über seine Verbreitung außerhalb Indiens im Dunkeln. Weder die Chinesen noch die Javaner noch die mohammedanischen Völker haben eine Überlieferung aufzuweisen, daß sie das Schattenspiel aus Indien empfangen hätten.

---

[1]) Was G. Jacob (Geschichte des Schattentheaters, S. 8) von Ceylon berichtet, beruht doch auf An-deutungen, die kaum auf einer sicheren Grundlage fußen.

Über die geschichtliche Stellung des Schattenspiels auf Java enthalte ich mich vorläufig jeder Meinung, da ich noch nicht Gelegenheit gehabt habe, die Quellen, auf denen sich die Ansicht vom hohen Alter des javanischen Spiels stützt, kritisch nachzuprüfen. Wenn es zutreffend sein sollte, daß es wirklich bereits der ersten Hälfte des elften Jahrhunderts angehört, so bleibt diese Gleichzeitigkeit des Auftretens mit dem Schattenspiel der Sung-Periode doch eine auffallende chronologische Tatsache, die nach meinem Empfinden für eine Abhängigkeit von China spräche, unter Berücksichtigung der sehr alten Handels- und Kulturbeziehungen zwischen beiden Ländern. Die sprachlichen Argumente für den einheimischen Ursprung sind nicht beweiskräftig; einheimische Namen für eingeführte Dinge aus der Fremde können überall entstehen. Man darf nicht außer acht lassen, daß in diesem Falle nur die Übernahme einer Technik in Frage steht; denn genau genommen darf man nicht von einer Wanderung des Schattenspiels reden, sondern man kann höchstens eine Wanderung der Figuren in Anspruch nehmen. Von einem hohen Altertum des Schattenspiels ist nirgendwo, selbst nicht in China, die Rede: die Zeit vom elften bis dreizehnten Jahrhundert kennzeichnet sich deutlich als die Periode seiner Entwicklung und seiner Ausbreitung über Asien. Es ist daher lediglich eine Kunst des Mittelalters; und naturgemäß wurden ihm überall, wohin es vordrang, die einheimischen Sagenstoffe untergeschoben. Es traf überall auf einen bereits angebauten Kulturboden. Eine so einfache Technik aber wie die der reizvollen, lebendigen und erfolgreichen Schattenfiguren mußte notwendig zur Nachahmung herausfordern. Daß jedes Volk seinen eigenen Stil und Kunstsinn in die Figuren hineinlegte und daß fortgesetzte Übung weitere Veränderungen oder örtliche Variationen geschaffen hat, ist nicht wunderzunehmen; daß aber in China und Java die Idee der Figuren unabhängig voneinander entstanden sein sollte, ist schlechterdings undenkbar. Schlegel (T'oung Pao, 1901, p. 203 und 1902, p. 43) hat die Abhängigkeit des javanischen vom chinesischen Schattenspiel behauptet, aber nicht bewiesen; daß Ma Huan nur das Wayang beber und nicht das Schattenspiel erwähnt, beweist natürlich gar nichts.

Wenn es berechtigt ist, die Heimat einer Sache da zu suchen, wo sie in größter technischer Vollkommenheit erscheint, so sprechen gewiß die chinesischen Schattenspielfiguren selbst eine beredte Sprache für die Ursprünglichkeit des Gedankens auf chinesischem Boden. Schon F. von Luschan (Int. Arch. für Ethn., Bd. II, 1889, S. 140) hat darauf hingewiesen, daß die chinesischen Figuren in der Berliner Sammlung sich nicht nur durch besonders sorgfältige, reinliche Arbeit, sondern auch durch künstlerische Behandlung auszeichnen. Dieses Urteil ist vollkommen richtig, denn es ist in der Tat die feine künstlerische Form, welche die Figuren der Chinesen von denen der Türken, Araber, Javaner und Siamesen vorteilhaft unterscheidet. Über diese Ansicht dürften sich wohl alle einig sein, die in unseren Museen Gelegenheit zu einer vergleichenden Betrachtung des einschlägigen Materials gehabt haben. Eleganz und Anmut der Formen, liebevolle Sorgfalt für die dekorativen Elemente des Kostüms, feinfühliger Farbensinn, solide Ausführung und unbegrenzte Beweglichkeit bestimmen den Charakter der chinesischen Schattenfiguren. Sie verdienen eingehendes Studium. Die javanischen sind von Serrurier in einer vornehmen Publikation beschrieben worden, die türkischen von F. von Luschan; über die ägyptischen verdanken wir P. Kahle eine wertvolle Abhandlung (Der Islam, Bd. I, 1910, S. 264—299 und Bd. II, 1911, S. 143—195, mit zahlreichen guten Abbildungen). Den Schattenfiguren dieser Völker sollen gewiß nicht ihre Vorzüge abgestritten werden, aber sie erreichen nicht das Ebenmaß und die Harmonie der chinesischen und noch weniger deren Beweglichkeit. Davon wird noch weiter unten die Rede sein.

In ganz anderen Bahnen als das Schattenspiel bewegt sich die Geschichte des Puppenspiels. Zu einer Zeit, als die kulturhistorische Methode noch wenig ausgebildet war und die Sinologen ihren chinesischen Vorbildern nacheifernd in gläubig-kindlichem Bestreben alle Erfindungen und Gedanken den Chinesen zuschrieben, kam die gelehrte Legende von einem hohen Altertum des Puppenspiels in China auf und Schlegel (Chinesische Bräuche und Spiele in Europa, S. 28) meinte sogar: „Es scheint somit nicht zu gewagt anzunehmen, daß sie auf demselben Wege wie die anderen Spiele nach Europa gekommen sind." Die ganze Beweisführung stützte sich auf zwei Anekdoten. Die eine derselben erzählt, daß, als der Hunnenfürst Mao-tun die von dem Han-Kaiser Kao-tsu (206 v. Chr.) verteidigte Feste P'ing oder Pai-téng belagerte, sein Ratgeber Ch'èn P'ing († 179 v. Chr.) den Entsatz der Festung herbeiführte, indem er auf der Stadtmauer eine schöne Frauenpuppe aus Holz tanzen ließ. Die List war auf die Eifersucht von Mao-tun's Gemahlin berechnet, die gefürchtet haben soll, daß der Khan nach dem Fall der Stadt diese angebliche Schönheit zu sich nehmen könne. Von dieser albernen Anekdote findet sich in den zeitgenössischen Quellen wie Se-ma Ts'ien's Annalen keine Spur; sie tritt zum ersten Male auf in einer kleinen Schrift über Musik und Tanz vom Ende des zehnten Jahrhunderts n. Chr., dem Yo fu tsa lu (p. 17b der Ausgabe in der Sammlung Schou schan ko ts'ung schu, vol. 66), d. h., wie wir weiter sehen werden, über dreihundert Jahre nach der wirklichen Einführung der Marionetten in China. Diese Geschichte von jungem Datum, die nur zu dem Zweck erfunden ist, um den Puppen ein hohes Alter und einen einheimischen Ursprung zuzuschreiben, ist ohne jeden Wert; es handelt sich in dieser plumpen Erfindung auch gar nicht um ein Puppenspiel. K'ang-hi's Wörterbuch, chinesische Enzyklopädien und eine Reihe von Sinologen bringen mit der Geschichte des Puppenspiels eine andere Anekdote in Verbindung, die sich in dem Buche des angeblichen Philosophen Lie-tse befindet. Danach soll zur Zeit des Königs Mu (etwa um 1000 v. Chr.) ein geschickter Mechaniker, Meister Yen, einen singenden und tanzenden Automaten verfertigt haben, der vor dem Könige eine Vorstellung gab; als der Automat schließlich mit den königlichen Frauen liebäugelte, geriet König Mu in Zorn und wollte Yen auf der Stelle töten; der zerstörte in seiner Angst das Kunstgebilde und zeigte, daß es nur ein Gefüge von Leder, Holz, Leim und Firnis sei (s. E. Faber. Naturalismus bei den alten Chinesen, S. 126; L. Wieger, Les pères du système taoiste, p. 145; Mayers. Chinese Reader's Manual, p. 276). Hier handelt es sich nicht um Puppen, sondern um eine ganz andere Sache, das Automatentheater. Bei Marionetten werden gewiß nicht die inneren Organe, Herz, Leber, Nieren dargestellt, wie sie hier beschrieben werden und die mit Mund, Augen und Füßen verbunden gewesen sein sollen. Berichte über solche Automatenfiguren gibt es ziemlich viele, besonders in der Geschichte der Tsin-Dynastie (265—419), aber ich kann an dieser Stelle nicht näher auf das Thema eingehen. Die Redaktion des unter Lie-tse's Namen gehenden Textes, in der uns überlieferten Gestalt, kann meines Erachtens nicht älter als das Zeitalter der Han (206 v. Chr. bis 220 n. Chr.) sein,[1] da das Buch von den in dieser Periode auftauchenden hellenistischen Einflüssen wimmelt. Zu diesen gehört auch das Automatentheater, das zuerst von den großen alexandrinischen Mechanikern, Philon und Heron. konstruiert und beschrieben worden ist (vgl. V. Prou. Les théâtres d'automates en Grèce au IIᵉ siècle avant l'ère chrétienne, in Mémoires présentés par divers savants 1886, pp. 117—274; und W. Schmidt, Heronis Alexandrini opera, Vol. I, pp. 338—453). Die erste zuverlässige Erwähnung von Marionetten unter dem noch gegenwärtig üblichen Namen k'ui-lei oder k'uei-lei ist erst im siebenten Jahrhundert unter den T'ang bezeugt und das Jahr 633 (siebentes Jahr der Periode Chêng-kuan) mag als das der ersten sicheren Erwähnung gelten. Damals war der Kronprinz dem Puppenspiel leidenschaftlich ergeben und hohe Staatsbeamte erhoben Einsprache gegen die Ausübung dieser neuen Kunst; der Vorsitzende des Ministeriums der öffentlichen Arbeiten, Tuan Lun, mußte seine Entlassung nehmen, weil er die Erlaubnis zur Herstellung hölzerner Marionetten durch einen geschickten Handwerker, namens Yang Se-ts'i, gewährt hatte. Wenn man den wahren Ursprung derselben bisher verkannt hatte, so lag dies daran, daß ihr alter Name, wie er in den Annalen der T'ang überliefert ist, nicht ans Licht gezogen war. Heute und seit den Tagen der Sung schreibt man den Namen k'uei-lei mit den Schriftzeichen 傀 儡. Da jedes dieser beiden Zeichen mit dem Klassenzeichen „Mensch" verknüpft ist, so hielt man natürlich das Wort für einheimisches Sprachgut. In der Geschichte der T'ang (Kiu T'ang schu, Kap. 29, p. 6b) jedoch finden wir zwei alte Schreibungen k'u-lei-tse 窟 石壘 子 und k'uei-lei-tse 魁 石壘 子, beide Bezeichnungen als ein Spiel mit singenden und tanzenden Puppen erklärt, das ursprünglich die Unterhaltung der Leute bei den Trauerfeierlichkeiten war. Von diesen beiden Namen ist der erstere sicher der ursprünglichere.

---

[1] Das Buch wird zuerst in den Annalen der Früheren Han als ein Werk in acht Kapiteln erwähnt (s. Ts'ien Han schu, Kap. 30, p. 12b).

da er rein phonetisch geschrieben ist und keinerlei Bedeutungselemente enthält. In der T'ang-Zeit wurde aber das Schriftzeichen, das jetzt k'u lautet, in der Regel kut ausgesprochen und lei wie le oder lai. Die Japaner nannten die Puppen und das Puppenspiel in der Fujiwara-Periode kugutsu (mit den modernisierten Zeichen wie oben geschrieben), wodurch erwiesen wird, daß ihnen das Wort von den Chinesen der T'ang-Zeit nicht als kut, sondern in der Form kuk überliefert worden ist; im Dialekt von Fu-kien hat dasselbe Wort noch auslautendes k. Somit gelangen wir zu der lautlichen Wiederherstellung kuk-le und dieses Wort ist nichts anderes als das mittelgriechische κούκλα „Puppe". Den Chinesen kam das Wort gewiß durch Vermittlung von Türken oder Iraniern Zentralasiens oder vielleicht von Kucha (s. w. unten) zu. Die Osmanen erhielten das Wort von den Byzantinern (W. Radloff, Versuch eines Wörterbuches der Türk-Dialecte. Bd. II, col. 898: osmanisch kukla „Puppe" und kuklači „Marionetten-spieler"); ebenso ist dasselbe in die slavischen Sprachen gedrungen (E. Berneker, Slavisches etymologisches Wörterbuch. S. 640: russisch und bulgarisch kukla; russisch kukolnik „Puppen-macher", früher „Hanswurst, Gaukler", kukolnaya komediya „Puppenspiel"). Das Puppen-spiel hat sich innerhalb der griechischen Welt bis zum Untergang von Byzanz erhalten (H. Reich, Der Mimus, Bd. I, S. 672—673) und Byzanz haben wir uns als Ausgangspunkt für seine Ver-breitung unter Slaven und Innerasiaten vorzustellen. Es ist auch bemerkenswert, daß sich im gegenwärtigen Pekinger Dialekt die alte anlautende Tenuis und überhaupt die alte Tradition des Wortes in der Form ku-li erhalten hat. Aber nicht nur das fremde Wort, auch die Sache haben die Chinesen der T'ang-Zeit übernommen. Das geht klar aus den alten Beschreibungen der Marionetten hervor, die an Fäden aufgehängt waren, noch jetzt eine in China übliche Form des Puppenspiels, zu der zwei jüngere Formen hinzukommen. Der chinesische Ausdruck huan sien k'uei-lei („an Fäden aufgehängte Marionetten"), der sich z. B. im Mong liang lu findet, liest sich wie eine Übersetzung des griechischen τὰ νευρόσπαστα. Diesem Terminus ent-spricht ferner das indische sūtraprota (sūtradhāra = νευροσπάστης) und das japanische ayatsuri („Puppenspiel", vom Verbum ayatsuru „die Fäden einer Gliederpuppe ziehen": K. Flo-renz, Geschichte der japanischen Literatur, S. 583 und Mitteilungen der Deutschen Gesellschaft Ost-asiens, Bd. IX, 1903, S. 275). Daß man in Hellas, Indien, Turkistan, China und Japan den-selben Typus, die gleiche Technik der Marionetten angewandt hat, kann nicht dem geringsten Zweifel unterliegen. Ebensowenig kann es eine Meinungsverschiedenheit darüber geben, daß diese Dinge historisch zusammengehören und daß sich von Hellas die Marionetten über Asien verbreitet haben. Die Wanderung des Wortes kukla nach Zentralasien und China verleiht dieser Ansicht eine sichere Grundlage. Ich bin weit von der Meinung entfernt, den Hellenen die Erfindung der Marionetten, wenn man überhaupt in diesem Falle von Erfindung sprechen kann, zuzuweisen oder den Ursprung des Marionettenspiels in Griechenland zu suchen. Ur-sprungsfragen werden wir niemals lösen. Vielleicht haben die Griechen das Spiel wie so vieles andere aus dem alten Orient empfangen, aber sicher ist das eine, daß wir die ältesten Nach-richten über das Spiel in griechischen Autoren besitzen (vgl. H. Reich, Der Mimus, Bd. I, S. 669—675; C. Magnin. Histoire des marionettes, pp. 6—50; E. Maindron, Marionettes et guignols, ist ein unkritisches Buch; mit vielen Aufstellungen dieser Gelehrten bin ich nicht ein-verstanden und eine wirkliche Geschichte der alten Marionetten bleibt ein Desideratum). Bei aller Begeisterung für das indische Altertum vermag ich leider nicht mit R. Pischel (Die Heimat des Puppenspiels. Halle 1900) Indien als das Stammland des Puppenspiels anzusehen; hier ist nicht der Ort für eine Kritik dieser Abhandlung, die ich in meiner ausführlichen Arbeit über diesen Gegenstand zu geben hoffe. Wenn ich an dieser Stelle auf dieses Thema eingegangen bin, so geschieht es deshalb, um auf den bisher nicht beachteten fundamentalen Unterschied in der Geschichte des Schattenspiels und Puppenspiels aufmerksam zu machen und ferner, um die Verschiedenheit zu betonen, mit der die Chinesen die beiden nur scheinbar ähnlichen Spiele verwandt haben. Wie schon die Annalen der T'ang berichten, bildeten die Marionettenspiele einen Bestandteil der Trauerzeremonien; ehedem wurden zwei Tage vor dem Begräbnis Theater-vorstellungen veranstaltet (Wieger und Davrout, Moral Tenets and Customs in China, p. 569). Erst die Gesetzgebung der Mandschu schaffte diese alte Sitte ab (De Groot, Religious System of China, Vol. II, p. 608). Wir hören dagegen nichts, daß das Schattenspiel je mit einer religiösen Feier verknüpft gewesen wäre; im Gegenteil, von den Tagen der Sung an gerechnet erscheint es als eine rein weltliche Unterhaltung. Wir haben also hier eine gegensätzliche Entwicklung: das Schattenspiel, das einen religiös-spiritistischen Ursprung nahm, hat sich bald verweltlicht und wurde zu einem Schauspiel, das dem Zeitvertreib, der Erheiterung und Belehrung diente; das Puppenspiel, das außerhalb Chinas von Hause aus nichts anderes als eine verkleinerte und zusammengedrängte, dem Vergnügen huldigende Bühne war, gestaltete sich zum Werkzeug einer religiösen Feier. Wenn sich in den letzten Jahrhunderten, unter den Ming und den Mandschu, beide Schaustellungen in Bezug auf ihr Wesen sowie in Programm und Inhalt ihrer Stücke stark genähert haben, so ist der Grund darin zu suchen, daß beide fruchtbare Anregungen von der großen Bühne empfingen und sich deren Stoffe zu eigen machten. Wir haben es hier demnach mit einer psychologischen Erscheinung zu tun, die wir gegenwärtig als Konvergenz zu bezeichnen pflegen. Der Schattenspieler hat sich aber stets von den Einflüssen der Bühne unabhängiger gehalten als der Puppenspieler. Das lehrt schon ein flüchtiger Blick auf beider Apparat. Wenn wir die mythologischen Figuren betrachten, die sowohl im Puppen- als im Schattenspiel eine große Rolle spielen, so sehen wir, daß der Puppenspieler nichts anderes zustande bringt als eine Nachahmung der Masken, welche die Schauspieler tragen. Gilt es z. B. einen bestimmten Tiergeist darzustellen, so wird das betreffende Tier in stilisierter Behandlung der Gesichtsmaske aufgemalt oder in verkleinerter Nachbildung aus Papier geformt und auf Stirn oder Scheitel der Maske aufgesetzt; in manchen Fällen werden beide Methoden vereinigt. Der Puppenspieler arbeitet getreulich nach diesem Muster; der Kopf seiner Marionette ist ein Miniaturabbild des Schauspielers. Anders der Schattenspieler. Ein technischer Grund zwingt ihn zunächst, auf die Schablone der Bühne keine Rücksicht zu nehmen; er kann seine Figuren nur im Profil ausschneiden, und Masken sind ihm folglich wertlos. Er schöpft daher unmittelbar aus der Quelle der Mythologie und stellt die Tiergeister in glücklicher Verbindung eines tierischen Körpers mit einem menschlichen Antlitz dar. Er erreicht so einen Realismus, der dem Puppenspiel und der großen Bühne versagt ist, und erschließt uns eine lebendige Quelle für das Studium der volkstümlichen Götterlehre. Der Geist der Muschel z. B. erscheint in den Schattenfiguren als eine zwei Schwerter haltende weibliche Fee, die von zwei Muschelschalen eingeschlossen ist und sich nach Belieben zwischen denselben bewegen kann; oder der Geist der Schildkröte ist als wirkliches Tier, aber mit menschlichen Armen und Beinen dargestellt, während die betreffende Puppenfigur ohne jedes besondere Merkmal nur die Gesichtsmalerei des Clowns oder Mimus trägt, eine nur aus chinesischen Anschauungen heraus verständliche Anspielung, da die Schildkröte als ein Symbol des Unanständigen und Unmoralischen gilt. Ein eingehender Vergleich der Schatten- und Puppenfiguren sowie der letzteren mit den Masken, besonders für die Dramen mythologischen Inhalts, würde eine lehrreiche Aufgabe bilden. Was das Verhältnis der von den Marionetten- und Schattenspielern in der Sung-Periode vorgetragenen Stoffe betrifft, so waren beiden damals die alten Historien gemeinsam: das Mong liang lu berichtet ausdrücklich, daß die Puppenspieler die Geschichtsbücher und die alten Angelegenheiten der Beamten und Generäle in chronologischer Reihenfolge vortrugen, und daß ihre Librettos sich mit denen der öffentlichen Geschichtenerzähler deckten. Der Puppenspieler scheint jedoch damals ein weiteres Feld als sein Kollege gepflegt zu haben, denn jener führte auch Liebes- und Intriguenstücke, mythologische Gegenstände, Kampfszenen und die später auf der Bühne und auch im Schattenspiel so beliebten Gerichtsverhandlungen auf. Inwieweit der Puppenspieler dabei auf Originalität Anspruch erheben darf, wieviel davon er dem Theater zu verdanken hat, inwiefern er den Schattenspieler beeinflußt hat, das sind Fragen, die sich zur Zeit noch nicht beantworten lassen.

Die vorliegende Sammlung gewährt uns einen trefflichen Überblick der reichen Auswahl an Stoffen, die dem chinesischen Schattenspieler der Gegenwart zur Verfügung stehen. Vom buddhistischen und taoistischen Legenden- und Zauberdrama mit seinem romantischen Reiz führt er uns in die Haupt- und Staatsaktionen der chinesischen Geschichte ein, in die Taten und Meinungen alter Paladine und Amazonen, denen ein Beigeschmack heldischer Vorzeit oder unseres höfischen Mittelalters anhaftet, um mit einer Reihe bürgerlicher Schauspiele und Schwänke abzuschließen, die uns ein Sittenbild Chinas von einer Treue und Wahrheit vor Augen führen, wie es keine europäische Feder je gezeichnet hat. Hier treten uns unverfälschte Urkunden des chinesischen Volkslebens entgegen: hier spricht zu uns das Volk in seiner eigenen Sprache und legt sein Fühlen und Denken bloß. Da sehen wir die Chinesen, wie sie in Wirklichkeit sind und sich selbst zeichnen, nicht als die Musterknaben oder Idealmenschen, zu denen sie gewisse Schwärmer gestempelt haben, noch als die Ungeheuer, zu denen sie die gefärbten Berichte anderer Tendenzschriftsteller gemacht haben. Wir sehen sie als Menschen von Fleisch und Blut, denen nichts Menschliches fremd ist, die wir aber nicht für besser zu halten brauchen als sie sich selbst: die Bestechlichkeit der Richter, die verkommene Beamtenwirtschaft, kalte Selbstsucht und Berechnung, herzlose Geldgier, grausame Behandlung von Kindern (vgl. Nr. XXXV. S. 321), die ganze Äußerlichkeit und Hohlheit der papiernen Moral sind traurige Wahrheiten, die der Schattenspieler ernst und eindringlich predigt. Gewiß fehlen diesem sozialen Bilde nicht die Lichtseiten: Gatten- und Kindesliebe, Güte und Edelmut, Wohltätigkeit und Aufopferung, Treue und Biedermannsart, Lernbegier und Bildungstrieb, strebsamer Ehrgeiz im Staatsdienst und andere Tugenden der chinesischen Gesellschaft werden mit gebührendem Nachdruck hervorgehoben.

Wer sich an eine ästhetische Beurteilung dieser Erzeugnisse wagen will, muß mit denkbar größter Vorsicht verfahren und sich beständig gewärtig halten, daß die Texte, während sie wohl in allgemeinen Zügen den Gang der Handlung durchsichtig erkennen lassen, keinen Ersatz für das lebendig gespielte Stück zu bieten vermögen. Man muß an unsere Opernlibrettos, namentlich die älteren des achtzehnten Jahrhunderts erinnern, und beim Lesen unserer Dramen darf man niemals den Gesichtspunkt außer acht lassen, daß erst Musik und Melodie der Handlung Leben und Farbe verleihen. Die Aufführung der Schattenspiele wird durchweg von einem kleinen Orchester begleitet; die Rezitative und eingestreuten Verse werden gesungen und zwar, wie meist von den Chinesen, mit der Kopfstimme (vgl. die Selbstverspottung des Schattenspielers auf S. 398, wo von einem häßlichen Manne gesagt wird: „Er hat eine Fistelstimme, als wäre er bei einem Schattenspieler in die Lehre gegangen"). Von der Mehrzahl der chinesischen Dramen ist ein feststehender Text überhaupt nicht zu erlangen, und von vielen Stücken sind mehrere voneinander abweichende Rezensionen vorhanden. In zwei Fällen (No. I. 4 und XXXIV) haben wir zwei Redaktionen desselben Stückes mitgeteilt. Dieses Schwanken der Textgestaltung hängt mit der Erscheinung zusammen, daß es eigentlich nicht der Dichter, sondern in weit höherem Grade der Schauspieler ist, der das Drama schafft. Der chinesische Schauspieler ist nicht verpflichtet oder gewohnt, sich streng an den Wortlaut einer auswendig gelernten Rolle zu binden: er schafft seine Rolle ziemlich frei, entfaltet ein großes Improvisationstalent, erfindet Witze und Lieder, macht Anspielungen auf zeitgemäße politische Fragen und redet zuweilen sogar die Zuschauer an. Manche Szenen werden der Eingebung der Spieler gemäß ausgedehnt, andere werden nach ihrem Ermessen gekürzt. Berechnung waltet dabei nicht immer vor, sondern mancherlei hängt von Umständen und Gelegenheit, z. B. von der Rücksicht auf die jeweilige Zusammensetzung des Publikums ab. Dieselben Stücke kann man daher auf recht verschiedene Art vortragen hören. Einige habe ich auf ihre Zeitdauer geprüft und verglichen. Das besonders bei den Frauen beliebte buddhistische Spiel von der Weißen Nägi (No. I), das ich öfters zu sehen Gelegenheit hatte, mag eine Zeitdauer von zwanzig Minuten bis zu einer Stunde und mehr beanspruchen. Nach einem eingehenden Studium der Texte ließ Herr Krebs in Peking eine Truppe Schattenspieler kommen und sich deren Repertoire vorspielen. Er bemerkt dazu: „Übrigens hielten sie sich nicht streng an den Text, sondern flochten aktuelle Tageswitze ein. In dem Lustspiel ‚die Drei Pantoffelhelden' z. B. sagte eine der Frauen: ‚Ich gehe jetzt ins Hotel Wagons-Lits [ein europäisches Hotel in Peking], wo ich eingeladen bin.'" Außer der Stegreifdichtung kommt dann noch das Darstellungstalent der Schauspieler oder Sänger in Betracht; der Erfolg eines Stückes hängt von der Lebendigkeit und Beweglichkeit ihres Spiels ab, und die Chinesen sind vorzüglichere Schauspieler als Dramatiker. Während China keinen großen Bühnendichter in unserem Sinne hervorgebracht hat, können sich seine Schauspieler getrost mit den unsrigen messen und übertreffen sie auf jeden Fall an Beweglichkeit. Keine Bühne der Welt führt Kampfszenen geschickter vor als die chinesische und ohne Hülfe eines Regisseurs bietet sie einen Reichtum künstlerisch schöner Bilder. Die Kürze und scheinbare Banalität mancher Texte darf uns daher nicht täuschen. Was uns beim Lesen fade oder trivial erscheint, mag sich auf der Bühne dennoch lebensvoll gestalten dank der Geschicklichkeit der Schauspieler. Die Verse des Textbuchs mögen uns herzlich unbedeutend vorkommen, ihr Inhalt ist auch unwesentlich, die Zuhörer erfassen kaum die Worte, aber Musik und Gesang mögen sie zur Begeisterung fortreißen. Die Schattenspieler extemporieren womöglich in noch höherem Grade als die Berufsschauspieler der großen Bühne. Ihr handschriftlicher Text, besonders was die Komödien und Schwänke betrifft, ist nur ein Leitfaden zur Einführung des Anfängers und eine gelegentliche Gedächtnisstütze für die Vortragenden, die meist den ganzen Text auswendig kennen. Ein bloßes Ablesen ihrer Rollen ist schon dadurch ausgeschlossen, daß sie ihre ganze Aufmerksamkeit der Handhabung der Figuren zuwenden müssen, die sich fast nie im Zustand der Ruhe befinden, sondern bei fast jedem Wort Leben und Bewegung bekunden.

Eine kritische Beleuchtung der Quellen, aus denen unsere Dramen geschöpft sind, wäre eine dankbare, aber im einzelnen auch recht schwierige Aufgabe. Es hilft uns vorläufig wenig, daß wir die religiösen und historischen Stoffe auf beliebte Romane oder literarische Bühnendramen zurückführen können, da wir weder eine kritische Geschichte des Romans noch des Dramas, dieser beiden von der Gelehrtenkaste verachteten Literaturgattungen, besitzen. In den Historiendramen wäre es nicht ausgeschlossen, daß die Schattenspieler noch eigene in die Sung-Zeit zurückgehende Traditionen besäßen; leider sind die Vorträge der Geschichtenerzähler der Sung, soviel wir wissen, niemals aufgezeichnet worden. Auf freierem Felde bewegen wir uns im Gebiete der Lustspiele und Schwänke, die unmittelbar aus dem Volksleben gegriffen sind und den eigentlichen Nährboden des Schattenspiels bilden. Da begegnen uns alle Züge, die Jacob als charakteristisch für das Schattenspiel bezeichnet hat. Hier mögen vor allem einige Worte über den Mimus am Platze sein. Sogleich im ersten Schwank (S. 359) erneuern wir eine alte Bekanntschaft, den Mimus in der Rolle des Kahlkopfs. Von alters her galt der Kahlkopf in China als eine komische Person, und „Kuo der Kahle" war ehedem eine volkstümliche Bezeichnung für die Marionetten. Schon das Fong su t'ung i, ein Werk über Sitten und Bräuche, das dem Ying Shao (zweites Jahrhundert n. Chr.) zugeschrieben wird, erzählt, daß sich alle Leute, die den Familiennamen Kuo tragen, für den Spitznamen „Kahlkopf" eignen, da es einmal in der Vorzeit einen gewissen Kuo gegeben habe, der durch Krankheit seines Haarwuchses verlustig ging und dann durch Liedervorträge und Tanzen seinen Unterhalt verdiente. In der antiken Welt war die Glatze das Hauptmerkmal gewisser komischer Typen (A. Dieterich, Pulcinella, S. 38). Ebenso treten der indische Vidûṣaka, der türkische Karagöz und andere Mimen kahlköpfig auf. Dies Moment beweist wenig oder nichts für einen historischen Zusammenhang, denn körperliche Gebrechen und ihre übertriebene Hervorhebung in der schauspielerischen Darstellung werden überall zur Erzielung komischer Wirkungen breit ausgenutzt (man vergleiche die grotesk-komische Schilderung des häßlichen Mannes auf S. 398 und des häßlichen Mädchens auf S. 404). Im allgemeinen entfaltet der chinesische Mimus eine enge Wesensverwandtschaft mit seinen indischen und westlichen Kollegen, aber ihn deshalb mit Haut und Haar aus dem Westen herzuleiten liegt kein Grund vor, ungeachtet des bekannten Buches von Reich über den Mimus, der das Wesen und die vielfache Verschlungenheit der einschlägigen Probleme völlig verkennt. Als Ganzes genommen ist das Werk von Reich gewiß eine achtunggebietende und ehrenvolle Leistung; seine Behandlung des griechischen Mimus selbst stellt eine ebenso gründliche und gelehrte als unterhaltende und anziehende Untersuchung dar. Ihre offenkundigen Mängel indessen liegen in den Irrtümern der von ihm angenommenen entwicklungsgeschichtlichen Methode, die nicht-zusammengehörende Erscheinungen gewaltsam miteinander verknüpft und ein bloßes Einteilungsschema zu einer Evolutionsreihe erhebt, sodann was Indien und Ostasien betrifft, in mangelhafter Information und unzureichendem Tatsachenmaterial. Die einseitige Methode verführt den Verfasser zu beweismatten subjektiven Konstruktionen, die in der Theorie gipfeln, daß die Mimen aller Völker und Zeiten aus griechischer Grundlage entsprungen seien, eine Halbwahrheit und maßlose Übertreibung; mit diesem Trank eines vorgefaßten Dogmas ist es natürlich leicht, Helenen in jedem Weibe und den hellenischen Mimus in jedem Witzbold der Welt zu sehen. Das schablonenhafte Schema, das Reich auf einer Tafel

III*

ausgearbeitet hat und daß die Entwicklung des Mimus bis zum Jahre 1902 veranschaulichen soll, ist kaum mehr als eine geistreiche Phantasie. Gewiß — und darin hat er vollkommen recht — sind Mimen gewandert, vom vorderen Orient nach Indien, von Indien nach China, sowohl zur See als auf dem Landwege über Turkistan, — das läßt sich an der Hand chinesischer Quellen und archäologischer Denkmäler zeigen, aber Reich hat kaum den Schein eines Beweises für seine Behauptung vorgebracht (man lese S. 698 sein Kapitel „Die Mimen wandern nach Indien", wo kein einziger Beleg für die These angeführt wird, oder die seltsame Beweisführung S. 691: „Da das javanische Puppenspiel schon Serrurier an den türkischen Karagöz erinnerte, so ist damit der Zusammenhang mit dem hellenischen Mimus hergestellt!"). Die Wanderung einiger oder sogar vieler Mimen beweist aber lange noch nicht, daß alles Mimische, das wir in Indien und China finden, auf Griechenland oder den hellenistischen Orient zurückgehen muß. Zunächst haben wir an der Tatsache festzuhalten, daß es überall auf dieser Welt zu Humor, Witz und Satire neigende Menschen gegeben hat, Clowns, Gaukler, Narren und Spaßmacher. Aus diesen haben sich allerwärts die lustigen Rollen bei religiösen Tänzen oder pantomimischen Aufführungen herausgebildet. In erster Linie handelt es sich um eine allgemein menschliche, rein psychische Erscheinung im Völkerleben, die zu jeder Zeit treibt und wirkt. „Die Neigung des Menschen zum Grotesk-Komischen oder zur komischen Karikatur ist so alt als irgend ein anderer Zweig des Komischen; ja es ist wahrscheinlich, daß er an Alter alle anderen übertrifft. Denn ehe der Mensch so gesittet wird, daß er das Fein- und Hoch-Komische erfinden oder an demselben Behagen haben kann, ist der Geschmack an dem übertriebenen oder grob Komischen lange vorhergegangen, weil sich dieser wie von selbst mit den rohen Sitten des ungebildeten Menschen am besten verträgt und natürlicherweise daraus entstehen muß", so eröffnet unser trefflicher Floegel seine Geschichte des Grotesk-Komischen. Die entlegensten Naturvölker haben uns so zahlreiche Beispiele dieser Erscheinung geliefert, daß wir sie mit Recht als universal betrachten dürfen. Es wäre z. B. interessant zu erfahren, wie sich Reich zu der Frage des unter den nordamerikanischen Indianern blühenden Mimus (vgl. die Übersicht bei M. W. Beckwith, Dance Forms of the Moqui and Kwakiutl Indians, Amerikanisten-Kongreß Quebec Bd. II, S. 79—109) stellen, ob er etwa die griechischen Mimen auch die Beringstraße überschreiten lassen würde.

Überall paßten sich die Rollen des Mimus gegebenen Verhältnissen an: wenn wir den konfuzianischen Gelehrten mit seinen komischen Anwendungen klassischer Zitate in dieser Rolle sehen (No. XLVIII und XLIX), so haben wir hier einen spezifisch chinesischen Typus des Spaßmachers vor uns. Ebenso sind der Bauer (No. XXXVIII und L), der Richter (No. XLVI und XLVII), der Eseltreiber (No. LI), der Bramarbas (No. LIII), die Aufschneiderin (No. LIV) und die Pantoffelhelden (No. LV) Charaktere aus dem chinesischen Volksleben, zu deren Erklärung keine Übertragungstheorie erforderlich ist. Als Quacksalber treffen wir den Mimus in den Pantomimen der Stelzentänzer (W. Grube, Zur Pekinger Volkskunde, S. 105). Räuber und Diebe gehören zu seinen bevorzugten Rollen; dagegen fehlt der komische Diener fast ganz. Die typische Erscheinung des chinesischen Mimus ist folgende: weiß getünchte große Nase, zwischen den Brauen und in den Augenwinkeln gemalte schwarze Streifen, ein stark nach oben gezogener Schnurrbart sowie blatternarbiges Kinn und Wangen; zuweilen trägt er einen aus Pferdehaar geflochtenen hohen Hut. Dies ist der übertriebene, eigentliche Narrentypus, der je nach der besonderen Rolle bedeutenden Abwandlungen unterliegt. Ein gutes Beispiel für die Altertümlichkeit des Mimus auf chinesischem Boden ist der bereits erwähnte Kahlkopf, ein anderes ist der Zwerg. Nach einem Bericht des Philosophen Wang Ch'ung, der im Jahre 82 oder 83 n. Chr. schrieb, fungierten schon in den Tagen des Altertums Zwerge als Spaßmacher (A. Forke, Lun-hêng, Vol. I, p. 473) und waren wie der zwerghafte Minister Yen Ying († 493 v. Chr.) durch Witz und Schlagfertigkeit ausgezeichnet (A. Tschepe, Histoire du royaume de Tch'ou, p. 149). Das Wei lio (drittes Jahrhundert n. Chr.) erzählt von einem Pygmäenreich im Nordwesten von Sogdiana mit zahlreicher Bevölkerung, wo Männer und Frauen nur drei Fuß hoch sind (Chavannes, T'oung Pao, 1905, p. 561 f.), und im achten Jahrhundert wurden Zwerge aus Sogdiana als Tributgaben an den chinesischen Hof gesandt (Chavannes, Documents, p. 136). Unter den Denkmälern der T'ang-Zeit finden sich gut modellierte Tonfiguren, die Zwerge, vielleicht Hofnarren, in komischen Aktionen darstellen. Eine derselben in meiner Sammlung (Field Museum, Chicago) verkörpert einen Zwerg von negerartiger Gesichtsbildung mit flacher Glabella, breiter Nase und aufgeworfenen Lippen, offenbar mit den K'un-lun ts'eng-k'i der chinesischen Berichte identisch (Hirth und Rockhill, Chau Ju-kua, pp. 149—150). Man sieht, daß China bodenständige und fremdrassige Zwerge gehabt hat, und daß der Zwerg als Mimus keine Berührung mit dem Hellenismus aufweist.

Dank den Forschungen von H. Lüders (Bruchstücke buddhistischer Dramen, Berlin 1911, und das Śāriputraprakaraṇa, ein Drama des Aśvaghoṣa, SBAW 1911, pp. 388—411) wissen wir jetzt, daß es in Indien bereits in alter Zeit eine buddhistische dramatische Literatur gegeben hat. Da die von Lüders studierten Handschriftenfragmente aus einem Höhlentempel westlich von Kucha stammen, so deutet dieser Umstand auf eine hohe Entwicklung der dramatischen Kunst im Kulturgebiet von Kucha hin. Dies wird durch die Nachrichten der Chinesen teils in den T'ang-Annalen teils in Yu yang tsa tsu, Yo fu tsa lu und anderen Werken bestätigt. Dankenswerte Nachrichten aus der chinesischen Literatur über die hohe Bedeutung von Musik und Tanz in Kucha haben M. Courant (Essai historique sur la musique classique des Chinois, pp. 192—197, Auszug aus Encyclopédie de la musique) und nach ihm S. Lévi (Le „Tokharien B", langue de Koutcha, JA 1913, Sept.-Oct., pp. 41—44) gegeben; in Lévis Abhandlung findet man auch eine sehr verdienstliche Zusammenstellung aller auf die Geschichte von Kucha bezüglichen Quellen. Ich habe oben auf die chinesische Entlehnung des Wortes kukla und der Marionetten in der Ära der T'ang hingewiesen; nach meiner Ansicht ist diese Übertragung von Kucha her erfolgt. Das arische Wort kukla wird sich gewiß noch in Tokharisch B (nach Lévi identisch mit der Sprache von Kucha) nachweisen lassen. Aber noch mehr. Unter den Tonfiguren der T'ang-Kunst finden sich prachtvolle Statuetten von Mimen und Schauspielern, die in ihrer Lebenswahrheit und überraschenden Pose und Geste wie griechische Bildwerke erscheinen (man wird dieselben im zweiten Teil der Chinese Clay Figures des Verfassers abgebildet und beschrieben finden). Diese Tonfiguren sind von chinesischer Hand geformt; ihre Gesichtstypen dagegen sind unchinesisch und ausgesprochen arisch. Meines Erachtens stellen sie Mimen und Schauspieler von Kucha dar, die bereits im sechsten und siebenten Jahrhundert gern gesehene und beliebte Gäste am chinesischen Kaiserhofe waren. Es kann also gar keinem Zweifel unterliegen, daß arisches Mimentum nach China eingedrungen ist und die chinesische Schauspielkunst beeinflußt hat. Der Grad und die Ausdehnung dieses Einflusses entziehen sich vorläufig unserer Kenntnis und werden sich erst dann bestimmen lassen, wenn uns Texte chinesischer Dramen aus der T'ang-Zeit vorliegen werden.[1]) Aus Indien sind solche Einflüsse weit früher, schon zur Han-Zeit, nach China gekommen; läßt sich doch die chinesische Gauklerkunst mit ihren Zauberkunststücken auf Indien, teilweise auch auf den hellenistischen Orient zurückführen. Hier harren noch manche wichtige Probleme ihrer Lösung, wie z. B. der Ursprung und die geschichtliche Eingliederung der chinesischen, japanischen und tibetischen Masken. Noch mehr als auf anderen Gebieten müssen wir hier jedoch gegen voreilige Verallgemeinerungen Verwahrung einlegen, wie den Welttraum, der alles Mimische aus Hellas herleiten möchte, und in der Untersuchung von fest umschriebenen und handgreiflichen Tatsachen ausgehen, um von da zur Synthese fortzuschreiten anstatt vorgefaßte Dogmen in den Tatbestand zu projizieren. Es kann auch keine Rede davon sein, daß erst die Berührung mit dem Westen den Chinesen den Mimus vermittelt hat, denn schon im alten China hat es Spaßmacher und Komödianten gegeben, die mit dem uralten und einheimischen Worte ch'ou bezeichnet werden (W. Grube, Zur Pekinger Volkskunde, S. 121). In seinem Shi ki (Kap. 126) gibt uns Se-ma Ts'ien Nachrichten von dem Leben zweier Komödianten Mong und Chan. Vom Kahlkopf und Zwerg ist schon oben die Rede gewesen.

Gegen de Groots Annahme (Religious System of China, Vol. IV, p. 88), daß das Schattenspiel, obwohl noch eine gelegentliche Belustigung in einigen Teilen des Reiches, niemals sehr

---

[1]) Wir haben gegenwärtig keine älteren Texte chinesischer Dramen als aus dem vierzehnten Jahrhundert und selbst bei diesen besteht die Wahrscheinlichkeit späterer Umarbeitungen. In Kozlovs Bücherfunden von Karakhoto befindet sich der Text eines Dramas mit dem Titel Liu chi yüan chuan, von Pelliot dem Jahre 1300 zugewiesen (JA 1914, Mai-Juni, p. 510).

volkstümlich gewesen zu sein scheint, hat sich schon Schlegel mit Recht gewandt (T'oung Pao, 2. Serie, Bd. III, 1902, S. 44), indem er auf die Popularität des Spiels in Amoy hinwies. Ich selbst bin seinen Spuren nachgegangen und habe im mittleren China Schattenspieler in Hang-chou, der Hauptstadt der Provinz Che-kiang, und in Han-k'ou am Yangtse, im nördlichen China in Peking, in Si-ngan, der Hauptstadt von Shen-si, und in Ch'eng-tu, der Hauptstadt von Sze-ch'uan, gefunden. In allen diesen Orten gibt es ständige, berufsmäßige Schattenspielertruppen, die bei alt und jung äußerst beliebt und mit dem denkbar größten Apparat ausgerüstet sind. Überall auch wird das Schattenspiel von Knaben zum Zeitvertreib geübt, und auf den Marktplätzen von Si-ngan und Ch'eng-tu kann man für einige Pfennige solche aus Papier hergestellte Spielzeug-figuren erstehen. Diese heißen in der Volkssprache 'Papiermenschen' (chi jên) im Gegensatz zu den 'Ledermenschen' (p'i jên), deren sich der Berufsschattenspieler bedient; ebenso heißen die Marionetten je nach dem Material 'Tonmenschen' (ni jên) oder 'Holzmenschen' (mu jên); die allgemeine Bezeichnung für solche Spielfiguren ist 'kleine Menschen' (siao jên, in Peking mit der Verkleinerungsendung êrh siao ji'r gesprochen). Das für die Schattenfiguren gewählte Material ist in den einzelnen Landesteilen verschieden: in Peking zieht man Eselsleder vor, in Hang-chou Schafsleder, in Han-k'ou und Ch'eng-tu Rindsleder. Das Leder muß fein gegerbt und durchscheinend sein. Die Figuren werden mit einem spitzen Messer aus freier Hand aus-geschnitten und allmählich fortschreitend auf beiden Seiten gleichmäßig gefärbt und gefirnißt. Diese Arbeit erfordert Zeit und Ausdauer, da mehrere Lagen Farbe notwendig sind und Schicht auf Schicht trocknen muß; auf gleichmäßige Verteilung der Farbe wird große Sorgfalt ver-wendet, da zu dickes oder zu dünnes Auftragen an einzelnen Stellen den Gesamteindruck er-heblich stören würde. Kopf, Arme und Beine werden in der Regel besonders ausgeschnitten; der Arm wiederum besteht aus Oberarm, Unterarm und Hand; die Hand ist meist aus zwei Teilen gebildet, Daumen und die vier übrigen Finger; die Gelenke sind durch Schnüre ver-bunden. Der Kopf wird in der Regel vermittelst Eisendrahts am Rumpfe befestigt. Da die Figuren auf beiden Seiten gleichmäßig gearbeitet sind, sowohl was die Hauptteile als auch die einzelnen Ornamente betrifft, so können sie beliebig von rechts oder links auftreten, was natür-lich für schnelle Aktion von großer Bedeutung ist. Im allgemeinen stellt jede Figur eine Ein-heit und eine bestimmte Persönlichkeit vor. In Peking gibt es jedoch auch zusammensetzbare Figuren, d. h. getrennte Köpfe und Körperformen, die nach Belieben oder je nach Bedürfnis verbunden werden können. Da der chinesische Künstler niemals einen Körper, sondern nur Bekleidung und Schmuck darstellt und damit nur das Vorhandensein eines Körpers andeuten will, so handelt es sich bei den Körperformen des Schattenspielers nur um Trachten; so stehen ihm Alltagskleider für Mann, Frau und Kind zur Verfügung, dann gestickte Kleider, Drachen-mäntel, historische Kostüme wie die der Ming- oder Mandschu-Dynastie, lokale Frauentrachten, die sich z. B. durch weite und eng anschließende Ärmel unterscheiden, Rüstungen, militärische Uniformen usw. Diesen werden dann, wie es die Gelegenheit erfordert, passende Köpfe auf-gesetzt und bei diesem Verfahren macht der Schattenspieler eine große Ersparnis, wo es gilt, Menschenmassen in Bewegung zu setzen und dem Bild der Bühne in rascher Folge neue Farben zu verleihen. Für die eigentlichen dramatis personae hat natürlich auch der Pekinger Schatten-spieler ständige Typen. Die Zahl der zur Bewegung der Figuren dienenden Stäbchen ist ge-wöhnlich drei. Sie bestehen aus einem Stück Eisendraht, der in einen Griff aus Rohr ein-gesetzt ist. Das eine der drei ist ein Standstäbchen, das an der Brust der Figur befestigt ist und die beiden anderen um das Doppelte an Länge überragt; diese sind Bewegungsstäbchen, die von den Händen der Figur ausgehen und mit denen der Spieler die Figur manipuliert; er kann zwei Figuren gleichzeitig handhaben und während er diese an den Vorhang anlehnt, neue auf die Szene bringen. Die Bühne des Schattenspielers ist ein großer, rechteckiger, weißer, durchsichtiger Vorhang, der zwischen zwei Stangen in einer Ecke des Zimmers möglichst straff aufgespannt wird. Er ist aus dünner Seidengaze oder Nesselfasern gewoben und wird von hinten durch Öllampen erleuchtet, während die Stube dunkel bleibt. Arme Schattenspieler (und das sind wohl die meisten) müssen selbst auf den Luxus der Lampen Verzicht leisten und sich mit einiger Kunst etwas Tageslicht stehlen. In Han-k'ou hörte ich im April 1903 eine Schatten-spielgesellschaft, die in einer elenden, halb verfallenen Bauernhütte untergebracht war, die

Wände zum Teil aus Lehm, zum Teil aus Brettern gefügt und das Dach mit Stroh gedeckt. Der Raum war fensterlos, aber um die Ausgaben für die notwendige Beleuchtung des szenischen Vorhangs zu sparen, hatte der Unternehmer eine Lichtquelle in die Ecke der Wand gegenüber dem Vorhang gebohrt. Hier wurde daher des Nachmittags gespielt; die Gäste, Bauern und Arbeiter, saßen auf rohen Bänken um Tische herum und wurden mit Tee, Tabakspfeifen und Nüssen bewirtet. Nach jedem Stück wurden Beiträge gesammelt, die sich im Durchschnitt auf drei bis vier Kupfermünzen (kaum einen Pfennig) für eine Person beliefen; Nichtzahler wurden höflich zum Verlassen des Lokals aufgefordert.

Die allgemeine Beliebtheit des Schattenspiels beruht auf einer Kombination von Erscheinungen. Zunächst ist es in der lebendigen Volkssprache, abgesehen von eingestreuten lyrischen Gesangspartien, abgefaßt. Der Rückschlag gegen den gelehrten und gekünstelten Bücherstil setzte im Zeitalter der Mongolenherrschaft ein und übte einen fruchtbaren Einfluß auf die Entwicklung des Dramas und Romans aus (W. Grube, Geschichte der chinesischen Literatur, S. 361); nur darf man sich nicht vorstellen, daß die literarische Wiedergeburt der Volksrede erst beides ins Leben gerufen hätte, oder daß das Kunstdrama und der Kunstroman überhaupt spontan in China entsprungen wären. Zu diesem Ergebnis haben mannigfach verkettete fremde und nicht zum wenigsten über Turkistan einströmende indische Einflüsse beigetragen, wie ich an anderem Orte ausführlicher darzulegen hoffe. Die eifrige Pflege dieser beiden Literaturgattungen hatte die Anerkennung der Rechte der Volkssprache zur Folge. Diese Erscheinung allein aber genügt nicht zur Erklärung der Volkstümlichkeit des Schattenspiels, da auch die große Bühne und das Puppenspiel die Volkssprache pflegen. Betrachten wir zuerst, was Schatten- und Puppenspiel gemeinsam haben und was sie vom öffentlichen Theater unterscheidet. Dieses ist in den meisten Fällen an ein geräumiges Gebäude, auf jeden Fall aber an die Bedingung einer Bühne gebunden; es setzt die Unterhaltung einer ansehnlichen Schauspieler- oder Sängertruppe voraus mit einer kostspieligen Ausrüstung von Kostümen und Requisiten. Ein solches Unternehmen verlangt Kapital und die erhobenen Eintrittsgelder sind nicht immer im Bereich des armen Mannes. Aber noch mehr: in China ist der Besuch des öffentlichen Theaters das Vorrecht des Mannes und der Halbwelt, während der ehrbaren Frau der Zutritt streng verwehrt ist. Ihr kommt das fahrende Volk der Schatten- und Puppenspieler zu Hülfe, die der Familie angehören; in den Straßen umherschlendernd sind sie jederzeit gern bereit, der Aufforderung eines Besuches im Familienkreise Folge zu leisten und Unterhaltung und Frohsinn in die einsamen Frauengemächer zu bringen. In dieser Erscheinung findet die erzieherische Richtung, die oft aufdringliche Lehrweisheit mancher Stücke unserer Sammlung ihre Erklärung; darin liegt die hervorragend soziale Bedeutung dieser volkstümlichen Spiele, die alle dem chinesischen Herzen teuren Tugenden der Familie in anziehender Weise vor Augen halten oder das Unschickliche und Unsittliche geißeln. Einige Stücke unserer Sammlung (No. XXXIX, XL) geben pädagogische Anweisungen, ein anderes (No. XLII) enthält einen Traktat über Musik, ein anderes (No. XLIII) belehrt über die beim Laternenfest gebräuchlichen Laternen. Ebenso wollen Schatten- und Puppenspieler in ihren Darstellungen aus der alten Geschichte nicht weniger belehren als unterhalten; sie fühlen sich in der Tat als Erzieher des Volkes und indem sie das Heiligtum der Familie, die Frauen und die Jugend erreichen, haben sie einen weiteren und verantwortlicheren Wirkungskreis als die öffentlichen Schauspieler. Schatten- und Puppenspieler tragen zu irgendeiner Zeit die sonnige Welt in das Haus oder ziehen die Menge auf dem Marktplatz an. Ihr leichtes Gerät macht sie beweglich und allgegenwärtig, ungebunden von Zeit und Ort. Nun hat aber das Schattenspiel bedeutende Vorzüge vor der Bühne und den Marionetten aufzuweisen. Wie bekannt, entbehrt die chinesische Bühne fast gänzlich der Ausstattung; einen auf Zeug gemalten Hintergrund habe ich nur in dem religiösen Drama Ta siang shan, der Göttlichen Komödie der Chinesen, und in den zu Ehren des kaiserlichen Geburtstags aufgeführten Festspielen gesehen (beide Bühnenszenen sind im Field Museum ausgestellt). Die chinesische Bühne verzichtet auf jeden äußerlichen Schmuck und überläßt es der Schaffenskraft des Schauspielers, den Zuhörern die gewünschten Täuschungen einzuflößen. Der Puppenspieler ist ebenso hilflos in Bezug auf realistische Mittel und hat nichts anderes als seine Marionetten zur Verfügung; auf dem guten Aussehen und der geschickten Handhabung derselben beruht sein Erfolg. Anders der Schattenspieler, der Realist und Romantiker in einer Person ist. Er scheut vor keiner Schwierigkeit zurück, das Leben seiner Bühne so deutlich und glänzend als nur möglich auszumalen. Er ist Bühnenkünstler. Er zaubert die Szene hin. Da ist Hochgebirg, Felsgeklüft, Baum und Wasserfall. Da sind Haus, Tempel, Pagode und Kiosk; teppichgeschmückte Böden, mit Vasen besetztes Gesims, mit Stickereien bedeckte Tische, mit Tigerfellen belegte Stühle. Die Helden reiten hoch zu Roß, die Frauen wiegen sich auf sanften Eseln oder werden in Palankinen getragen und maultierbespannte Karren eilen an uns vorüber. Richten wir den Blick aufwärts, so gewahren wir auf Kranichflügeln dahingetragene Genien und in den Wolken schwebende Götter. Die Darstellung einer Überschwemmung des Yangtse ist für den Schattenspieler keine unüberwindliche Aufgabe; er setzt Wind und Wellen in Bewegung und läßt schreckliche Ungeheuer aus den Fluten emportauchen, Riesenfische, Muscheln, Krabben, Frösche, Schildkröten, Schnecken und die Drachengötter, taoistische Umformungen des indischen Nāga. Der Donnergott schlägt mit dem Hammer auf seine Trommeln, die Göttin des Blitzes läßt ihre Metallspiegel tönen, der Regengott stößt sein Schwert in die Wolken und gießt Wasser aus einem Gefäß herab, die Windgöttin reitet auf einem Tiger durch die Luft und öffnet den Sack, in dem sie die Winde gefesselt hält. Der Schattenspieler meistert die Elemente und beherrscht die Tierwelt. Niemand kann mit größerer Naturwahrheit als er die Bewegungen von Schlangen und Drachen dem Auge vorzaubern oder das Gähnen und Recken des Löwen, das Schleichen und den Sprung des Tigers. Er erfaßt mit der Sicherheit des chinesischen Malers das Leben in der Bewegung, er ist Bewegungskünstler. Vom Leben führt er uns durch die Welt zur Hölle, zu Yama, dem strengen Totenrichter, seinen ochsen- und pferdeköpfigen Trabanten und seinen gehörnten Teufeln. Die Qualen der Verdammten ziehen im Bilde an uns vorüber, bis der milde Blick der gnadenreichen Göttin Kuan-yin die Erlösung ankündet. Im finstern Reich der Schatten ist das Schattenspiel von doppelt packender Wirkung. Den Reiz, den es auf das Gemüt der Chinesen ausübt, kann man dem Kulturmenschen der Gegenwart vielleicht nicht besser erklären als durch die Anziehung, welche die photographischen Bewegungsbilder auf uns ausüben. Auch dann bleibt noch ein unbestimmbarer Rest rein gefühlsmäßiger Dinge zurück, die wir etwa verstehen können, wenn wir uns in die Seele des alten Kaisers Wu versetzen, da er die zarte Gestalt seiner verstorbenen Gemahlin im Schattenbilde wiedersah. Dies Gefühl des Schauderns, das unser Dichter der Menschheit bestes Teil nannte, ist an der Wirkung des Schattenspiels seit den Tagen des Spiritismus haften geblieben und bildet das wesentliche Element, das zu seiner Erhaltung und allgemeinen Beliebtheit beigetragen hat. Man vergesse nicht, daß die Aufführung eines Schattenspiels einen dunkeln oder halbdunkeln Raum verlangt, während das chinesische Theater stets gleichmäßig hell (oder den Umständen nach gleichmäßig schlecht) erleuchtet bleibt und auch dank dem lauten und ungezwungenen Verhalten der Zuschauer keinerlei Illusion aufkommen läßt. Das Schattenspiel ist die vollkommenste Illusion, welche das chinesische Drama hervorgebracht hat und hervorbringen kann. Es ist die einzige chinesische Bühne, welche in dem Zuschauer Stimmung und Weihe aufkommen läßt und die ihn mit romantischem Zauber umfangen hält. Es bezeichnet daher die künstlerisch höchste Stufe, welche die dramatische Darstellung in China erreicht hat.

124

纳柯苏的摩梭语手稿

THE
GEOGRAPHICAL REVIEW

VOLUME I

JANUARY–JUNE, 1916

PUBLISHED BY

THE AMERICAN GEOGRAPHICAL SOCIETY

BROADWAY AT 156th STREET

NEW YORK

THE NICHOLS MO-SO MANUSCRIPT

By BERTHOLD LAUFER, Ph.D.

Field Museum of Natural History, Chicago

That charming book, "Through Hidden Shensi," made the late Mr. Francis H. Nichols widely known to the public, but few are acquainted with his high aspirations in attempting to acquire the Tibetan language by leading a solitary life for a considerable time in the lama monastery of Wei-si in the prefecture of Li-kiang, Yün-nan Province.[1] On arriving at Mandalay, Burma, on September 20, 1904, he sent the following note: "My greatest treasure I enclose. It is a book in the Ton-ke, the original language of the Mo-so tribe, who now speak and write Tibetan. Books like this one are exceedingly rare and difficult to obtain. I consider myself fortunate in having secured this manuscript. Part of it I have been able to translate."

It is a pleasure to respond to the request of the American Geographical Society for some comment on this rare and valuable Mo-so manuscript which he presented to the Society. The designation Ton-ke, which is used by Mr. Nichols, is composed of *ton,* or what other travelers style *tong-pa,* the name for the shamans, or sorcerers, of the Mo-so, and the word *ke* (Tibetan *skad,* pronounced *kä*), which means "language." The hope that the partial translation of the manuscript might be preserved has unfortunately not been fulfilled. Mr. Nichols' mother, Mrs. William C. Nichols, of Evanston, Ill., states that this translation has not been found in his papers, or has never reached her.

At present the Mo-so inhabit the Chinese prefectures of Li-kiang and Ho-k'ing in the province of Yün-nan. They represent the remnants of a once powerful and more extended tribe (or even several tribes), which belong to the Tibeto-Burmese stock of the large Indo-Chinese family. Their idiom, as far as we can judge at present, shows decided affinities with Tibetan and Lo-lo; it is, however, not merely a Tibetan dialect, but an independent language with peculiar characteristics. The name Mo-so is applied to them by the Chinese and presents the combination of two tribal names, Mo and So. They style themselves Na-shi, and are called by the Tibetans Jang. The latter term appears in Kara-jang, the medieval designation for the province of Yün-nan, applied to the latter by Marco Polo and the Persian historian Rashid-eddin.

For all that we know thus far about the history of this people we are indebted to the records of the Chinese. The Mo-so are first mentioned under the T'ang dynasty (608-916). Toward the end of the eighth century they

---

[1] An account of lamasery life at Wei-si from Mr. Nichols' diary was published in the *Bull. Amer. Geogr. Soc.,* Vol. 47, 1915, pp. 100-114, where further details of his travels in China may be found.

274

were subjugated by Yi-mou-sün, ruler of the kingdom of Nan-chao in Yün-nan, which was composed of members of the Tai or Shan stock, the ancestors of the subsequent Siamese. Under the Sung dynasty, Mong-ts'u, the chief of the Mo-so, captured the town of Li-kiang. The Mongols, after their conquest of Yün-nan, established two departmental districts in the Mo-so territory, called Tsaghan-jang and Kara-jang, that is, "White and Black Jang."

In former times the Mo-so were very warlike tribesmen, fighting many a battle with their Tibetan neighbors, and were even bold enough to invade the southeastern portions of Tibet. The memory of this heroic age is preserved in the great national epic poem of the Tibetans, the Gesar Saga, divided into three sections, one of which is devoted to the struggle of the Tibetan warriors with the Mo-so.

There is a Chinese chronicle devoted to the history of the chiefs of Li-kiang, who from the year 1382 bore the family name Mu. This work has been translated and interpreted by E. Chavannes.[2] It contains an interesting genealogy of the Mo-so chieftains from the beginning of the twelfth to the first part of the seventeenth century. To the industry of P. Cordier, the eminent bibliographer of the literature relating to Eastern Asia, we owe an excellent summary of what has been written on the Mo-so.[3] This study also contains a Mo-so vocabulary.

What renders the Mo-so particularly interesting is a peculiar pictographic system of writing and a religious literature composed in it and extant in a number of manuscripts. T. de Lacouperie, the versatile orientalist, was the first to take an interest in this matter and to publish facsimiles of two Mo-so manuscripts procured by Captain W. Gill and Father A. Desgodins.[4] He made several very ingenious observations on the script, but owing to lack of information he was unable to make a statement as to the contents of the manuscripts. The beginning of the same manuscript acquired by Desgodins was reproduced also by G. Devéria,[5] likewise without explanation. Prince Henri d'Orléans brought back five Mo-so manuscripts from his journey in 1895-96 and figured the first three pages of two of them, accompanied by a translation and transcription of one of the texts.[6]

The first scholarly contribution to Mo-so literature was made by Monsieur Charles-Eude Bonin, at present Consul-General of France in Montreal, Canada.[7] In 1895 he was charged with a mission along the borders of

[2] Documents historiques et géographiques relatifs à Li-kiang, *T'oung Pao*, 1912, pp 565-653; also reprinted in the work of J. Bacot: Les Mo-so, pp. 127-215 (Leiden, 1913).

[3] *T'oung Pao*, 1908, pp. 663-688 (compare also his article on the Mo-so in *Journ. des Savants*, 1911, pp. 129-134).

[4] T. de Lacouperie: Beginnings of Writing in Central and Eastern Asia, pp. 45-50, Plates I-III (London, 1894). The manuscripts were first published in the *Journ. Roy. Asiatic Soc.*, 1885.

[5] La frontière sino-annamite, p. 166 (Paris, 1886).

[6] In his book "Du Tonkin aux Indes," pp. 364-367.

[7] Note sur un manuscrit Mosso, *Actes du onzième Congr. Intern. des Orientalistes*, 2d section, pp 1-10, Paris, 1898. This article is reprinted and generally accessible in the author's book "Les royaumes des neiges: États himalayens," pp. 281-296 (Paris, 1911).

Tibet and had the opportunity of traversing, as the first white man, the territory of the Mo-so. He visited also its capital, Li-kiang, and was fortunate enough to secure a Mo-so manuscript from a native priest, a so-called *tong-pa* (in Chinese, *to-pa*). The latter are the shamanistic sorcerers representing the aboriginal form of religion that obtained in times prior to the introduction of Buddhism among the Mo-so people from Tibet. A certain affiliation still exists between the Buddhism of Mo-so land and the Lamaist church of Lhasa. Capable Mo-so men are recruited by the Tibetan lamas for the ecclesiastic army and sent to Lhasa in order to receive the benefit of a religious and theological education. Despite this official conversion to Buddhism the adherents of the primitive native religion still hold their ground, and it is their principal ritual that is contained in the book secured by M. Bonin. The poor *tong-pa* who transmitted it to him secretly made a translation of the first six pages into Chinese, which was recorded by a Chinese interpreter, and it is on this Chinese version that M. Bonin's French rendering is based. The original manuscript, unfortunately, is not reproduced by him, but we trust that M. Bonin, in the interest of science, will be induced to publish it in the near future. The description given of the manuscript shows that in its technical make-up and appearance it comes very near to the manuscript of Mr. Nichols. Like the latter, it is an album of oblong size, comprising twelve folios of very strong paper, the verso and recto of each leaf being inscribed, save the last page, which is decorated with designs of red flowers. The two sides of the first folio are occupied with representations of deities and objects of the cult. The ten following folios are covered with colored hieroglyphs peculiar to the writing of the *tong-pa*, and on each page distributed over three horizontal lines, each line being divided by vertical strokes into two or three sections. These form a series of rectangular enclosures, the characters of each quadrangle representing a phrase, so that the vertical lines assume the function of our punctuation. M. Bonin insists on the strictly hieroglyphic character of Mo-so writing, saying that nearly all characters directly picture the object for which they are intended. Likewise he lays emphasis on the indigenous origin of the system, which appears not to contain any elements borrowed from the script of the Lo-lo or the Chinese. The notable foreign elements seem to him rather to be derived from India, as, for instance, the design of the svastika, which is expressive of the idea of bounty. According to the vocabulary of J. Bacot[8] the symbol of the svastika would mean a "remedy." The manuscript of M. Bonin is the only one thus far known in which pigments are applied to the characters: that of Mr. Nichols and all others are written exclusively in black Chinese ink, and, apparently, not with a brush, but by means of a pointed wooden stylus, such as is in vogue among all present Tibetan tribes and as was employed also by the ancient Chinese in times prior to the invention of the writing-brush. In

---

8 Les Mo-so, p. 103.

the Bonin manuscript the colors seem to bear out a symbolic significance, the idea of "earth," for instance, being conveyed by a blue band covered with bunches of grass.

It is difficult to form an opinion on the translation of the Mo-so ritual, as it is offered by M. Bonin. He himself admits that it should not be regarded as definite, but simply as an essay to make out a coherent concept of the ideographic characters, which merely convey notions of realistic objects without being capable of expressing the verb. A great antiquity is ascribed to his text, regarded as the product of indigenous religious thought, the original being traced to an epoch when the Mo-so were not yet in contact with Chinese or Mongols, and under their chief Mong-ts'u conquered the territory of Li-kiang. This conclusion is tempting, but it can be accepted only with certain strictures. First of all, the Indian-Buddhist influence transmitted through the medium of Tibet is quite apparent in this brief text, as particularly shown by the thirty-three spirits of heaven, which are mentioned twice. Every one familiar with Buddhist literature knows that these are the thirty-three gods, headed by Indra, of the Brahmanic heaven. Among the offerings made to the gods by the family we note "tobacco," and this is a recent element giving rise to suspicion. To be sure, it may be a subsequent interpolation smuggled into a text of older date, but a sacred ritual that is thus impaired and modernized can hardly lay claim to that purity of native atmosphere with which it is credited.[9] Be this as it may, the great merits of M. Bonin as a pioneer worker in the field of Mo-so literature remain uncontested.

An essential progress in the advance of Mo-so studies is signaled by the recent work of J. Bacot.[10] This energetic traveler and explorer has made not only most interesting contributions to the life and religion of the people but has furthered a great deal also our knowledge of their language and systems of writing. He is the first to have revealed the interesting fact that the Mo-so have not, as had hitherto been believed, merely one system of writing, but that they have at least two. Besides the pictograpic method there is a syllabic script; and the latter, again, appears in combination with the former, as the Japanese combine the syllabary of their Katakana or Hiragana with the Chinese characters. In its style and outward appearance the marks of the Mo-so syllabic system bear a certain resemblance to

---

[9] Among the many curiosities of the Tibetan-English Dictionary, published in 1902 by Sarat Chandra Das, the well-known Bengali student of Tibetan and explorer of Tibet, we read (p. 565) that the evil drug tobacco (in Tibetan *tha-ma-kha*) appeared in ancient time about one hundred years after the death of Buddha, which would yield the date 377 B. C., and that mention is made of tobacco also in the writings of a lama who is dated in the twelfth century A.D. In 1908 I enjoyed the privilege of spending several months in "Lhasa Villa," the house of Das in Darjeeling, and when one evening, in the course of a learned conversation with him, I ventured to draw his attention to this chronological anomaly and the post-Columbian introduction of tobacco into Europe and Asia, he replied, "This is your tradition, and that (pointing to the passage in his dictionary) is our tradition; and our traditions certainly are as good as yours."

[10] Les Mo-so: Ethnographie des Mo-so, leurs religions, leur langue et leur écriture, par J. Bacot. Avec les documents historiques et géographiques relatifs à Li-kiang par E. Chavannes (Leiden, 1913, 41 plates and a map).

FIG. 1.

278

FIG. 3.

FIG. 4.

FIG. 3—Recto of the second folio (page 3).

FIG. 4.—Recto of the fifth folio (1

279

the Önmun alphabet of the Koreans; and the reason presumably is that, like the latter, it may be traceable to the Indian Devanāgarī or an allied alphabet.[11] It is needless, however, to speculate on this point, as we are ignorant at present of the origin and history of the Mo-so syllabary. M. Bacot has published seven plates containing so-called Dhāranī (mystic prayer forms, spells, and charms) composed in this style of writing; and if the syllabary is with predilection chosen for this class of Buddhist literature, we may presume with some degree of probability that this script received its impetus from India. On pages 68-117 of his book a very useful list of Mo-so words is given with their pictographic and syllabic characters. As a foundation it is excellent, though it is certainly not complete, as we meet in the Mo-so texts numerous characters not yet explained. Our knowledge of the Mo-so grammar being only of the most rudimentary kind (the dictionary remains as yet to be made), it is impossible to decipher Mo-so manuscripts at home in the study. Such researches, naturally, can be pursued only in the field, in the midst of the Mo-so, with the assistance of native scholars or learned priests. It is hoped that some of the French missionaries who applied themselves so successfully to the study of the Lo-lo will soon attack also the Mo-so problem.

As regards the origin of the pictograph system of writing, M. Bacot reiterates the very interesting theory which had already been advanced by T. de Lacouperie.[12] He supposes that it is identical with the script which is said to have been in vogue among the Tibetan shamans prior to the seventh century. Forgotten in Tibet, it would have survived among the Mo-so, while its traces may still be recognized in certain pictorial charts employed by the Tibetans for purposes of magic. It is very likely that the realistic and partially conventionalized designs now serving the expression of ideas were in their origin religious symbols utilized in magic rites and devil exorcisms, or even purely ornamental motives. Certain it is that, in the same manner as the peculiar writing of the Lo-lo, the pictographs of the Mo-so bear no relation to those of the Chinese and are entirely independent in their origin and development. The Chinese Chronicle of Li-kiang relates that Nien-pao A-tsung, the son of the ancestor of the Mo-so chiefs, invented a system of writing for his country in the twelfth century, but we are left in ignorance as to what this writing was.

The Mo-so manuscript of Mr. Nichols consists of nine oblong folio leaves, measuring 28.3 x 9.7 cm. and sewed together on the left-hand margin by means of treble yellow-silk threads. The peculiar size is derived from that of Tibetan books, which, on their part, imitate the palm-leaf manuscripts of India. The writing, done by means of black Chinese ink, is apparently accomplished with a bamboo or wooden stylus. The paper is stiff and heavy and has assumed a yellowish brown color. Dr. C. F. Millspaugh, Curator

---

[11] Compare W. G. Aston: Writing, Printing, and the Alphabet in Corea, *Journ. Roy. Asiatic Soc.*, 1895, p. 510.

[12] Beginnings of Writing, p. 48.

f Botany in the Field Museum, who has a wide experience in the study
f Oriental papers, has been good enough to examine the paper of this
manuscript, with the following result: "The paper employed in the fashion-
ng of this book," he remarks, "is manufactured from the bark of *Streblus
sper,* a rigid shrub or small, gnarled tree, belonging to the same botanical
amily (*Moraceae*) that includes the paper mulberry. The tree is found
n the dryer parts of India from Rohilkhand eastward and southward to
Travancore, Penang, and the Andaman Islands. It is prevalent also in
Cochin-China, Siam, and the Malayan Islands. In Siam it forms the pulp
or the famous *Laos* paper. There the bark is known as *Khoi* bark. In Siam
he bundles of bark are soaked in water for two days, then steamed with
ime for the same period, shredded, placed in earthen jars and steeped in
lear water and lime for a few days, washed free of lime, beaten to a pulp
with wooden mallets, floated in water where the flakes of unmacerated bark
and knots are picked out, and finally floated out, squeezed, and layered into
he thickness desired for pressure and drying into paper."

Four pages of the manuscript, which consists of nine folios of two pages
each, are reproduced herewith.[13] The title of the work is written in the
center of the front cover (Fig. 1), occuying two lines enclosed by rectangles,
he symbols running from the left to the right, as is the case also in the
Tibetan alphabet.

The verso of the first folio (Fig. 2) contains three mythological representations which stand outside of the frame of the text proper, that begins
with the second folio. They are, so to speak, miniatures intended as book-
decoration and selected with due regard to the contents and purport of
he book. As will be seen, they even afford an important clue to reveal to
us its subject-matter. The idea of placing miniatures in front of the book
s borrowed from the Tibetans, who on their part have been inspired by
traditions inflowing from India. The Lama artists are masters in the
execution of exquisite miniature painting in colors,—gold, silver, or black,—
spread over the covers or the pages of a book. The animal on the left-
hand side of the Nichols manuscript is a tiger, well characterized by the
shape of the head and the stripes of the skin. This identification is con-
irmed by a note of Mr. Nichols himself, who has accompanied it with a
pencil-mark reading *tag.* The latter word (written *stag*) is the general
Tibetan designation for the tiger and probably is also the corresponding
Mo-so word for it.[14] In the shamanism of Tibet (as well as in the shaman-
stic religions of Siberian peoples, particularly the Amur tribes) the tiger
plays the rôle of a powerful deity. The central figure of Figure 2 represents a subject of purely Indian mythology, the bird Garuda, the companion
and riding-beast of the god Vishnu. He is the inveterate enemy of the

[13] Students desiring more information may consult the manuscript in the Society's library.—EDIT. NOTE.
[14] M. Bacot (*loc. cit.*, p. 108) gives *la* as the Mo-so term for the tiger, which is also the word of the Lo-lo.
D'Ollone (Langues des peuples non chinois de la Chine, p. 64, Paris, 1912) gives the Mo-so word in the form
a which is the same as Nichols' *tag.*

serpent-demons, the Nāga, and constantly engaged with them in a struggle
of destruction. The Mo-so representation is perfectly in accordance with
indeed, an exact reproduction of what we meet in Tibetan art, and familia
to all students of Tibetan iconography. The bird is outlined *en face* wit:
outspread wings, soaring above the clouds, with triangular hooked beak, seiz
ing in its claws a snake whose head droops toward the left. The Nāga i
an inhabitant of the water, and accordingly we see a sketch of wave-line
along the lower margin, being symbolic of the watery element. The Garud
is supposed to have wrested the Nāga from its abode; and the latter, a
indicated by the downward position of its head, makes a supreme effor
to be restored to its former home. Mr. Nichols has made here a pencil-not
"phœnix," which may be an erroneous explanation given by his Chines
interpreter.

The most interesting of all is the figure on the right-hand side of thi
page. We note what appears as a male deity clad with a panther's skin an
boots of Chinese style, and crowned with a five-lobed diadem. He stand
with straddling feet over a nude figure which lies on its back. In his righ
hand he brandishes a rattle-staff (known in Sanskrit as *khakkara*), and i
his left a tambourine (*damaru*) or cymbal. This figure is well known a
that of the great Indian magician Padmasambhava, the "lotus-born," wh
introduced a degenerate form of Buddhism into Tibet toward the middl
of the eighth century. He conjured and overpowered the native demon
but assimilated all of them to the pantheon of Çivaitic gods, which he intr(
duced. He became the founder of a sect known as the Old Church, i
distinction from the subsequent Reformed Church, organized by Tsong
kha-pa. The later generations canonized him as the most efficient sain
and the second Buddha; and the common people, particularly of Sikkiı
and eastern Tibet, still worship him directly as a god. An immense litera
ture, written in Tibetan and Lepcha, has crystallized around this interesi
ing personage: a bulky biography, teeming with miracles and romanti
adventures, and numerous other books have been produced in his honoi
Only a small fraction of these works has thus far been made accessible i
translations: E. Schlagintweit, A. Grünwedel, and the present writer hav
contributed several monographs to the subject.

In the work of J. Bacot (Pl. xvi) we find the reproduction of a Mo-s
painting representing the same personage. He occupies the center of th
picture, being surrounded by four smaller effigies of himself, differing solel
in the posture of the hands and feet. Beneath we see a tiger facing a bul
the two being separated by a censer from which burning incense flame
up and on which six death's heads are figured. On the top is shown
Garuda standing in clouds and holding a serpent in its beak and betwee
its claws. This picture, accordingly, comprises all essential elements coi
tained on the inner title-page of our Mo-so book. M. Bacot has terme
this image Tumbashera, who, according to him, is identical with Sheral

ni-vo, the mythical founder of the Tibetan Bon sect. As seen also by
M. Bacot, however, the Mo-so legends connected with Tumbashera contain
vestiges of the Buddha legend (as, for instance, the story of his birth from
his mother's armpit), and, on the other hand, show a strong resemblance
to the tradition of Padmasambhava. The case is such that the whole
Buddha legend has been plagiarized and transferred to the revered saint
Padmasambhava. The Tibetan people in general do not care much for the
life of Buddha, but they are keenly interested in that of his prophet; and
thus the conclusion is justified that also the Mo-so, who are surrounded by
Tibetan tribes adhering to Padmaism rather than Buddhism, are driven to
the same tendency. There is positive evidence forthcoming from the
Nichols manuscript for the fact that this book deals with a legend taken
from the cycle of Padmasambhava: it is adorned with his portrait on the
page preceding the text, and this is a safe criterion to warrant the conclusion that the text can have only Padmasambhava as its subject. If therefore his portrait, as assured by M. Bacot, is styled Tumbashera by the Mo-so,
this would mean that Tumbashera is the Mo-so designation for Padmasambhava. Thus the not unimportant result is assured that the Nichols manuscript represents a chapter or a story from the legendary biography of this
saint, a book celebrated all over Tibet and Mongolia.

It is evident also that many other Mo-so manuscripts are nothing but
detached chapters from the same work in Mo-so translation. To these
belongs the manuscript published by T. de Lacouperie,[15] as proved by the
image of Padmasambhava outlined in the beginning of the first page. Here,
in correspondence with numerous Tibetan statues and paintings, he holds
in his left hand a skull-bowl supposed to be filled with blood. The title of
the same manuscript opens with the well-known Indian design of a lotus-
blossom surrounded by leaves, and it is plausible that it may express the
first element of the name of the great saint, *padma,* the "lotus." The
technique and the style of writing of this manuscript are exactly the
same as those of the Nichols book. Also M. Bacot's manuscript[16] manifestly comes from the same source. It is, however, not complete, or not
completely illustrated, only four pages being reproduced. The first page,
at least, is missing; and the long legend given in translation on pages 18-20
of his work obviously corresponds to far more than four Mo-so pages
with only sixteen short lines.

The specific Buddhist contents of Mo-so literature are strongly borne
out by the Buddhist images and emblems with which the manuscripts are
adorned: M. Bonin has encountered in his manuscript a sketch of the
Buddhist lion, further the traditional picture of the Bodhisatva Avalokiteçvara, and to his right the design of the conch-shell trumpet, an ancient
symbol of the Buddhist church, the call of which still summons the lamas

---

[15] Beginnings of Writing in Central and Eastern Asia, Pls. i and ii.
[16] Les Mo-so, Pls. xviii and xix.

to their prayer-meetings. Finally, what M. Bonin describes as "the chie of the evil genii, whose robe, head-dress, and boots remind us of those stil worn by the itinerant lamas of Tibet," in my opinion, can be nobody bu Padmasambhava. In the title-page of de Lacouperie's manuscript we note besides the lotus above referred to, the symbol of the so-called *çrīvatsi* ("lucky knot"), the mystic emblem of the god Vishnu and his adherents

The second folio of the Nichols manuscript (Fig. 3) opens with a desigr of waves common to the art of the East in general. It fulfills a purely ornamental function; likewise the following spiral pattern, which is ver frequently found in the beginning of Tibetan books. The first symbols ar the sun and the moon or month. The striking characteristic of Mo-so writ ing is the naïve and refreshing realism manifested by the representation of animal pictures. We see many birds on the wing, well-outlined heads o mammals, and even complete beasts like a horse, a stag, and a tiger (Fig. 4) In de Lacouperie's manuscript (Pl. iii, Nos. 101 and 102) we note an ell and a yak of surprising naturalness; and M. Bacot (p. 79) gives tw sketches of ravens, which breathe a truly artistic spirit. Of animal heads the goat chewing herbage (*ibid.*, p. 77) and the mule (p. 96) are remarka ble. Others are less good, and could hardly be guessed without commen tary. The ethnologists interested in the decorative art and picture-writin; of primitive man will find here ample and interesting material for study In the human face we find the front as well as the profile represented i' the Nichols manuscript (Fig. 3, line 4). As a rule, the human figure i strongly abridged, characterized only by a circle to which lines are attached marking arms and feet. In the first line of Figure 3 we observe the outline of a house, in the interior of which are standing two persons, the second being distinguished by a head-dress. This design is many times repeated in our text. M. Bonin had already drawn attention to it, interpreting it a the expression of the idea of the family, man and wife living in their house M. Bacot (p. 91) adopts another explanation, which is, "host who receives master of the house." This being the case, the person honored by the head dress would be the guest. Headgear, presumably a helmet, is met also in connection with warriors brandishing a weapon (Fig. 3, line 2), a symbo which according to Bacot (p. 71) means "to fight." Accordingly, tw stages are notable in Mo-so writing,—full pictures of natural objects and conventionalized shortenings.

But this is not all; we encounter even abstract symbols of geometri character, especially for conveying abstract and verbal notions, and some times even denoting living beings. Thus the horse is represented by square #, very much like the Chinese character signifying a "well.' A "road" is expressed by a multiplication sign plus a dot in the lowe part (X) If a horizontal line is added above (X), the symbol signifie a "king." It is difficult to discuss the majority of these symbols withou having reference to reproductions of them in the text.

In the writing of the Lo-lo the pictographic characteristics, if they ever existed, have disappeared; and the symbols are composed of decidedly geometric elements. They are pre-eminently ideographic, while Mo-so has a pictographic substratum with a tendency to strive for simplification. There is no similarity or coincidence between Lo-lo and Mo-so characters.[17]

[17] The best means for the study of Lo-lo writing are the books of Father P. Vial, "De la langue et de l'écriture indigènes au Yun-nan" (Paris, 1890), "Les Lolos" (Shanghai, 1898), and "Dictionnaire français-lolo" (Hongkong, 1909). A Lo-lo manuscript of twenty folios has been reproduced in facsimile by Professor F. Starr of Chicago (in fifty copies).

125

中国的人种学传说

AUFSÄTZE

ZUR

KULTUR- UND SPRACHGESCHICHTE

VORNEHMLICH DES ORIENTS

ERNST KUHN

ZUM 70. GEBURTSTAGE AM 7. FEBRUAR 1916

GEWIDMET VON

FREUNDEN UND SCHÜLERN

MÜNCHEN 1916

BRESLAU, VERLAG VON M. & H. MARCUS

1916

Ethnographische Sagen der Chinesen.

Von

Berthold Laufer (Chicago).

Ursprünglich waren die Chinesen ein Bauernvolk des Binnen-
landes, das mit dem Meere in keinerlei Verbindung stand. Erst
in allmählicher Ausdehnung ihrer Macht nach Osten und Südost
gewannen sie Fühlung mit den Küstengebieten, und gegen Ende
des dritten vorchristlichen Jahrhunderts unter den Ts'in regte sich
der Drang, unbekannte Lande im östlichen Ozean zu entdecken.
Unter den Han nahm diese seewärts gerichtete Expansion eine
ungeahnte Entwicklung, während sich gleichzeitig Zentralasien und
Indien den Chinesen zu erschließen begannen und ihr reger
Handelsgeist sie mit den Ländern des hellenistischen Orients in
Verbindung brachte. China nahm damals seinen Platz in der
Welt ein und wuchs sich zu einem Weltreich in dem Sinne aus,
daß alles zu jener Zeit bekannte Gute bereitwillig Aufnahme fand
und fremde Kulturelemente die einheimischen Gedanken be-
fruchteten und vertieften. Zu diesen charakteristischen Entlehnungen
aus dem Westen gehören nicht in letzter Reihe die sagenhaften
Motive der Wundervölker und der sich an diese anschließenden
Wundermären. Die in der chinesischen Literatur über diesen
Gegenstand handelnden Quellen haben schon vielfach den Scharfsinn der Sinologen in Anspruch genommen, ohne daß ein allgemein befriedigendes oder annehmbares Ergebnis erzielt worden
wäre. Vor allem hat G. Schlegel (Problèmes géographiques, Les
peuples étrangers chez les historiens chinois, Leiden 1892; zuerst
in den Bänden III—VI des T'oung Pao) mit einem Aufwand
großer Gelehrsamkeit, aber mit mangelnder Kritik sowohl in der
Behandlung der Quellen als in ihrer Interpretation, eine Reihe

jener Wundervölker von geographischen und ethnographischen Gesichtspunkten behandelt und systematisch klassifiziert. Von dem Glauben erfüllt, daß die einschlägigen chinesischen Nachrichten aus der Anschauung und Beobachtung wirklicher Zustände hervorgegangen seien, versuchte er dieses angebliche Wirklichkeitsbild zu rekonstruieren und ein festes geographisches Schema zu zeichnen, in dem fast alle gegenwärtig lebenden Völker des nordöstlichen Asiens, wie die Ainu, Kamtschadalen, Korjaken und Tschuktschen, den alten Chinesen bekannt gewesen seien. Über den allgemeinen Mißerfolg dieses Versuches hat wohl niemals Meinungsverschiedenheit geherrscht. Einmal können die meisten Texte, auf denen sich Schlegels Anschauungen gründen, keine geschichtliche Glaubwürdigkeit beanspruchen, sondern gehören der taoistischen, zu einem geringeren Teile der buddhistischen Wunderliteratur an. Sodann ist es uns längst bekannt, daß die ethnographischen und kulturhistorischen Verhältnisse im nordöstlichen Stillen Ozean in alter Zeit weit von denen der Gegenwart verschieden waren, so daß wir die heutigen Zustände nicht in alte Berichte hineinlesen können. Was aber das Wichtigste ist, die Interpretation dieser Dinge ist ganz anderswo zu suchen als in der vergeblichen Bemühung, sie mit geographischen und ethnographischen Namen gleichzusetzen. Wer die chinesischen Berichte mit einiger Aufmerksamkeit und Kritik liest und wer mit dem Bilde der Wundervölker vertraut ist, wie es unter indischem Einfluß die griechischen Berichterstatter über Indien, Skylax, Hekataios, Ktesias und Megasthenes den Hellenen vermittelt haben, wird bald zu der Überzeugung gelangen, daß hier auffällige Übereinstimmungen vorwalten, die sich in einer Anzahl von Fällen auch als historische Zusammenhänge nachweisen lassen. Wie die Kunde der indo-hellenischen Wundervölker durch Vermittlung des Alexanderromans sowie des Plinius und seiner Nachfolger das gesamte Abendland durch das Mittelalter bis in den Beginn der Neuzeit mächtig angezogen und die Sagen der seefahrenden Araber beeinflußt hat, so hat sie auch auf die Phantasie der Chinesen eingewirkt. Die folgenden Ausführungen wollen nur als ein Programm gelten; sie sind der Auszug aus einer größeren, noch unveröffentlichten Abhandlung, in welcher das Thema auf breiter Grundlage dargestellt ist.

Um die Abhängigkeit ethnographischer Sagen der Chinesen
von westlichen Vorstellungen zu erweisen, wird es am besten
sein, von fest umschriebenen Tatsachen auszugehen. Als ge-
wichtigstes Beweisstück bietet sich die Sage von den Pygmäen
und ihrer Kämpfe mit den Kranichen dar (Ilias III 3; Hekataios,
fr. 266; Herodot II 32; Ktesias § 11, p. 81b; Aristoteles, Hist.
animalium VIII 12 = Aubert und Wimmer, Aristoteles Tierkunde,
II, p. 151; Megasthenes bei Strabo XV 1 § 57; Plinius VII 2
§ 26, X 30 § 58; vgl. W. Reese, Die griechischen Nachrichten
über Indien p. 101 f.; O. Keller, Antike Tierwelt II, p. 188). Tu
Yu (735—812) erzählt in seinem zwischen den Jahren 766 und
801 verfaßten T'ung tien (T'ai p'ing yü lan, Kap. 796, p. 8) die
folgende Version dieser Legende: „Die Zwerge leben im Süden
von Ta Ts'in [der hellenistische Orient]; ihr Wuchs beträgt nur
drei Fuß. Während sie der Bestellung ihrer Felder nachgehen,
fürchten sie von den Kranichen verschlungen zu werden. Jedes-
mal, wenn Ta Ts'in ihnen in ihrer Bedrängnis Beistand leistet,
vergelten die Zwerge diesen Dienst durch eine Sendung wertvoller
Waren[1]." Tu Yu hatte einen Verwandten namens Tu Huan, der
im Jahre 751 von den Arabern in der Schlacht bei Talas ge-
fangengenommen wurde, dann Zentralasien und arabisches Gebiet
bis zum Mittelmeer durchwanderte und endlich im Jahre 762 auf
dem indischen Seewege nach Kanton in China zurückkehrte. Dort
schrieb er ein Werk über seine Abenteuer, das leider verloren
gegangen ist, von dem aber Fragmente im T'ung tien seines Verwandten Tu Yu aufbewahrt sind. Da wir in diesem Buche
arabische Überlieferungen finden, die mit guten Gründen auf

---

[1] Hirth (China and the Roman Orient p. 87; 202) zitiert diese Sage
nach dem Wen hien t'ung k'ao des Ma Tuan-lin vom Jahre 1319, der
jedoch in diesem Falle das ältere T'ung tien wörtlich ausgeschrieben hat.
F. W. K. Müller (Z. f. Ethnologie 38, 1906, p. 750) bezieht sich aus-
schließlich auf die japanische Ausgabe von 1713 der chinesischen Enzyklopädie San ts'ai t'u hui vom Jahre 1607, ohne selbst Ma Tuan-lin's zu
erwähnen. Dieser Passus samt der schwachen Illustration war übrigens
lange vorher von F. de Mély (Le »de Monstris« chinois, Revue archéol.
1897, p. 353) veröffentlicht worden. Es bedeutet natürlich wenig, hellenische
oder andere westliche Stoffe in der chinesischen Literatur nachzuweisen;
die Hauptsache bleibt, dieselben auf ihre ältesten Quellen zurückzuführen und
zu zeigen, wann und auf welchen Wegen sie nach China eingedrungen sind.

Rechnung der unfreiwilligen Reise des Tu Huan gesetzt werden (Chavannes, T'oung Pao 1904, p. 486), so dürfen wir wohl auch die Sage von den Pygmäen und Kranichen auf dieselbe Quelle zurückführen und annehmen, daß Tu Huan sie von den Arabern gehört hat. Daß die griechische Sage den letzteren geläufig war, ersehen wir aus Qazwīnīs Kosmographie (J. Ansbacher, Die Abschnitte über die Geister und wunderbaren Geschöpfe aus Qazwīnīs Kosmographie p. 31)[1]). Nun ist es aber wahrscheinlich, daß schon vor Tu Yu dieselbe Sage aus anderer Quelle nach China gewandert ist; denn schon in den im Jahre 629 verfaßten Annalen der Liang hören wir von Zwergen des Landes Ta Ts'in (Hirth, China and the Roman Orient p. 48), und im Schên i king lesen wir folgendes: „Im Gebiete des Westlichen Meeres (Si hai, das Mittelmeer) liegt das Land der Kraniche (hao kao), wo Männer und Frauen nur sieben Zoll hoch sind und nach dem natürlichen Sittengesetz leben. Sie halten gern Beratungen ab und knien nieder zum Zeichen des Grußes[2]). Sie scheinen zu fliegen, wenn

---

[1]) Qazwīnī erzählt noch eine andere Version der Sage, die nach ihm in Damīrī's Lexikon der Tiere zitiert wird (A. S. G. Jayakar, Ad-Damīrī's Ḥayāt al-Ḥayawān, Vol. II, pt. 1, p. 450). Danach wurde ein Mann von den Leuten von Rūmīyah (Rūm) bei einer Reise im Meere von Zanj durch Sturm auf eine Insel verschlagen und kam zu einer Stadt, deren Einwohner etwa eine Elle hoch und die meisten einäugig waren. Er wurde gefaßt und vor den König gebracht, der ihn in einen Käfig einsperren ließ. Eines Tages sah er sie Vorbereitungen zu einem Kampfe treffen und erkundigte sich nach der Ursache. „Wir haben einen Feind, der gewöhnlich zu dieser Zeit zu uns kommt," wurde ihm zur Antwort gegeben. Bald darauf wurden sie von einer Schar Störche angegriffen. Die Einäugigen unter ihnen hatten ihr Auge dadurch verloren, daß diese Vögel es ausgestochen hatten. Der Mann von Rūm machte einen heftigen Angriff auf sie mit einem Stock, worauf sie fortflogen und verschwanden, und wurde darob von den Zwergen hoch geehrt. M. Gaster (Folk-Lore 26, 1915, p. 202) hat jüngst eine ganz übereinstimmende Geschichte aus dem Eschkol Kakofer mitgeteilt, einem polemischen Werke, das ein in Konstantinopel lebender keraitischer Schriftsteller, namens Judah Hadasi, im Jahre 1148 verfaßte; auch hier landet ein gewaltiger Mann von Konstantinopel im Lande der Pygmäen und hilft ihnen, ihre geflügelten Feinde zu vertreiben. Gaster hält wohl mit Recht diese Erzählung für den Vorläufer oder gar die Quelle von Swifts Gulliver and the Lilliputians.

[2]) Nach Ktesias folgten die indischen Pygmäen indischen Gesetzen und waren Biedermänner.

sie gehen, und legen an einem Tage tausend Meilen (*li*) zurück.
Sie sind alle gebildet und scheuen davor zurück, einander zu
kränken. Die einzigen Geschöpfe, die sie fürchten, sind die
Kraniche, die vom Meere her kommen. Diese Kraniche, die mit
einem einzigen Fluge tausend Meilen machen, können sie verschlingen; doch die Leute, die drei Jahre alt sind, erliegen nicht
ihrem Angriff." Das Schên i king gehört zu den schwer datier-
baren Werken. Die Verknüpfung mit dem Namen des Tung-fang
So (zweites Jahrhundert v. Chr.) ist nichts anderes als eine
taoistische Legende; die Annahme, daß das Werk im vierten oder
fünften Jahrhundert n. Chr. entstanden sei, ist wahrscheinlich, und
das Alter unseres Textes wird durch seine Aufnahme in die im
Jahre 983 vollendete Enzyklopädie T'ai p'ing yü lan (Kap. 797,
p. 9) gewährleistet. Im dritten Jahrhundert berichtet Yü Huan
in seinem Wei lio von einem Zwergenreich im Nordwesten von
Sogdiana (K'ang-kü), wo Männer und Frauen drei Fuß hoch sind
(Chavannes, T'oung Pao 1905, p. 561). Im T'ung tien des Tu Yu,
der diesen Text kopiert hat (T'ai p'ing yü lan, Kap. 796, p. 7b),
ist der Zusatz enthalten, daß dieses Land kostbare Edelsteine
sowie die in der Nacht leuchtenden und die mondhellen Perlen
hervorbringe; bei dieser Vorstellung handelt es sich um eine den
Chinesen aus dem hellenistischen Orient vermittelte Tradition, die
Verfasser in seiner Abhandlung 'The Diamond, a Study in Chinese
and Hellenistic Folk-lore' eingehend dargestellt hat. Yü Huan
hat überdies den Chinesen das Bild der antiken Kentauren oder
vielmehr Hippopoden zugeführt: „Greise unter den Wu-sun er-
zählen, daß sich das Land der Leute mit Pferdeschenkeln bei den
nördlichen Ting-ling befinde; ihre Sprache gleiche dem Ruf wilder
Gänse und Enten; oberhalb der Knie hätten sie einen menschlichen Körper und Kopf; unterhalb der Knie hätten sie Haar-
wuchs, Pferdeschenkel und Pferdehufe; sie stiegen nicht zu Pferde,
sondern liefen schneller als ein Pferd; sie seien tapfere und kühne
Leute im Kampf[1]." Wenn sich nun diese nur aus dem Helle-

---

[1] In Anbetracht der Tatsache, daß die Pygmäen, Amazonen, Hundsköpfe und andere Wundervölker von der chinesischen Sage in den östlichen
Ozean versetzt werden, ist es interessant, daß in der Enzyklopädie San ts'ai
t'u hui (1607), die einen Katalog der Wundervölker enthält, auch die Ting-
ling bewohnenden Pferdefüßler im Meere angesiedelt werden. Man sieht an

nismus zu erklärende Vorstellung im Schan hai king befindet, worauf Chavannes hinweist, so ist es ohne weiteres klar (und dies wird durch viele andere Texte bestätigt), daß diese merkwürdige Kosmographie nicht das fabelhafte Altertum beanspruchen kann, das ihr chinesische Gelehrte und in ihrem Bann befangene europäische Sinologen zuschreiben; in der uns vorliegenden Textgestalt ist das Buch zur Han-Zeit redigiert worden und hat, was die Wundervölker betrifft, starke Einflüsse vom Hellenismus erfahren. Wenn die Chinesen und Sinologen an dem Versuch gescheitert sind, seinen Inhalt und sein Wesen zu deuten, so liegt dies daran, daß ihnen diese Zusammenhänge fremd geblieben sind: nach meiner Auffassung ist das Schan hai king nur aus dem Hellenismus zu erklären und geradezu als hellenistisches Produkt aufzufassen. In demselben Werke begegnen wir auch einer Andeutung betreffs eines Pygmäenlandes, das jenseits des östlichen Meeres in einer großen Einöde projiziert wird. Schlegel (T'oung Pao IV, 1893, p. 324) bemerkt, daß die chinesischen Annalen diese Zwerge des äußersten Nordostens nicht erwähnen; dies ist indessen ein Irrtum, denn im Nan shi (Kap. 79, p. 3b), das die Geschichte Chinas vom Jahre 420 bis 589 behandelt und das in der ersten Hälfte des siebenten Jahrhunderts von Li Yen-schou verfaßt worden ist, wird im Anschluß an einen Bericht über Japan unter dem Namen *Chu-žu*[1]) ein Land der Pygmäen erwähnt, die

---

diesem Beispiel, wie hier die Legendenbildung nach einer Schablone arbeitet und Geographie und Ethnographie fabriziert. — Nach Solinus gab es ein Volk Hippopodes im Skythenlande; die Hereford-Karte gibt eine Abbildung mit der Bemerkung „equinos pedes habent" und verlegt ihren Wohnsitz in die Nähe des kaspischen Meeres (vgl. L. E. Iselin, Der morgenländische Ursprung der Grallegende p. 127).

[1]) Eine allgemeine Bezeichnung für Zwerge (Legge, Chinese Classics V, p. 422; 424). Nach den Bambusannalen (d. h. auf Bambustafeln geschriebenen Annalen) soll ein Häuptling der Zwerge sich bereits dem mythischen Kaiser Yao unterworfen und im Wasser untersinkende Federn als Tribut dargebracht haben (Legge, Chinese Classics III, Prol. p. 112). Diese Federn sind an sich recht verdächtig. Die Bambusannalen, die im Jahre 299 v. Chr. in einem Grabe geborgen und im Jahre 281 n. Chr. ans Licht gezogen wurden, sind, wie Chavannes (Mémoires historiques de Se-ma Ts'ien V, p. 446—479) in einer ausführlichen Studie nachgewiesen hat, ein authentisches Werk, aber wie Chavannes selbst zugibt, ist die Möglichkeit späterer Umarbeitungen nicht ausgeschlossen. Die Pygmäen als Interpolation aufzufassen, liegt um so mehr Grund vor, als dasselbe Werk von zwei anderen Wundervölkern be-

vier Fuß lang sind. Schlegel hat diese Pygmäen mit den Koro-
pokguru identifiziert, die in den Sagen der Ainu eine Rolle spielen,
und aus der Kraft seiner Phantasie eine Pygmäenrasse geschaffen,
die in unterirdischen Wohnungen gelebt und die Küsten des
Japanischen Meeres vom Amur, die Inseln des Japanischen und
Gelben Meeres sowie die Kurilen bis Kamtschatka bewohnt haben
soll. Diese „Rasse" hat natürlich nie existiert, wie Verfasser
bereits (Zentralblatt für Anthropologie 1900, p. 321—330) gezeigt
hat. Der vagen Anweisung der Pygmäen im nordöstlichen Meere
kommt ebensowenig ein geographischer oder ethnographischer
Wert zu als z. B. den Pygmäen, die Odoric von Pordenone an
den Yang-tse verlegt (Yule, Cathay, neue Ausg., Vol. II, p. 207).
Es handelt sich vielmehr um einen Nachhall hellenistischer Sage,
denn nicht nur die Pygmäen, sondern auch die Amazonen und
Hundemenschen sind mit jenen von den Chinesen in den nordöstlichen Ozean versetzt worden. Der Fall der Hundemenschen
ist besonders schlagend.

Sieht man von einigen zusammenhanglosen Notizen bei Hesiod,
Simmias von Rhodos und Aischylos sowie von Herodots (IV 191)
etwas vagen nordafrikanischen κυνοκέφαλοι ab, so ist Ktesias in
seinen Indica (§ 20, 22—23) der erste, dem wir eine ausführ-
liche Beschreibung eines indischen Volksstammes unter dem Namen
der Hundeköpfe verdanken, die Klauen wie Hunde haben, aber
länger und runder, und noch größere Zähne. Männer und Frauen
haben einen Schwanz über den Hüften wie ein Hund, aber größer
und dicker, und begatten sich nach Art der Vierfüßler. Sie besitzen keine Sprache, sondern bellen wie Hunde und verständigen
sich auf diese Weise. Im Verkehr mit den übrigen Indern verstehen sie diese zwar, sie selbst aber können nicht sprechen,
sondern machen sich durch Gebell und Gebärdensprache ver-

---

richtet, den Menschen mit durchbohrter Brust und den Langbeinen (Legge,
p. 109), von denen sonst erst zur Han-Zeit die Rede ist (A. Forke, Lun-
hêng, pt. II, p. 263 und Chavannes, La sculpture à l'époque des Han p. 79).
Der Name der von Forke erwähnten Tan-êrh, deren Ohrlappen bis auf die
Schultern herabhängen, ist gewiß ein Nachhall der Ἐνωτοκοῖται des Ktesias
und Megasthenes, der Karṇaprāvaraṇa, Karṇika, Lambakarṇa, Mahākarṇa usw.
der Inder. Im San ts'ai t'u hui (1607) erscheint ein Volk der Langohre,
die ihre bis auf den Gürtel herabhängenden Ohren auf dem Marsch mit den
Händen festhalten.

ständlich, wie Taubstumme; im Verkehr sind sie bieder. Dann
folgt eine kurze Schilderung ihrer Kultur, die sehr wohl einem
Wirklichkeitsbilde entspricht. J. Marquart hat in seinem hervorragenden Werke „Die Benin-Sammlung des Reichsmuseums für
Völkerkunde in Leiden" (p. CC—CCXIX) die gesamte antike
Überlieferung von den Hundemenschen mit gewohntem historisch-
kritischem Scharfsinn behandelt. Mit Recht weist Marquart auf
den Einfluß des Alexanderromans auf die Ausgestaltung der Sage
hin, besonders was das Motiv betrifft, daß nur die Knaben Hundemenschen, die Mädchen dagegen mulieres pulcherrimae magni
sensus werden[1]. Für unseren Zweck ist besonders die Stelle
bei Adam von Bremen (Gesta Hammaburgensis ecclesiae ponti-
ficum 4, 19) von Bedeutung, der über das Frauenland (Terra
Feminarum) auf der östlichen Seite der Ostsee berichtet: Cumque
pervenerint ad partum, si quid masculini generis est, fiunt cyno-
cephali, si quid feminini, speciosissimae mulieres. Ferner weist
Marquart auf ein im Jahre 555 geschriebenes syrisches Verzeichnis
süd- und nordkaukasischer Völker hin, in welchem jenseits der
hunnischen Völker die Pygmäen und Hundsmenschen sowie die
Amazonen aufgeführt werden, die sich mit den ungeschlachten
Hros paaren. Auf die Gedankenwelt unseres Mittelalters haben
die der Antike entlehnten Wundervölker einen tiefen Eindruck
gemacht, wozu vor allem der vor dem Jahre 1177 verfaßte Brief
des Priesters Johannes an den byzantinischen Kaiser Emanuel
(F. Zarncke, Der Priester Johannes I, p. 85—91) beitrug, in dem
sie alle nach Indien verlegt werden; hier sind pigmei und ceno-
cephali aneinander gerückt. So konnte sich auch Konrad von
Megenberg (1309—1374) diesem Einfluß seiner Zeit nicht entziehen und nahm mehr seinen Freunden zu Gefallen, als weil er
selbst Geschmack an diesen Fabeln gefunden hätte, ein Kapitel
über die Wundermenschen in sein anmutiges Buch der Natur auf
(F. Pfeiffer in seiner Ausgabe p. XXXIII).

Die früheste Nachricht über ein Frauenreich im nordöstlichen
Ozean findet sich in den Annalen der Späteren Han (Hou Han
schu, Kap. 115, p. 4b) als Anhang eines Berichts über das öst-

---

[1] Der Kannibalismus der Hundsköpfe, wie ihn z. B. Qazwīuī hat
(J. Ansbacher a. a. O. p. 31), ist auch auf den Alexanderroman zurückzuführen (A. Ausfeld, Der griechische Alexanderroman p. 82).

liche *Wo-tsü*, eines Stammes au der Ostküste Koreas: „Man erzählt
ferner, daß sich im Meere ein Frauenland befindet, wo es nur
Weiber, aber keine Männer gibt. In diesem Lande gibt es einen
wunderbaren Brunnen: wenn die Frauen in diesen hineinschauen,
so gebären sie alsbald Kinder." Es ist von großem Interesse,
daß Ma Tuan-lin in seinem Bericht über Ta Ts'in (Hirth, China
and the Roman Orient p. 84) von einem Amazonenreich im
Westen spricht und anfügt, daß die Frauen dort unter dem Einfluß des Wassers Kinder gebären. Diese Tradition muß also
wohl, wie auch Hirth (p. 200) folgerichtig annimmt, unmittelbar
aus Ta Ts'in geflossen sein und in China bereits zur Han-Zeit
bestanden haben, wie der analoge Text der Han-Annalen zeigt
und es war bereits in der Han-Periode, daß unter dem Einfluß
der taoistischen Wundermänner die hellenistischen Wundervölker
in den nordöstlichen Ozean festgelegt wurden. Zum Teil, wie
wir noch sehen werden, setzten diese Bestrebungen schon unter
der vorausgehenden Ts'in-Dynastie im dritten Jahrhundert v. Chr.
ein. In der von Klaproth (Aperçu général des trois royaumes
p. 149) nach der chinesischen Reichsgeographie übersetzten Beschreibung Koreas wird jene Sage einem Greise von Wo-tsü in
den Mund gelegt, der noch hinzusetzt, daß diese auf einer Insel
im Meere der Wo-tsü lebenden Frauen nur mit einem Gewande
aus Zeug bekleidet die See durchschwimmen, daß sie in ihren
körperlichen Eigenschaften Chinesinnen gleichen und daß die
Ärmel ihrer Kleider drei Klafter lang seien. Im Ling wai tai ta
des Chou K'ü-fei vom Jahre 1178 wird eine Amazoneninsel im
Osten von Java erwähnt, wo die Frauen, wenn der Südwind weht,
sich entkleiden und vom Winde konzipieren, aber nur Mädchen
gebären (Pelliot, BEFEO IV, 1904, p. 302; Hirth und Rockhill,
Chau Ju-kua p. 151). Brunnen- und Windmotiv vereinigt erscheinen im San ts'ai t'u hui. Ebenso erzählt Qazwīnī von einer
Fraueninsel im Chinesischen Meere, wo die Frauen vom Winde
empfangen und nur Mädchen zur Welt bringen, während sie nach
anderen vom Genuß einer Baumfrucht schwanger werden (G. Ferrand, Textes relatifs à l'Extrême-Orient p. 311). Sodann findet
sich diese Form der Legende bei Japanern und Ainu (B. H. Cham-
berlain, Language, Mythology etc. of Japan viewed in the Light
of Aino Studies p. 22; B. Piłsudski, Materials for the Study of

the Ainu Language and Folklore p. 91, Veröffentliehung der Kais. Akademie von Krakau). Übrigens hat die Sage von der Fraueninsel nichts mit der von Pilsudski mitgeteilten Ainu-Legende Nr. 5 zu tun, wie er meint, noch ist die von ihm vorgetragene medizinisch-pathologische Erklärung auf letztere anwendbar; es handelt sich hier vielmehr um eine ainu-giljakische Entlehnung eines nordwestamerikanischen Mythus, den schon F. Boas (Indianische Sagen von der Nord-Pazifischen Küste Amerikas p. 24, 66) behandelt hat und der unter dem Titel vagina dentata den Amerikanisten in vielen Varianten bekannt ist. F. de Mély (Revue archéol. III, 1897, p. 372) hat wohl mit Recht bemerkt, daß die Vorstellung von der Konzeption durch den Wind in der chinesischen Tradition auf den Einfluß griechischer Sage zurückzuführen ist; dasselbe gilt wohl auch von dem Brunnenmotiv. In dem bereits erwähnten Nan schi (Kap. 79, p. 4) finden wir nun die Hundemenschen unmittelbar an die Amazonensage angeschlossen, die der berüchtigte Śramaṇa Huei Schên wieder in Umlauf setzte. Wie bekannt, soll derselbe im Jahre 499 nach China zurückgekehrt sein und einen Bericht über das fabelhafte Land Fu-sang erstattet haben, der zu der sensationellen Hypothese einer chinesisch-buddhistischen Entdeckung Amerikas Veranlassung gegeben hat (s. zuletzt T'oung Pao 1915, p. 198). Huei Schên war nichts anderes als ein literarischer Betrüger, wie sie zu jener Zeit in China blühten, der längst bestehende Traditionen zusammenschweißte und durch nicht einmal sehr geistreiche Umformungen und Umdeutungen einen ethnographischen Roman zusammenbraute, den nur mit gleicher Phantasie begabte Geister für bare Münze hinnehmen konnten. Der Text des Nan schi lautet folgendermaßen: „Über tausend Meilen (li) östlich von Fu-sang liegt das Frauenreich. Diese Weiber haben regelmäßige Körperformen und eine weiße Hautfarbe, doch ihr ganzer Körper ist mit Haaren bewachsen, und ihre Kopfhaare sind so lang, daß sie auf die Erde herabfallen. Im zweiten oder dritten Monat baden sie in einem Flusse und werden davon schwanger; im sechsten oder siebenten Monat gebären sie Kinder. Diese Weiber haben keine Brustwarzen, sondern Haare hinten am Nacken; die weißen Haare enthalten eine Flüssigkeit, mit der sie ihre Kinder säugen. Hundert Tage nach der Geburt können die Kinder gehen, so daß sie im

dritten oder vierten Jahre erwachsen sein dürften. Beim Anblick
von Männern ergreifen sie entsetzt die Flucht, denn sie verab-
scheuen jeden männlichen Verkehr. Wie wilde Tiere leben sie
von salzhaltigen Pflanzen (Genus Salsola), deren Blätter denen
der Pflanze *sie hao* (Seseli libanotis; G. A. Stuart, Chinese Materia
Medica p. 405) gleichen, die angenehm duften, aber salzig von
Geschmack sind. Im sechsten Jahre der Periode Tʻien-kien (507
n. Chr.) der Liang-Dynastie war ein Mann von Tsin-ngan (in
Tsʻüan-chou, Provinz Fu-kien), der das Meer durchfuhr und von
einem Sturm auf eine Insel verschlagen wurde. Als er seinen
Fuß aus Land setzte, traf er auf die Bewohner, unter denen
Frauen und Männer sich scharf unterschieden: nur die Frauen
glichen denen von China, redeten aber eine unverständliche Sprache.
Die Männer besaßen einen menschlichen Körper, dagegen den
Kopf eines Hundes, und ihre Stimme tönte wie Hundegebell.
Kleine Bohnen bildeten ihre Nahrung, und ihre Kleidung schien
aus Zeug gemacht zu sein. Aus gestampfter Erde waren die
Wände ihrer Wohnungen verfertigt, die in Kreisform erbaut waren
mit höhlenartigem Eingang." Hier haben wir also dieselbe klare
Scheidung zwischen menschlich geformten Frauen und hundeköpfigen Männern innerhalb einer sozialen Gemeinschaft wie bei
Adam von Bremen. Diese auffallende Übereinstimmung kann
nicht auf Rechnung eines blinden Ungefährs gesetzt, sondern sie
kann nur so erklärt werden, daß der europäische und chinesische
Chronist aus einer gemeinsamen Quelle geschöpft haben, und dies
wird eine aus dem Alexanderroman geflossene oder damit in Verbindung stehende hellenistisch-orientalische Quelle gewesen sein.

In dieser Form ist die Sage den Chinesen bis ins dreizehnte
Jahrhundert lebendig geblieben, denn der armenische König
Haithon erzählt, daß es jenseits von Khatai ein Land gebe, wo
die Frauen menschliche Form haben und mit Vernunft begabt
sind, wo dagegen die Männer die Gestalt von Hunden haben,
vernunftlos, groß und behaart sind. Diese Hunde erlauben
niemandem, ihr Gebiet zu betreten; sie liegen der Jagd ob und
nähren sich wie die Frauen von dem erbeuteten Wilde (vgl.
Ktesias). Die Knaben, die aus der Verbindung dieser Hunde mit
den Frauen entstehen, gleichen Hunden, die Mädchen aber Frauen [1]:

---

[1] Vgl. Klaproth, Aperçu des entreprises des Mongols en Géorgie et
en Arménie (Nouveau J. A., Vol. XII, 1833, p. 287, wo Klaproth auf die-

Daß das Volk der Hundsköpfe auch der lamaischen Geschichtschreibung bekannt ist, ersehen wir aus Huths Geschichte des
Buddhismus in der Mongolei (II, p. 33). Hier werden sie mit den
Amazonen Tsug-te und den Wundermenschen mit einem Auge auf
der Brust aufgezählt. Die letzteren stammen gleichfalls aus dem
Alexanderroman: „Von da kamen wir an einen Ort, wo Menschen
ohne Kopf waren. Sie hatten Augen und Mund auf der Brust
und sprachen nach Menschenart, waren zottig, in Felle gekleidet,
Fischesser" (A. Ausfeld, Der griechische Alexanderroman p. 83).

Im dritten Jahrhundert v. Chr., unter der Regierung des
Kaisers Ts'in Schi, lesen wir zum ersten Male in den chinesischen
Annalen von Drei Inseln der Seligen, P'êng-lai, Fang-chang und
Ying-chou, im östlichen Ozean als dem Aufenthalt der Unsterb-
lichen, wo Vierfüßler und Vögel weiß, Paläste und Tore aus Gold
und Silber sein und vor allem ein Kraut wachsen sollte, das den
Tod verhindert und ewiges Leben verleiht. Unter den Han-
Kaisern, besonders unter Wu (140—85 v. Chr.), als die Alchemisten
am Kaiserhofe großen Einfluß erlangten, nahm der Glaube an
die Wirklichkeit dieser Inseln festere Formen an und rang sich
zu einer kunstvollen Darstellung in der den Toten gewidmeten
Keramik durch (Laufer, Chinese Pottery p. 191 ff.). In der legenden-
haften taoistischen Literatur werden die Wunder dieser Inseln in
breiter Weitschweifigkeit verherrlicht. Schlegel hat auch diese
sagenhaften Inseln unter dem Titel 'Les trois îles enchantées'
(T'oung Pao, VI, 1895, p. 1—64) vom geographischen Standpunkt
beleuchtet und alles Wunderbare nach rationalistischer Methode
fein säuberlich gedeutet; unter andern Wundern wird uns hier
der Narwal, der nur in den hohen arktischen Meeren lebt, als
ein den Chinesen wohl bekanntes Geschöpf aufgetischt (p. 23 f.).
Man braucht aber nur E. Rohdes geistreiche Darstellung der
hellenischen Vorstellungen von den Inseln der Seligen (Der
griechische Roman 3, p. 213 ff.) zu lesen, besonders die „Heilige

selbe Sage bei Plano Carpini hinweist und den rezenten Text des San ts'ai
tu hui über das Land der Hunde übersetzt) und E. Dulaurier, Les Mongols
d'après les historiens arméniens (J. A. 1858, p. 472). In seinen kurzen
Notizen zu Marco Polos hundeköpfigen Andamanern verwechselt Yule (Marco
Polo II, p. 312) zwei ganz verschiedene Dinge, hundeköpfige Menschen und
Stammessagen von einem Hundevorfahren. Vgl. auch Rockhill, Journey of
Rubruk p. 36.

Urkunde" des Euhemerus, in dessen Gruppe glücklicher Inseln
sich drei besonders auszeichneten, um zu der Überzeugung zu
gelangen, daß der chinesische Gedankenkreis nichts anderes als
ein Reflex griechischer Ideen ist, die durch den hellenistischen
Orient und Zentralasien vermittelt wurden. Dem hellenischen
Vorbild getreu haben die Chinesen die Wunderinseln und Wundervölker in den östlichen Ozean gelegt, nur daß naturgemäß für
sie, was den Hellenen der Indische Ozean war, der Stille Ozean
werden mußte. Diese geographische Verschiebung aber, die gleichzeitig mit der Ausdehnung chinesischer Macht im Pazifischen
Meere in Verbindung steht, darf uns nicht verhindern an der Er-
kenntnis der wirklichen historischen Zusammenhänge, die den
Hellenismus mit dem Chinesentum verknüpfen. Die reale Grundlage für diese Untersuchungen liefern uns die auf das Land Ta
Ts'in, den hellenistischen Orient, bezüglichen Texte der chinesischen Annalen: sie zeigen uns vor allem, daß mit den Handels-
artikeln des vorderen Orients und Indiens auch hellenistisch-
orientalische Sage nach China eindrang und daß insbesondere die
Kenntnis der Edelsteine und ihrer Eigenschaften von dort ein-
geführtes Gut ist. Die chinesische Mineralogie (von einer solchen
kann nur in beschränktem Maße die Rede sein), Alchemie, Astronomie und Astrologie (wie insbesondere F. Boll gezeigt hat)
wurzeln im Hellenismus; ebenso aber zahlreiche Fabeln, Wunder-
geschichten und religiöse Vorstellungen, wie Verfasser in einer
Reihe von Einzelstudien zu zeigen hofft. Die gesamte frühe
taoistische Literatur mit ihren Mirakeln, Zaubern, Naturwundern,
seltsamen Menschen und Monstern ist stark vom Hellenismus
angehaucht: nicht nur äußerlich, sondern auch innerlich ist sie
der alexandrinischen Literatur der Wunderbücher, den θαυμάσια,
παράδοξα und ἰδιοφυῆ eng verwandt (F. Susemihl, Geschichte
der griechischen Literatur in der Alexandrinerzeit I, p. 463 ff.).
Auf dem Gebiete der Mechanik und technischen Erfindungen läßt
sich ein gleiches Maß hellenistischer Einflüsse nachweisen: die
Glasur der Tongefäße, die allmählich zur Herstellung des Porzellans
führte und die dafür notwendige Vorbedingung schuf, ist aus dem
vorderen Orient nach China gelangt, ebenso das Glas. Die
Mechanik und Ingenieurkunst der Han-Zeit steht völlig im Bann
der großen Meister von Alexandria.

126

西夏语言：一项印–汉文字学研究

T'OUNG PAO

通報

ou

ARCHIVES

CONCERNANT L'HISTOIRE, LES LANGUES,
LA GÉOGRAPHIE ET L'ETHNOGRAPHIE
DE
L'ASIE ORIENTALE

Revue dirigée par

Henri CORDIER
Membre de l'Institut
Professeur à l'Ecole spéciale des Langues orientales vivantes
ET
Edouard CHAVANNES
Membre de l'Institut, Professeur au Collège de France.

VOL. XVII.

LIBRAIRIE ET IMPRIMERIE
CI-DEVANT
E. J. BRILL
LEIDE — 1916.

THE SI-HIA LANGUAGE,

A STUDY IN INDO-CHINESE PHILOLOGY.

BY

BERTHOLD LAUFER.

————••⋅⟨⊃Ӡ⟩⋅••————

CONTENTS.

1

INTRODUCTORY.

"After you have travelled thirty days through the Desert, as I have described, you come to a city called Sachiu, lying between north-east and east; it belongs to the Great Kaan, and is in a province called Tangut. The people are for the most part Idolaters, but there are also some Nestorian Christians and some Saracens. The Idolaters have a peculiar language, and are no traders, but live by their agriculture." Thus MARCO POLO opens his chapter on the Great Province of Tangut.[1] According to YULE, the "peculiar language," as Neumann supposes, may have been Tibetan; while, on the other hand, he was inclined to think that the language intended by Polo may have been a Chinese dialect. At present we might confidently say that Polo certainly visualized the Si-hia language, and, judging from the fragments of it now at our disposal, and to be discussed on the pages to follow, that he was quite right in styling the language "peculiar." It is in fact a peculiar language, not, however, plainly Tibetan, as believed by Neumann and several successors,[2] but an independent and peculiar idiom within the great family of Tibeto-Burman languages, fundamentally evincing decided affinities with the Lo-lo and Mo-so group.

In designating the language as Si-hia I follow the established convention, although it is not logical to speak of a Si-hia language or people, as Si Hia 西夏 is a purely Chinese term pertaining to the dynasty of the kingdom. The people were descendants of

---

[1] *The Book of Ser Marco Polo*, ed. of YULE and CORDIER, Vol. I, p. 203.

[2] DEVÉRIA (*L'Ecriture du royaume de Si-hia ou Tangut*, p. 28 of the reprint from *Mémoires présentés par divers savants à l'Académie des Inscriptions*, first series, Vol. XI, pt. 1, Paris, 1898) designated the language as possibly being a Tibetan dialect. M. G. MORISSE (*Contribution préliminaire à l'étude de l'écriture et de la langue Si-hia*, p. 37, reprint from *Mém. prés. par divers savants à l'Académie des Inscriptions*, first series, Vol. XI, pt. 2, Paris, 1904, pp. 313—379) concluded from certain analogies of syntax that "la langue tangoutaine était apparentée au tibétain."

the Tang-hiang 党項, in particular of the *gens* Tᶜo-pa 拓拔, forming one of the eight *gentes* of that tribe.[1] They were accordingly members of the Tibetan family. The word *Tang* seems to have resulted in the Mongol name *Tangut* (-*ud* being the ending of the plural), employed by Marco Polo and Rashid-eddīn.[2] The term *Tangut*, however, antedates the Mongol epoch, for, as already observed by DEVÉRIA,[3] it is found as early as the year 734 in the Sino-Turkish inscription of Kosho-Tsaidam.[4]

A considerable literature must have been extant in the Si-hia language. Chinese books were eagerly sought in the kingdom, and were not only brought there from China proper, but were also printed in the country itself, sometimes even at the initiative of the rulers of the Si-hia dynasty.[5] Confucian literature was cultivated under the reign of Li Jên-hiao (1140—93), when schools were erected after the Chinese model in all departmental and district towns, swelling the number of students to three thousand. All children of the royal family, from their seventh to their fifteenth year, were obliged to frequent a college installed in the palace. The king and the queen did not think it beneath their dignity to impart instruction there. In 1145 a learned academy was founded after the Chinese model. In 1150 Wa Tao-chung was appointed professor of Chinese and Si-hia literatures; he translated the Analects of Confucius, and provided them with a commentary in thirty books.

---

[1] See the interesting observations of E. CHAVANNES, *Dix inscriptions chinoises de l'Asie centrale*, p. 13 (reprint from *Mém. prés. par divers savants*, etc., first series, Vol. XI, pt. 2, 1902, pp. 193—295).

[2] See an interesting note on Tangut, in W. W. ROCKHILL, *Land of the Lamas*, p. 73; and *J.R A.S.*, 1891, p. 189.

[3] *L'Ecriture du royaume Si-hia ou Tangut*, p. 21.

[4] Compare W. RADLOFF, *Alttürkische Inschriften der Mongolei*, I, p. 58; and V. THOMSEN, *Inscriptions de l'Orkhon*, pp. 123, 178.

[5] P. PELLIOT, *Les documents chinois par la mission Kozlov* (*Journal asiatique*, 1914, mai-juin, p. 504).

Besides, he wrote a treatise on divination. His works, composed in the national system of writing, were propagated in the kingdom. At the time when it was overthrown by the Mongols (1226), Chinese literature was in a flourishing condition, and numerous books were carried away with the booty of the conquerors.[1] In 1154 Li Jên-hiao requested permission from the Kin to purchase from them books relating not only to Confucius, but also to Buddha. The Buddhist Tripiṭaka has been entirely translated into the Si-hia language, and an edition of this translation was prepared in print in 1294.[2] The interest in Buddhism awoke under Li Yüan-hao (1032—48), who procured a collection of Buddhist works from the Chinese Court in exchange for fifty horses. He constructed a monastery and high Stūpas east of his capital, where the sacred canon was deposited. Uigur monks were called to this establishment to study and interpret these texts and to render them into the Si-hia language. The Uigur, a Turkish tribe bordering the Si-hia Kingdom on the west, furnished the missionaries and translators of Buddhism to the Si-hia in the same manner as at a later date to the Mongols. The Saddharma-puṇḍarīka-sūtra translated into Si-hia has been made the object of an ingenious study by M. G. Morisse,[3] who availed himself of a manuscript (incomplete, in three volumes), written in gold on stiff blue-black paper (a style familiar also in Tibet). The Si-hia rulers likewise had annals kept in their language and writing, but no historical works seem to have as yet come to the fore.

The Si-hia system of writing was constructed in 1037 by Li Yüan-hao. It is perhaps the most complicated system ever invented by a human mind,—ideographic, like Chinese, the single characters being

---

[1] G. Devéria, *L'Ecriture du royaume de Si-hia*, pp. 23—26.

[2] Pelliot, *l. c.*, p. 518.

[3] *Contribution préliminaire à l'étude de l'écriture et de la langue Si-hia (l. c.).*

composed of a bewildering mass of irregular lines, in which no method or principle has as yet been disclosed. Only a small fraction of characters has been read; of others, the meaning only is known, but not the phonetic value. Devéria is presumably right in tracing the origin of Si-hia writing to that of the Khitan, which, on its part, was derived from Chinese script in 920. Apart from books, manuscripts, and scrolls, we have but few monuments of Si-hia culture in the form of lapidary inscriptions, coins, and seals. In the celebrated inscription of Kü-yung-kuan (in the Nan-kʻou Pass, on the road from Peking to Kalgan), dated 1345, four sections are composed in Si-hia, two remaining undeciphered; the two others present the literal phonetic transcription in Si-hia characters of Sanskrit Dhāraṇī. The majority of the latter characters have been identified by A. WYLIE [1] and E. CHAVANNES. [2] It is to the merit of G. DEVÉRIA to have established the fact beyond doubt, that the inscriptions in question were really of Si-hia writing, through comparison with an inscription of 1094 erected in the Temple of the Great Cloud at Liang-chou in Kan-su Province, and studied by him in the publication above cited. S. W. BUSHELL [3] examined twelve Si-hia coins, and succeeded in deciphering the sense of some forty characters, without being able, however, to assign to them their phonetic value. Reproductions of some Si-hia seals have been published in the Chinese journal *Shên chou kuo kuang tsi*, No. 10. King Li Yüan-hao received from the Chinese Court a plated silver seal on which were engraved four characters, reading, "seal of the lord of the country of Hia" 夏國主印.[4]

---

[1] *On an Ancient Buddhist Inscription at Keu-yung kwan* (*J.R.A.S.*, N. S, Vol. V, 1870, pp. 14—44).

[2] *Documents de l'époque mongole, édités par le prince Roland Bonaparte*, Plate X.

[3] *The Hsi Hsia Dynasty of Tangut, their Money and Peculiar Script* (*J. China Branch R. A. S.*, Vol. XXX, 1895—96, pp. 142—160).

[4] *Sung shi*, Ch. 485, p. 8 b.

The only available material for a study of the Si-hia language is a brief manual, unfortunately incomplete, found, together with a large number of other Si-hia books, in 1908, by Gen. Kozlov, in a Stūpa not far from Karakhoto,[1] and published by Professor A. IVANOV.[2] This booklet was prepared by an author of Si-hia descent, Ku-le or Ku-lo (Ku-ro, Gu-ro) 骨勒, with the intention of facilitating for his compatriots the study of the Chinese language. The work, written in 1190,[3] consists of a vocabulary, arranged according to subject-matter, in Chinese fashion, — heaven, earth, man, body, costume, implements, fauna, flora, minerals, nutrition, abstract nouns, measurements, adjectives, verbs, and numerals. Each word is given in Si-hia script (third column); it is accompanied by the pronunciation of the corresponding Chinese word in Si-hia characters (first column), by the Chinese rendering (second column), and by Chinese characters indicating the reading of the Si-hia word (fourth column). Professor Ivanov, to whom we are under great obligation for having rendered this important material accessible, has added a few Chinese, Mongol, and Tibetan words for purposes of comparison, but the real analysis of the language remains to be made. Besides the vocabulary in question, I availed myself of a number of Si-hia words recorded in the Chinese Annals (Nos. 46, 138—142).

As regards the transcription of the Chinese characters reproducing the Si-hia words, Mr. Ivanov is generally correct in what he remarks on this point (p. 1225). I think, however, that Northern Mandarin should be preferred to the Peking dialect. Thus

---

[1] The ruined town of Karakhoto, first visited and explored by Gen. Kozlov in 1908—09, was investigated again by A. Stein on his recent expedition to Central Asia, with good results: abundant remains of Buddhist manuscripts and prints, both in Tangut script and Tibetan, and miscellaneous records on paper in Chinese, Tangut, and Uigur, were brought to light by him (see his report in *The Geographical Journal*, 1915, pp. 408, 409).

[2] *Zur Kenntnis der Hsi-hsia Sprache* (*Bull. de l'Acad. imp. des sciences de St.-Pétersbourg*, 1909, pp. 1221—1233).

[3] The correct date has been established by CHAVANNES (*T'oung Pao*, 1910, p. 151).

Si-hia 七吾 ("salt") doubtless is *ts'i-ñu*, not *ch'i-wu*, and Si-hia 皆 ("gold") is *kie*, *kiä*, *k'ä*, not *chieh*. A point chiefly to be noted is that modern Chinese, lacking the sonants, resorts to the corresponding surds in transcribing the sonants of foreign words; while there is every reason to believe that Si-hia, like its congeners, possessed the sonants *g, b, d*. Only close comparison with cognate forms of speech can teach us whether the sonants or surds have to be adopted for Si-hia; and in many cases, this remains doubtful. For example, 芭不 ("daddy") is not to be rendered *pa-pu*, but *pa-bu*, the element *bu* being the same as the diminutive ending *bu* of Tibetan; 則移 ("water"), as shown by linguistic comparison, is *dzei, jei*, not *tse-yi*. Whether *l* or *r* is to be read for Chinese *l*, can be decided solely by comparative methods; even then the decision is not always easy, as *l* and *r* are to a great extent interchanged in Tibeto-Burman. In a single case, Chinese *l* corresponds to Si-hia *n*: 羅 *lo* ("disease"), being identical with A-hi Lo-lo *no*, Nyi Lo-lo and Tibetan *na*, should be read as Si-hia *no*.

The comparative material brought together in elucidation of the Si-hia onomasticon has been arranged on as broad a basis as possible, covering, so far as this can be done at present, the Indo-Chinese languages in their entire range. It is thus hoped that this essay will be interesting, not only to students of Si-hia and Tibetan, but also to those engaged in the pursuit of other Indo-Chinese branches and to philologists in general. It will be seen from the following pages that the Si-hia language is more than an isolated phenomenon of local interest, and that it even has a fundamental value for the understanding of the speech history of the Indo-Chinese family, and in many cases largely contributes to a correct appreciation of its genetic growth. The ethnologist and historian of culture may also find here some new facts of importance, particularly with reference to the utilization of metals and cultivated plants. It was not feasible, in each and every case, to state the source from which a word has been derived; only in specific cases acknowledgment has been rendered, author's names have been cited in parenthesis wherever it seemed that responsibility must solely rest with them. In general, only well-authenticated material has been adopted, each item being drawn and verified from first-hand sources. It is chiefly to the following works that I am under obligation. My knowledge of the Lo-lo languages is based on the fundamental studies of Fathers P. VIAL (*Les Lolos*, Shanghai, 1898; and *Dictionnaire français-lolo*, Hongkong, 1909) and A. LIÉTARD (*Notions de grammaire lo-lo, Bull. de l'Ecole française*, Vol. IX, 1909, pp. 285—314; *Notes sur les dialectes lo-lo, ibid.*, pp. 549—572, an achievement of principal value on account of the copious comparative vocabulary; *Notions de grammaire lo-lo, Toung Pao*, 1911, pp. 627—662; *Essai de dictionnaire lo-lo français, dialecte A-hi, ibid.*, pp. 1—37, 123—156, 316—346, 544—558; *Vocabulaire français lo-lo, dialecte A-hi, Toung Pao*, 1912, pp. 1—42; *Au Yun-nan, les Lo-lo-pᶜo, Bibliothèque Anthropos*, Vol. I, No. 5, Münster, 1913; this work contains a solid grammar of the Lo-lo-pᶜo dialect, also several texts with translation). Vial and Liétard are the founders of Lo-lo philology, and their phonetic reproduction of the language may be regarded as fairly exact; at any rate, it is infinitely superior to what is offered us in the lists of words haphazardly gathered by casual travellers. I feel the more grateful toward the work of the two fathers, as the present investigation could not have been made but for their efforts, one of my main results being the proof of a close affinity between Si-hia and Lo-lo. The materials of BONIFACY (*Etude sur les langues parlées par les populations de la haute Rivière Claire, Bull. de l'Ecole française*, Vol. V, 1905, pp. 306—327; and *Etude sur les coutumes et la langue des Lo-lo et des La-qua du Haut Tonkin, ibid.*, Vol. VIII, 1908, pp. 531—558) have also proved useful. Occasional references have been made to the Lo-lo collectanea of E. C. BABER (*Travels and Researches in Western China*, *Roy. Geogr. Soc. Suppl. Papers*, Vol. I, 1886, pp. 73—78), and H. R. DAVIES (*Yün-nan, the Link between India and the Yangtze*, Cambridge, 1909), but only in cases where they offer material not listed by Vial and Liétard. Next to Lo-lo, the language of the Mo-so has proved most important for my purpose. The interesting work of J. BACOT (*Les Mo-so, ethnographie des Mo-so, leurs religions, leur langue et leur écriture*, Leide, 1913) has furnished most of the Mo-so material (with occasional quotations from the work of Davies). M. Bacot's transcription of Mo-so is not made after rigid phonetic principles, but according to the French alphabet. The comparative method, however, brings out the fact that his mode of writing is generally correct. It has, of course, been reduced to the phonetic system adopted by me. It will be noticed that the phonology and vocabulary of Mo-so are closely allied to Lo-lo, and hence to Si-hia, and that even in some cases Mo-so furnishes the sole clew to a Si-hia word. Thus new light falls on the Mo-so language itself. These results, together with other observations on the language, I hope to set forth in the near future. It hardly requires mention that ample use has been made of that admirable linguistic encyclopædia, the *Linguistic Survey of India*, edited by G. A. GRIERSON, or that B. H. HODGSON's *Essays on the Languages, Literature, and Religion of Nepal and Tibet* (London, 1874), have been laid under contribution for the languages of Nepal, or the Gazetteer of Upper Burma and the Shan States for the Karen dialects. It is regrettable that since Hodgson's day our knowledge of Nepalese idioms has so little advanced; this immense field merits a most serious exploration. Newārī remains the only language of this group of which we possess a somewhat superficial knowledge, thanks to the pioneer labors of A. CONRADY (*Das Newārī, Grammatik und Sprach-*

proben, *Z.D.M.G.*, Vol. XLV, 1891, pp. 1—35; and *Ein Sanskrit-Newārī Wörterbuch*, *ibid.*, Vol. XLVII, 1893, pp. 539—573). My acquaintance with Lepcha is founded on G. B. MAINWARING, *A Grammar of the Rong (Lepcha) Language* (Calcutta, 1875), and *Dictionary of the Lepcha Language*, compiled by G. B. MAINWARING, revised and completed by A. GRÜNWEDEL (Berlin, 1898). The majority of Bunan words are derived from the article of H. A. JÄSCHKE (*Note on the Pronunciation of the Tibetan Language*, *J.A.S.B.*, Vol. XXXIV, pt. 1, 1865, pp. 91—100); an exhaustive study of this interesting language would be a primary desideratum and of greatest service to Indo-Chinese philology. Kanaurī or Kanāwarī words are taken from Pandit T. R. JOSHI, *Grammar and Dictionary of Kanāwarī, the Language of Kanāwar* (Calcutta, 1909), and from T. G. BAILEY, *Kanaurī Vocabulary* (*J.R.A.S.*, 1911, pp. 315—364). The sources of Tibetan do not need specification. The Tromowa and Sikkim dialects are quoted from the important paper of E. H. C. WALSH (*A Vocabulary of the Tromowa Dialect of Tibetan spoken in the Chumbi Valley*, Calcutta, 1905), partially also from G. SANDBERG (*Manual of the Sikkim Bhutia Language*, Westminster, 1895). Materials relating to the East-Tibetan dialects, Jyaruṅ and Gešitsᶜa, are derived from the writer's own collectanea made in the field. Words of the Tibetan dialect of Yün-nan are partially due to Father TH. MONBEIG (*Bull. de l'Ecole française*, Vol. IX, 1909, pp. 550—556), partially to the book of Davies above cited. As regards the Tᶜai languages, I hold myself under greatest obligation to the work of H. MASPERO (*Contribution à l'étude du système phonétique des langues Thai*, *Bull. de l'Ecole française*, Vol. XI, 1911, pp. 153—169); from many another study of this eminent philologist I have derived as much instruction as pleasure (compare *J.R.A.S.*, 1915, p. 757). As to Ahom, I am indebted to the excellent vocabulary of G. A. GRIERSON (*J.R.A.S.*, 1904, pp. 203—232; and *Z.D.M.G.*, Vol. LVI,

1902, pp. 1—59). Pa-yi words are culled from F. W. K. MÜLLER's study of this language (*T'oung Pao*, Vol. III, 1892, pp. 1—38). Other material relative to the T'ai languages of Yün-nan is drawn from DAVIES (*l. c.*), G. W. CLARK (*Kwiechow and Yün-nan Provinces*, Shanghai, 1894), S. R. CLARKE (*Among the Tribes in South-West China*, London, 1911), and in some instances from T. DE LACOUPERIE (*Les langues de la Chine avant les Chinois*, Paris, 1898, used with reserve; this also holds good for D'OLLONE, *Langues des peuples non chinois de la Chine*, Paris, 1912). I am further indebted to B. HOUGH-TON for his *Kami Vocabularies* (*J.R.A.S.*, 1895, pp. 111—138), his *Southern Chin Vocabulary* (*ibid.*, pp. 727—737), and his *Tibeto-Burman Linguistic Palæontology* (*ibid.*, 1896, pp. 23—55), a first attempt of culture-historical and philological study in the Indo-Chinese field, to which I owe a great deal of inspiration and encouragement; to T. C. HODSON (*Thādo Grammar*, Shillong, 1905; and other monographs quoted in the proper place); and to STEN KONOW (*Zur Kenntnis der Kuki-Chin Sprachen*, *Z.D.M.G.*, Vol. LVI, 1902, pp. 486—517). Kachin is drawn upon from H. F. HERTZ, *Handbook of the Kachin or Chingpaw Language* (Rangoon, 1895; 2d ed., 1902). It is a matter of course that the standard works of A. JUDSON, PALLEGOIX, McFARLAND, etc., have been consulted for Burmese and Siamese, and that the synthetical studies of E. KUHN, A. CONRADY, H. MAS-PERO, and others, have been duly appreciated. To these and other contributions special acknowledgment is given in every instance.

THE NUMERALS.

In the vocabulary published by Mr. Ivanov we find only the numerals for 1, 2, 4, 5, 8, 10, 100, and 10,000. These numerals, while in general reflecting Tibeto-Burman types in the widest sense of the word, reveal far-reaching deviations from Tibetan proper, and show that Si-hia is not a Tibetan dialect, but is an independent idiom, which takes its place beside Tibetan.

Si-hia 1, *a* 阿, deviates from the standard Tibetan form * či-k* (written *gčig*) and *ti-k* (Jyaruṅ *k-ti*, *k-ti-k*; Bunan *ti*; Lo-lo *tᶜi*; Burmese *taϑ*, spoken language *tīt-ta*; hence *či-k* seems to be evolved from *t-ši-k, which is confirmed by Lepcha *ka-t*, Chinese *yi-t*, Milčaṅ and Chuṅ-kia *i-t*; Tibetan *g-či-g*, therefore evolved from **ga-* or *ka-t-ši-k).[1] The only western Indo-Chinese language known to me, in which the element *a* appears as the numeral 1, is Aka in Assam.[2] On the one hand, however, the habitat of Aka is so far removed from that of Si-hia, with no missing links discoverable between the two, that a correlation of the two forms, at least for the present, would seem a hopeless venture. On the other hand, Mr. T. C. HODSON, to whom we are indebted for a very important and ingenious study of the numeral systems of the Tibeto-Burman languages, inclines toward the belief that Aka *a* is the survival of fuller forms, like *aka, akhet, ekhū, ākū, hak*, encountered in other Assam languages; and traced by him to Indo-Aryan *eka*.[3] Moreover, Mr. Hodson makes the very appropriate observation that "the freaks and fantasies of phonetic growth and decay in this area are such that seemingly identical forms may be evolved out of totally distinct original forms."[4] Whatever the origin of Aka *a* may be, therefore, its ancestry might be entirely distinct from that of Si-hia *a*. It may be suggested that the latter is possibly related to Tibetan *ya* ("one of two things forming a pair, one of two op-

---

[1] It has not yet been pointed out that Tibetan *čiy* is related to Cantonese *čık, ček*, Hakka *čak*, North Chinese *či* 隻 ("single").

[2] A numeral *a* ("one") appears in Miao and Yao of Yün-nan (H. R. DAVIES, *Yun-nan*, p. 339), classified by Davies in the Mon-Khmêr family. In Mon-Khmêr, however, the typical form of the numeral 1 is *mui*.

[3] *J.R.A.S*, 1913, p. 320. According to STEN KONOW (*J.R.A.S.*, 1902, p. 129), who writes *ū*, Aka *a* corresponds to Meithei *a-mū*, Kachin *ai-mū*.

[4] *L. c.*, p. 335.

ponents"), and possibly to Chepaṅ *ya* ("one").[1] The reason why Si-hia rejected the common type may be explained by another suggestive opinion of Mr. Hodson, "There can be no doubt that the use of the numeral for 'one' as an affix to indicate singularity has led to the disappearance of the original numeral in several cases and the employment of some word of different origin in its place." [2] Unfortunately, it is not known what the suffix of singularity was in Si-hia; but some such factor, as indicated by Mr. Hodson, seems to have been at play in that language.

Si-hia 2, *nŏṅ*, *nõ̃*, *noṅ* (or perhaps *loṅ*) 能 , is different from the common Tibetan type *ñi*, *ñis* (*g-ñis*, from *ga-* or *ka-ñi-s*; Lepcha *ñat*; Lo-lo and Mo-so *ñi*). The nearest forms of relationship that I am able to trace are Chinese *liaṅ* (Cantonese *lŏṅ*) 兩 ("both, a pair") and Chuṅ-kia *s-loṅ* (the prefix *s-* presumably being the survival of *sam* of the T‘ai group). Perhaps also Hai-nan and Swatow *no* belong to this series. Further relationship may exist with the base *niṅ* of Miju Mišmi, Laluṅ, and Garo Jalpaiguri.[3]

Si-hia 4, *le* 勒 , represents the typical base of Tibeto-Burman *le*, *li*, *ri*, *lu*: Burmese *le*, Kachin *ma-li*, Haka *pa-li*, Tauṅtha *p-li*; Kami *ma-li*, *m-le*; Lepcha *fa-li* (from *ba-li*), Lušai *pa-li*; Jyaruṅ *k-b-li* (from *ka-ba-li*); Miao-tse *p-lu*, Mo-so *lu*, Chuṅ-kia *s-le*, Lo-lo *š-le*; Tibetan *b-ži*, evolved from *ba-ži*, *ba-ši*, or *bi-ži*: Bunan *bi* ("four"), the base *ži*, *si*, being identical with Chinese *sẹ* 四 and T‘ai *si*.

Si-hia 5 is handed down to us in the transcription *ku-yü* 骨魚 . As the standard type of the numeral 5 in Tibeto-Burman is *ña*, *ñu*,

---

[1] B. H. HODGSON, *The Phœnix*, Vol. III, p. 46; or *Essays on the Languages, Literature, and Religion of Nepal and Tibet*, pt. 2, p. 52.

[2] *L. c*, p. 321.

[3] HODSON, *l. c.*, p. 322, who justly remarks that the closing *ṅ* may, on further investigation, prove to be only a nasalization. Hence I have added above also the transcription *nõ̃* (see Phonology, § 14).

ño, we have doubtless to assume the prevalence of the former initial *ñ* in *yü* 魚 , and to restore the Chinese transcription to Si-hia *k-ñü* or *k-ñu*. An analogous case of transcription occurs in *ku-yü-mo* 骨魚沒 ("heaven"), which in my opinion should be restored to Si-hia *k-ñum*, and be correlated with Tibetan *d-guñ* and *g-nam* (see No. 35). As to the prefix *k-*, Si-hia coincides with Jyaruṅ *k-mu*. The fact that this prefix is the remnant of an independent base with the meaning "five," we may glean from Lo-lo *gha*, Chuṅ-kia and Tᶜai languages *ha*; hence Si-hia *k-ñu* appears to have arisen from **ka-ñu*. The second element *ñu* is identical with Chinese *ñu* 五 , Lepcha *fa-ño*, while other languages have the vowel *a*: Burmese *ña*, Tibetan *l-ña*, Lušai *pa-ña*, Mo-so and Gešitsᶜa *wɹ*.

Si-hia 8, *ye* 刵 (possibly also *ya*), again agrees with the typical Tibeto-Burman series, particularly with Dhimal *ye*; Kami *ta-ya*, *te-ya*, or *ka-ya*; Lo-lo *e*; Gešitsᶜa *r-ya*, *vɑ-r-ya*; Jyaruṅ *v-r-yat* or *v-r-yä-t*; written Tibetan *b-r-g-ya-d* (now sounded *gyat*; *b-* evolved from **va*).

The Si-hia numeral 10 is mysterious. It is transcribed 奄 , which Mr. Ivauov renders *yen, am*; probably we have to read *an, en*. Whatever transcription we may choose, however, this form remains isolated, and defies identification with any type of Tibeto-Burman or any other Indo-Chinese group known to us. The consonantal basis for the numeral 10 is *s, š, č,* and *t*: Tibetan *b-čù* (from **ba-t-šu*), Jyaruṅ *š-či*, Lepcha *ka-ti*; Gešitsᶜa *z-ra* (from **si-ra*), and *s-ka* (from **si-ka*) in multiplication; Kami *ka-su*; Siamese *si-p*; Chinese **ši-p, ži-p* 十 .[1] Owing to the isolation of the Si-hia form, it is justifiable to regard it as a loan-word, perhaps from Turkish *on*.

Si-hia 100, *i* or *yi* 易 , corresponds to the Tibeto-Burman base *ya*, with the remarkable modification of *a* into *i* (see Phonology,

---

[1] For further information see HODSON, *l c*, pp 327–331.

§ 1): Burmese *ta-ya*, Karen *ta-ye*, other Burmese languages *a-ya*;[1] Jyaruṅ *b-ri-yä̇*, Gešits͑a *r-yo*, *r-ya*, Milčaṅ *ra*, Bunan *gya*, Lepcha *gyo*, Tibetan (written language) *b-r-g-ya*. Si-hia *i*, *yi*, is perhaps preserved in Siamese *roi*, if the latter should have been evolved from *ro-i*, *ro-yi*.

Si-hia 10,000, *k͑o* 刻, is an isolated form (Tibetan *k͑ri*, *t͑i*; Lo-lo *t͑i-va*; Mo-so *mö*; T͑ai group *mün*, *mōn*). It is notable that the word is identical with Si-hia *k͑o* ("foot"), which is reproduced by the same Chinese character (see No. 51). Hand and foot, as is well known, belong to the primitive means of counting; and the words for these bodily parts, as well as for fingers and toes, have left their imprint in the numerals of many peoples.

ANALYSIS OF THE SI-HIA ONOMASTICON.

GROUP I.

Words of Common Indo-Chinese Type.

In this group are treated words of general Indo-Chinese character, which bear the same or similar features, or a peculiar form in Si-hia.

1. Si-hia *ṅo* 我, silver. There is only one metal the designation of which is common to all Indo-Chinese, and that is silver. In the eastern group we encounter Amoy *gin*, Cantonese *ṅön*, Hakka *ṅyin*, Fukien *ṅüṅ*, Ningpo *ṅyiṅ*, Wen-chou *ṅiaṅ*; Siamese *ṅön*, Ahom and Shan *ṅün*, Black Tai *ṅön*, White Tai *ṅün*, Dioi *ṅan*;[2] Chuṅ-kia *ṅan*, Hei Miao *ṅi*, Hua Miao *ñai*, Ki-lao *ñin*. In the western group we have Burmese *ṅwe*; Tibetan (written language) *d-ṅul* (from *de-ṅul*, compare Mo-so *de-gu*), K͑ams *γ-ṅul* (Purig and Ladākhī *š-mul*; *š-* = Bunan *ši*, "white;" Ladākhī also *mul*), Jyaruṅ *po-ṅï'*, Central Tibetan *ṅü-l*, Sikkim and Lhoke *ṅü*, Yün-nan Tibe-

---

[1] *Gazetteer of Upper Burma and the Shan States*, Part I, Vol. I, p. 646.

[2] Compare MASPERO, *Bull. de l'Ecole française*, Vol. XI, p. 168.

tan *ṅo-l*; Mo-so *ṅö*; and corresponding to the last-named we have Si-hia *ṅo*, that accordingly stands closer to the western than to the eastern branch of Indo-Chinese.[1] The coincidence of Si-hia with Mo-so, in particular, is notable; and in all probability we have to recognize in *ṅo* the base of the word that, with the vocalic variations *ṅu*, *ṅü*, *ṅö*, assumes in Sino-Tᶜai a suffix -*n*, and in Tibetan a suffix -*l*. In some languages the initial guttural nasal has been transformed into the surd or sonant of the guttural series: Mo-so *de-gu* corresponds to Chö-ko *a-ko* and Lepcha *ko-m*. Tibetan, Bunan, and Kanaurī *mu-l*, Milčaṅ *mi-l* and *mu-l* (parallel to Mi-ñag [*Baber*] *mwe*), Murmi *mui*, and Lepcha *ko-m*, in all probability, represent an independent type, which has not arisen from *d-ṅul*, but which is formed from the base **mu*, *mi*, with suffix -*l*. This is demonstrated by two facts. First, the Lepcha word *ko-m* consists of the two elements *ko* + *m*, both of which cannot well have been evolved from the same root. Second, we may examine the Mongol-Tungusian languages that have adopted the Indo-Chinese word for "silver." In these, the element *mo*, *mö*, *me*, appears coupled with *ṅün*, *ṅö*: Mongol *möṅün*, Buryat *möṅö*, *möṅön*, *möṅün*; San-čᶜuan Tᶜu-jen (ROCKHILL, *Diary*, p. 377) *miengo*; Niüči *meṅguwen*, Manchu *meṅgun*, Tungusian *meṅün*, Gold *muṅgu*; Yakut *maṅuni* (Tungusian loanword, that occurs beside the Turkish word *kömüs*). The latter series forms a well-defined group, diverse from what is found in the Turkish and Finno-Ugrian languages: Uigur and Djagatai *kümüš*, Kirgiz *kümüs*, Yakut *kömüs*;[2] Suomi *hopea*, Esthonian *hōbe*, Votyak *opea*, Livonian *öbdi*, Chudian *hobet*; Magyar *ezüst*, Syrjän *ezyś*, Votyak *azveś* (probably "white copper;" -*veś* = Suomi *vaski*, "copper"). The Mongol-Tungusian loan-words which must be of ancient date

---

[1] In other Tibeto-Burman languages, "silver" is expressed by "white:" Nyi Lo lo *šla* ("white" and "silver"); Li-su *pᶜu* ("silver"), *pu-pu*, *i-pᶜu*, and *yu-pᶜu* ("white").

[2] RADLOFF, *Wörterbuch der Türk-Dialecte*, Vol. II, col. 1527, 1528.

hint also at the fact that in Northern Chinese, in the same manner as in the south, the word for "silver" might formerly have been *n̄ūn* or *n̄ön*, until it was turned into the modern form *yin*. [1] They further demonstrate that besides *n̄ün*, *n̄ön*, an element *mo, mö*, with the meaning "silver," appears to have existed within the northern or north-western range of Indo-Chinese, and the survivals of this ancient word we encounter in Tibetan and Kanaurī *mul*, Murmi *mui*, Lepcha *ko-m*.

2. Si-hia *k'ä, g'ä* [2] 皆, gold. B. HOUGHTON, in his very interesting and suggestive essay "Tibeto-Burman Linguistic Paleontology," justly says that there seems little doubt that gold was practically unknown to the Burmans and Tibetans before they divided. [3] Indeed, Burmese *hrwe* (spoken Burmese *šwe*) and Tibetan *gser* are distinct types. Moreover, the Indo-Chinese languages in general do not have a common word for "gold." Chinese practically has no word proper for it, but styles it "yellow metal," *huan kin* 黄金. The same meaning underlies the Tibetan term *g-ser*, which is derived from the adjective *ser* ("yellow"), and, as shown by the corresponding Mo-so word *ke-se*, [4] is to be analyzed into **ke-ser* (*ge-ser*). The element *ke*, probably related to Chinese **gim, kim* 金 (Cantonese *köm*, Fukien *kin̄*, Japanese *kin, kon*), [5] appears to have had in its origin the significance "metal." It is exactly the same word that we meet in Si-hia *k'ä, g'ä*, which, accordingly, mirrors

---

[1] As to the change of *n̄* to *y*, compare 雁 Fukien *n̄an̄*, Cantonese and Hakka *n̄an*, Tibetan *n̄an̄* ("wild goose"), Northern Chinese *yen*, Si-hia *ya*; 贋 Cantonese and Hakka *n̄an* ("false"), Tibetan *n̄an* ("bad"), Northern Chinese *yen*, Si-hia *wen*.

[2] The accent following a consonant is intended to express the palatalization of the latter.

[3] *J.R.A.S.*, 1896, p. 34.

[4] Thus given by J. BACOT, *Les Mo-so*, p. 46. H. R. DAVIES (*Yün-nan*, p. 360) imparts *ha* as Mo-so word for "gold;" this appears merely as a variant of *ke-, ge-*.

[5] Siamese *gam* (*k'am*), Ahom and Shan *k'am* (MASPERO, *Bull. de l'Ecole française*, Vol. XI, p. 168); Kachin *gyā*, "gold" (*Gazetteer of Upper Burma and the Shan States*, Part I, Vol. I, p. 662), in Karen dialects *kam* and *kyam* (*ibid.*, pp 650, 651).

2

the same state of affairs as Chinese *kim*, *kin*, inasmuch as the original meaning "metal" passed into that of "gold." In like manner we find in Miao-tse *ko* ("gold"). Si-hia *k'ā*, therefore, is preserved in the first element of Mo-so *ke-se* and Tibetan *g-ser*. The latter word is still articulated *γser* in Balti and K͏cams, but *ser* in all other dialects. Since this element *ser* has the same tone (the high tone) as the word *ser* ("yellow"), the Tibetans simply designate gold as "the yellow one." [1] The same condition is reached in Lo-lo: Nyi Lo-lo *še*, A-hi Lo-lo *ša*, Lisu *ši*, P͏cu-p͏ca *si*, Chö-ko *a-si* ("yellow" and "gold," akin to Tibetan *ser*); further, Ngačań *se*, Lisu *šö*, A-k͏ca *su*.

3. Si-hia *šań* 伺, iron. There is no general Indo-Chinese word, either, for "iron." The working of this metal became known in eastern Asia at a comparatively late period. Chinese *t͏cie* 鐵 (formerly *t͏cit*, *t͏cet*, *det*; Japanese *tetsu*) is apparently an ancient loan-word, somehow connected with the Old-Turkish forms: Uigur *tämūr*, Orkhon inscriptions *tümir*, in other dialects *täbir* or *timir*, Mongol *tümūr* (*tümür*). The Si-hia word *šań*, as suggested also by Mr. Ivanov, may be akin to Tibetan *l-čag-s* (Lepcha *ča*, Central Tibet *čak*, *ča*);

---

[1] It is curious that so eminent a philologist as JÄSCHKE (*Tibetan-English Dictionary*, p. 590) could connect Tibetan *γser* with Persian *zer*. Persian *zer* (Pahlavī *zar*, Kurd *zer*, *zir*, Afghan and Baluči *zar*) goes back to Avestan *zairi*, *zaranya* (Vedic *hari*, *hiraṇya*; compare P. HORN, *Neupersische Etymologie*, No. 554); and Persian *zerd* ("yellow"), to Avestan *zairi*, *zaray*. In order to make Tibetan *γser* (from *ke-ser*, *ge-ser*) dependent upon Persian *zer*, it would be required to prove also that Tibetan *ser* ("yellow") is traceable to Persian; this assumption, however, is disproved by the Lo-lo forms *ša*, *še*, *ši*, *si*, and Kachin *si-t*, which show that Tibetan *ser* is derived from a base *se*, followed by a suffix *-r*. The Lepcha words for "gold," *jer* and (in the legends of Padmasambhava) *zar*, are singular. Again, it seems to me that the coincidence of Lepcha *zar* with Pahlavī *zar* is accidental, for an influence of this language on Lepcha can hardly be assumed. In view of Kanauri *zań*, *za-ń* ("gold"), it seems to me conceivable that Lepcha *ze-r* and *je-r* are variations or perhaps older forms corresponding to Tibetan *zer*. Compare (under No. 199) Tibetan *zań-s* ("copper"), developed from *jań*, where Lepcha, Mo-so, and Yün-nan Tibetan have initial *s*.—Regarding gold as the yellow metal compare Phrygian γλουρός ("gold") = Greek χλωρός ("yellowish"); Gothic *gulþ*, Slavic *zlato*, Lettic *zelts*, from Indo-European *g̑helto-s* ("yellow"), etc.

Hor-pa *ču*. The alternation of the palatal surd *č* with the palatal sibilant *s* is not unusual in Tibeto-Burman languages, and, in the present case, is confirmed by Jyaruṅ *šo-m*, Mo so *šu*, *šo*, Bodo *šu-r*, and Manyak *ši* ("iron"). The final guttural nasal is met also in Raṅkas *čyaṅ*, Dārmiyā *ni-jaṅ*, Chaudangsī *na-jaṅ*. [1] There is an exceedingly large variety of words for "iron" in Indo-Chinese. Suffice it to note here that Nyi Lo-lo *re* (compare Ahom and Shan *lik*; Kachin *p^cri*; Kanaurī and Kanāshī *ron*), A-hi Lo-lo *hŏ*, *ho*, Lisu *ho* (*huo*), are unrelated to Si-hia and Tibetan. [2]

4. Si-hia *wu* 悟, cow = Shan *wo*, *wuw* (Karen *pu*), Ki-lao *wu*, Ahom *hu*, Khamti *ṅo*, Laos *ṅoa*, *ṅ-woa*, *ṅ-wau*; Chuṅ-kia *rai*, *ṅai*; Ya-č'io Miao *ṅuei*; Cantonese *ṅau*; Burmese *nwā* (spoken *nwau*); Thādo *bo-ṅ*; Tibetan *ba*. B. HOUGHTON (*J.R.A.S.*, 1896, p. 36) has combined Burmese *nwā* with Tibetan *nor* ("cattle"); this is partially correct, inasmuch as Tibetan *nor* possibly inheres in the prefix *n-*, the first part of the Burmese word, which has arisen from *no-wa,—*no* answering to Tibetan *nor*, and *wa* to Tibetan *ba*. Tibetan *nor* itself is *no* + *r*, *r* being an affix; the root *no* survives in Lo-lo-p'o *no-ñi* ("ox"), [3] in Southern Chin *no* and *no-n* ("cattle"), in Kami *pan-no* and *ma-na* ("buffalo"), and in Lisu *a-ṅa* ("buffalo"). In the Jyaruṅ dialect "cattle" is called *nuṅ-wa*, *nu-ṅ* being the equivalent of *no-r*, the guttural nasal being substituted for *r*.

5. Si-hia *žu* 汝, fish = Mi-ñag *zö*; Tibetan *ña*; Lepcha *ṅo*; Newārī *ṅā*; Lo-lo-p'o *ṅo*, P'u-p'a *ṅa*, A-hi Lo-lo *ṅo-zo* and *a-ṅo*, Nyi Lo-lo *gha*, Hua Lisu *ṅ-wa* (Black and White Lisu *wa*); Mo-so *ṅi*; Chinese *ṅi*, *yü* 魚; Burmese and Kachin *ṅa*. In its phonetic formation the Si-hia word is isolated, and seems to answer only to

---

[1] *Linguistic Survey of India*, Vol. III, pt. I, p. 538.

[2] Khamba and Rāi *sel*, Bāhiṅ *syal*, Kirānti *syel*, curiously remind one of Manchu *sele*.

[3] The element *ṅi* is identical with Lisu *a-ni* ("ox" and "cow").—Chinese *ṅu*, *wu* 悟 ("wild ox") should be added to the above series.

Mi-ñag *zö* and the element *zo* of Lo-lo *ṅo-zo*, rather than to the typical Indo-Chinese series.

6. Sia-hia *žou* 手, louse = Nyi Lo-lo *ši-ma*; Kachin *tsi*; Tibetan *ši-g*, *ši-k*, Lepcha *ša-k*; Annamese *sa-t*, Chinese *ši-t* 虱, Fukien *sai-k*; Southern Chin *hai-t*, *hĕ-k*; Kami *χē-t*; Bunan *šrig*, evolved from **ši-rig*, for in Kanaurī *rig*, *ri-g*, means "louse" (compare Ahom *rin*, "flea;" Shan *hin*, "sandfly;" Ahom *rau*, *raw*, Shan *haw*, "louse;" and Ahom *rai*, "kind of louse or mite," Shan *hai*, "lice of animals and fowl"). Kanaurī *š-pōg* (*š-püg*), "flea," appears to have originated from **ši-pōg*; the association of flea and louse is evidenced by Tibetan *k⁀yi-šig* ("flea," literally, "dog-louse") and *a-ji-ba*, *l-ji-ba* (flea"),[1] the stem *ji* representing the same base as *ši* (compare also Nyi Lo-lo *c̓e-ši-ma*, "dog-louse, flea," and Ahom *bat*, "a kind of louse found on the body of a dog," and Shan *mat*, "flea").[2]

7. Si-hia *laṅ-nöṅ* 浪能, camel. In order to understand the Si-hia word for "camel," it is necessary to remember the history of this domestic animal and its designations in the cognate languages. In ancient times the Indo-Chinese group of peoples was not acquainted with the camel. It was unknown to the ancient Chinese; at least there is nothing on record about the matter. The camel, together with the donkey and the mule, belonged to the "strange domesticated animals" (奇畜) which the Chinese for the first time encountered among the Hiung-nu.[3] The name for the camel then

---

[1] The prefix *l-* originated from *li*; we find *li* ("flea") in Minbu and Sandoway (HOUGHTON, *J.R.A.S*, 1895, p. 734).

[2] Likewise in the Mon-Khmêr languages the flea is known as the louse of the dog. It is also a fact of great interest that this family, unrelated to Indo-Chinese, has a common word for "louse" (see E. KUHN, *Beiträge zur Sprachenkunde Hinterindiens, S.B.A.W.*, 1889, p. 215). The louse belongs to the very oldest ingredients of Indo-Chinese and Mon-Khmêr cultures.

[3] *Shi ki*, Ch. 110, p. 1; *Ts'ien Han shu*, Ch. 94 A, p. 1. In view of this fact,

was $t^{c}o$-$t^{c}o$ 橐佗, and in this form we find it written also in the contemporaneous documents of the Han period. [1] The vacillating modes of writing the word in the Annals of the Han 宅, [2] 佗, and 駝, show well that an effort toward reproducing a foreign word was made; $t^{c}o$-$t^{c}o$ appears to have been the Hiung-nu name for the camel. [3] How the latter was articulated we do not know exactly, but the restoration of $t^{c}o$-$t^{c}o$ to the older phonetic stage with initial non-aspirated sonant—that is, da-da or do-do—is legitimate. [4] An exact parallel to this word is no longer preserved, but it seems to be traceable in Mongol adan, Turkish atan ("gelded camel"), further in the general word for "camel:" Uigur töbe, töbek; Djagatai töve, töye, tüye; Altaic and Teleutic tō (from tögö), also tebe, tebege; Soyot täbä; San-čᶜuan Tᶜu-jen time; Taranči tȳgä; Osmanli däwä (deve); Serbian deva; Albanian deve; Ossetian tewa, täwa; Magyar teve; Old Chuvaš täwä, Cumanian tova; Mongol tämäγän, tümān (Guiragos in 1241 wrote the word thaman); Niüči teo, Manchu temen, Gold tyme, Solon temuγe. The Taranči word tȳgä means also "a sepulchral mound," and Djagatai täbä (Altaic töbö, Osmanli tebe, Chuvaš tübe) signifies "summit, mound," likewise; in Mongol we meet dobo ("hill") and in Manchu and Tungusic dube ("summit"). The word for "camel," accordingly, seems to be based on that conspicuous property of the animal, the hump. For this organ the Chinese have preserved a special word fui (from *buñ), formerly

---

BRETSCHNEIDER's generalization (*Mediæval Researches*, Vol. I, p. 150) that "the Chinese were acquainted from remote times with the camel of Mongolia," is difficult to understand.

[1] CHAVANNES, *Documents chinois découverts par Aurel Stein*, p. 74, No. 319.

[2] Thus still in *T'ang shu*, Chs. 170, p. 3 b; 217 B, p. 1.

[3] P. P. SCHMIDT (*Essay of a Mandarin Grammar*, p. 175, in Russian, Vladivostok, 1914), as far as I know, is the only one who justly states that "in all probability the Chinese word $t'o$ is of Central-Asiatic origin."

[4] The Japanese and Annamese reading of 駝 is da. Regarding $t'o = da$ see particularly PELLIOT, *Bull. de l'Ecole française*, Vol. VI, 1906, p. 372.

written 封, [1] and subsequently 峯 ("peak of a mountain"). [2] We first meet it in the designation of the dromedary or single-humped camel of Ngan-si (一封橐駝). This word *fuṅ* (*buṅ*) is traceable in the second element of the Tibetan expression for the camel *rṅa-boṅ* (Balti and Purig *śṅa-boṅ*, Spiti *ṅa-boṅ*, West Tibetan *ṅa-moṅ*). The Chinese-Tibetan *buṅ*, *boṅ*, however, is again an ancient loan-word received from Turkish-Mongol languages: Mongol *bükü* or *büküṅ* [3] and Manchu *boḥoto* ("camel-hump"); Djagatai and Kirgiz *bykyr* ("hump"); [4] Djagatai *buṅra*, Kirgiz *būra*, Osmanli *buṅur*, [5] Mongol *bughūra* ("camel-stallion").

At a later period [6] the Chinese word *tꞌo-tꞌo* was altered into *lo-tꞌo* 駱駝. At that time *lo* was still possessed of a final *k* (compare Japanese *raku*); and this word **lok*, in my opinion, is the reproduction of some Turkish word, the remnants of which have survived in Djagatai and Kirgiz *lök* لوك ("single-humped camel"). [7]

In this connection, the Lepcha word for the camel, *lum-daṅ*, deserves special consideration. MAINWARING and GRÜNWEDEL, in their excellent dictionary of the Lepcha language (p. 354), have added to this word the comment "Chinese" in parentheses. But *lum-daṅ* cannot well be a mere Chinese loan-word in Lepcha; for

---

[1] See *Tsꞌien Han shu*, Ch. 96 A, p. 6 b.

[2] The so-called "wind-camel" (*fung tꞌo* 風駝, see BRETSCHNEIDER, *l. c.*), that is, a swift-footed camel for the despatch of a special courier, is probably nothing but a substitute suggested by punning upon the loan-word *fuṅ* (**buṅ*).—From a semasiological point of view, compare Persian *kohān* کوهان ("hump of a camel"), derived from *koh* کوه ("hill").

[3] The former is given in Kꞌien-lung's Polyglot Dictionary (Ch. 31, p 62), the latter in the Mongol dictionary of I. J. Schmidt. Kovalevski and Golstunski also have recorded the former in their Mongol dictionaries.

[4] RADLOFF, *Wörterbuch der Türk-Dialecte*, Vol. IV, col. 1878.

[5] RADLOFF, *ibid.*, col. 1806, 1807, 1817.

[6] Without attempting here to define its limits exactly, it may be pointed out that the word *lo-tꞌo* occurs in *Sung shu*, *Liang shu*, and *Tꞌang shu*

[7] RADLOFF, *l c*, Vol. III, col. 755. Compare also Turki *lōkun*.

the initial sonant of *daṅ* would indicate that the word was at least borrowed at a time when the present Chinese *t͡ʻo* was still articulated *da*, which would carry us at least as far back as the T͡ʻang period. We know nothing, however, about a possible contact of the Lepcha with the Chinese in that era, nor is the very existence of a Lepcha tribe attested at that date. Yet the fact remains that the Lepcha, in the persistent isolation of their mountain-fastnesses, have preserved the word in that original phonetic state which not only we are bound to assume for the Chinese, but which we find also in Mongol *adaṅ* (corresponding to Turkish *atan*, "gelded camel"). Again, it cannot be denied that the first element of the Lepcha word bears a certain relation to Chinese *lok,* although the final *m* in place of the guttural surd remains singular; it was perhaps suggested by the Lepcha word *fyam* or *a-fyam* ("hump"); compare *lum-daṅ on fyam* ("camel's hump").

The Si-hia word for the camel, *laṅ-nöṅ*, is composed of the two elements *laṅ* and *nöṅ.* For the first element, the alternative between *laṅ* and *raṅ* remains, the former being preferable in view of Chinese *lo* and Lepcha *lum-daṅ*, the latter in view of Tibetan *rṅa (-boṅ).* [1] Tibetan *r-ṅa* should be conceived as having originated from **ro* (or *ra)-ṅa,* [2] and Si-hia *laṅ* in like manner from **lo* (or *la)-ṅa,* the word *ṅa* referring to the hump of the animal (compare the derivative *r-ṅo-g*, "hump;" Siamese *hnòk*, "hump of zebu"). [3] The accent operated differently in the two languages: in Tibetan, the strong accentuation of the ultima, **ro-ṅá*, finally resulted in *r-ṅa;* in Si-hia, the accent was thrown upon the first syllable,

---

[1] The complement *boṅ* is not essential to the word. It may be dropped at least in the formation of compounds, for instance, *rṅa-rgod* ("wild camel;" Polyglot Dictionary of K͡ʻien-lung, Ch. 31, p. 3), *k͡ʻal-rṅa* ("loaded camel;" *ibid.*, p. 47).

[2] A similar development is found in *rṅa-ma* ("tail"), from **raṅ-ma* or *ra-ṅa-ma;* Ahom *raṅ* ("tail"). Tibetan *ma*=Thādo *me*, Kami *a-mai*, Lo-lo *ma*, *mử* ("tail").

[3] *Gye-gu-čan* ("provided with a hump") is a Tibetan synonyme of "camel."

*ló-ṅa, lí-ṅa, and ultimately brought about the monosyllabic product laṅ, under the pressure of the dissyllabic compound laṅ-nöṅ. It is well known that in Tibeto-Burman the ending of a dissyllabic word may be dropped when it enters into composition and thus forms a new dissyllable.

The second part of the Si-hia word laṅ-nöṅ cannot be explained from Tibetan boṅ or moṅ. The element nöṅ being written 能, and the same character serving for the transcription of the numeral "two" (nöṅ), it seems probable that the same meaning attaches to the word nöṅ in the compound laṅ-nöṅ. It would thus have the sense "two-humped camel" (*lo-ṅa-nöṅ, literally, "camel-hump-two"). This implies that also the single-humped camel or dromedary was known to the Si-hia. The Annals of the T͑ang Dynasty attribute to Tibet single-humped camels capable of running a thousand li a day. [1] The Si-hia country appears to have abounded in camels. Tribute gifts to the Chinese Court of three hundred camels are mentioned in the Chinese Annals; and Rashid-eddīn states that Chinggis, after subjugating part of Tangut, drove off many camels which formed their wealth.

8. Si-hia liṅ-lo, riṅ-lo, riṅ-ro 領羅, horse. To all appearances this word presents a compound formed by two synonymous terms. As to the first element, liṅ, riṅ, we may compare it with the widely diffused word raṅ, which in Kanaurī is the designation for "horse." [2] This type, further, occurs in Bunan ɣ-raṅ-s, Chepaṅ se-raṅ, Burmese m-raṅ (from *mo-raṅ), Wa ma-röṅ, Riaṅ ma-raṅ; Palauṅ b-raṅ, Rumai r-b-raṅ. The fact that the second element of

---

[1] 獨峯駞日馳千里 (T͑ang shu, Ch. 216, p. 1b). This item is not given in the corresponding passage of the Kiu T͑ang shu (Ch. 196 A, p. 1b).

[2] To compare Si-hia liṅ with Kanaurī gi-liṅ-ta ("horse") would be a fallacy. The latter is a Tibetan loan-word and composed of gi-liṅ (Jäschke: "a strong-bodied, durable horse") and r-ta ("horse"). The former, in all probability, is nothing but a transcription of Chinese k͑i-lin (*gi-lin, in Tibetan also gi-lin and gyi-liṅ).

the Si-hia compound, *lo* or *ro*, is an independent base with the meaning "horse," is well evidenced by the Si-hia language itself, which offers the word *lo-i* or *ro-i* 羅依 ("saddle"), literally, something like "horse-covering" (see Morphological Traits, § 2).

The same word for "horse" is met in Tᶜo-ču *ro*, Newārī *sa-la* (the element *sa* being identical with the above *š-* and *se-* of Bunan and Chepaṅ; Southern Chin *ši*), Pahrī *so-ro*; Kachin *kum-ra*; Jyaruṅ *mo-rú* and *bo-rú*, Manyak *bó-ro*, *bro*, and in the prefix *r-* of Tibetan *r-ta*, which appears to have been evolved from *ro-ta, and Tibetan *r-kyaṅ* ("wild horse"), from *ro-kyaṅ. [1]

An old theory, already pointed out by Tomaschek, is to regard the Indo-Chinese word-types for "horse" in general as borrowings and to render the Mongol-Tungusic *morin* responsible for all phenomena of the kind in the Indo-Chinese languages. In view of the history of the domestication of the horse, it may indeed be possible that the word *morin* has a certain share in these formations, but certainly it is not capable of accounting for all the manifold variations that we meet in Indo-Chinese. It was W. SCHOTT [2] who wisely cautioned against too wide an application of Tomaschek's theory. A. CONRADY, [3] nevertheless, adopted the latter in its entire range, but the proof given in support of his opinion is not wholly convincing. There is no resemblance, for instance, between Tibarskad *šuṅ*, Kanáurī *šaṅ*, Southern Chin *ši*, Singpᶜo *gūm-raṅ*, Taṅkᶜul *sa-puk*, Karen *tᶜi*, *ka-tᶜi*, etc., and Mongol *morin*. According to the doctrine of the loan-theory, the word *morin* was dissected by the Indo-Chinese into *mo-rin*, and the first element *mo* (*ma*) is found in Chinese, Lo-lo (*mo*, *mu*), and in the Tᶜai languages. [4] The

---

[1] The element *kyaṅ presumably is related to *r-god-pa* ("wild").

[2] *Über einige Tiernamen*, p. 12 (*A. Ak. W. B.*, 1877).

[3] *Indochin. Causativ-Bildung*, p. XII.

[4] Regarding the latter see MASPERO, *Bull. de l'Ecole française*, Vol. XI, p. 165.

second element -*rin* we should then meet in our Si-hia word *riṅ*. In Conrady's opinion the prefix *r*- should represent the survival of the root-word in Tibeto-Burman, and be reducible to the Mongol prototype *morin*. [1] This view conflicts with Si-hia and also other Indo-Chinese formations. If we infer (and perhaps justly so) that *morin* yielded to Si-hia the word *riṅ*, it is difficult to realize that this language should have drawn upon *morin* twice,—first, to adopt the syllable *rin*, and second to distil an element *lo* or *ro* from the *r* of the same word. On the contrary, *lo* or *ro*, as we have seen, represents another Indo-Chinese base conveying the notion "horse," and independent of *morin*. We recognize this also from Rumai *r-b-raṅ* and Jyaruṅ *mo-ro*. The latter, certainly, is not merely a variation of *morin*, but is composed of *mo* (the Chinese-Lolo-Tᶜai word, possibly derived from the first syllable of *morin*) and *ro*, an entirely distinct word for "horse." Remnants of this word *lo* or *ro* in the Ural-Altaic languages possibly are encountered in Magyar *ló*, Irtish-Ostyak *t-lau-χ* or *t-lo-χ* (from **ta-lo-χ*; regarding the word *ta* see below), Mongol and Kirgiz *o-lo-ṅ* ("girth in harness"). [2]

In the Gešitsᶜa language (Hodgson's Hor-pa), a peculiar Tibetan dialect spoken in the territory stretching from Dawo to Kanze in the northwestern part of Sze-chᶜuan, we meet as the word for "horse" *ṛyi* (*r'i*). As in Buryat the word for "horse" shows the same palatalization of *r*, *moryeṅ* (*mor'eṅ*) and *morye* (*mor'e*), similarly in Gold *mor'ă*, it seems likely that the Gešitsᶜa word is traceable to the former contact with a Mongol-Tungusic dialect in which this process

[1] It is hardly possible to suppose with Conrady an older form **mor*. Such can be inferred neither from Mongol-Tungusian, nor from Chinese or any other Indo-Chinese speech. Chinese *ma* never had a final consonant.

[2] RADLOFF, *Wörterbuch der Türk-Dialecte*, Vol. I, col. 1086. As Altaic *koloṅ* corresponds to Kirgiz-Mongol *oloṅ*, it may be permissible to refer also to Altaic, Kirgiz, etc., *kulun* ("foal;" *ibid.*, Vol. II, col. 979).

of palatalization had taken place. [1] By means of this suggestion, we may account for the peculiar Si-hia word *yiṅ* 迎, used for the horse in the cycle. Tentatively I would propose that it may have been derived from a supposed form *moryiṅ* (*mor'iṅ*), analyzed into *mor-yiṅ*. The existence in Indo-Chinese of the type *mo* on the one hand, and the type *riṅ, raṅ, ryi, ri*, on the other hand, raises to a high degree of probability the supposition of a loan from *morin*.

The second element of the Tibetan word *r-ta*, in my opinion, is of Turkish origin, and should be connected with the word *at* ("horse"), common to all Turkish idioms. The phonetic combination *at* does not exist in Tibetan, for which the metathesis *ta* was required. [2] There is even reason to believe that in some ancient Turkish dialect a word for "horse" of the form *ta* seems to have existed. As is well known, the stem *at* appears in Mongol in *adaγun* (*adaγu, aduγu*), "herd, herd of horses" (Buryat *adūχuṅ*, Tungusian *adugun*, Manchu *adun*, "herd;" Manchu *ad-u-la*, "to graze"), *adaγ uči* ("herdsman tending horses"), *adaγusun* ("beast, domestic animals"). [3] W. RADLOFF [4] has added to this group Turkish *atan* ("gelded camel") and Mongol *adaṅ tïmäγän* of the same meaning. These equations bear out the fact that in the Turkish-Mongol languages the word *at* originally did not have the significance "horse," but conveyed the general meaning "gregarious or domestic animal," and subsequently was differentiated into the specialized categories "horse" and

---

[1] In Hodgson's list of Sok-pa words the horse is *ma-ri*. Considering the exuberant number of Mongol loan-words in Sok-pa, this word may well have been directly borrowed from Mongol. *Mari* is likewise the Korean word for "horse."

[2] A similar metathesis prevails between Tibetan and Bunan: Tib. *k'a* ("mouth")— Bunan *ag*, Tib *r-tsa* ("vein")—Bunan *sta*. Compare also Ladākh *rgun-drum* ("grape")— Balti *urgun*.

[3] This relationship was first pointed out in 1836 by W. SCHOTT (*Versuch über die tatarischen Sprachen*) and repeated by him in *Über einige Tiernamen* (*A. Ak. W. B.*, 1877, p. 12). Mongol *ajirga*, Manchu *ajirχa*, Tungusian *ad'irga* (CASTRÉN, *Tungusische Sprachlehre*, p. 72), Yakut *atyr* (PEKARSKI, *Dictionary of the Yakut Language*, in Russian, col. 201), "stallion," may be added to this series.

[4] *Wörterbuch der Türk-Dialecte*, Vol. I, col. 441, 454.

"camel." If *atan* in Turkish and *adaŭ* in Mongol refer to the camel, it is obvious that also the stem *tü-*, *tö-* (see p. 21), applied to the camel, is connected with the stem *at* for the designation of the horse. Moreover, as pointed out by Schott, the horse is styled *tau*, *tav*, in the language of the Irtish-Ostyak. And *tai* in Uigur, Osmanli, Djagatai, and many other dialects, means a "young horse;"[1] Djagatai *tatu*, according to VÁMBÉRY,[2] is "a strong horse of medium stature, a cart-horse," according to Sulejman Efendi "a stallion."[3] The Mongol word *taki* relates to the wild horse.

9. Si-hia *siŭ* (character not given), heart, mind (occurs in the phrase *siŭ-le*, "to think") = Chinese *siem*, *sim* 心; Tibetan *sem-s* ("mind, soul"), *sem-pa* ("to think"); White Tai *sam*; Kachin *sin-tu* ("heart").

10. Si-hia *r-ni* (*lu-ni*) 六 尾, ear. This word represents a compound formed by the two synonymes *r* + *ni*, each of which has the meaning "ear." It corresponds to Tibetan *r-na* (from **re-na*), except the peculiar vowel change from *a* into *i* (see Phonology, § 1). The vowel *i* occurs also in Hakka *ni*, Pahri (Nepal) *ni-sab-ne*, Rai (Nepal) *ni-čo*. The element *r-* is preserved in Chinese *r* 耳, in Bunan *rĕ-tsi* ("ear"), Chun-kia *re*, *reo*, Ya-čʿio Miao *b-re*. In Jyaruŋ *d-r-nä* ("ear") *r* still forms a syllable with vocalic value. In the majority of Tibeto-Burman languages we find only the base *na*, *no*, in combination with various elements: Nyi Lo-lo *na-po*, A-hi and Lo-lo-pʿo *no-pa*, Lisu *na-bo*, Pʿu-pʿa *na-be-tla*, Chö-ko *na-ku*; Burmese *nā*, Kʿchin *na*; Kʿyeñ *ma-nho*; Southern Chin *a-hno*; Guruŋ *nha*; Sunwar *no-pʿa*; Magar *na-kep*; Balti and Purig *s-na*, etc. Gešitsʿa *ña* (book-language *s-ñan*) and Lepcha *a-ñor* are connected

---

[1] RADLOFF, Vol. III, col. 765. Perhaps also Mongol and Manchu *to-χo* ("to saddle"), Mongol *to-χo-m*, Manchu *to-χo-ma*, Tungusian *to-ku-m* ("saddle-cloth"), have to be associated with this root.

[2] *Čagataische Sprachstudien*, p. 11.

[3] I. KUNOS, *Sulejman Efendi's Čag.-osman. Wörterbuch*, p. 42.

with another stem, ñan ("to hear"). The element č̕o in Rai ñi-č̕o seemingly corresponds to the second part of Ladākhī nam-č̕ok (from *r-na-ba-č̕og); the same formation might eventually be recognized in Si-hia r-ni č̕añ-ni 六尾長尾 ("exterior part of the ear"). Ni is a Si-hia suffix, and Si-hia final ñ may correspond to Tibetan final g (see Phonology, § 27). On the other hand, the expression č̕añ-ni, written in the same manner, means also "to fly," and the term r-ni č̕añ-ni, after all, may be of independent Si-hia origin, without Tibetan affinity.

11. Si-hia ni or ñi 你, nose = Mo-so ñi-ma; Nyi Lo-lo na-bi, Chö-ko na-mo, Lo-lo-pᶜo no-bi, A-hi Lo-lo no-bo, Lisu na-pe, na-kŏ (Kami na-baun; Ṭōṭo na-ba); Tibetan s-na, Gešitsᶜa s-ni and s-na, Jyaruṅ te-š-no-s; Burmese nhā (Newārī nhā-sa), Kuki-Lušai hnā, Southern Chin h-nu-t-tō; Lepcha tuk-n-om; Sunwār (Darjeeling) ne. Besides the i-forms of Mo-so and Gešitsᶜa, we have nhi-se in Pahrī (Nepal) and a-nhi-č̕a in Angao-Nāgā.

12. Si-hia li 力, moon. This is a peculiar variation of the general type la, le, lo, that we meet in Tibeto-Burman. The form li, however, is not entirely isolated, but occurs in Ya č̕io Miao and Hua Miao li, in Leṅ-ki Miao ka-li, and in Miao-tse ka-h-li. The more important representative forms are: Burmese la, Lepcha la-vo, Nyi Lo-lo š-la-ba, Tibetan z-la-ba (Central Tibetan da-va, Kᶜams and Ladākhī lda-va); Bunan h-la; A-hi Lo-lo h-lo, h-lo-bo, Kᶜyeṅ and Southern Chin k-h-lo, Kuki lha, Haka k-la, Thuluṅ k-h-le, k-h-lye; Kachin ša-ta. Tib. z-la is evolved from *za-la, as shown by Lahūl la-za.

13. Si-hia tsᶜi-ñu 七吾, salt = Tibetan tsᶜwa (tsᶜua), tsᶜa, tsᶜo; Nyi Lo-lo tsᶜa, A-hi Lo-lo and Lo-lo-pᶜo tsᶜo; Lisu tsᶜa-po; Pᶜu-pᶜa sa-m, Chö-ko tso-m; Mo-so tse; Hua Miao ndse; Newārī če-lu ("salty taste"); Manyak če; Lepcha č̕a; Kanaurī č̕a; Burmese č̕ā; Karen i-sē, Yintale i-sā. The Si-hia form with the vowel i occurs only in Thādo č̕i (T. C. Hodson, Thado Grammar, p. 99), Chin dsī

and *tsĭ*, and Hei Miao *śie*. The corresponding Chinese word is *tsᶜo* 醝 (Canton *čᶜa*, Fukien *čᶜwa*, Sino-Annamese *sa*). The second element *ńu* in the Si-hia compound seems to be an independent base, likewise with the meaning "salt." It corresponds to Kuei-chou Chuň-kia *ku* and Kuang-si Chuň-kia *giu* ("salt:" S. R. CLARKE, *l. c.*, p. 310), Ahom *k-lu* (from **ku-lu*), Shan *küw*, Siamese *k-lua*.

14. Si-hia *ńo* 遏, 1 = Chinese *ńo* 我; Tibetan *ńa, ńo*; Lo-lo *ńa*, Lisu *ńwa*, Mo-so *ńa, ńö*; Burmese *ńa*, Kachin *ńai*.

15. Si-hia *wei* 爲, to do = Chinese *wei* 爲, Tibetan *byé-d* (*v'e, j'e*), Mo-so *be, pe*.

16. Si-hia *si* 悉, to die = Chö-ko *si-pü*, Pᶜu-pᶜa *se-poa*; Kachin *si*; Lo-lo *šö*; Mo-so *še*; Tibetan *ši*, Yün-nan Tibetan *šö*; Digaru, Dafla, and Miri (Assam) *sĭ*; Chinese *se* 死.

17. Si-hia *wen* (character not given), bad = Chinese *ńan (yen)* 贋 ("bad"); Tibetan *ńan*; Ahom *ńam* ("false, falsehood").

GROUP II.

Words directly Related to Lo-lo or Mo-so, or Both Idioms, and bearing a Further Relationship of the Second Degree to Tibetan or Chinese, or to Both Languages.

18. Si-hia *kou, gou* 苟, ant = Nyi Lo-lo *kau-ma*, A-hi Lo-lo *ka-vu* (*vu* is perhaps akin to Chinese *wei* 蝟, "wingless insects"); Ahom *kau* ("spider"), Shan *kuň-kau*; Bhūtan *kyo-ma*; Kirānti (*saču-*) *ka-va*; Kachin *ka-gyin*. The element *-ma* of *kau-ma* meets its counterpart in Tibetan *grog-ma*.[1] The latter, stem *grog*, as shown by Jyaruň *go-rok* (Sog-pa *kᶜo-rok-we*, Hor-pa *kᶜro*), should be analyzed

---

[1] In all probability the element *ma* in this case is not a mere suffix, but an independent base meaning "ant;" compare Ahom *mau, ma-t, mu-t* ("a kind of ant"); Shan and Siamese *mo-t*, Pa-yi *mu-t*; Thādo *si-mi* ("ant")

into *yo-rog or *gi-rog (rog, "black"). The base *yo, gi is related to Si-hia kou, gou, on the one hand, and to Chinese ṅi (Japanese gi) 蟻 on the other hand. The ant is called bo-yo in Lo-lo-pᶜo, bu-ku in Pᶜu-pᶜa, and bu-ma in Chö-ko. This word bu seems to me to be identical with Tibetan ɑbu ("insect, worm;" combining the significance of Chinese čᶜuṅ 虫) and Burmese pü:.

19. Si-hia pan-bu, ban-bu (perhaps with labial assimilation pam-bu, bam-bu) 板哺, butterfly. The second element bu occurs in Nyi Lo-lo bu-lu-ma, A-hi Lo-lo bu-hlo, Lo-lo-pᶜo bo-lu ("butterfly"), and possibly in Chinese fu (from *bu) 蚨 ("water beetle;" fu-tᶜie, "butterfly"). The first element may be in some relation with Tibetan pᶜye-ma of pᶜye-ma-leb ("butterfly"), Ladākhī pe-ma-lab-tse; Tromowa pᶜi-ma-laq, pim-lab, Sikkim byam-lap; Mo-so (written language) pᶜe-le;[1] Magar whā-mā; Kachin pa-lam-la; Thādo peṅ-pu-lep. B. HOUGHTON (J.R.A.S., 1896, p. 37) has identified Tibetan pᶜye-ma-leb with Burmese lip-pyā (the members of the compound being transposed), explaining pyā as meaning "to fly." This is interesting, and in some measure is corroborated by the Bunan verb pan-čum ("to fly;" -čum is a verbal ending), so that Si-hia pan-bu would mean "flying insect." It seems doubtful, however, that, as suggested by Houghton, the element lip, leb can be interpreted through Tibetan leb, which means "flat." This conception of the matter is contradicted by the variant lab. It seems more appropriate to take the element leb or lab in pᶜye-ma-leb as *le-bu, la-bu, and to identify *bu with the base bu in the Si-hia and Lo-lo words, and *le, la, perhaps with Lo-lo lu, lo.

20. Si-hia moṅ-tsi 夢積, fly. This is a compound formed by two synonymes. The first element moṅ is apparently identical with Chinese moṅ 蝱 ("gadfly"), moṅ 蠓 ("flies"); Siamese, Shan and

---

[1] The primeval forms presumably are *be, bʹe, ba, bʹa, while ma, -m, may be suffixes or euphonic insertions. In this way we arrive at a satisfactory explanation of Si-hia pam-bu, bam-bu, by analyzing the latter into pa-m-bu, ba-m-bu.

Laos *meṅ* ("insect"); Pa-yi *myăn-moṅ* ("fly"); Ahom *mliṅ* ("white ant, firefly"), *mŭk* ("mosquito"); Lepcha (HODGSON) *maṅ-koṅ* ("mosquito"); Tibetan *boṅ* in the names of small insects (*rgyas-poi boṅ-bu,* "sugar-mite;" *boṅ-nag,* "dung-beetle"), and Tibetan *muṅ-ba, mug-pa* ("moth, worm"). The second element *tsi* is comparable with Nyi Lo-lo *je-mu,* A-hi Lo-lo *yi-mu,* Lo-lo-pᶜo *ya-mu* (possibly further relationship with Tibeto-Chinese *yaṅ, yiṅ,* see No. 21).

21. Si-hia *mou-ṡuai, mou-ṡu* 謀率, bee. The first element *mou* is met with in Kanaurī *mö-khăr* ("beehive"), in Chuṅ-kia *mo-vei* ("bee"), and Mo-so *mba-me.* According to J. BACOT (*Les Mo-so,* p. 29), this word literally means "honey-mother," so that *mba* would signify "honey" (*me,* "mother"). At the same time we have in Mo-so *ba-ler* and in the written language *mbar* in the sense of "fly." Mo-so *mba,* therefore, seems to be evolved from **mo-ba.* This analysis is confirmed by Chepaṅ (Nepal) *tu-mba* ("bee") and *tu-m* ("honey"),[1] which shows that Mo-so *mba* (*mo-ba), in the same manner as Chuṅ-kia *mo-vei* and Si-hia *mou,* has also the meaning "bee." The element *vei* (= Mo-so *ba*) appears in Nyi Lo-lo *d-la-vu-kᶜiā* ("wild bee"). It is noteworthy that in the Indo-Chinese languages the notions "insect, fly, bee, honey" inhere in the same roots. There is the remarkable parallel: Tibetan *buṅ-ba* ("bee"), Chinese *fuṅ* (from *buṅ) 蜂, Siamese *pᶜŭṅ* (from *bŭṅ: MASPERO, *Bull. de l'Ecole française,* Vol. XI, p. 158), Ahom *pᶜrŭṅ.* The Si-hia word, of course, bears no direct relation to this series, or to Lo-lo-pᶜo *byo* ("bee"). The latter is akin to Tromowa *byo-mo, bya-mo,* and Sikkim *byam* ("fly"). These forms are contractions of **bu-yo, bu-ya,* which, as a matter of fact, we meet in West Tibetan *bu-yaṅ* ("humble-bee"). The latter is by no means a corruption of *buṅ-ba,* as asserted by JÄSCHKE

---

[1] The element *tu* is related to another word for "fly" in the Mo-so written language, *ndu,* apparently coincident with A-hi Lo-lo *do* ("bee"). The same base is found in Mo-so *bu-tu* ("insect"), Lo-lo-pᶜo *hyo-to* ("mosquito"), and Nyi Lo-lo *dla-ma* ("bee"), from **do-la.*

(*Tibetan Dictionary*, p. 393 a), but an independent formation: *yaṅ* is a root-word with the meaning "insect," that is found in Chepaṅ and Burmese *yaṅ* ("fly"), Kachin *či-yoṅ* ("mosquito"); Kanaurī *yaṅ* ("fly, bee"); Newārī *yaṅ-kela* ("bee"), Tibetan *s-b-raṅ* ("fly, bee"), Lepcha (*sum-*)*b-r-yoṅ*, Ahom *jṅ*, *jiṅ* ("dragon-fly"), Shan *yiṅ*; and Chinese *yiṅ* (Wen-chou *yaṅ*) 蠅, "fly." Tibetan -*raṅ*, that as independent word occurs in West-Tibetan *raṅ-ṅu* ("fly") and *raṅ-si* ("honey"), is merely a phonetic variation of *yaṅ*, for in Sikkim we have *se-byam* and in Tromowa *se-byom* as equivalents of common Tibetan *sbraṅ*.[1] It is therefore obvious that the latter has been evolved from *se-bu-raṅ* = *se-bu-yaṅ*. Lepcha *sum-bryoṅ* is developed from *su-bu-ryoṅ*, with a euphonic insertion of *m* between *su* and *bu*. The base *bu* seems to be associated with Tibetan *a-bu* ("worm, insect") and *s-bu-r* ("beetle"); it is widely disseminated also in the languages of Nepal, in Lo-lo, and Si-hia (see Nos. 18, 22). The second element of the Si-hia word *mou-šu* (probably from *mou-su*), in my opinion, may be correlated with the above Tibetan-Lepcha bases *se, su, Sikkim se-byam, Tromowa se-byom. This s-base appears also in Lohoroṅ *bᶜu-su-na* ("mosquito"), Magar *bᶜu-s-na*, Kuswar *bᶜu-n-si*.

22. Si-hia *mo-lu* (possibly pronounced *m-lu, m-ru, b-ru* 沒魯), worm, snake. In order to understand this formation, it may be well to proceed from an analysis of Tibetan *sbrul* ("snake"). This complex word (at present articulated *sbrul* only in Purig; Central Tibetan *ḍul*), as follows from a comparison with the facts of the cognate languages, presents a triple compound evolved from *sa(se)-bu-ru-l, sa(se)-bu-lu-r. 1. The base *sa, se, si, is widely spread in

---

[1] *Ya-ta* and *ra-ta* (that is, the subscribed letters *y* and *r*) interchange within Tibetan: *abras* ("rice")—Tromowa *bya* (WALSH, *Vocabulary of the Tromowa Dialect of Tibetan*, p. vi), *k'rag* ("blood")—Trom. *k'yag, abros-pa* ("to flee")—Trom. *byo-po, ak'rid-pa* ("to guide")— Trom. *k'yi-ko, abrog-pa* ("herdsman")—Trom. *byo-ko, brag* ("rock")—Trom. *byag, gro* ("wheat")—Trom. *gyo*, etc.

3

Indo-Chinese: Limbu *o-se-k*; [1] Gyami *š-re*; Pᶜu-pᶜa *a-si-ma*, Chö-ko *sie-na*, Nyi Lo-lo *še-pᶜai*; Mo-so *jö*; Chinese *džie, že, še, šö, ša, sia* 蛇. As to Tibetan, it is found again in Central Tibetan *sa-gu-tse* ("worm") and in the prefix *s-* of *s-braṅ* (see No. 21), *s-rin* ("worm, insect"), and *s-bu-r* ("beetle"). 2. The base **bu* (*mu, mo*) occurs in Tibetan *a-bu* ("worm, insect;" *a-bu-riṅ*, "snake"); Tibetan *ąbruɡ*, Balti *bluɡ* ("dragon, thunder"), evolved from **ąbu-ruɡ, ąbu-luɡ* (**luɡ* = Chinese *luṅ*, "dragon"); Tromowa *bu*, Sikkim *bi-u* ("serpent"), Bunan *de-bu* ("serpent"), Lepcha *bu* ("serpent"), Kirāntī *pu*, Takpa *mrui* (from **mo-rui*), Manyak *bru* (from **bu-ru*); Murmi *pu-ku-ri*; Guruṅ *bᶜu-ɡu-ri*; Magar *bu-l* (Mi-ñag *bu-r*); Kachin *la-pu*; Bhrāmu *pai*; Lo-lo (*še-*)*pᶜai*, Lo-lo (BABER) *vu*, Lisu *fu*; Kami *pu-wi* (besides, *me-kᶜwi, m-kᶜwe, ma-kᶜui*); Khambu dialects *pu*. 3. The base **lu, ru*, is encountered as independent word in Mo-so *lu* ("worm"), Pa-yi *low* ("python"), in the second element of Lo-lo-pᶜo *bo-lu* and Nyi Lo-lo *bu-lu-ma* (both "butterfly"); in Sokpa *tᶜo-le* ("snake"); in Manyak *bru* ("snake"), from **bu-ru*; Tapka *mrui*, from **mo-rui*; further in Burmese *kᶜrū* (also Burmese *mrve*, "serpent," from **mo-ru-ve*) and Tibetan *k-lu* ("cobra, serpent-demon"), evolved from **gu-lu, ku-lu, ku-ru* (the first element appears in Central Tibetan *sa-gu-tse*, "worm," Thādo *gᶜū-l*, "snake;" Ahom *ku*, "worm;" Murmi *pu-ku-ri*, Guruṅ *bᶜu-gu-ri*, [2] Tᶜočŭ *bri-gi*, "snake;" the latter from **bu-ri-gi*, instead of **ru-gu*), and finally in Tibetan **ru-g, lu-g* of *ąbruɡ, bluɡ*, previously cited, and in *s-bu-r* ("beetle"), from **se-bu-ru*. As demonstrated by Si-hia and Mo-so *lu* on the one hand and by Tibetan *k-lu* and *s-b-rul* on the other hand, the final *g* of **ruɡ, luɡ* is not inherent in the stem, but is either a terminating affix (**ru-ɡ, lu-ɡ*),

---

[1] Kanaurī *sa-pös* and Milčaṅ *sūbūs* appear to be loan-words from modern Indo-Aryan *sapa, sap* (from Sanskrit *sarpa*).

[2] Corresponding to Milčaṅ and Bunan *gur-gu-ri*, "thunder" (A. CUNNINGHAM, *Ladák*, p. 403), which I conceive as a gemination (**gu-ri-gu-ri*), and to Kanaurī *gur-gur* ("thunder").

or the survival of the above base *gu; in this case ᶏbrug would represent a triple compound, contracted from *ᶏbu-lu-gu. For this reason I conceive also Chinese luᶇ ("dragon") as lu-ᶇ, that is, stem lu + affix ᶇ; this opinion, from the viewpoint of Chinese, is confirmed by the form löe for lu-ᶇ in the dialect of Wen-chou. In Ladākhī and Lahūl, "snake" is called rul; possibly rul has arisen from a contraction of ru-lu, that is, a gemination of the base *ru, lu. On this assumption, sbrul would even be a quadruple compound (*se-bu-ru-lu), each element having the significance "worm, snake." It is conceivable, however, that the final -l of sbrul, rul, is merely an affix on the same footing as -g, -n (compare spre-l, "monkey," derived from spre). There can be no doubt of the fact that Si-hia mo-lu is composed of the two bases dealt with under 2 and 3. What may be questioned is solely the correct articulation of the Si-hia speakers, as there is a somewhat wide range of possibilities in the allied languages. Magar bu-l, for instance, might tempt one to restore a Si-hia form *mu-l; but considering the close affinity of Si-hia with Lo-lo and Mo-so, where we have the base lu, I believe we are justified in adhering to a Si-hia word of the type mo-lu, m-lu.

23. Si-hia wei (perhaps wöi, wö) 崑, dragon = Mo-so lö, l-wö, from *lö-wö (compare Wen-chou löe for luᶇ 龍). The Mo-so element lö, l-, has perhaps survived in Tibetan ᶏbrug ("dragon, thunder"), Balti b-luᶇ, possibly evolved from *ᶏbu ("worm, snake")- lug (*lug = Chinese luᶇ, "dragon"). It is therefore not necessary to regard A-hi Lo-lo lo as a mere loan-word from Chinese luᶇ, as proposed by LIÉTARD, [1] but Lo-lo lo may very well be anciently allied to Mo-so lö, Tibetan *lug, and Chinese luᶇ. There is a further possibility that this root is connected with the base lu ("worm") men-

---

[1] *Bull. de l'Ecole française*, Vol. IX, p 557.

tioned afore, if *lug and luṅ be lu-g and lu-ṅ; that is, if the finals
g and ṅ should not be inherent in the stem, but merely terminating
suffixes. This is confirmed by Wen-chou löe.

24. Si-hia ri-ṅ or possibly rĭ (with nasalized i) 領, bear = A-hi
Lo-lo rö-mo, Lo-lo-pᶜo vö-mo (LIÉTARD, Au Yun-nan, p. 214); Hua
Lisu wo; Lo-lo (BABER) wo; Mi-ñag (BABER) re. Further relationship seems to exist with Tibetan dred ("yellow bear") to be analyzed into d-re-d, so that the initial d- would have to be regarded
as a prefix somehow related to Tib. dom ("black bear," Ursus tibe-
tanus, Sanskrit bhallūka).

25. As we note in the preceding example an alternation of r
and v in the Lo-lo dialects, I am inclined to connect Si-hia ro 勒
("wolf") with Nyi Lo-lo ve and A-hi Lo-lo vö-mo.

26. Si-hia dzei, zei (zöi) 則夷, panther = Lo-lo ze, zö; Mo-so
ze; Tibetan g-zi-g; Lepcha syi-čak.

27. Si-hia wo 訛, hog, swine = A-hi Lo-lo vye, Nyi Lo-lo and
Lo-lo-pᶜo ve, Pᶜu-pᶜa and Chö-ko va; Lisu a-ve; Mo-so bu, bo; Miao
ba; Newārī pᶜā ("boar"); Gešitsᶜa va; Tibetan (written language)
pᶜa-g (in many dialects pᶜa); Kanaurī fa-g; Burmese va-k (vet);
Thādo vō-k; in Khambu dialects ba-k, pᶜa-k, ba, bᶜa, bo, po, pa. [1]
In view of Mo-so bu, bo, the series mu of the Tᶜai languages and
Chuṅ-kia possibly may belong to the same root. [2]

---

[1] Linguistic Survey of India, pt. III, Vol. I, p. 344.

[2] The coincidence of the type vak, bak, etc., with Old Javanese wòk, Kawi wĕk,
Sumban we, Malayan babi ("hog"), etc., is hardly fortuitous. One of the great centres
where the domestication of the pig (Sus indicus, traceable to the wild form Sus vittatus)
was brought about in a prehistoric age was located in southeastern Asia, inclusive of Java.
This species gradually extended to Yün-nan, Sze-chᶜuan, southwestern Kan-su, and farther
into Tibetan territory. In Sze-chᶜuan and Kan-su it meets with the species of northern
China, which, as far as we can judge at present, is the product of an ancient Chinese
domestication from indigenous wild material. These zoölogical results may account also
for the fact that the Chinese nomenclature relative to swine is perfectly independent, being
without parallels in Tᶜai and Tibeto-Burman.

28. Si-hia *k᷋ū*, *č᷋ü* 屈, dog = Mo-so *k᷋ö*, *k᷋e*, *kü*; A-hi Lo-lo *k᷋i*, Nyi Lo-lo *č᷋e*; Lo-lo-p᷋o *č᷋ö*; Mili Si-fan (Davies, *Yün-nan*, p. 360) *č᷋ö*; Northern Lo-lo (Davies) *ksö* (from **ki-sö*?); Yün-nan Tibetan (Davies) *ts᷋ö*, (Monbeig) *tsö*; Ya-č᷋io Miao and Hua Miao *k-le* (from **ki-le*; Hei Miao *la* ("dog"); Bunan *k᷋yu*; Tibetan, in general, *k᷋yi* (K᷋ams *k᷋ye*, Jyaruṅ and Spiti *k᷋i*, Gešits᷋a *d-ga* and *k-ta*); Digaru *n-kwī*, Dafla *i-kī*, Miri *e-kī*; Burmese *k᷋wē*, Kachin *gwi*. The primeval form seems to be **gi*, *g᷋i*. I am not convinced that Chinese *kou*, *ku* 狗 belongs to this series.

29. Si-hia *tsai-šu* (*tsai-šuai*), *dzai-šu* 宰率, rat. The first element appears to be related to Mo-so *dse* ("rat"), Tibetan *tsi*, *tsi-tsi* ("mouse, shrew"), Bunan *myu-tsi* ("rat, mouse"), [1] Burmese *čwet* (written *krvak*). The word seems to be imitative of the animal's voice (compare Malayan *tikus*, "mouse," Javanese *tjit*, "piping of the mouse;" Sikkim *tiñ-rjiñ*, "shrew;" Turkish *syčkan*, *syčan*, Kirgiz *čičkan*, "mouse;" Magyar *cziczkány*, "shrew"). Lo-lo *e* and Chö-ko *a-i* (stem *i* with prefix *a*) correspond to Tibetan *byi*. The second element of the Si-hia compound is presumably identical with Chinese *šu* 鼠 ("rat, rodent").

30. Si-hia *lo-wo* 勒訛, hare = Lo-lo *lo*: Lo-lo-p᷋o *a-lo* (*lo* with prefixed *a*), *ti-h-lo*, and in the cycle *ta-lo*; [2] Nyi Lo-lo *a-š-la*; Mo-so *to-le*. The element *wo* of Si-hia *lo-wo* possibly is akin to the second element of Tibetan *ri-boñ*, *ri-voñ* (Tromowa and Sikkim

---

[1] In Kami, *myu* means "rat" (Houghton, *J.R.A.S.*, 1895, p. 137).

[2] Liétard, *Au Yun-nan*, p. 230; the element *ta*, as shown by the phonetic variations *ti* and *to* in the cognate idioms (perhaps also Siamese *kă-tay* belongs to the series) is indigenous. Lo-lo-p᷋o *ta-lo*, accordingly, bears no relation to Mongol *taulai*, *tolai*, Kalmuk *tūlai* (from **tabulai*, Old Mongol *tablga*), Yakut *tabisχan*, Orkhon Inscriptions *tabyšγan*, Osmanli *tawšan*, Khitan *t᷋ao-li* 淘裏 or 陶里 (K. Shiratori, *Sprache des Hiung-nu Stammes*, p. 44) The case of Nyi Lo-lo *t᷋o š-la*, used in the duodenary cycle for the year of the hare, is different. As rightly observed by P. Vial (*Dict. français-lolo*, p. 201), this *t᷋o* is of Chinese origin (*t᷋u* 兔), as is likewise Siamese *t᷋o* employed in the cycle.

ri-goṅ); but I do not feel certain of this. Tibetan *-boṅ* is comparable with Ahom *paṅ*, Shan *paṅ-lai* ("hare"). [1]

31. Si-hia *la* 粹, stag = Nyi Lo-lo *la* ("chevrotain;" *la-re*, "musk"); Lo-lo of T͟ung River (BABER) *lö* ("musk-deer"); Mi-ñag (BABER) *lie* (*id.*); Tibetan *g-la* (*id.*); Chinese *lu-k* 鹿 ("stag"). Lo-lo *re* ("musk") seems to tally with Chinese *šö, žö* 麝.

32. The designation for the domestic fowl in Si-hia is *wo-yao* 訛要. This is a compound formed by the two synonymes *wo* and *yao*, each of which has the general meaning "bird." As shown by the Si-hia form and by what is found in other Tibeto-Burman languages, Tibetan *bya* ("bird, fowl"), which at first sight appears as a primitive stem-word, has resulted from a contraction of **ba* (*va*) or *bo* (*vo*) + *ya*. In Balti and Purig a glide is still audible between the initial sonant and *y* (*b^e ya-po*).

Si-hia *wo* and Tibetan **ba, va*, occur in Nyi Lo-lo *va* ("bird"), used only in the written language. [2] In the idioms of Nepal, *wa* is the generic of birds of the fowl kind; Chepaṅ (*mo-*)*wa* ("bird") and *wa* ("fowl"); T͟oču *mar-wo* ("bird"); Limbu *wa-bha-le*; Magar *gwha-bha* (*bha* corresponding to Tib. *p͟o*, "male," in *bya-p͟o*, "rooster") and *gwa-ja* (correctly analyzed by Hodgson into *g-wa-ja*; regarding *g-* see No. 33); Murmi *hwā-bā*; Sunwār (Darjeeling district) *wo-a*; Khambu *wa-pa* ("rooster"). Kami *ka-va, ta-va* (*a-bwi*, "rooster," *va-ā, wa-ā*, "to crow"); [3] Lepcha *fo*; Kachin *wu, u* ("fowl").

Si-hia *yao* and Tibetan **ya* are found in Nyi Lo-lo *ye* ("fowl"), *ye-p͟u* ("rooster"), *ye-p͟ä* ("hen"); A-hi Lo-lo *ye* (*ye-p͟u*, "rooster;"

---

[1] It is possible that Tibetan *ri-boṅ* means "wild ass" (*ri*, "mountain, wilderness;" *boṅ-bu* or *boṅ-bo*, "ass"). The long ears form the point of comparison between hare and donkey: compare Nyi Lo-lo *a-žla-la-mu* ("donkey;" literally, "hare-horse"); Persian χergöš ("hare;" literally, "donkey's ear").

[2] P. VIAL, *Dictionnaire français-lolo*, p. 78 (see *chauve-souris*). Baber gives for Lo-lo of the T͟ung River *wo* ("fowl").

[3] B. HOUGHTON, *J.R.A.S.*, 1895, pp. 116, 118. As to *ka* in *ka-va* compare No. 33.

$p^c u$ = Tib. $p^c o$); Lo-lo-$p^c o$ yi-$p^c o$; $P^c u$-$p^c a$ ya-$p^c u$-ma; Hua Lisu ai-ya, White Lisu a-$y\ddot{o}$-r; Mo-so $\bar{a}$, a-me (compare $P^c u$-$v^c a$ e-ma and Chö-ko e-me, beside za-$p^c o$-ma; Thādo $\bar{a}$, "bird," $\bar{a}$-$p\bar{i}$, "hen"); Murmi nam-ya, Guruṅ nem-ya; Chin $p^c a$-$y\mathring{a}$. As Tibetan bya was formerly taken for a stem-word, it was subjected to erroneous comparisons; it cannot directly be correlated, for instance, with Newārī $j^c a\dot{n}$-gal, as proposed by CONRADY (*Causativ-Bildung*, p. 105), in which the element gal coincides with Sog-pa $t^c a$-kol, and $j^c a\dot{n}$ with Kirānti $\check{c}o\dot{n}$-$w\bar{a}$. On the other hand, we meet in Newārī and also in Pahrī a form $j^c a$-$\dot{n}a$, which seems to be the antecedent of the contraction $j^c a\dot{n}$. As $j^c a$ corresponds to ya in other languages of Nepal, we may equalize the two forms, which answer to *ya, with the second element of Tibetan bya; Central Tibetan $j^c a$, $j^c ya$, however, is a recent affair, developed from bya, — a specific Tibetan home-affair which has nothing to do with Newārī $j^c a$. [1]

33. Si-hia ku-$ki\ddot{a}$, gu-$gi\ddot{a}$, gu-$g'\ddot{a}$ (ku-$kiai$) 姑皆, phœnix (in Buddhist texts presumably for the designation of the Garuḍa). The literal meaning of this compound is "the bird gu (ku);" for the second element is identical with Nyi Lo-lo ge ("bird, fowl"), Black Lisu a-ke ("chicken"), Hor-pa gyo, and Magar g-wa-ja (see No. 32). This base appears also in Mi-ñag ge-ji ("bird") and Mo-so gi-$\bar{a}$ ("duck;" $\bar{a}$ borrowed from Chinese ya 鴨), and further points to the T^{c}ai languages: Shan and Dioi kai, Siamese and Laos $k^c ai$, White Tai $k\hat{e}i$ ("bird, chicken"), developed from *gai (Annamese ga), Chuṅ-kia and Hua Miao kai; and to Chinese *gi, ki 鷄 ("fowl,

---

[1] The Si-hia word for "egg" is unfortunately not given. It is a felicitous suggestion of B. HOUGHTON (*J.R A.S.*, 1896, p. 36), in view of radically different words for "egg" in Tibetan (sgo-$\dot{n}a$) and Burmese (u), taken with the diversity of the name for the domestic fowl in the two languages, that the fowl was not kept by the Burmans in prehistoric times prior to their separation from the Tibetans. Even in the Lo-lo dialects the words for "egg" are at variance: Nyi $\check{z}la$, A-hi $t'o$, Lo-lo-$p^c o$ fu, Black Lisu a-le-fu, Hua Lisu ai-ya-$k^c u$. Tibetan sgo-$\dot{n}a$, $sgo\dot{n}$-$\dot{n}a$, or $sgo\dot{n}$, appears in the written language of the Mo-so as $g\ddot{o}$, to which a-ku, a-kvu, of the oral language, is perhaps related.

rooster"), Cantonese and Hakka *kai*, Kuei-chou *gi*. As to formations analogous to the Si-hia compound, compare Nyi Lo-lo *k^cïñ-ge*
("pheasant"). The first element *gu*, *ku*, appears to be identical with
Tibetan *go-bo* ("eagle;" stem *go*), Tibetan *g-laj* ("eagle," from **go-
laj*), Kanaurī *g-ol-t^cŏs* ("vulture"), Mo-so *ko-n*, *ko-n-do* ("eagle"),
Lepcha *ko-juk ge-bo fo* ("a species of eagle"). The last named cannot
be identified with Tibetan *k^cyab-aju g dge-ba*, as proposed by GRÜN-
WEDEL (*Lepcha Dictionary*, p. 26), but Lepcha *ko-juk* is the phonetic equivalent of Tibetan *k^cyuñ*, evolved from **go-juñ*, *go-yuñ*, the
element **go* being identical with *go* ("eagle"). Tibetan *k^cyuñ* was
heretofore known to us solely as a rendering of Sanskrit *garuḍa*,
but, as demonstrated by Kanaurī *k^cyuñ pyā* ("eagle") [1] and by the
very phonetic development of the word itself, it originally had the
significance "eagle." The Si-hia compound, accordingly, means
"eagle-bird."

34. The Si-hia word for "heaven" is *mo* 沒, and that for
"sun" is *mo* 墨. As these two Chinese characters represent a
sound-combination *mo* of the same tone (the entering lower one),
the two words, separated in the glossary, apparently are identical,
the single stem *mo* combining the two significances "heaven" and
"sun." This state of affairs is confirmed by Lo-lo, where we meet
the same base *mu* in the same tone and with the same duplicity
of meaning, as well as by Mo-so *mu*, *mŏ*. It is obvious that the
Si-hia word for "sun" has nothing in common with Tibetan *ñi-ma*,
which Mr. Ivanov has added to it, or with Chinese **ži-t*, *ñi-t* 日,
coincident with Tibetan *ñi*. The affinity of Si-hia, in this case,
decidedly points to Lo-lo and Mo-so In Mo-so, "heaven" is *mu*,
mŏ (written language also *mu-n*); in A-hi Lo-lo *mu*; in Nyi Lo-lo,
the vowel is but dimly sounded, so that P. VIAL writes *m(u)*, while

---

[1] T. G. BAILEY, *J R A.S.*, 1911, p 335.

LIÉTARD [1] transcribes the corresponding word in P^{c}u-p^{c}a m^o. This m enters into composition with $k^c e$, $k^{c\prime} \ddot{a}$ in Nyi Lo-lo m^u-$k^c e$, m^u-$k^{c\prime} \ddot{a}$ (Hua Lisu mu-kua), and with ti in Chö-ko (m-ti-ma). The former presents an exact analogon with Tibetan m-$k^c a$ ("heaven"), which accordingly is evolved from *mu-$k^c a$. [2] In Jyaruṅ we have te-mu ("heaven"). The Lo-lo-p^{c}o dialect has combined the base $m\bar{o}$ with $\tilde{n}i$-mo (= Tib. $\tilde{n}i$-ma, "sun") into $m\bar{o}$-$\tilde{n}i$-mo ("heaven"), and employs the same compound $m\ddot{o}$-$\tilde{n}i$ in the sense of "sun." The base $\tilde{n}i$ becomes ni in P^{c}u-p^{c}a (ni-$z\ddot{o}$-ma) and Chö-ko (ni-ma). [3] A-hi Lo-lo has mu-$\tilde{n}o$ for "sun," besides li-ki; [4] Mo-so (written language) $m\ddot{o}$ ("sun"). In Black Lisu mu-$ts^c a$ and White Lisu $m\ddot{o}$-$\breve{c}^c a$ the meaning is perspicuous, being "heat of heaven" ($ts^c a$, "hot" = Lo-lo and Tibetan $ts^c a$). The same root covers much ground also in the Burmese languages: Karen $m\hat{a}$, ta-$m\hat{a}$, Yintale ta-$m\ddot{u}$-n, other dialects $t\ddot{u}$-mu, $m\ddot{u}$ $m\ddot{u}$-$\tilde{n}$; [5] Kachin la-mu.

35. Another Si-hia word, although I cannot trace it in Lo-lo or Mo-so, may be discussed in this connection, as it belongs to the same semasiology. This is transcribed in Chinese $ku(gu)$-$y\ddot{u}$-mo 骨魚沒, and is translated by Mr. Ivanov "supreme Heaven." The character mo, as, for instance, also in lo-mo 羅沒 = lom = Tibetan $k^c ron$-pa, $k^c rom$-pa (see No. 114), serves to denote a final -m. The combination ku-$y\ddot{u}$, written in the same manner, is intended also for the numeral "five," which may be restored to

---

[1] *Bull. de l'Ecole française*, Vol. IX, p. 550.

[2] These comparative considerations, again, refute the unwarranted opinion that Tibetan $mk^c a$ should be a Sanskrit loan-word (see *T'oung Pao*, 1914, p. 101); on the contrary, it is a dissyllabic compound framed from Indo-Chinese elements. The element $k^c a$ appears in Kami as $k^c \hat{a}$, $k^c u$, $k^c u$-$su\tilde{n}$, and $k^c au$-$sani$.

[3] The Lo-lo dialects make a far-reaching use of the stem mu in the terminology relative to atmospheric phenomena: mu-ho and a-mu-ho ("rain"), mu-$hl\ddot{o}$ and a-mu-$\check{s}i$ ("wind"), mu-$d\ddot{o}$ ("thunder"), mu-llo ("lightning"), etc. Compare also E. HUBER, *Bull. de l'Ecole française*, Vol. V, 1905, p. 324.

[4] The element ki perhaps coincides with the prefix ke- in Jyaruṅ ke-$\tilde{n}i$.

[5] *Gazetteer of Upper Burma and the Shan States*, Part I, Vol. I, p. 630.

k-ńü, k-ńu (see above, p. 13). In the same manner, the transcription *ku-yü-mo* leads to the restitution *k-ńum* or *g-ńum*, which in my opinion may be correlated with Tibetan *d-guń* and *g-nam* ("sky, heaven") and Burmese *koń-kań*.

36. Si-hia *mo* 沒, fire = A-hi Lo-lo *mu-tö, mö-te*, Nyi Lo-lo *mu-tu*, Pᶜu-pᶜa *mi-to* (Lo-lo-pᶜo and Lisu *a-to*; that is, stem *to* with prefixed *a-*), Chö-ko *bie-tu*; [1] Mo-so *m'ö, mi*; Tibetan *mö, me*; Lepcha *mi*; Newārī *mī*; Khambu dialects *mi*.

37. Si-hia *tsei, dsei, jei* 則移, water = Nyi Lo-lo *je*, A-hi Lo-lo *yi-j'e*, Lo-lo-pᶜo *vi-dye* (*dye* from **je*); [2] Mo-so *jie* (beside *gi*); Lisu *a-čia, i-čia*; Dhimal *či*: Limbu *čua*; Mišmi *m-čⁱi-n*; Jyaruń *ti-či* (a dissyllabic split of the primitive form; also in Chinese *šwi* 水 we have the *i*-vowel); Li-fan Tibetan *tse*. The last two forms are nearer to the Si-hia and the Lo-lo series than *čᶜu* of the Tibetan written language, which represents a later stage of development; but also in Tibetan we have the base *či* (*tši*) in *g-či-n* ("urine"); [3] compare Kachin *n-sin* ("water"). Lepcha *ji-t* ("urine"), derived from the same root, is certainly identical with the Si-hia and Lo-lo type for "water," *jei, je*. There can be no doubt also that the aspirate of Tibetan *čᶜu* is secondary, and that *čᶜu* has been evolved from **ču, ju* (*t-šu, d-žu*), still preserved in the derivation *b-ču-d* ("juice, sap"); [4] Newārī *čo* ("urine"). Tibetan *čᶜab*, the word for "water" in the respectful and elegant language, seems to be a

---

[1] As indicated by Miao-tse *tö* ("fire"), the element *tu, to, te*, in Lo-lo, is an independent word meaning "fire," so that the above compounds consist of two synonymes.

[2] Considering Karen *tö*, in cognate languages *ti* and *tai* (*Gazetteer of Upper Burma and the Shan States*, Part I, Vol. I, p. 652), Kami *twi, tu*, and *tui*, Thādo *tui*, Jyaruń *ti-či*, Miao-tse *t-le*, Kanaurī *ti* (*Linguistic Survey of India*, Vol. III, pt 1, p 423), we probably have to assume a primeval form **d-že, t-še*, from which the type *je, či*, etc., was evolved.

[3] In view of the Chö-ko form *zi* ("urine"), it seems admissible to associate the Lo-lo series of the same meaning (A-hi *zo*, Pᶜu-pᶜa *a-zo*, Lo-lo-pᶜo *še-vi*) with Tibetan *g-či-n*.

[4] Also in *b-ču, b-ču-s* ("to scoop water").

compound, presumably formed by contraction of $\check{c}^c u + bab$ ("to flow;" compare $abab$-$\check{c}^c u$, "river, rain"), as we have $snabs$ ("mucus, snivel"), from *sna-bab-s (literally, "nose-flowing"). [1] Also Tibetan $\check{s}wa$ ("flood"), $\check{c}^c ar(-pa)$, "rain," and m-$\check{c}^c i$-l-ma ("saliva") seem to belong to the same root.

38. Si-hia *wei, wö, bō* 為, snow = Mo-so (written language) *be*, (coll.) *m-be* (from *mu-be); Nyi Lo-lo *va*, A·hi *uo, wo*; Shan and Laos *nam m-we* (from *mu-we); Lisu *wa*; Bunan *mu*; Chepañ *če-pu* ("ice"); Newārī *čvā-pom* (from *$\check{c}e$-va-pom); Tibetan $k^c a$-ba, $k^c a$-wa. This case bears out the interesting fact that -*ba*, the second element of the Tibetan word, is not the affixed particle *ba*, but is an independent base meaning "snow;" as in West-Tibetan "snow" is called $k^c a$, the combination $k^c a$-ba ($k^c a$-wa) presents a compound consisting of two synonymes. Moreover, Tibetan $k^c a$ is evolved from *ga: the latter base occurs in $ga\dot{n}s$ ("snow, ice, glacier"); $ga\dot{n}s$ is to be analyzed into $ga\dot{n}$-s (-s being a suffix), and $ga\dot{n}$ is a compound contracted from *ga-$a\dot{n}$, for in Bunan we have the base $a\dot{n}$ as independent word for "ice." The latter notion is expressed in Milčañ by *pam, pañ*, which accordingly presents a contraction of *ba-$a\dot{n}$, *ba being identical with the above base *be, wa, wo* for "snow." The Milčañ word sheds light also on Newārī *čvā-pom* (from *$\check{c}e$-va--pom), which is a triple, and eventually even a quadruple compound. Finally the question may be raised whether Chinese *piñ* (from *$bi\dot{n}$, Annamese *bañ*) 冰 ("ice") does not belong to the same group.

39. Si-hia *lō, lo*, or *ro* 勒, wind = Nyi Lo-lo *mu š-le*, A-hi Lo-lo *mu h-lo* (or *h-lö*); [2] Burmese *le*, Kachin *la-ru* ("storm"), [3] Ahom *rau, raw* ("air, atmosphere"). The Si-hia stem-word has survived in the prefix *r-* of Tibetan *rluñ* ("wind") and *rdzi* ("wind"), which

---

[1] CONRADY, *Indochin. Causativ-Bildung*, p. 107.

[2] The form *h-lo* is evidently a contraction of *he-lo, as shown by Lisu *me-hei, mi-hi* ("wind").

[3] Kachin *m-boñ* ("wind"), evolved from *mu-$bo\dot{n}$ (*mu* = Lo-lo *mu*).

accordingly are developed from *ro-luṅ and *ro-dzi. Comparison with Bunan laṅ, Burmese luṅ, Ahom and Shan lum, lōm, Siamese lom, Chuṅ-kia rum, röm ("wind"), bears out the fact that Tibetan *ro-luṅ is a compound formed by two synonymes, each conveying the notion "wind." Mo-so ör [1] is obviously identical with Tibetan ur ("roar of a tempest"), to which possibly also the element ru of Kachin la-ru belongs; presumably Tibetan ur is only a metathesis of ru = *ro ("wind").

40. Si-hia le, lo 勒, earth, field = Mo-so le ("field"), Hua Miao liai, Hei Miao li, Ya-čʿio Miao lie (S. R. CLARKE, Among the Tribes in South-West China, p. 312); Lušai and Lai lo, Raṅkʿol loi, Kami la, le, Meitei and Thādo lau, Shö lai, Burmese lay-yā (le); Red Karen lyā, kē-lā; Leṅ-ki Miao le ("earth"); Khamti la-ṅuin ("earth"); [possibly Shan and Ahom na ("field"); Siamese and Chuṅ-kia na;] Chepaṅ b-lu ("cultivated field": HODGSON, The Phœnix, Vol. III, p. 46); Tibetan k-lu-ṅ-s ("cultivated land, field, a complex of fields").

41. Si-hia lu 盧, stone. This word occurs in the compound lu-yi ("mineral coal;" yi 乙 means "charcoal"), corresponding in sense to Tibetan rdo-so. [2] The word lu answers to Mo-so lu, lö, lu-n, lu-pa; Nyi Lo-lo lu-ma, A-hi Lo-lo lo-mo or lo-po, Lo-lo-pʿo lo-di, Pʿu-pʿa lo-ka, Chö-ko lo-ma; Black Lisu lu-ti, White Lisu lo-ti; Karen lå, Yintale lo-n, Manö lü, in other dialects of this group lo, lo-m, lo-m-tu; [3] Kachin n-loṅ; Khambu dialects luṅ; [4] Lepcha laṅ, luṅ; Mišmi m-p-la; Newārı lo-ho; Tʿočʿu γo-lo-pi, Sok-pa čʿi-lo;

---

[1] Transcribed ḣʿeur by M. BACOT (Les Mo-so, p. 52).

[2] Coal is nowhere found in Tibet. It is known there, however, as a produce of China, as noticed at Si-ning and other marts and border-places.

[3] Gazetteer of Upper Burma and the Shan States, Part I, Vol. I, pp. 650, 651. Possibly also Siamese pʿloi ("jewel") belongs to this group. The element tu is related to do in Tibetan r-do.

[4] Linguistic Survey of India, Vol. III, pt. 1, p. 346.

Jyaruṅ *ru-gu*; Tibetan *r-do*, from *lu-to, *lu-do*, *ru-do*, or *ra-do* (Bunan *ra*, Milčaṅ *ra-g*, *ra-k*; Ya-čˤio and Hua Miao *re*, "stone").

42. Si-hia *si*, *zi* 西, grass = Mo-so *zi*; Bunan, Milčaṅ and Kanaurī *či*; A-hi Lo-lo *hi*; Nyi Lo-lo *še*, Lo-lo-pˤo *šö-ba*; Lo-lo of Tˤung River (Baber) *jih-pa*; Meng-hua Lo-lo *šo*; Yün-nan Tibetan (Davies) *su*, (Monbeig) *tsoa*; Central Tibetan *tsa*; Ladākhī and Lahūl *sa*; Balti and Purig *r-tswa* (*tsu̯a*), *s-tswa*; Tibetan written language *rtswa*; Chinese *tsˤao* 草, Hai-nan *šau*.

43. Si-hia *fu* 縛 (probably to be restored to *bo*, *vo*; *fo*?), flower = Mo-so *bo-bo*, *ba-ba*; A-hi Lo-lo *vi-lo*, Nyi Lo-lo *vi-lu*, Lo-lo-pˤo *ve-lu*; White Lisu *su-wei*, Hua Lisu *su-wye* (Newārī *s-woṅ*); Kirānti (*buṅ-*)*wai*; Gyami *kˤwā*; Milčaṅ *u*; Karen *pˤō*; possibly also Chinese *hua*, *hwa*, *fa* 花, as proposed by Mr. Ivanov.

44. Si-hia *čˤi-ma* 吃麻, orange. This word corresponds to Nyi Lo-lo *čˤu-se-ma* and Tibetan *tsˤa-lum-pa*. According to a phonetic law (see Phonology, § 1), Si-hia *čˤi* is the equivalent of Tibetan *čˤa*, *tsˤa*. It is reasonable to suppose that the term in the three languages is a loan-word, but its history is still obscure. In A-hi Lo-lo, the name for the orange is *hua-ko*, explained by Liétard [1] from Chinese *huaṅ kuo* 黃菓 ("yellow fruit").

45. Si-hia *tsu-ni* 卒尼, man, homo (*ni* being a suffix) = Nyi Lo-lo *tsˤo*, A-hi Lo-lo *tsˤu*, Lo-lo-pˤo *tsˤa*, Black Lisu *tsˤou-tsa*, White Lisu *tsˤo-tsa*, Hua Lisu *la-tsˤu-n*; Pˤu-pˤa *čö*, Chö-ko *u-čö*; Mo-so *zu-ču* (*zu* = A-hi *zo-pˤo*, Lo-lo-pˤo *u-čöl-me*, Pˤu-pˤa *za*, Burmese *sû*, *θo*, "vir"); Leṅ-ki Miao *tsi-ne*. Further relationship may exist with Manyak (Hodgson) *čˤo* ("man"), Tibetan *tsˤo* ("number, host"), plural suffix of living beings and pronouns, and possibly with Chinese *tsu-t*, *tswo-t* 卒 ("servant, retainer, soldier," etc.). In Lepcha,

---

[1] *Bull. de l'Ecole française*, Vol. IX, p. 557.

the males of some animals are expressed by *a-tsu*, as, *luk* ("sheep"), *luk-tsu* ("ram"); *món* ("pig"), *món-tsu* ("boar"). [1]

46. Si-hia *gu-tsu* (*ñu-tsu, wu-tsu*) 吾 祖 or 兀 卒, king. This is the Si-hia term for "king," handed down in the Annals of the Sung Dynasty (*Sung shi*, Ch. 485, p. 8). This title was officially adopted in 1032 by King Li Yüan-hao 李 元 昊, [2] and is said to have the meaning of *khagan* 可 汗. The reasons why I prefer the reading *gu* are prompted by the results of linguistic comparison, as stated below, and by the fact that Rashid-eddīn calls the King of Tangut تنكقوت Lung-šādir-ghū لونك شـادرغـو. [3] Lung-šādir appears to be intended for his name, while *ghū* is his title ("king"). Parallels to the Si-hia title we meet in Nyi Lo-lo *tsᶜie ko-tsᶜo* ("*chef de village*") [4] and *o-ko-tsᶜo* ("*homme de la tête*" = *chef*). [5] The Lo-lo word *tsᶜo* means "man," and as the corresponding Si-hia word is *tsu* (No. 45), we are perfectly justified in identifying with it the second element of *gu-tsu*. The attribute "the man, the male" in the royal title is very similar to the *pᶜo* in ancient Tibetan *rgyal-pᶜo*. [6] Si-hia *gu* belongs to the base **go* ("head, chief") that we find in Tibetan *m-go* ("head"), *ạ-(m-)go-pa* ("head man, chief, alderman"), *go-ñ-ma* ("a superior, emperor"); Nyi Lo-lo *ge-mu* ("king, emperor"), A-hi Lo-lo *ñö-ma, rö-mu, wo-mö* (these variants illustrate very well that Si-hia *gu* could have been sounded also *ñu, wu; gu*, at any rate, represents the older form, and *ñu, wu*, a subsequent development; at the same time, these phonetic variants of A-hi

---

[1] G. B. Mainwaring, *A Grammar of the Rong (Lepcha) Language*, p. 25.

[2] G. Devéria, *L'Ecriture du royaume de Si-hia*, p. 17.

[3] F. v. Erdmann, *Uebersicht der ältesten türkischen Volkerstämme nach Raschid-uddin*, p. 62 (Kasan, 1841); E. Blochet, *Inscriptions turques de l'Orkhon*, p. 51.

[4] P. Vial, *Dict. français-lolo*, p. 344 (see *village*). Regarding *tsᶜic* ("village") see No. 47.

[5] *Ibid.*, p. 79.

[6] *T'oung Pao*, 1914, p 102, note 2

testify to the correctness of the equation here proposed); and Mo-so *gi-bu* ("king").

47. Si-hia *tsö-ni*, town. This word is known to me only from the note given by M. PELLIOT in *Journal asiatique* (1914, mai-juin, p. 506). Mr. Ivanov has equalized the Si-hia word with a Tibetan term transcribed by him in Russian *dzon*; this is intended for *dzoñ*, written language *r-dzoñ*. Mr. Ivanov certainly does not visualize the Tibetan word *groñ*, as supposed by M. Pelliot; and Mr. Ivanov, as will be seen presently, is quite right in his conception of the matter. First of all, the Si-hia word *tsö-ni* (stem *tsö*, *ni* being a suffix) corresponds to Nyi Lo-lo *ts^cie* ("village") and to the first element in Mo-so *je-nua* ("village"); Ahom *če* ("town"), Shan *če* ("province"); Tibetan *ts^co* in *yul-ts^co*, *groñ-ts^co* ("village"). A-hi Lo-lo *k^cye* ("village"), in all probability, is derived from a different base; and, as shown by the compound *k^cye-ra-mo* (LIÉTARD: "*bourg*"), is connected with Tibetan *k^cyer* in *groñ-k^cyer* ("town"). Tibetan *k^cyer*, *k^cye-r*, is a contraction of **k^cye-ra*, exactly corresponding to A-hi *k^cye-ra-mo*, and consists of the bases *k^cye* ("house," from which also the word *k^cyi-m*, "house," is derived; primeval form **gi*, *g'i*) [1]

---

[1] Tibetan *k'yim* ("house") has been brought together with Burmese *im* by B. HOUGHTON (*J.R.A.S.*, 1896, p. 44). In the Kuki-Chin languages, with the exception of Mei-tei, it is *im* or *in*. The Mo-so word for "house" is given by M. BACOT in the form *gi*. In Bunan, "house" is *gyum* (Lepcha *k'yum*, Abor *e-kum*, and Meitei *yum*, show likewise the *u*-vowel). Mo-so and Bunan have preserved the original initial sonant; and in view of Moso *gi* and Burmese *im*, we are bound to presume that Tibetan *k'yim* is evolved from **gi-im* or *gi yim* (Bunan from **gi-yum*, compare Meitei *yum*), the aspirate surd being a subsequent development. Newārī *č'em* and Thādo *čen* ("house") have arrived at the same stage as Tibetan *k'yim* Tibetan **gi*, further, corresponds to Chinese **gia*, *kia* 家 and Hei Miao *gie* ("house"). In regard to the change of initial guttural sonant into surd compare Chinese **gŭ* (*kŭ*) 懼 ("to fear"), Bunan *gyar-čum* with Ahom *kŭ*, *küw* (Shan *kuw*) and Tibetan *s-krag-pa*. In the same manner as *k'yim*, also Tibetan *lam* ("road") is a compound formed from **la-am*; for in Newārī we find *la* ("road") beside *lam*, and in Bunan *am*, *am-tsi*, *om*, Milčaṅ *om* ("road"); Tibetan **la* is presumably identical with *la* ("mountain-pass"), so that *lam* would mean "road leading over a pass," practically the only kind of road in Tibet.

and *ra* (*ra-ba*), "enclosure, wall, pen, fold." The word *kᶜyer*, accordingly, signifies an "assemblage or block of houses." [1] Si-hia *tsö* (probably to be restored to *dzö, jö*) is finally found also in Tibetan *rdzoñ*, which doubtless is divisible into *r-dzo-ñ* and evolved from *ra-dzo-ñ (*ra = *ra*, "enclosure"). The usual pronunciation in Central and Eastern Tibet is *joñ*. [2] There is, in all probability, further relationship with Chinese **dzoñ, joñ* (*dziañ, ziañ*) 牆 ("wall"). Rashid-eddīn narrates that the people of Tangut have dwelt in towns and steppes since oldest times.

Two other Tibetan words merit consideration in this connection, — *groñ* ("town") and *kᶜrom* ("market, bazar"). Both apparently represent parallels developed from the same base, *gi-roñ (*rom*), *roñ* and *rom* being identical. Both the primeval form *gi and the later development *kᶜi are encountered in the cognate languages: No-su Lo-lo *gi kao*, Kuei-chou Chuñ-kia *gei*, Kuang-si Chuñ-kia *heo* ("market;" S. R. CLARKE, *Tribes in South-West China*, p. 310); Ya-čᶜio Miao *ki*; Nyi Lo-lo *kᶜe*, A-hi Lo-lo *čᶜö*; Ki-lao *kᶜö*, Hua Miao *kᶜü* ("market"); Kuki-Chin: Lai-kwa, Rañkᶜol *kū*, Thādo and Lušai *kᶜua*, Meitei *kᶜū-l* ("village"). Tibetan **roñ, rom*, may be compared with Kachin *ma-reñ, ma-re* ("village;" *ma* being a prefix), Burmese *rwā*.

48. Si-hia *či* 直, flesh = Mo-so *ši, še, šö* (written language *šü, še, ši*); Mi-ñag (BABER) *ši*; Tibetan *ša*; Kachin *ša-n*; Burmese *a-sâ'*; Chinese *žou*, **yu-k, ñu-k* (Japanese *šiku*) 肉 ; Lo-lo (Nyi *ra*; A-hi *ho, po-ho*) is different, but Lo-lo of Tᶜung River (BABER) *ši-ni*. Note

---

[1] From what has been said about the development of *k'yim* it follows that the primeval type of *k'ye-r* must have been **g'e-r, ge·r* (*ge·ra*). This **ge·r* I believe I recognize in the name of the border-fortress *Liger* لیکر, which, according to Rashid-eddīn, was situated in the realm of Tangut, and was destroyed by Chinggis.

[2] Mr. WALSH (*J R.A S*, 1915, p. 466) is quite right in saying that "a *dzong* in Tibet is not merely a fort, but a district, of which the fort is the headquarters of the administration under the Jong-pön." As shown by the etymology of the word, it originally designated a fortified ("walled," *ra*) place.

that both Si-hia *či* and Chinese *žou* have the deep tone, while Tibetan *ša*, owing to the change of the initial, is high-toned. Chinese *ki, či* 肌 ("flesh"), on account of being high-toned, cannot come into question for comparison.

49. Si-hia *la* 梓, tongue = A-hi Lo-lo *lo*; Black Lisu *la-č^cue*; Ahom *li*, Shan *li-n*; Guruṅ, Murmi, and Sunwār *le*; Magar *le-t*, Ṭōṭo *le-be*, Vāyu *li*, Lepcha *a-li* (with prefixed *a*), Lušai *lei*; Gešits^ca *v-le*; Bunan *l-he*; Burmese *lhyā*, Karen *p-li*. [1] Tibetan *lče*, as demonstrated by the facts of the cognate languages, is a compound evolved from **le + če* (*je*), each of these components signifying "tongue." The second element, *če*, is encountered in Mo-so *či*, Sharpa (Darjeeling) *če-lak*, Thāmi (Darjeeling) *či-le*, Yün-nan Tibetan *j'e-le*, [2] Jyaruṅ *de-ž-mí*, Chinese **džiet, žiet, še* 舌 (Hakka *še-t*, Canton *šu-t*, Fukien *sie-k*), Lo-lo-p^co *še-ve*; Lo-lo of T^cung River (BABER) *šie*.

Another Tibetan word for "tongue" is *ljags*, now assigned to the respectful style of speech (*že-sai skad*). Its origin is now clearly indicated by the phonetic writing *l-ja-g-s*, evolved from **le-ja* (**ja* = *če, je*), to which the consonants *-g-s*, terminating nouns, are affixed. The prefix *l-* in the verb *l-dag-pa* ("to lick") has doubtless sprung from the base *le* ("tongue").

50. Si-hia *la* 腬, hand = Mo-so *la*; Nyi Lo-lo *le-p^ce*, A-hi Lo-lo *lye-pö*, Lisu *le-pe* (Hua Lisu *la-kua*); Kachin *la-ta, la-pan*; Pahrī *lā*; Newārī *lā-hā*; Tromowa and Sikkim *la-ko*; Tibetan *la-g*; Taungyo (Karen language) *la-k*.

51. Si-hia *k^co* 刻, foot = Mo-so *k^cö, k^cu* [A-hi Lo-lo *k^ci-bye*?]; Hor-pa (HODGSON) *ko*, Sok-pa (HODGSON) *k^co-il*; Li of Hai-nan *k^co-k*; Kami and Shö *k^co* (also *a-k^co* and *a-k^cu-t*); Meitei *k^co-ñ*; Thādo *k^cu-t* ("finger"), *ke-ñ* ("foot"); Lušai *ke*; Burmese *k^cre*; Kachin *la-go-ñ*.

---

[1] Newārī *me* is isolated.

[2] The two forms *či-le* and *j'e-le* are merely inversions of Tibetan **le-če*.

4

The primary base is *go, preserved in Kachin, further in Kanaurī *gu-d* ("hand, arm"), and in Old Chinese *gu-k 脚 ("foot"); it further appears in Tibetan *a-gro* ("to go, walk"), *gro* from *go-ro (*ro also in *a-grul*, from *go-ru-l), and in Mišmi *mgro* ("foot"), from *me-go-ro.

52. Si-hia *k͑o-i* 刻移, boot. Probably a derivation from *k͑o* ("foot").

53. Si-hia *mei*[3] 每, eye = Lo-lo-p͑o *me-du* (Bhūtān *mi-do*, Sokpa *nū-tu*); Mo-so *mö*, *m͑ö*; Lisu *mie-su*; Jyaruṅ *d-mye*; Newārī *mi-k͑ā*; Chinese *mu-k* 目; Tibetan *mi-g*; Kirānti *ma-k*; Burmese *mya-k*, *mye-t*; Thādo *mi-t* (*mū*, *mū-k*, "to see"). The base is *me, mi, etc. (compare also Tibetan *me-loṅ*, "mirror," Takpa *me-loṅ*, according to HODGSON, "eye;" Hor-pa *mo*, Manyak *mni*, from *mo- or *me-ni*; Kachin *myi*; Chinese *mou* 眸, "pupil of the eye," and *mou* 瞀, "near-sighted," derived from the same base). In the Si-hia word for "lungan," *wöi-mei*, which means "dragon's-eye," being a literal translation of the Chinese term *luṅ-yen*, the word *mei* ("eye") is written *mei*[2] 梅. It is not, however, a difference in tone which is here intended, but it seems that the classifier 木 was added to the phonetic element merely for the purpose of indicating to the eye the botanical character of the term.

54. Si-hia *mo*[4] 墨 eyebrows = Chinese *mei*[2], *mi*[2] 眉; Tibetan *s-mi-n-ma*; Newārī *mi-sa*. The Lo-lo and Mo-so terms are not on record. On p. 1232, Mr. Ivanov states that "eyebrows" is *mo-ma*, so that we should have the same affix as in Tibetan.

55. Si-hia *si* 息, liver = Nyi Lo-lo *se*; Mo-so *se-r*; Jyaruṅ Tibetan *te-že*; Tibetan written language *m-č͑i-n*, from base *či; Kanaurī *či-pur* and *śi-ṅ*; Newārī *sya-lā*.

56. Si-hia *č͑i* 吃, gall = Nyi Lo-lo *če*, A-hi Lo-lo *i-ki*; Tibetan *m-k͑ri-s*; Lepcha *k͑i-bo*; Kachin *ža gri*.

57. Si-hia *mo* 沒, lip = Lo-lo-p‘o *me-čo*; Tibetan *m-č‘u*, evolved from **me-* or *mo-č‘u*. Nyi Lo-lo *ñi-p‘u* and A-hi Lo-lo *ni-p‘ye* are independent words.

58. Si-hia *č‘ui-ko* 垂箇, tooth. This word represents the combination of two synonymes, each with the meaning "tooth." Si-hia *č‘ui* is not related to Tibetan *so*, as stated by Mr. Ivanov, but is akin to Nyi Lo-lo *č‘e-ma*, Hua Lisu *ts‘e-ču-r*, Lo-lo of T‘ung River (BABER) *ji-ma*; Shan *k‘iw*, Ahom *k‘riw*; Tibetan *ts‘e-m-s*, Burmese *ꙃwä*, Chinese *č‘i* 齒. The element *ču* of Hua Lisu answers to A-hi Lo-lo *ča-rö* or *ča-ho*, and to the second element in Lo-lo-p‘o *so-čo*. The latter compound plainly shows that the series with initial sibilant is distinct from that with initial palatal surd. The element *so* in Lo-lo-p‘o *so-čo* certainly is identical with Tibetan *so*, Gešits‘a *šo*; it further occurs in P‘u-p‘a *su*, Chö-ko *su-ma*, Mo-so (written language) *še*, and in the languages of Nepal: Gurung *sa*, Murmi *s-wä*, from **sa-wa* (Newārī, Takpa, and Pahrī *wä*), Magar *śyäk*, Ṭōṭō *si*. The word *ko*, forming the second part of the Si-hia compound, is met in Mi-ñag (BABER) *fu-k‘wa* ("molars"), Karen *ku-kö*, Manö *ku-ki*, Yintale *ta-kai*;[1] Siamese *k‘iau*, Shan *k‘io*, Laos *kiu*.[2]

59. Si-hia *wo-wei* 訛味, stomach, abdomen. This is a compound formed by two synonymes, that are found in the same manner in La-hu Lo-lo (DAVIES) *wop-pe* (for *wo-pe*). The first element, *wo*, occurs in Tibetan *p‘o-ba* (*p‘o-wa*), Tromowa *p‘o*, Yün-nan Tibetan *a-po*; Lepcha *pu-p*, *ta-fu-k* ("abdomen"); Red Karen *p‘ū*, Burmese *wam-pü-k*, *po-k*; Kuki-Chin group: Meitei *pu-k*, Lušai *pu-m*, Lai *på*. The second element, *wei*, is found in Chinese *wei*[4] 胃, "stomach" (note the identity of tone with *wei*[4] 味), Nyi Lo-lo *e-pi* ("abdomen"), in other Lo-lo dialects *vi-mu*, Thādo (Kuki-Chin) *wai*, *oi*.

---

[1] *Gazetteer of Upper Burma and the Shan States*, Part I, Vol. I, p. 648.

[2] *Ibid.*, p. 630; and MASPERO, *Bull. de l'Ecole française*, Vol. XI, p. 167. Perhaps further relationship exists with Amoy *ge*, Ningpo *ꙇo*, and Cantonese-Hakka-Fukien *ꙇa*, *ň'a* 牙.

60. Si-hia *ṅo* 蒡, back. This formation is isolated, as far as the initial guttural nasal is concerned, which in the cognate languages answers to *g*: Mo-so *gu-dse, gö-se*; Lepcha *ta-gu-m*; Tibetan dialects *g'a-p, g'e-p*, written language *r-gya-b*; Guruṅ *g'o*.

61. Si-hia *wu-yi, u-yi, u-i* 勿移, backbone. The first element is presumably related to No. 60; in this case -*i* is a suffix (see Morphological Traits).

62. Si-hia *no* (*lo*) 羅, disease = A-hi Lo-lo *no*, Nyi Lo-lo and Tibetan *na* (Tibetan also *na-d*), Burmese *nâ*.

63. Si-hia *tsö-wei* (*wö*) 則胃, hatchet. The nearest approach is BABER's Mi-ñag *wo-tsu* ("axe") with inverted members; for the Lo-lo of T°ung River he gives *wu-ma*; compare Chö-ko *sa-va*, Kachin *niṅ-wa*. In Nyi Lo-lo *ra-tsu* ("hache"), *ra* corresponds to Si-hia *wei*; and *tsu*, to *tsö* (regarding the interchange of *w* and *r* compare Nos. 24 and 25); A-hi Lo-lo *ö-ts°ö*, Lo-lo-p°o *a-tso*, P°u-p°a *sa-č̌u*. Compare further Tibetan *tog-rtse, tog-tse, tog-tsö* ("hoe"), Yün-nan Tibetan *tar-rö* (common Tibetan *sta-re*); Thādo *tū-tsā* ("hoe"). Chinese *pu, fu* 鉄, 斧 ("axe") may be grouped with Si-hia *wei*, Mi-ñag *wo*, Lo-lo *wu*.

64. Si-hia *le, lo* 勒, heavy = Lo-lo-p°o *li*;[1] Kachin *li*; Mo-so *li, lia*; Black and White Lisu *a-ke-li*, Hua Lisu *li*; A-hi Lo-lo *h-lö*; Nyi Lo-lo *l-je*; Tibetan *l-č̌i*, evolved from **le-* or *li-č̌i*; Bunan *li-ko*; Kanaurī *li-k, li-g*; Milčaṅ *li-hig*.

65. Si-hia *mi* 迷, high = Mo-so *me-ša-ša*, Lo-lo of T°ung River (BABER) *a-mo-so*; Nyi Lo-lo *mu*, A-hi Lo-lo *mo*; La-hu Lo-lo (DAVIES) *mwa*; Tibetan *m t°o*, from **me-t°o* (the vowels of these restored elements naturally remain uncertain); Burmese *mraṅ, myen, t'wā*; Yün-nan Tibetan *t°wa*. The Mo-so and Lo-lo elements *ša, so*, are comparable with Ahom *šu-ṅ*, Shan *hsu-ṅ* ("high"), Thādo *a-sāṅ*

---

[1] A. LIÉTARD, *Au Yun-nan*, p. 222.

("high"). E. H. Parker (*Up the Yang-tse*, p. 273) notes a Miao-tse word *ša* ("high").

66. Si-hia *na-kü*, *na-gü* 那 局, at night. The second element seems to correspond to Mo-so *mu-ku*, *me-k^cu* ("night"), *mö-kö* ("evening"), *u-ko* ("midnight"); and Black Li-su *mu-k^ce* ("night"); Ahom *k^cü-n*. The Mo-so element *me*, *mu*, answers to Lo-lo-p^co *a-mo-čo* ("night"); otherwise the Lo-lo languages have different words: A-hi *so-vu*, P^cu-p^ca *na-si-lya* (*na* possibly related to the first element of the Si-hia word), Chö-ko *si-pa-i*. The equation of Si-hia *na-kü* with Tibetan *nam-guñ*, proposed by Mr. Ivanov, is untenable; for *nam-guñ* means "midnight" (*nam*, "night;" *guñ*, "middle"), while in the Si-hia term *kü*, *gü*, refers to "night," and *na* might well be a pronominal or adverbial element. We meet the latter again in Si-hia *na-lo* ("to-morrow"); compare further Milčan *na-sam*, "to-morrow" (*sam* = Tibetan *sañ*, "to morrow;" Bunan *zañ-ma*, "day," *bar-sañ*, "year").

Group III.

Words Related to Lo-lo and Mo-so, without Equivalents in Tibetan and Chinese.

67. Si-hia *lo* 勒, tiger = A-hi Lo-lo *lo*, Nyi Lo-lo *la*, Lo-lo-p^co *lo-mo*, Lisu *la-ma*; Mo-so *la*; Manyak (Hodgson) *lêphê*. This series is independent of Tibetan *stag*, Chinese *fu*, *hu* 虎, and the type *su*, *sü*, encountered in all T^cai languages (Siamese and Black Tai *süa*, Shan *sü* [*süw*], Ahom *su*, *shū*, Khamti *sü*).[1] A base with initial *s* occurs likewise in Kačari *mo-sa*, *ma-sa*, and Lepcha *sa-t^cañ* or *sa-t^coñ* ("tiger").[2] This word simultaneously shows us that Tibetan *stag* is a compound, which has arisen from *sa-tag, sa-d^cag. The

---

[1] Maspero, *Bull. de l'Ecole française*, Vol. XI, p. 167.

[2] Perhaps also in Lušai *sa-kei*, and Kachin *si-roñ*; -*kei* answers to Meitei *kei* and Burmese *kyā* ("tiger").

element *sa* survives in Tibetan *g-sa* ("*Felis irbis*"), the prefix *g-* being identical with that in *g-zig* ("*Felis leopardus*"). The second element, **tag*, *dᶜag*, corresponds to Lepcha *tᶜaṅ*, *tᶜon*; Newārī *dᶜū*, *dᶜuṁ* (at present *tuṅ*); [1] Wāliṅ *dᶜin(a)-ra*; Kami *ta-ka-i* (or *ta-ke-i*); [2] and language of Chᶜu *wu-tᶜu* 烏菟 or *tᶜu* (from **du*, *dᶜu*) 虍兔. [3]

Also in Tibetan we find a form with initial aspirate surd: *tᶜuṅ-ṅa* ("a three-years-old tiger" 三歲虎), [4] which seems to have arisen from *tᶜu-ṅa* (**dᶜu-ṅa*). While the ancient language of Chᶜu is thus connected with Tibetan and still more closely with Nepal, the *l*-base of Lo-lo, Mo-so, and Si-hia is wedged in between them. This state of affairs is curious, and may raise the question as to whether it may be due to an outside influence upon these languages. In the Mon-Khmêr family we have also an *l*-base for the designation of the tiger: Kuy *kᶜo-la*, Khasi *la* and *kᶜla*, Mon and Bahnar *k-la*, Khmêr *kᶜ-lā*, Stieṅ *klah*; Kolh *ku-la*; [5] Palauṅ *la-wai*, Wa *ra-woi*. A derivation of Lo-lo *lo* from this quarter, of very ancient date, seems to me quite possible.

68. Si-hia *ye* 野, sheep, mutton, wether = Mo-so *yo*, *yu*, *yü*; Nyi Lo-lo *jo*, A-hi Lo-lo *ju*; Hua Lisu *a-ju*, Black and White Lisu *a-čᶜö*; Min-kia, Mi-ñag and Lo-lo of Tᶜung River *yo*; Jyaruṅ *ke-yó*; Thādo *yao*, *yā-m*. [Possibly related to Hok-lo *yo*, Amoy *ye* 羊 (*yaṅ*).] [6]

69. Si-hia *tan*, *dan* 怛, mule. This word certainly is not related to Tibetan *rta* ("horse," see No. 7), as suggested by Mr. Ivanov,

---

[1] A. CONRADY, *Z.D.M.G.*, Vol 47, 1893, p. 565; and *Causativ-Bildung*, p. 106.

[2] The *k*-base denoting "tiger" occurs in Tᶜoču *kᶜoh*, Gyami *kᶜu*, Jyaruṅ *koṅ* (HODGSON).

[3] LEGGE, *Chinese Classics*, Vol. V, pp. 117, 297, T. DE LACOUPERIE, *Les Langues de la Chine avant les Chinois*, pp. 18, 19; CHAVANNES, *Quatre inscriptions du Yun-nan* (*Journal asiatique*, 1909, juillet-août, p. 31).

[4] Thus explained in Kᶜien-lung's Polyglot Dictionary (Ch. 31, p 5). JÄSCHKE, following I. J. Schmidt, says that it means "three years old, of animals."

[5] Compare E KUHN, *Beiträge zur Sprachenkunde Hinterindiens* (*S.B.A.W.*, 1889, p. 213).

[6] An interesting equation is presented by Tibetan *tᶜoṅ-pa* ("a young ram in the first and second year") and Chinese *tᶜuṅ* 撞 ("a young ram").

nor to Tibetan *dre* ("mule"), Jyaruṅ *dar-ke*. The latter word apparently is associated with the verb *a̱-dre-ba* ("to be mixed"), and accordingly means "mongrel." The Chinese word *lo-tse* 騾 子 has been widely disseminated over the Indo-Chinese area (for instance, A-hi Lo-lo *lo-dse*, Nyi Lo-lo *a-la-sa*; White Lisu *a-mu*["horse"]-*lo-tzu*; Shan *lă*, Siamese *la*, Laos *luwa*, etc.), and has transgressed the Great Wall (Mongol *lo-sa*, Niŭči *lao-sa*, Manchu *lo-sa*). Mo-so (written language) *ku* appears to be an independent word; but the first element of Mo-so *ten-ja* ("donkey") seems to be in some relation with our Si-hia word; Mo-so *ja* means "horse."

70—71. The Si-hia language possesses two words for "year,"— *kou* 荀 and *wei* 韋. Si-hia *kou* = Mo-so *kᶜu*, *kvu*; A-hi Lo-lo *kᶜu*, Lo-lo-pᶜo *kᶜo*; Black and White Lisu *kᶜo*, Hua Lisu *čie-kᶜo*; Lo-lo of Tᶜuṅg River (BABER) *koa*; Pᶜu-pᶜa *kwe-so-mo*, Chö-ko *ta-ko*; Burmese *ku*. Si-hia *wei* (perhaps *bei*) = Chuṅ-kia of Kuei-chou *bi*, of Kuang-si *bei*; Ahom *pĭ*, Siamese and Shan *pi*; Li of Hai-nan *po*, Tai of Phu-qui *suay*; [1] Old Chinese *s-wai*; Fu-kien *s-wui* 歲 (as phonetic *wei*); Ya-čᶜio Miao *sö*, Hua Miao *su-ṅ*.

72. Si-hia *tsu* 祖, winter = Mo-so *tse-lu* (*lu* = Si-hia *lu*, "season"); Nyi Lo-lo *tsᶜe*, A-hi Lo-lo *jye*; Hua Lisu *mu-tsᶜu*.

73. Si-hia *nöṅ*, *nö̆* 能, spring = Nyi Lo-lo *nä*; A-hi Lo-lo *ni*; Mo-so *ṅi*.

74. Si-hia *čᶜiṅ-ni*, *čᶜi-ni* 頃 尼, summer = Mo-so *je*; Nyi Lo-lo *ši*; Hua Lisu *mö-ši*.

75. Si-hia *nie*, *ṅe* 烈, mouth = Nyi Lo-lo *ṅi-ṅa*, A-hi Lo-lo *ni-pᶜye*, Hua Lisu *me-ne*, Mo-so *nö-ta* (compare also Magar *ṅer*, Ṭōṭō *nui-gaṅ*); perhaps also Ahom *na*, Shan *na* ("mouth, face"). [2] This element has shrunk into a prefixed consonant in Pᶜu-pᶜa *n-to* and Chö-ko *n-ku*. The element *to* in Pᶜu-pᶜa *n-to* is apparently identical

---

[1] MASPERO, *Etudes sur la phonétique hist. de la langue annamite*, p. 45.
[2] E. H. PARKER (*Up the Yang-tse*, p. 272) notes a Miao-tse word *ngha-niou* ("mouth").

with the second element in Mo-so *nö-ta*, with Pahrī *to*, and Newārī *mhu-tu*. The Lo-lo-pᶜo word *me-ku* is different from that in the other dialects, the element *ku* being identical with the first element in Black and White Lisu *kᶜua-pᶜei* (perhaps also Chinese *kᶜou* 口 and Tibetan *kᶜa*).

76. Si-hia *wu-ki* 勿卽 (probably *wu-gi* or *bu-gi*), kidney = Nyi Lo-lo *ju-ghe*. The element *wu, bu*, is possibly related to Jyaruṅ *po-ta* ("kidney").

77. Si-hia *k'ä-i, č'ä-i* 皆移, genuine, true = A-hi Lo-lo *če*, Nyi Lo-lo *če, je*, Mo-so *čo-n*; Ahom *te* ("truth"). Mr. Ivanov has added to this word the observation, "Tibetan *kie*." Such a Tibetan word is not known to me.

78. Si-hia *na* 那, dog (used only in the duodenary cycle, as far as we know at present) = Lo-lo *a-na, a-no* (see No. 175).

79. Si-hia *tsō, dsō, jö* 則, mountain = Ki-lao *dse*; Lo-lo-pᶜo *u-tsye-bo*, Hua Lisu *wa-či-la-ku*, Mo-so *ji-na-me*. The element *bo* of the Lo-lo-pᶜo word occurs as independent word for "mountain" in Nyi Lo-lo *pö*, A-hi Lo-lo *po*, Black Lisu *wa-pᶜö*; further, in Pᶜu pᶜa *bo-mi*, Chö-ko *ha-pu-ma*, Chuṅ-kia *bo*, Hei Miao *bao, biei*, Ya-čᶜio Miao *bie*, and seems to be allied to Siamese *bᶜu (pᶜu)*, etc.; [1] Thādo *mo-l.* It is fossilized in Li-fan Tibetan *b-se* (from **bu-se*).

80. Si-hia *dsei, jei* 則移, south = Nyi Lo-lo *šle-če*, A-hi Lo-lo *hli-ki* (the words *šle* and *hli* mean "wind"); Mo-so *i-čᶜi-me* (Black Lisu *yi-mö*).

81. Si-hia *la* 粹, north; possibly = Mo-so *lo* in *huṅ-gu lo*. The Lo-lo base is *ma* (A-hi *hli-ma*, Nyi *šle-ma*).

82. Si-hia *wu* 勿, east; possibly = Lo-lo of Tᶜung River (BABER) *bu-du*. The element *du* refers to the sunrise, as shown by Nyi Lo-lo

---

[1] MASPERO, *Bull. de l'Ecole française*, Vol. XI, p. 159.

će-du ("sun rising" = east).   Whether Newārī *wam*(-*tā*) may be utilized for comparison seems doubtful. [1]

GROUP IV.

T'ai Affinities.

83. Si-hia *liao* 料, blood.   As far as I know, this word meets its counterpart only in Siamese *lüet* ("blood"). [2]

84. Si-hia *k'o* 客, rice = Siamese *k'ao*, Shan and Ahom *k'au*, White Tai *k'ou*; [3] Leň-ki Miao *kia*; [4] Mo-so *k'ia*; Meng-hua Lo-lo *sa-k'ao*; La-hu *sa-k'a*. [5]

85. Si-hia *mo*, possibly *ma* 魔 ("forest"), meets no parallel in Tibetan or Chinese, or in Lo-lo (Nyi *se-šlai*, A-hi *le-bö*, Lo-lo-p'o *sö-dzö-li*) or Mo-so (*bi-na*, *nau* = Tibetan *nags*), but it is comparable

---

[1] The designations for the four quarters in Si-hia—*la* ("north"), *jei*, *dsei* ("south"), *wu* ("east"), *liu*, *rin* ("west")—do not agree with the corresponding names in Tibetan (*byaň*, *lho*, *šar*, *nub*). This is not surprising, in view of the fact that there is no coincidence along this line between Tibetan and Chinese and among the Indo-Chinese languages in general.   This group of words was evidently formed after the separation of the Indo-Chinese tribes   Tibetan and Burmese have only the word for one quarter of the compass in common; namely, that for the west, Tibetan *nub*, Burmese *nok*, proposed by B. HOUGHTON (*J. R. A. S.*, 1896, p. 34), provided this combination be correct.   For Si-hia *liň* 嶺 ("west") I have only one parallel to offer, Milčaň *niň* (A. CUNNINGHAM, *Ladák*, p. 403).

[2] Mr. Ivanov transcribes the character in question *hsieh*, and accordingly connects the word with Chinese *hüe* 血 .

[3] MASPERO, *Bull. de l'Ecole française*, Vol. XI, p. 167.

[4] T. DE LACOUPERIE, *Les langues de la Chine avant les Chinois*, p. 45.

[5] The last two words are contributed by H R. DAVIES (*Yün-nan*, p. 361).   Davies proposes also to connect this type with the Mon-Khmêr languages: P'u-man *n-k'u* (compare Kachin *n-gu*), Wa *n-gou*, Palauň *la-kou*, Khmêr *aň-ka*, K'a-mu *un-k'o*, Annamese *gao*, La *kao*.   While I do not share the view of Davies that Mon-Khmêr and Indo-Chinese are mutually related, his proposition notwithstanding is highly suggestive, inasmuch as the word may be a very ancient Mon-Khmêr loan-word in the T'ai languages (or *vice versâ?*).— Note that the T'ai word is employed only in some Lo-lo dialects: A-hi *ta-mi* and Lo-lo-p'o *te-mi* have the Chinese word *mi* 米; Nyi Lo-lo has an apparently independent word, *ts'i-se* ("*riz non decortiqué*") and *tsa* ("*riz cuit*"), where the vowel-change for the expression of differentiation in meaning is interesting.

with Leṅ-ki Miao *ma-le*, [1] and possibly with the series *pa* of the T͑ai languages. [2]

86. Si-hia *ma* 麻 ("tree"), which seems to root in the same base as the preceding word, likewise points to connection with the T͑ai family: Ahom and Shan *mai* ("wood, tree"), Siamese *mai* ("wood, tree"). Father Th. Monbeig [3] imparts a Yün-nan Tibetan word *ma-den* ("trunk of a tree"): the second element, *den*, is obviously identical with Tibetan (written language) *s-doṅ*, "stem" (coll. Tibetan *šiṅ-doṅ*, "tree;" Murmi *d͑oṅ*, "tree"), so that *ma* would have the meaning "tree." In other Tibetan dialects such a word *ma* is not known to me, and if it should occur only in the Tibetan of Yün-nan, it might well be a loan-word due to Shan influence. In Lo-lo and Mo-so we have a root *se, sö, so, sü*: Nyi *se*, A-hi *se-ts͑e*, Lo-lo-p͑o *so-dso*, P͑u-p͑a *su-ma*; Mo-so *se, sö, so, sa* ("wood"), *ndsö, se-n-dsen* ("tree"). In view of Chö-ko *si-ma* (in which the Shan element *ma* may be visible), a relationship of the Lo-lo and Mo-so series with Tibetan *šiṅ* ("wood, tree") is possible; also in Tibetan dialects, the final guttural nasal is eliminated, for instance, in Jyaruṅ *še*. The coincidence of Chö-ko *si-ma* with Newārı *si-ma* ("tree") is very striking.

87. The Si-hia word for "fruit," *ma* 麻, is evidently the same as that denoting "tree," and likewise meets its counterpart in Siamese *ma* ("fruit") and Chuṅ-kia *lek-ma* (*lek* is numerative). [4] In this case, however, we have a missing link in Nyi Lo-lo *ma* ("fruit").

88. An interesting T͑ai element in Si-hia is presented by the word *kun* ("man;" Ivanov, p. 1233, Chinese transcription not being

---

[1] T. de Lacouperie, *l. c.* The element *le* is related to *le-bö* of A-hi and *li-dú-ma* of Chö-ko (the *ma* of the latter perhaps is identical with Miao *ma-*).

[2] Maspero, *l. c.*, p. 159.

[3] *Bull. de l'Ecole française*, Vol. IX, p. 551.

[4] P. Vial, *Les Lolos*, p. 34.

given). In the T͡ʻai languages we have Siamese *kʻon*, Shan *kon*, Laos *kʻon*, Ahom *kūn, kun* ("man"). Perhaps also Tibetan *kun* ("all") belongs to this series.

89. Si-hia *kia, kʻa* 假, duck, wild duck = Hei Miao *ka* (S. R. CLARKE, *Tribes in South-West China*, p. 309). The relationship with Chinese *ya, ap* 鴨 ("domestic duck"), proposed by Mr. Ivanov, is difficult to admit.

90. Si-hia *wei-ma* 韋麻, pear (*ma*, fruit). The word *wei* may be connected with Ya-č͡ʻio and Hua Miao *ra*, Hei Miao *za* ("pear;" S. R. CLARKE, *l. c.*, p. 312; the *z*, according to Clarke, is "a rough initial sound, indescribable, and must be heard to be appreciated;" these difficult sounds, of which we have as yet no exact phonetic description, are recorded in a state of embarrassment as *z* or *r*; we found several examples where these correspond to Si-hia *w*, see Nos. 24, 25). Lo-lo has independent words: Nyi *se-če-ma*, A-hi *sa-li* (*li* Chinese loan-word); *se-ndu* (CLARKE).

GROUP V.

Chinese Affinities.

91. Si-hia *wu* 勿, father = Chinese *fu* 父.

92. Si-hia *ma* 麻, hair = Chinese *mao* 毛. Further parallels are in Guruṅ (Nepal) *mui*, in Limbu (Nepal) *mū-rĭ* ("hair of body"), in Saṅpaṅ (a Khambu dialect) *mwa* (in other dialects of the same group *mi, muṅ, muwa, mua, mui, māa*).

93. Si-hia *pa* 巴, palm of the hand = Chinese *pa* 把, to grasp with the hands; *pa-čaṅ* 巴掌, palm.

94. Si-hia *kwaṅ-niṅ* 光寧, neck. The first element of this compound seems to be related to Chinese *kōṅ*, Hakka *kiaṅ* (Korean *kiöṅ*) 頸 ("neck, throat"); but it is very striking that the Si-hia form closely agrees with Buṅan *koaṅ-gul, kwaṅ-gul* ("neck"), where

gul answers to Tibetan *m-gul*. The second element is perhaps comparable with Chinese *liṅ* 領 ("throat") and Lepcha *tuk-liṅ*, *tuṅ-liṅ* ("neck").

95. Si-hia *wu-ni* 勿 你, wild animals; possibly = Chinese *wu(t)* 物 ("creature, animal"); Chepaṅ *sva* ("quadruped"). Perhaps also Mo-so *go-ge* ("animal") belongs here.

96. Si-hia *tu*[1] 瀆, bean = Chinese *tou*[1] 荳 (*tu* is possibly a Chinese loan-word in Si-hia).

97. Si-hia *ya* 牙, goose = Northern Chinese *yen* 雁 (Fukien *ṅaṅ*, Cantonese and Hakka *ṅan*; Tibetan *ṅaṅ*).

98. Si-hia *maṅ*, serpent (sixth year of the cycle, but presumably the general word for "snake")[1] = Chinese *maṅ* 蟒, python.

99. Si-hia *ćai-ni* 窜 尼, fox; possibly = Chinese *ćʻai* 豺. In Lo-lo and Mo-so we have a base with initial *d*: Nyi *o-du-ma*, A-hi *a-dö*, Mo-so *ndra*, and Mo-so written language *da*; *pa*, another word for "fox" used in the latter, apparently is identical with Tibetan *wa* (also *wa-tse*), Bunan *goa-nu*, *g-wa-nu* (Tromowa *am*, *a-mu*; Sikkim *am*); perhaps also Ahom *ma*, Shan *mā-lin*.

100. Si-hia *šaṅ-wei* 尙 嵬 (translated by Mr. Ivanov "shirt");[2] the first element *šaṅ* = Chinese *šaṅ* 裳 ("clothes on the lower half of the body").[3] The second element *wei* refers likewise to an article

---

[1] This may be inferred from the fact that Si-hia offers a word of its own for "dragon," which is used also in the Si-hia cycle for the designation of the dragon-year, while *maṅ* strictly refers to the year of the serpent. If *maṅ* in Si-hia would denote the dragon, I should not hesitate to regard it as a loan-word. As such we find *maṅ*, for instance, in Niŭči *maṅ-lu-wen*, being the equivalent of Chinese *maṅ-luṅ* 蟒 龍, in opposition to the native word *mudur* (Manchu *muduri*).

[2] Probably it is "undergarment." The Tibetan tribes and their congeners have no shirts. "Le Lolo ne porte pas de chemise" (P. VIAL, *Dict. français-lolo*, p. 79). The coincidence of Chinese *šam* 衫 ("shirt") with Tibetan *šam*, *g-šam* ("lower part of a thing"), *šam-gos*, *šam-tʻabs* ("lower garment, skirt"), is interesting.

[3] The character transcribing the Si-hia word is the phonetic element employed in the character 裳.

of clothing, for in the chapter on the Si-hia, embodied in the Annals of the Sung, this word *wei*, reproduced by the same character, is explained as "a red, knotted ribbon hanging down from behind the button of the official cap." [1] This certainly is a rather specialized significance, which does not fit the case of *śaṅ-wei*, but we are perhaps allowed to infer from this example that *wei* had also a more general meaning with reference to attire. Further, Si-hia *wei* may be compared with Lo-lo *bä*, *bi* ("coat"): Nyi Lo-lo *šlo-bä* ("*habit*"), *šla-bä* ("*pantalon, c'est-à-dire habit enveloppant la cuisse*"); A-hi Lo-lo *a-bi*, *ka-bi* ("*habit*"), *lo-bi*, *lu-bi* ("*pantalon*") [Mo-so *ba-la*, "*habit*;" Bu-nan *p꜀os*, "garment, dress"?]; Chuṅ-kia *bu*, Hei Miao *u* ("clothes"); Meitei *p꜀i*, Kachin *m-ba*, Burmese *a-wat*. The combination of *śaṅ* and *wei* leads me to conclude that *śaṅ*, in the same manner as *wei*, is a genuine old Si-hia word, not merely a Chinese loan-word. Chepaṅ *sum-ba* ("lower vest") [2] is an analogous formation, though I am not convinced that the element *sum* is the phonetic equivalent of Si-hia *śaṅ*.

101. Si-hia *ño, o* 餓, lake (Ivanov: "sea;" compare Tibetan *mts꜀o*, "lake," *rgya-mts꜀o*, "sea"). Possibly related to Chinese *hu* 湖, Cantonese *u, wu*.

GROUP VI.

Tibetan Affinities.

102. Si-hia *lu* 六, body = Tibetan *lu, lu-s*.

103. Si-hia *wu* 吳, head = Tibetan *wu, u* (written *dbu*); Mi-ñag (BABER) *we-li*. The same word occurs in the first element of A-hi Lo-lo *o-ko*, Lo-lo-p꜀o *u-di*, and Mo-so *wu-k꜀ua*. The element *ko* in A-hi *o-ko* and Chö-ko *i-ko* (further Mo-so *ku-lä, ku-lö*) is apparently identical with Tibetan *m-go* ("head").

冠頂後垂紅結綬自號嵬名 (*Sung shi*, Ch. 485, p. 5 b).

[2] HODGSON, *The Phœnix*, Vol. III, p. 45; or *Essays*, pt. 2, p 51.

104. Si-hia *wu* 悟, centre, central = Tibetan *wu, wu-s* (written *dbus*).

105. Si-hia *niṅ* 寧, heart = Tibetan *s-ñiṅ*.

106. Si-hia *o* (*wo, ṅo*)-*diṅ* 訛丁, neck (Ivanov), but more probably throat, windpipe = Tibetan *'o-doṅ* (written also *'o-ldoṅ*). The first element of the latter word, *'o*, appears also as *'og* (*'og-ma*, "throat, neck;" *'og-ajol*, "gullet") and *'ol* (*'ol-mdud*, "larynx"). It is further equivalent to *l-kog, r-kog* (*l-kog-ma*, "gullet, windpipe, throat, neck"), which is related to *m-gu-r, m-gu-l, m-g-rin* ("throat, neck"); that is to say, all these variations are derived from a common base **gu, go*, further developed into **ko* and *'o*. There is possible relationship with Chinese *hou* (in dialects *u, wu, ho, hau*) 喉 ("throat, gullet"), Siamese *kᶜo*, Pᶜu-pᶜa *ko-bya*, and Mo-so *ki-pa*; Black Lo-lo *ko*; Burmese *kup-jak* (*gok-zet*); [1] compare further Kanauri *gol·öṅ* and Bunan *sta-gor-wa*.

107. Si-hia *miṅ, mĭ* 名, man = Tibetan *mi*.

108. Si-hia *miṅ, mĭ* 名, not = Tibetan *mi*; Si-hia *mo* 沒 ("not") = Tibetan *ma*. Compare Canton, Hakka, and Shanghai *m* 唔 ("not").

109. Si-hia *ma šuo* 麻說, river, that is, the River = Tibetan *r-ma čᶜu*, the Yellow River (Huang ho).

110. Si-hia *pᶜu* 普, a place of higher altitude = Tibetan *pᶜu*, upper part of an ascending valley (Si-hia *pᶜu* is perhaps Tibetan loan-word).

111. Si-hia *kᶜia* 恰, magpie = Tibetan *skya ka, skya·ga*.

112. Si-hia *lo-tsei* (*tsö*) 羅賊, species of antelope; *lo* = Tibetan *lug* ("sheep"); *tsö* = Tibetan *gtsod* (*tsö*), *gtso, btso* ("*Pantholops hodgsoni* Abel").

113. Si-hia *po* 孛, species of antelope = Jyaruṅ *po, ba*; Kanauri

---

[1] E HUBER, *Bull de l'Ecole française*, Vol. V, 1905, p 325

$p^c o$ and $f\check{o}$ ("deer"); Ahom $p\bar{u}$ ("a fallow deer"); Tibetan written language r-go-ba, d-go-ba (Central Tibetan go-a), "*Procapra* or *Gazella picticaudata*." [1]

114. Si-hia *lom*, *rom* (*lo-mo*) 羅沒, spring, well = Tibetan k^c-ron-pa (k^c-rom-pa), Yün-nan Tibetan *lom-ba* (*Bull. de l'École française*, Vol. IX, p. 550), Lepcha *ram* (compare No. 35).

115. Si-hia *lu* 路, season = Tibetan *lo*, year.

116. Si-hia *na* 那, cereals, barley = Tibetan *na-s*, Tromowa and Sikkim *na*.

117. Si-hia *na* 納, deed, action; possibly = Tibetan *la-s*, Bunan *len*.

118. Si-hia *go*, *ko* 各, form; possibly = Tibetan *go-po* (written *sgo po*), "the body with respect to its physical nature and appearance." (Ahi Lo-lo *gö-mo*, Lo-lo-p^c o *gö ćö*, P^c u-p^c a and Chö-ko *gu-mo*, "*corps*".)

119. Si-hia *liṅ*, *riṅ* 令, great; perhaps = Tibetan *riṅ*, "long, high, tall" (or rather = Ahom and Shan *luṅ*, "great, large"?). This word *liṅ* doubtless occurs in the two titles of Si-hia officials, *niṅ-liṅ* and *mo-niṅ-liṅ* (G. DEVÉRIA, *L'Écriture du royaume Si-hia*, p. 18).

120. Si-hia *sie-niṅ* 薛寧, the day after to-morrow. The first element, *sie*, signifies "following, coming," as shown by the phrase *sie wei* 斜韋 ("next year;" *wei*, "year"). *Niṅ*, therefore, is a noun, and identical with Tibetan *naṅ(-mo)*, "morning," especially "the following morning;" *naṅs-par*, *naṅ-la*, written language *g-naṅ*, "the day after to-morrow." In like manner Si-hia *yiṅ* 盈 ("light, in weight") corresponds to Tibetan *yaṅ-po*, Lepcha *kyaṅ-bo*. Regarding Si-hia *i* = Tibetan *a*, see Phonology, § 1.

---

[1] See M. DAUVERGNE, *Bull. du Muséum d'histoire naturelle*, Vol. IV, 1898, p. 219. Mr. Ivanov translates the Si-hia word by "yellow sheep" as the literal rendering of Chinese 黄羊; but the latter term corresponds to Tibetan *rqo-ba*, Manchu *jeren*, Mongol *dsegēre*, *dsēre* (K'ien-lung's Polyglot Dictionary, Ch. 31, p 14).

121. Si-hia *tu* 潰, fruit; possibly = Tibetan *tᶜo-g* ("produce, fruit"), *lo-tog* ("annual produce, harvest"). Or the same as *tu* ("bean")? See No. 96.

122. Si-hia *ma-mo* 麻沒, mother = Tibetan *ma* ("mother"), *ma-mo* ("grandmother"). No equivalents in cognate languages are known to me for Si-hia *čᶜŏň* 成 ("mother"). Si-hia *nir (ni-lo)* 你羅 ("relatives") possibly is identical with Tibetan *ňer* (*ňe, ňen, g-ňen*). Si-hia *pᶜu-bu* (characters not given; certainly it is not *pᶜu-pu*, as transcribed by Mr. Ivanov), "the ancients," may be akin to Tibetan *pᶜu-bo*, *pᶜo-bo* ("elder brother").

GROUP VII.

Dubious Cases.

123. Si-hia *tsu-ni* 足尼, rain (-*ni* being suffix), has no correspondent equivalent in Lo-lo and Tibetan (Tibetan *čᶜar* belongs to the base *či, ju, ču*, see No. 37). Whether *tsu* can be correlated with Mo-so *šŏ*, Lepcha *so*, seems doubtful. It is striking, however, that we find an element *tsu* ("rain") in the Karen languages: Yintale *kan-tsu*, Manö *ka-ču*, Karen *ke-tsi* (in other dialects *kan, kam, kyan, kă, ka-le*). [1]

124. Si-hia *yiň-na* 迎那, agate. The first element *yiň* may be identical with the Si-hia word *yiň* 迎 ("star"). Whether the element *na* may be identified with Chinese *nao* 瑙, is questionable. The Tibetans transcribe Chinese *ma-nao* as *ma-nahu*; the Mongols have it as *manu, mānu*; the Niŭči as *ma-nao*.

125. Si-hia *yi, i* 移, woman = Kachin *yi* ("female"), Ahom and Shan *ī* ("the youngest of several, young girl"), perhaps also Shan *yiň*, Ahom *ňiň* ("female"). The character 移 has the same tone (even lower) as *i* 姨 ("a wife's sister, mother's sister"); but owing

---

[1] *Gazetteer of Upper Burma and the Shan States*, Part I, Vol. I, pp. 652, 653.

to the difference in meaning, relationship of the two words seems to me doubtful. The types of words expressing the notion "woman" are widely varying in Indo-Chinese, and even Tibeto-Burman has no word in common for it. A-hi Lo-lo *i-mo* ("female") presents merely a seeming coincidence with Si-hia, for, as shown by *i-po* ("male"), the base *i* means "man" (*-mo* = Tibetan *mo*, "woman, female;" Mo-so *a-mu*), and answers to *ya* in Lo-lo-pᶜo.[1] Lepcha *yu*, again, is a different word from Si-hia.

126. Si-hia *tsan* 拐, lungs (no relation to Chinese or Tibetan), possibly = Newārī *som*; Nyi Lo-lo *tsᶜe-pᶜu-ma*. Other comparative material is not available.

127. Si-hia *tsan* 拐, Chinese. Mr. Ivanov, who transcribes *tza*, suggests relation to Tibetan *rgya*, *rgya-mi*. This is difficult to prove. It is true that *rgya* is sounded at present *gya*, *g'a*, *jya*, and that the character in question was sounded in the Tᶜang period *ja* and *ča*;[2] but we cannot base our Si-hia readings on the Tᶜang phonology, still less is there any evidence that the Tibetans of the Tᶜang period articulated the word *rgya* as *ja* or *j'a*. The most probable assumption is that they pronounced it *gya*. The same Chinese character, that above and No. 126, is employed also for writing the Si-hia word *tsan* ("autumn"), which I am unable to explain through comparative analysis.

128. Si-hia *kwan* 幹, shoulder. This word, possibly, might be connected with Mo-so *kwa-pi* and Chinese *kien* 肩 ("top of shoulder"). On the other hand, Mo-so *koa-pi* may point to Chinese **kiap* 胛 ("part under and between the shoulder-blades"); the case, therefore, is still dubious. There are no corresponding forms in Tibetan or Lo-lo. The element *bu*, *bo*, *po* in Lo-lo (Nyi *bu-kᶜiä*, Pᶜu-pᶜa *na-po*,

---

[1] Compare Hua Lisu *mu-tsᶜu* ("woman"), from *mu* ("female") and *tsᶜu* ("homo").

[2] F. W. K. MÜLLER, *Uigurica*, II, p. 101. It renders *sa* in Sanskrit words (A. VON STAËL-HOLSTEIN, *Kien-chᶜui-fan-tsan*, p. 185).

5

Chö-ko *na-bo-ma*) may be related to Siamese *ba*, Shan *ma*. [1] A-hi Lo-lo *p̒a-ñi* may point to Tibetan *p̒ra-g*; and Chinese *pan* 髈 may be associated with Tibetan *d-pun*, Lepcha *tuk-pun*.

129. Si-hia *pu*, *bu* 不, spleen. Related to Newārī *al-pe*? The word is not recorded for Lo-lo and Mo-so; it is independent of Tibetan *m-č̒er(-pa)*, and Chinese *p̒i* 脾, which has the lower tone, while *pu* 不 is in the high tone.

130. Si-hia *no-čui-ni* 蔓追尼, chair. *Ni* is a suffix. The element *no* seems to be the same as the word *no* ("back"), transcribed by the same character (No. 60).

131. In the Indo-Chinese languages we meet a common word for "elephant" only in the eastern branch, T̒ai and Chinese: Siamese *čan*, Shan *san* or *tsan*, Khamti *čan tsan*, Laos *tsan* (Mo-so *tso*, *tso-n*), Palaun *san*, Ahom *tyan* (Lepcha *tyan-mo*); Cantonese *tsön*, Hakka *sion*, Fukien *č̒ion* (Japanese *šō*, *dso*), Northern Chinese *sian*. [2] In the western branch of Indo-Chinese there is no common word for "elephant." The Tibetans evidently made the elephant's acquaintance only when they came in contact with India, as shown by their term *glan-č̒en* or *glan-po-č̒e* ("big bull"). [3] In Nyi Lo-lo we have *a*, in A-hi Lo-lo *ro* (distant relationship with Lepcha *ran-mo?*); Mo-so *tso*, *tso-n*, as indicated above, points to the T̒ai group. In view of this diversity of words in the western branch of Indo-Chinese, it is not surprising to find a seemingly independent name for "elephant" in Si-hia *mu* 暮. Perhaps this word is related to some such form as Lisu *a-mu*, Chin *mwie*, Kachin *magwi*; but this is doubtful.

---

[1] MASPERO, *Bull. de l'Ecole française*, Vol. IX, p. 553.

[2] Mongol *dsān* (Buryat *zan*), written *dsagan* (*aga* being a graphic expedient to indicate the length of the vowel: compare *Begē-dsin* = *Pei čin* 北京 [Peking]; *togor* ["peach"] = *t̒ao'r* 桃兒; *lagūsa* or *lagūse* ["mule"] = *lo-tse* 騾子) is apparently a Chinese loan-word.

[3] Also *ba-lan* (from *bal glan*), that is, ox of Nepal.

132. Si-hia *č̓ui* 垂, nit. The Chinese word corresponding in meaning is *ki, či* 蟣; the Tibetan, *s-ro-ma*; the Kachin, *tsi-ti*. Unfortunately the word is not known to me in other languages, and I therefore hesitate to establish a relationship between the words named, although this seems not to be impossible. As to Tibetan *s-ro*, the element *s-* appears to be related to Lepcha *sa-* in *sa-fyat* ("flea"); and the element *-ro*, to Ahom *rau, raw*, "louse" (see No. 6).

133. Si-hia *liṅ* 令, tortoise. The word has no equivalent in Chinese or Tibetan (*rus-sbal*, "bone frog"). The following are merely tentative suggestions: Kami *ta-lī*, Newārī *kāp-li*, Kanaurī *ri-hōnts*, Kachin *tau-b-ren* (Ahom, Shan, Pa-yi, and Siamese *tau*, "tortoise").

134. Si-hia *ku-ni, gu-ni* 穀尼, flour. Possibly connected with Bunan *k̓u-g* ("meal of roasted barley").

135. Si-hia *maṅ* 疕, white. This word has no counterpart in Lo-lo, Mo-so, Tibetan, Chinese, or T̓ai. Perhaps Burmese *màṅ*, Thādo *boṅ*, and Chepaṅ *p̓am(-to)*, may be enlisted for comparison.

136. Si-hia *fu* (*bu, bo*)*-sai* 縛腮, lotus. The first element appears to be identical with the Si-hia word *fu* ("flower"), see No. 43; but it should be pointed out that Ahom *bū* (Shan *wuw, muw*) means "lotus, water-lily." See also p. 95.

137. Si-hia *miṅ* 名, lower as to position or altitude. Mr. Ivanov gives another word *miṅ* ("low, small"), but accompanied by a character reading *liṅ* 令. He further compares the latter with a Tibetan word, printed in Tibetan letters as *smeṅ*, and transcribed *me*. I presume that this is intended for Tibetan *smad*, which may indeed be pronounced *me*; this being the case, a relationship of Si-hia *miṅ* with the Tibetan word is hardly apparent. As some confusion has here arisen, judgment should be held in abeyance till we hear again from Mr. Ivanov himself. An application of the word *miṅ* doubtless occurs in the official title *ki-miṅ*, the lowest grade in the Si-hia official hierarchy (G. Devéria, *L'Écriture du royaume Si-hia*, p. 18).

GROUP VIII.

Words Peculiar to Si-hia.

There are naturally a number of words in Si-hia, which cannot be traced in cognate forms of speech, and which, at least for the present, must be characterized as peculiar features of this language. Those preserved in the Chinese Annals may first be passed under review.

138. Si-hia *se* 廝, sorcerer, shaman, priest (西夏語以巫 爲廝也。 *Liao shi*, Ch. 115, p. 3). As far as the Indo-Chinese languages are concerned, this word may stand isolated. [1] The Lo-lo and Mo-so use different words for their native medicine-men (Mo-so *to-pa, tuṅ-pa*; Lo-lo *pi-mo*). Chinese **bu, wu* 巫 ("shaman;" probably connected with **bu-k, pu-k* 卜, "to divine") belongs with Tibetan *ɋba* ("sorcerer;" *ɋba-g,* "mask;" *ɋbo-g,* "to sink down in a fainting-fit;" *smyo-ɋbo-g,* "madness"), Ahom *mo* ("a learned man, a Dēodhai or Ahom priest"), and possibly Burmese *rwa* ("witchcraft").

139. Si-hia *wei li* 嵬理. This was the name borne by the Si-hia King Nang Siao or Li Yüan-hao (1032—48) during his childhood, as stated in *Sung shi* (Ch. 485, p. 5 b). According to the interpretation there given, the word *wei* means "to regret, to pity" 惜; and the word *li*, "rich and of high rank" 富貴. These words are not given in our Si-hia glossary.

140. Si-hia *wei* 嵬 or 威 (compare No. 100), a red, knotted ribbon hanging down from behind the button of the official cap (*Sung shi, l. c.*; and *Liao shi*, Ch. 115, p. 2 b). The first of the characters given is employed in the *Sung shi*, the second in the *Liao shi*.

---

[1] EITEL, in his *Cantonese Dictionary* (p 419), gives a term "廝也 *sz-ma,* a witch" (GILES, No. 7871, "necromancers"). The history of this term is not known to me; perhaps this Chinese *se* is somehow connected with the Si-hia word.

This duplicity is striking, as the one has the even lower, the other the even upper tone. The reading presented by the Sung Annals, which is the older work, merits preference.

141. Si-hia *niṅ liṅ ko* 甯令哥. This was the juvenile name of King Li Liang-tso (1049—67), oldest son and successor to Li Yüan-hao (*Sung shi*, Ch. 485, p. 9). The word *niṅ* is explained in the Annals as meaning "to rejoice in what is good" (歡嘉). The word *liṅ* is said to be the name of the river Liang-ch'a 兩岔河. This may well be the case, but it is not plausible that the designation of a river should enter the personal name of a royal child. As *ko* means "elder brother" (our Si-hia glossary gives this word in the form *a-ko*), it would seem more probable that *liṅ* is identical with the adjective *liṅ* (No. 119), "great," *liṅ-ko* being the "great brother."

142. Si-hia *to-pa* 拓跋, title of the Si-hia sovereigns, said to signify "king of the earth" (G. DEVÉRIA, *L'Écriture du royaume Si-hia*, p. 15). It is doubtful, however, whether this etymology is correct; the word in question, in all probability, is not of Si-hia origin.

143. Si-hia *lo* or *ro* 勒, fir-tree. In Tibetan we have *som* or *gsom*, that has been compared with Chinese *suṅ* 松 by SCHIEFNER; [1] but *suṅ* is evolved from **zuṅ, dzuṅ, duṅ* (Sino-Annamese *tuṅ*, Cantonese *ts'uṅ*), and is rather related to Tibetan *t'aṅ*, Burmese *t'aṅ-rū*, Lepcha *duṅ-šiṅ* or *tuṅ-šiṅ* (*šiṅ*, "tree"); Newārī *t'a-sim* (*sim*, *si-mā*, "tree"); Mo-so (written language) *ton*, *t'o*; Nyi Lo-lo *t'o-se*, Lo-lo-p'o *ta dsö*. [2] See also p. 95.

144. Si-hia *yiṅ* 迎, star. The relation with Chinese *siṅ* 星, proposed by Mr. Ivanov, is not convincing: the Si-hia word, as indicated by the Chinese transcription, is in the lower tone, while

---

[1] *Mélanges asiatiques*, Vol. I, p. 340.

[2] LIÉTARD, *Au Yun-nan*, p. 216. The word *dsö* means "tree." The Polyglot Dictionary (Ch 29, p 17) justly identifies Tibetan *t'aṅ-siṅ* with *suṅ* 松.

siṅ has the upper tone, which makes for a net differentiation of the two words; the assonance *iṅ* is not conclusive, the initial sounds are the decisive factors in Indo-Chinese, and a correspondence of Si-hia *y* to Chinese *s*, for the present at least, cannot be established. In my opinion, therefore, Si-hia *yiṅ* is to be regarded as an independent Si-hia word.

145. Si-hia *yao* 要, day. In opposition to Tibetan and Lo-lo *ñi*, Chinese **ži-t, ñi-t* 日.

146. For Si-hia *la-nu* 粹怒 ("the four stars of the Dipper"), no equivalent in another language is known to me.

147. Si-hia *ko-ni* 葛[1] 尼, owl. The stem *ko* looks like an inversion of Tibetan *ug*; but this means nothing along the line of mutual relationship, as the two words may be conceived as independent formations mimetic of the bird's cry (compare Sanskrit *ulūka*, Latin *ulucus, ulula*; Mongol *ugūli, ūli*; Persian *kokan*; Kachin *u-kᶜu*; Ahom *kaw*).

148. As to other names of birds, *tsŏ-ni, jŏ-ni* 則尼 ("wild goose"), *mo-ni* 莫你 ("cuckoo"), *kᶜiṅ* 慶 ("pigeon"), *ta-yaṅ, da-yaṅ* 打檬 ("swallow"), *yaṅ-hei* 檬黑 ("raven;" *hei*, "black"), *taṅ-laṅ* 党俍 ("quail"), seem to be word-formations peculiar to Si-hia.

149—150. Si-hia *saṅ* 桑 ("male") and *tu, du* 瀆 ("female") appear to be words peculiar to Si-hia. For the latter I find a slight comparative indication in Miao-tse *tsᶜu-to* ("female"). [2] As *tsᶜu* means "man" in general, Miao-tse *to* must have the meaning "female." There is, further, *na-tǎ* ("women") in Chin. [3] The word *saṅ* is

---

[1] Mr. Ivanov (p. 1229) prints the character *ko* 各; but in the plate attached to his paper, where a facsimile of two pages of the glossary is given, the character appears as above.

[2] D'OLLONE, *Langues des peuples non chinois*, p 66.

[3] Kanauri has a feminine suffix *-de* joined to verbal forms ending in *či* or *ži*: *čᶜug-ži-de* ("a lady visitor"), *lan-či-de* ("she who wears"); but I do not feel sure that this *-de* is connected with Si-hia *tu, du*.

utilized in *wei-san* 崔桑 ("sparrow"), literally, "male bird;" *wei* ("bird") is inferred from the phrase *čʿan-ni wei* ("flying birds").

151. Si-hia *yi* 乙 ("charcoal") is isolated. In Tibeto-Burman we find chiefly two stems, one with initial *s* and another with initial *ts*, for expressing the notion "coal:" Tibetan *so-l(-ba)*, *soʾ*, *söʾ* (Yün-nan Tibetan *se-a*); Burmese *miː-swe*; Nyi Lo-lo *tse-se*, Lo-lo-pʿo *se-ñi*, A-hi Lo-lo *tsa-sa*; Mo-so *še*. [1]

Other Si-hia words not traceable to allied languages are:

152. *lan-to* or *-do* 浪多, younger brother.

153. *čʿön* 成, mother.

154. *ki* 即, brain.

155. *tʿien-čo* 天捉, loam-house.

156. *yi* 移, ladle, spoon (hardly related to Tibetan *kʿyem-bu, skyog*).

157. *kʿu* 枯, saw.

158. *tsu* 足 / *mo* 末 } drum.

159. *čio, tsio, kio* 爵, spider (related to Chinese *či* 蜘?).

160. *yŭ* 玉, silk.

161. *čʿou-na* 抽那, coin.

162. *yin* 盈, mark, characteristic.

163. *tsin, dsin* 精, law.

164. *mei* 每, virtuous.

165. *sie* 寫, wise.

166. *to, do* 多, true.

167. *nian* (Chinese character not given), word.

168. *liu* (Chinese character not given), world.

---

[1] The word *millò*, added by J. BACOT (*Les Mo-so*, p. 34), is composed of *mi* ("fire") and *lo* ("stone"). Likewise Nyi Lo-lo *lu-ma* and A-hi Lo-lo *lo-mo* ("mineral coal") are formed with *lu, lo* ("stone"), in the same manner as Si-hia *lu-yi* (see No. 41).

GROUP IX.

Technical Terms.

The Si-hia names for the animals of the duodenary cycle merit a special discussion. In this list we meet only two names identical with those for the real animals given in the vocabulary,—*lo* ("tiger") and *wŏ* ("dragon"), both genuine Si-hia words. In eight cases the names of the cyclical animals vary from the ordinary names for these animals; [1] in two other examples ("serpent" and "monkey"), we cannot decide the status, as the words for the animals are not contained in the vocabulary. Also in the Tibetan cycle, two animals appear with specific terms reserved for the cycle only,—*yos* ("hare;" common word *ri-boň*) and *tsᶜa* (or *mtsᶜa*)-*lu* [2] ("fowl;" common word *bya*).

169—170. In the same cases Si-hia offers likewise unusual names, —*tᵹŏ, dsŏ, jŏ* 則 ("hare;" common word *lo-wo*, No. 30) and *čᶜaň-ni* 長尾 ("fowl;" common word *wo-yao*). If we were positive that at the end of the twelfth century Tibetan *yos* already had the modern pronunciation *yŏ*, it might be permissible to correlate Si-hia *jŏ* with this Tibetan form. The expression *čᶜaň-ni* is a Si-hia formation: we find in the vocabulary *čᶜaň-ni wei* (*wŏi*) in the sense of "flying birds;" *čᶜaň*, accordingly, means "to fly," and with the suffix -*ni*, "the flying one."

171. The year of the rat is called *hi²* 㩗, while the rat ordinarily is styled *tsai-šu* (see No. 29). The former word seems to be identical with Chinese *hi²* 鼶, which in a poem of Hu Yen 胡儼 of the Ming period appears as name for the year of the rat. [3]

---

[1] This is likewise the case in Siamese (see CHAVANNES, *T'oung Pao*, 1906, pp. 52 and 53, note 2).

[2] This word survives in Jyaruň *pu-tsᶜa* ("chicken").

[3] CHAVANNES, *T'oung Pao*, 1906, pp. 56, 57.

172. The year of the ox is *mo* 没, while "cow, cattle," is *wu*. There is indeed in Indo-Chinese a stem *ma*, *mo*, with the meaning "cattle" or "ox." In discussing the Si-hia word *wu* (No. 4), reference was made to Kami *ma-na* ("buffalo"). There is, further, in Chuṅ-kia, a word *mo-tlaṅ* ("ox"), [1] the element *tlaṅ* being composed of *t-laṅ*, *laṅ* being identical with the base *laṅ* in Tibetan *g-laṅ* (Central Tibetan *laṅ*, Lepcha *loṅ*, Yün-nan Tibetan *lon*, Lo-lo *lo*, Balti *χ-laṅ*; as loan-word in Paśai *gō-lāṅ* [2]), and the element *t-* surviving in Ladākhī *χ-laṅ-to*, and Tibetan *po-to* ("bullock"), *be-to*, *be-do* ("calf"); so that the Chuṅ-kia word seems to be evolved from *mo-to-laṅ. The prefix *g-*, *χ-*, of Tibetan *g-laṅ*, likewise was once a full word, which is preserved in Mo-so *gö*, *na-gö*, [3] and in Siamese *kᶜo* (from *go); hence Tibetan *g-laṅ* from *go-laṅ. [4] Moreover, the *m-* type for "ox" appears in Lo-lo *ṅi-mu* (*ṅi* = Chinese *ṅi*, *niu* 牛; Pᶜu-pᶜa *ṅü*, Chö-ko *ṅö*), Miao-tse *ṅi-ma*, *s-ṅi-ma*. Hence it is legitimate to regard Si-hia *mo* as an indigenous word with the meaning "ox, cow, cattle."

As to *yiṅ* ("horse") see above, p. 27.

173. As regards *mo* 没, the year of the sheep or goat, it is curious that this word is identical with *mo* ("ox"); the vocalic timbre of the two words perhaps was different. The corresponding Siamese year is *māmä* or *mome* (the latter likewise in Cambodjan), year of the goat. I am inclined to assume a connection (that is, linguistic, not historical) between this and Si-hia *mo*, which would accordingly mean "goat," not "sheep," like the corresponding Tibetan and Mongol years. Compare also A-hi Lo-lo *kᶜi-mo*, Lo-lo-pᶜo *a-čö-mo*, Pᶜu-pᶜa *tsö-ma-la* ("goat").

---

[1] P. VIAL, *Les Lolos*, p. 36.

[2] GRIERSON, *Piśāca Languages*, p. 65.

[3] J. BACOT, *Les Mo-so*, p. 32.

[4] For this reason the above Paśai form is not a corruption of Tibetan *g-laṅ*, as assumed by GRIERSON (*l. c.*), but is a conservation of the original form of the Tibetan word.

174. The year of the monkey is termed *wei* 韋. Such a word for "monkey" is not found within Tibeto-Burman. [1] It therefore seems permissible to read the word *bei*, *bä*, and to derive it from Turkish *bäčin*, which in the Orkhon inscriptions is used for the year of the monkey [2] (Uigur *bičin*; [3] Mongol *bečin*, *mečin*; Niüči *mo-nen*; Manchu *mo-nio*, *bo-nio*). In HODGSON's Sokpa vocabulary we find as word for "monkey" *meči*, which is assuredly a Mongol loan-word. There are many of these in the same idiom: *tavoso* ("salt") from Mongol *dabusun*, *usu* ("water") from Mongol *usun*, *t'umar* ("iron") from Mongol *tämür*, *č'agan* ("white") from Mongol *tsagān*, etc. BABER's Mi-ñag word for monkey, *mi*, might be derived from the same source, unless it be related to White and Black Lo-lo *miu*, Müng *miau*, Burmese *myok* (*myauk*).

175. The year of the dog is *na* 那, while the general word for "dog" is *k'ü*. *Na* is a Si-hia word for "dog," for we encounter the same stem-word in Lo-lo with prefixed *a-*: Li-p'a and Li-su *a-na*; Lo-lo-p'o, Kö-sö-p'o, Ke-sö-p'o and Li-p'o *a-no*. [4] It further occurs in Murmi (HODGSON) *na-ñi* and Gurun *na-gyu*.

---

[1] My first impression was to connect Si-hia *wei* with Siamese *wok*, year of the monkey (the common word for "monkey" in Siamese is *liṅ*); but on account of the final *k*, which is absent in Si-hia, I have abandoned this theory.

[2] RADLOFF, *Wörterbuch der Türk-Dialecte*, Vol. IV, col. 1625.

[3] Aside from its Turkish ending -*čin*, this word certainly is not of Turkish origin; *bä-*, *bi-*, may have been derived from Greek πήθων, πίθηκος (from which also Old Slavic *pitikü* comes) at the time of the dissemination of the Hellenistic cycle over Central Asia. J. HALÉVY (*T'oung Pao*, 1906, p. 294) derives the Turkish word from Persian *püzineh* or *büzneh*, which does not sound very probable.

[4] A. LIÉTARD, *Bull. de l'Ecole française*, Vol. IX, pp. 563—567. The form *a-no* is indicated also by D'OLLONE, *Langues des peuples non chinois*, p. 60. Side by side with *a-no*, the word *č'ö* (corresponding to *k'i* of A-hi, *k'yi* of Tibetan, and *k'ü* of Si-hia) is employed in Lo-lo-p'o (LIÉTARD, *l. c.*, p. 551); in the Lo-lo-p'o cycle, however, the word *č'o* is utilized (LIÉTARD, *Au Yun-nan*, p. 230). The Lo-lo-p'o cycle, as given by A. LIÉTARD (*l. c.*), contains ten genuine Lo-lo words for the animals, and two borrowed from Chinese,—No. 6 *ya* from Chinese *yaṅ* 羊 ("sheep" in Lo-lo is *ju*, *ju*, and this word is used in the Nyi Lo-lo cycle; "goat," *k'i*, *ts'i*; final Chinese *ṅ* and *n* are dropped in Lo-lo [*Bull. de l'Ecole française*, Vol. IX, p. 557; and *Au Yun-nan*, pp. 201—204]); and No. 7 *myo*, year of the

176. The year of the hog in Si-hia is *yŭ* 玉 ("hog" in general being *wo*; the corresponding Lo-lo word *ve* is applied to the cycle). The word *yŭ* shows no relation to any equivalent in the cycle of outside languages, nor can I trace its existence in other Indo-Chinese idioms. It may be, therefore, an autochthonous Si-hia formation.

177. As to the Si-hia designations of the Ten Cyclical Symbols (*ši kan* 十干), we are confronted with a puzzle. The Tibetans have rationally transcribed the Chinese names as follows: *gya* (*g'a*), *yi, biň, tiň, wu, kyi* (*k'i*), *giň, zin, žim* or *žiň, gui.* Si-hia, however, offers the following series: *nai, liň* (*riň, ri*), *mi, wei, wei, tsˁi, lai, kˁo, nai, nu.* There are two homonymes in this series, 1 and 9 both being *nai* 乃, and 4 and 5 both being *wei* (*wö*) 嵬, certainly to the disadvantage of the system. All that can be said for the present is that the Si-hia people seem to have exerted their own ingenuity in framing this series, that shows no resemblance to the Chinese prototypes on which it is based.

GROUP X.

Loan-Words.

A. Chinese Loan-Words.

178—179. There are a certain number of Chinese loan-words in articles of clothing, in names of fruits and other objects imported from China. These, and others also, have been indicated by Mr. IVANOV: [1]

---

monkey, where Nyi Lo-lo has the regular word *nu.* I am inclined to regard *myo* as being derived from Chinese *mao, myao, myo* 猫 ("cat"). Lo-lo-p'o and A-hi apparently possess no word for "monkey;" Nyi has *a-nu,* P'u-p'a *a-no,* Chö-ko *a-si* (*Bull., l. c.,* p. 552),— evidently mere descriptive epithets of the animal. *Myo,* at any rate, is not a Lo-lo word.

' In a few cases, Mr. Ivanov has drawn also upon Mongol for the explanation of Si-hia words denoting bodily parts. At the outset, an influence of Mongol upon Si-hia could not be disavowed, though we are ignorant of the state of the Mongol language in that time when our Si-hia vocabulary was edited (1190) Si-hia *k'o* ("foot") is connected by Mr. Ivanov with Mongol *kül*; in fact, however, we have a stem *k'o* meaning "foot" within Indo-Chinese: Mo-so *k'ö, k'u,* etc. (see No. 51). It is therefore not necessary to

for instance, *tiṅ* or *diṅ* 丁 ("lamp") from *töṅ* 燈;[1] *kiai-i* (presumably *χ'ai-i*) 皆夷 ("shoe, slipper") from *χai* 鞋 (also in Chuṅ-kia *hai*, Hei Miao *ha*, Tibetan *χai* and Mongol *χei*). It seems unnecessary to me, however, to derive Si-hia *k'o-i* ("boot") from Chinese *hüe* 靴, as proposed by Mr. Ivanov. Aside from the phonetic difficulties of the case, Si-hia *k'o-i* appears as a legitimate derivation from Si-hia *k'o* ("foot"), No. 51, the suffix *i* being endowed with the same function as in *ro-i* ("saddle"), from *ro* ("horse"). Also in the Lo-lo idioms, the word for "boot" is based on that for "foot:" Nyi Lo-lo *č'e-nö* (literally, "*pied-exhaussement*"), A-hi Lo-lo *k'i-no*, Lo-lo-p'o *k'ye-no*. The combination of Si-hia *k'u* 枯 ("saw") with Chinese *kü* 鋸 seems to me doubtful, and a derivation of Si-hia *tsu* 足 ("drum") from Chinese *ku* 鼓 appears to be out of the question.[2]

180. Si-hia *yao* 藥 ("furnace") is hardly connected with Chinese *tsao* 竈, as conceived by Mr. Ivanov, but rather with Chinese *yao* 窰 ("kiln, furnace"), — a word adopted also by Mongol in the form *yō*.

181. The fact that the Chinese loan-words in Si-hia are not written in every case with the proper Chinese characters for these words, but that different symbols are chosen for them, does not militate against the conclusion that they are nothing but loan-words; for it is not necessary to assume that their status as loan-words rose into the consciousness of the author of our vocabulary or of

---

assume that Si-hia *k'o* is a Mongol loan-word. Si-hia *wu-i* ("backbone") is compared by Mr. Ivanov with a Mongol word written by him *ui*; KOVALEVSKI, in his *Mongol Dictionary* (Vol. I, p. 551) has a word *üye* ("joint, articulation"). This combination is not very convincing either; and as *-i* is a suffix that occurs in Si-hia, I prefer to think that *wu-i* is an original word-formation of this language (No 61). Si-hia *wu* ("cow") certainly bears no relation to Mongol *ükür* and Turkish *ut*, as proposed by Mr. Ivanov, but is a genuine Indo-Chinese word and the legitimate equivalent of Tibetan *ba* (see No. 4).

[1] This word was adopted by several other languages. In P'u-p'a we find it in the form *a-teṅ* with prefixed *a* (*Bull. de l'Ecole française*, Vol. IX, p. 554) In Buryat it occurs as *diṅ* with the meaning "candle" (SCHIEFNER in Castrén, *Burjat. Sprachlehre*, p. XIII). In Malayan we have *tiṅ* ("lamp").

[2] If a loan-word at all, Si-hia *tsu* would rather seem to come from Chinese *ču* 柷.

the Si-hia speaking population in general. This is well evidenced by the transcription *mao'r* 貌兒 ("cat"), which, as justly recognized also by Mr. Ivanov, certainly is a Chinese loan-word and the equivalent of 貓兒. [1]

182. Of color-designations, *hei* (*ha, ho*) 黑 ("black") has been received from Chinese.

183—184. Among fruits, we find *hn̈* 杏 ("apricot"), *Prunus armeniaca*, a native of China; [2] and *šwi-ma* 水麻 (*Diospyros kaki*), from Chinese *ši* 柿, [3] transcribed in Tibetan *se*; possibly also *tu* ("bean") from *tou* (see No. 96).

185. Si-hia *č'u-liň, č'u-li* 出令, plum. As *č'u* has the upper tone, it may be permissible to correlate it with *čo'* in Lo-lo-p'o *se-čo'-dsö* ("prunier"). [4] The element *se* of the latter word occurs in Nyi Lo-lo *se-ka-ma* ("prune") and Hua Lisu *se-li*. The second element of the latter word is indeed identical with Chinese *li* 李 ("plum"), and for this reason I am inclined to regard also Si-hia *liň, li*, as a Chinese loan-word. We find the latter also in Mongol (*lise*).

186. Si-hia *č'u-li* or *č'u-ri* 出梨 ("vinegar"), as justly observed by Mr. Ivanov, is borrowed from Chinese *ts'u* 醋, in the same

---

[1] The example is interesting for another reason: we note that *érh* was used as a diminutive suffix in Chinese at the end of the twelfth century.

[2] A. DE CANDOLLE, *Origin of Cultivated Plants*, p. 217; BRETSCHNEIDER, *Bot. Sin*, pt 2, No 471.

[3] Mr. Ivanov gives as translation "*hakki*," a word which does not exist, and which I believe is a misprint for *kaki*. In regard to another plant I cannot make any statement, as Mr. Ivanov does not specify the meaning exactly. This is *ts'üan-ni-na* 全尼那, a plant for which only the botanical term *Solanum esculentum* is given as equivalent. This term is now obsolete, and was formerly applied to the potato, the tomato, and the egg-plant. Since the first two are strictly American plants, they cannot come into question here; thus the supposition remains that the egg-plant (in Chinese *k'ie* 茄) may be intended. I hope that Mr. Ivanov will clear up this point in his further studies of Si-hia.

[4] A. LIÉTARD, *Au Yun-nan*, p. 216. The word *dsö* means "tree."

manner as Tibetan *ts͑uu*. The determinative element *li* or *ri* seems
to be a Si-hia addition. If *č͑u* is a Chinese loan-word, the supposition advanced by Mr. Ivanov (p. 1225) is not justified, that
č͑u-li should have terminated in -*l* or -*r*, "especially as in Tibetan
'acidity' is styled *skyur*." Si-hia *č͑u-li*, however, has no connection
with this Tibetan word which means "sour" (note what Jäschke
remarks on the unfamiliarity of Tibetans with vinegar), and which
presents a contraction from *skyu-ru*, *skyu-r* (compare Tromowa
kyu-pu, *kyum*; Sikkim *kyum*; Bunan *šu-ri*). [1] If the Si-hia word
were to be read *č͑ur*, an element *lo* or *lu* would have been chosen
for the transcription of *r* (compare *nir*, written *ni-lo*, No. 122),
not, however, *li*; and this hypothetical *č͑ur* could certainly not be
correlated with Chinese *ts͑u*. It seems quite reasonable to read
č͑u-ιi or *č͑u-ri*, and to look upon *č͑u* as a Chinese loan-word.

187—188. Si-hia *ts͑un* 寸 ("inch") and *šöñ* 聖 ("holy") are
likewise derived from Chinese.

B. Sinicisms.

Besides Chinese loan-words, we meet in Si-hia with certain terms
presenting literal translations into Si-hia of the corresponding Chinese
terms. Such are found in the field of astronomy.

189. Si-hia *mo-jei* 沒則移 ("galaxy") is composed of *mo*
("heaven") and *jei* ("water, river"), being a rendering of Chinese
t͑ien ho 天河, while Tibetan has a seemingly native term in
dgu-ts͑igs or *dgu-ts͑igs skya-mo*; Nyi Lo-lo has *če-k͑a* ("road of dew"),
and Lepcha *lóm toñ* ("highroad") or *lum toñ lum-ba*.

190—193. The Si-hia names of the planets *k͑ä yiñ* 皆迎
("gold star," Venus), *jei yiñ* ("water star," Mercury), and *mo* 沒

---

[1] The Polyglot Dictionary (Ch. 27, p. 12) has coined an artificial Tibetan word, *skyur-ẏu*
("sour water, vinegar"). Manchu *jušun* is likewise borrowed from Chinese; and so is
Ordos Mongol *oču*.

yiñ ("fire star," Mars), are modelled in imitation of Chinese *kin siñ* 金星, *šui* 水 *siñ*, aud *huo* 火 *siñ*. [1] The name for "Jupiter," *si yiñ* 西迎, forms au exception, *si* being a transcription of Chinese *sŭ* 戌 or *sui* 歲 with following *siñ* 星.

194. Si-hia *mo-wo* 沒訛 (*Anas galericulata*, the mandarin duck) is composed of *mo* ("heaven") and *wo* ("bird"), and looks like an imitation of Chinese *t'ien ngo* 天鵝 ("wild swan").

195. Si-hia *wei*(*wŏ*)-*mei* 嵬梅 (*Nephelium longan*) means "dragon's eye," being a literal rendering of Chinese *luñ-yen* 龍眼. [2] In the same manner the Tibetans have translated the Chinese term as *ạbrug-mig*; the Mongols as *lū-yin nidu*. [3]

C. Tibetan Loan-Words.

196. Si-hia *šu*(*šuai*)-*k'uai* 率塊, coral. The character *šuai* is used in the transcription of Si-hia words on two other occasions,— in *mou-šu* (No. 21) and *tsai-šu* (No. 29). As the second element, *k'uai*, is written 塊, it may be nothing but this Chinese word meaning "a piece." The Mongol word *širu* (written also *bširu*), Kalmuk *šuru* and *šūr*, and Manchu *šuru* ("coral"), are justly regarded as loanwords based on what is written in Tibetan *byi-ru* [4] or *byu-ru*, but articulated *j'u-ru*, *šu-ru*; there is, further, *šo-lo* ("coral") in Mo-so, and *šu-li*, *šu-lig*, in Kanaurı. There is thus good reason to assume that also the Si-hia word for coral was sounded *šu*, not *šuai*, and is derived from Tibetan.

Si-hia *p'u* ("place of higher altitude") might be regarded as a loan from Tibetan *p'u* (see No. 110).

---

[1] In the same manner, the Manchu have translated from Chinese *aisin usiha* (Venus), *muke usiha* (Mercury), *tuwa usiha* (Mars).

[2] BRETSCHNEIDER, *Bot. Sin.*, pt. 3, No. 285.

[3] *Polyglot Dictionary*, Ch. 28, p. 52.

[4] Derived from Sanskrit *vidruma* through the medium of some Indian vernacular.

D. Indian Loan-Words.

197. The peacock is termed in Si-hia *wo* (*vo*)-*lo*, or *wo-ro* 訛 勒 . The first element, *wo*, as already stated (No. 32), is the Si-hia word *wo* ("bird, fowl"). The second element, *lo* or *ro*, in my opinion is derived from the last syllable of Sanskrit *mayūra* ("peacock"), transcribed in Chinese *mo-yu(yū)-lo* 摩 由 (裕) 邏 , from which Manchu has distilled the form *molo-jin* (*-jin* being a Manchu ending). [1]

Sanskrit *mayūra* has affected also other Indo-Chinese languages. In Tibetan *rma-bya* (*bya*, "fowl"), the element *rma* (*r-* being silent in all dialects), in my opinion, is nothing but a reproduction of the first syllable in the Sanskrit name; also *yuṅ* in Lepcha *muṅ-yuṅ* and Ahom *yuṅ*, Siamese *ma-yūṅ* or *nok-yūṅ*, Pa-yi *nuk-yuṅ*, might be traceable to the same type. [2]

---

[1] K'ien-lung's Polyglot Dictionary (Appendix, Ch. 4, p. 6) has the series Manchu *molo-jin*, Tib. *ma-yu-ra*, Mongol *mayara*, Chinese *mo-yu-lo*. Manchu, moreover, has the following designations for "peacock:" *kundu-jin*, from Chinese *k'uṅ tu hu* 孔 都 護 (*ibid.*, where Tibetan *yo-bya* and Mongol *kündügüri* are added); *to-jin*, from Mongol *tōs* (written *tagus*, *togos*, to express the length of the vowel), doubtless conveyed through an Iranian language, Persian *ṭawus* طاووس (Turkī *ta'us*), that, together with Greek ταϝῶς, Hebrew *tuki*, *tūki*, has been traced to Tamil-Malayāḷam *tokei*, *togei*, by R. CALDWELL (*Comp. Grammar of the Dravidian Languages*, 3d ed., pp. 88—89); *yo-jin*, from Chinese *Yüe niao* 越 鳥 ("bird from Yüe"), Tibetan *yo rma-bya*, Mongol *yogon* (*ibid.*, Ch. 4, p. 6); and *ju-jin*, given as equivalent of Chinese *nan k'o* ("guest from the south"), Tibetan *lhoi rma-bya* ("peacock of the south"), Mongol *emürči*. All these are artificially coined book-words without legitimate life.

[2] Tibetan *rma-bya* is hardly intended to convey the meaning "*oiseau aux plaies*," as stated by J. HALÉVY (*Journal asiatique*, 1913, nov.-déc, p. 713); the Tibetan word *rma* certainly means "wound," but we cannot etymologize on this basis in Indo-Chinese languages. Nor can it be positively asserted, "le nom indien n'a jamais quitté la péninsule gangétique;" it has indeed wended its way to China, and has resulted in Manchu *molo-jin*. The dependence of Tibetan *rma* on Sanskrit *mayūra* is attested also by the fact that in cognate forms of speech we have a variation *mo*, *mu*, corresponding to modern Indo-Aryan *mōr*, *mūr*. In Kanaurī we have a duplicity of words,—first, *mōrös*, *mōres* (feminine *mōrī*), directly derived from Indian; and, second, *mom-za*, leaning toward Tibetan *mobza* ("peacock's feather"); Newārī *hmu-sa-khū*.

E. Iranian Loan-Words.

198. The Si-hia word for the lion is *ko-čen* 葛正. In all probability this is to be transcribed phonetically *go-čẽ* or *ko-čẽ* (that is, initial *č*, followed by a nasalized *e*; compare Phonology, § 14). The element *go* or *ko* seems to have the function of a prefix, and the element *čẽ* appears as the stem. The word *čẽ* appears to have arisen from *šẽ;[1] Si-hia *č* sometimes is developed from original *š*, as shown by Si-hia *či*, "flesh" = Tibetan *ša* (No. 48). The form *šẽ, without doubt, is identical with the same Iranian word as resulted in Chinese *ši* 獅. The latter has been derived from Persian *šēr* شیر.[2] This, however, is not quite satisfactory, since at the time when the first lions were sent to China by the Yüe-či (Indo-Scythians), in A.D. 88, the language, styled properly Persian, was not yet in existence. It seems that toward the end of the first century the word was transmitted to China through the medium of the Yüe-či, and that it originally hailed from some East-Iranian language, where it appears to have been known in the form *šē* or *šī*,[3] as Chinese *ši* (*šʹi) 師 has no final consonant.

On the other hand, if we adhere to the reading *čẽn*, a connection may be sought with Sanskrit *siṁha*, *siṅga* (Tibetan *seṅ-ge*, Newārī *sim*, Lepcha *suṅ-gi*).[4] The change of *s* to *č*, however, is difficult

---

[1] Compare Tibetan *šṅ-tse*, "lion" (SCHIEFNER, *Çdkyamuni*, p. 96), apparently a transcription of Chinese *ši-tse*.

[2] WATTERS, *Essays on the Chinese Language*, p. 350. The Persian word is traced to Avestan *xšaθr-ya* (= Sanskrit *kshatriya*), "imperiosus, lord, ruler" (C. BARTHOLOMAE, *Altiranisches Wörterbuch*, col. 548; SALEMANN, *Grundr. iran. Phil.*, Vol. I, pt. 1, p. 273; HORN, *ibid.*, pt. 2, p. 34, and *Neupersische Etymologie*, No. 803). It is perhaps under the influence of this signification that the Chinese chose the word *ši* 師 ("master") for the transcription of the Iranian word. This character was formerly employed for the designation of the lion without the classifier (see HIRTH, *J.A.O.S.*, Vol. XXX, 1910, p. 27).

[3] Tokharian A *šišäk*.

[4] The Lepcha form *suṅ-gi* is surprising. Still more curious is it that this variation meets an analogy in the languages of Madagascar: Madagasy *soṅ-ombi*, *šuṅ-ũmbi*; Merina *suṅg-ũmbi* (written *songombi*). This compound means literally "lion-ox" (a fabulous animal

to explain, and for this reason I prefer the explanation as outlined above. As to the meaning of the first element in the Si-hia
term, I have no definite opinion. From a Si-hia point of view the
syllable *ko* might be an epithet characterizing the lion, as, for instance, we have in Shina *guma šēr* ("a fiery lion"). [1] In view of
the affinities of Si-hia with Lo-lo, it may be pointed out that *ko*
is a Nyi Lo-lo word meaning "wild, savage" (A-hi Lo-lo *ku*), used
especially with reference to non-domesticated animals. [2]

199. Si-hia *lo*, *ro* 羅, copper. This word is not Indo-Chinese,
but it is doubtless derived from an Iranian language; for it is
identical with Persian *rō*, *rōi* روی, "brass, copper" (*royīn* رویین,
"brazen, bell-metal"); Yidghal and Khowar, two Hindukush dialects,
lo; Pahlavī *rōd*, Sogdian *rωδ* (from **rōδ*), [3] Baluči *rōd*), which
belongs to the well-known series: Sanskrit *loha*, Latin *raudus*, *rōdus*,
rūdus ("piece of ore"), Old Slavic *ruda* ("ore, metal"), Old Icelandic
raude ("red iron-ore"), — a group of words possibly connected with
or derived from Sumerian *urudu* ("copper"). The Si-hia form *ro*
shows that it is derived from Persian, not from Pahlavī. [4]

---

with the trunk of an ox or horse; also a strong, brave man). G. FERRAND (*Essai de
phonétique comparée du malais et des dialectes malgaches*, p. 298), who enumerates these
words among the Sanskrit elements in Madagasy, explains the vowel *u* or *o* from the effect
of vocalic attraction prompted by the *u* of *ūmbi* (*siña-ūmbi*, *siñ-ūmbi* becoming *suñ-ūmbi*).
In Javanese *siñā*, Malayan *sīña*, the Sanskrit word is preserved in its pure state. In
view of Lepcha *suñ-gi*, it is conceivable that in some language of India such a variant with
u vowel was already developed, especially as also in Cambodjan the lion is called *soñ*.

[1] G. W. LEITNER, *Languages and Races of Dardistan*, p. 26.
[2] See P. VIAL, *Dict. français-lolo*, pp. 256, 298.
[3] R. GAUTHIOT, *Essai sur le vocalisme du sogdien*, p. 101. See also P. HORN, *Neupersische Etymologie*, No. 635.
[4] Iranian loan-words exist in Tibetan also, and merit a special investigation. The
following may be called to mind: Tibetan *p'o-lad* ("steel"), from Persian *pūlād* پولاد (Pahlavi
pūlāfat, Armenian *polovat*, from **pavilavat*), a word widely diffused (Ossetian *bolat*,
Turkish *pūlād*, Grusinian *p'oladi*, Russian *bulat*, Mongol *bolot*, etc.); Tibetan *sag-lad* ("a
textile"), from Persian *saglāt* سقلات (see *J.A.S.B.*, 1910, p. 266); West Tibetan *ču-li*,
čo-li, Kanaurī *čul* ("apricot"), from the Pamir languages: Minjan and Galcha *čeri* (Dardu and
Shina *juru*). Ladākhī *a-lu-ča* ("plum"), from Persian *āluča* الوچه; Tibetan *deb-t'er*, *deb-*

In other Tibeto-Burman languages this Iranian word seems not to exist; at least, it has not yet been traced. There is a uniform Indo-Chinese word for "copper" pervading the chief members of the group, besides others covering limited areas. We have:

(1) Tibetan *zaṅ-s* (from **džaṅ*), [1] Lepcha *soṅ*, Yün-nan Tibetan *son*, Mo-so *soṅo*; Chinese *t͑uṅ* 銅 (from **dṅṅ, džuṅ, dsuṅ*), and Ahom *tåṅ* ("brass"), Shan *tåṅ* ("copper"), Siamese *t͑oṅ* ("metal, gold;" *t͑oṅ deṅ*, "copper"), from **doṅ*; Hei Miao *deo*, Ya-čio Miao *de*, Hua Miao *duṅ*.

(2) A-hi Lo-lo *ji*, Nyi Lo-lo *je*, Lo-lo-p͑o *jŏ*, Chö-ko *dsi*, Mo-so *šŏ-tu* ("bronze").

(3) Tibetan *k͑ro* and *ak͑ar* ("brass, bronze"), Burmese *k͑ë*. [2]

F. West-Asiatic Loan-Words.

200. Si-hia *po-lo* 字羅, radish (*Raphanus sativus*). Mr. Ivanov identifies this word with Chinese *lo-po* 蘿蔔. The case, however, is not such that the Si-hia would have inverted the Chinese term, but Chinese *lo-po* (**la-buk*) and Si-hia *po-lo* (from **buk-lo*) go back to two West-Asiatic names. The prototype of the Si-hia term is furnished exactly by Aramaic *fuglo* פּוּגְלָא ("radish"); [3] and we meet the same word in Grusinian *bolo-ki*, Ossetian *būlk͑*, Kabardian *belige*, all of which refer to the radish. [4] Whether this word occurs

gter, deb-ster ("book, document, record"), from Persian *dabtär* دفتر (derived from Greek διφθέρα). Tibetan *sur-na* (Chinese *so-na* 鎖吶), "hautboy, flageolet," from Persian *surnā* سرنا.

[1] The ancient initial *j* is preserved in the languages of Almora, Raṅkas, Dārmiyā, Chaudāṅsī, and Byāṅsī (see *Linguistic Survey*, Vol. III, pt. 1, pp. 538, 539), where *jāṅ* means "gold," answering to Kanaurī *zaṅ* ("gold"). In Bunan, *mal*, a typical word for silver (No. 1), means "gold." Confusion between the words for "gold" and "copper" obtains in several languages; for instance, Yakut *altun* ("copper"), Turkish *altun* ("gold"). Tibetan *zaṅ* is transcribed in Mongol *tsaṅ, čaṅ*.

[2] Proposed by B. HOUGHTON, *J R.A.S.*, 1896, p. 51.

[3] I. Löw, *Aramaeische Pflanzennamen*, p. 309.

[4] W. MILLER, *Sprache der Osseten*, p. 10.

also in Iranian is not known to me. [1] It is perfectly conceivable
that the Nestorians who were settled in the Si-hia kingdom [2] brought
with them the plant and the word. It is notable that the next
item introduced by our Si-hia vocabulary is the term *tsan po-lo*;
that is, "Chinese radish." Consequently the plain term *po-lo* must
have designated another species or variety which apparently was
non-Chinese.

Further, we find in Aramaic and Syriac *lafto* לִפְתָּא [3] (Arabic
laft لفت), derived from Greek *ράπυς* or *ράφυς* (Latin *rāpa* or *rāpum*). [4]
This type, in coalition with *fuglo*, seems to have conspired in forming
Chinese *la-buk*. T. WATTERS [5] insisted on "a suspicious resemblance
of the Chinese word to *rapa* and the kindred terms in Latin and
Greek." On the one hand, however, we cannot fall back on Greek
directly; and, on the other hand, the word *lafto* cannot fully explain
the Chinese term, but at best solely the first element *la*; while
Chinese *po* must be traced to *buk*, and the latter to *fuglo*. According

---

[1] A Persian word for the radish has been transmitted to Turks, Mongols, and Tibetans:
Persian *turma* ترمه and *turub, turb, turf* ترب; Balkar (PRÖHLE, *Keleti szemle*, Vol. XV,
p. 263) and Mongol *turma*; West Tibetan *se-rak turman*, written also *dur-sman* ("carrot"),
se-rak from Persian *zardak* زردك. The Polyglot Dictionary (Ch. 27, p. 18) writes the
Mongol word *turma*, and gives it as synonyme of *lobang* (Chinese *lo-po*; Manchu *mursa*).
G. A. STUART (*Chinese Materia Medica*, p. 371) holds the opinion that our English word
"turnip" is probably derived from Persian *turub*. This would not be so bad, if our turnip
had really come from Persia. There is, however, no trace of evidence to that effect; on
the contrary, our turnip is a very ancient European cultivation, being indigenous everywhere in temperate Europe (A. DE CANDOLLE, *Origin of Cultivated Plants*, pp. 36—38;
J. HOOPS, *Waldbäume und Kulturpflanzen*, pp. 351, 467). The *-nip* of "turnip" doubtless
goes back to Anglo-Saxon *nǣp* (Middle English *nepe, neep*; Old Norwegian *nǣpa*), from
Latin *nāpus* (French *navet, navette*); the usual explanation of the first part (from *turn* or
French *tour*, in the sense of turned, round) is hardly satisfactory, and, for myself, I do
not believe it.

[2] As related by Marco Polo (see YULE's note in his edition, Vol. I, p. 207).

[3] I. Löw, *l. c*, p. 241. This name in particular refers to the species *Brassica rapa*.

[4] Regarding the names of the European languages, see J. HOOPS, *Waldbäume und
Kulturpflanzen*, p. 350.

[5] *Essays on the Chinese Language*, p. 332.

to Bretschneider,[1] the word *lo-po* or *lo-p'o* first appears in Chinese books of the ninth century, and was originally used in the state of Ts'in (Shen-si and eastern Kan-su). The name accordingly makes its début in the T'ang period when numerous new species of cultivated plants were introduced into China from the West, and its first appearance in the border-land of Turkistan is likewise suggestive of a foreign origin. The various earlier designations of the plant, *lai-fu* 萊服 (in the *T'ang pên ts'ao*) and *lu fei* 蘆菔 or *lu fu* 服 in Kuo P'o, are independent of *lo-po*, referring to an apparently indigenous cultivation; Bretschneider assumes that the radish, as it is mentioned in the *Erh ya*, has been cultivated in China from remote antiquity. Since numerous varieties of this genus are under cultivation, not only in China, but also in Europe and India as well (and there also of uncontested antiquity), it is conceivable that a new variety might have been introduced into China from Western Asia through Turkistan under the T'ang to receive the foreign name *la-buk*. Bretschneider's observation, that "from China the cultivation of the radish spread over the neighboring countries, where the people generally adopted also the Chinese name of the plant," is somewhat too generalized. A. DE CANDOLLE[2] has remarked that "for Cochin-China, China, and Japan, authors give various names which differ very much one from the other." A specific instance may be cited to the effect that an Indo-Chinese nation received the name of the radish from India. The Sanskrit word for the radish is *mūlaka* (from *mūla*, "root"), also *mūlābha*. It is commonly cultivated in Western India and in the Panjab. In the latter territory it is styled *muñ-ra*, in Bombay *mogri*, in Hindustānī *mugra*.[3] An Indian form of the type *muñ-ra* appears to have resulted in Burmese *mun-lā*.

[1] *Bot. Sin.*, pt. 2, p. 39.

[2] *Origin of Cultivated Plants*, p. 31.

[3] WATT, *Dictionary of the Economic Products of India*, Vol. VI, p. 394.

This shows that not all Indo-Chinese words for the radish are traceable to Chinese, and that Burmese *mun-lā* has no relationship with Tibetan *la-p^c ug* (Ladākhī *la-bug*), as assumed by B. HOUGHTON. [1]

201. Si-hia *si-na* 悉那, mustard. This word is neither related to Chinese *kai, kiai* 芥, nor to Tibetan *yuǹs*; nor can any relationship with Nyi Lo-lo *o-na-se* be asserted. According to the explanation of P. VIAL, [2] *o* means "vegetable," *na* "nose," and *se* "congestive." As "nose" is *ni* in Si-hia, we should expect at least *si-ni*, if the Lo-lo and Si-hia words were identical. Si-hia *si-na* reminds one of Greek *sinapi* (σίναπι or σίνχπυ), which found its way into Latin *sinapis* (in Plautus), Gothic *sinap*, Anglo-Saxon *sënep*, Italian *senape*, French *sanve*, etc. However startling this derivation may seem at first, it is nevertheless possible. Si-hia *si-na* denotes a species different from Chinese *kiai*: the former relates to *Sinapis* or *Brassica alba*, the latter to *Sinapis juncea*. The home of the white mustard (*Sinapis alba*) is in southern Europe and western Asia. [3]

It first appeared on the horizon of the Chinese in the T^c ang period, being described under the name *pai kiai* 白芥 ("white mustard") by Su Kung, the reviser of the *T^c ang pên ts^c ao*, and said by him to come from the Western Jung (*Si Jung* 西戎). Under the term *Hu kiai* 胡芥 it is noted in the *Pên ts^c ao* of Shu 蜀 of the middle of the tenth century. It then was abundant in Shu (Sze-ch^c uan), and for this reason received also the name *Shu kiai* ("mustard of Shu"). [4] It is therefore logical to identify Si-hia *si-na* with the

---

[1] *J.R.A.S.*, 1896, p. 42. Moreover, the first part of the Burmese word is not traceable to the tribal name Mon, as supposed by the same author.

[2] *Dictionnaire français-lolo*, p. 227.

[3] A. ENGLER in Hehn, *Kulturpflanzen*, p. 212 (8th ed.); WATT, *Dictionary of the Economic Products of India*, Vol. I, p. 521.

[4] *Pén ts'ao kang mu*, Ch 26, p. 12. G. A. STUART (*Chinese Materia Medica*, p 408) states that the white mustard was introduced into Sze-ch^c uan from Mongolia; it is safer to say "Central Asia." Also in Tibetan a distinction is made between *ske-ts'e* = Sanskrit *rājikā* ("black mustard") and *yuǹs-kar* or *yuǹs-dkar* = Sanskrit *sarshapa* ("white mustard"),

white mustard of the Chinese, introduced under the Tͨang from a region of Central Asia; and it is not surprising, that, with this new cultivated species, also its Hellenistic name (sinapi) was diffused over the Asiatic Continent. It remains to determine through which language the transmission took place. Iranian may come into question, but Persian sipand سپند (also sapandān, sipandān, sipandīn, saped-dān) surely is not related to sinapi.

202. Si-hia liu-na 流那, cabbage. In attempting to account for this word, it is necessary to survey to some extent the field relating to the cultivation of the genus Brassica. The Tibetan language shows us best the various historical probabilities with which we are confronted. The Tibetans, not much given to the growing of vegetables, have three words for designating cabbage, that are borrowed, and curiously enough, each from one of those three great centres of culture with which Tibet was in contact on its western, southern, and eastern frontiers. In West Tibetan we meet the word kram,[1] which is traceable to Persian karanb كرنب [2] or kalam كلم, Arabic kiranb, kurumb, kromb, Hindustānī káramkállá كرم كلّا; Siamese ka-lam; Greek κράμβη, Latin crambē, Aramaic keruba כְּרוּבָא.[3] We further meet in West Tibetan gobi, Sikkim and Bhūtan kobi, an Indian word derived from Hindustānī kōbī كوبى, Bengālī kōpī, Gujerati kōbiā.[4] The last, in consequence of the

both terms being listed in the Mahāvyutpatti. The latter term is artificial, and plainly shows that the white mustard was foreign to the Tibetans likewise; for yuṅs, yuṅ, or ñuṅ relates to the turnip. Tibetan ske-tse is a transcription of Chinese kai ts'ai 芥菜; the two terms are equalized in the Polyglot Dictionary (Ch. 27, p. 19).

[1] H. RAMSAY (Western Tibet, p. 15) remarks that the Tibetan word is used for cabbage, but really means "a kind of spinach." According to Jaschke it only means "cabbage."

[2] The word appears as early as the latter part of the tenth century in the pharmacological work of Abū Mansūr (HORN, Grundr. iran. Phil., Vol. I, pt. 2, p. 6).

[3] I. Löw, Aramaeische Pflanzennamen, p. 213.

[4] The word is found in Singalese (kōvi), in the Dravida languages (kob, kobī, etc.), n Malayan and Javanese (kōbis, kūbis) See S. R. DALGADO, Influencia do vocabulário português em linguas asiáticas, p. 65 (Coimbra, 1913).

introduction of cabbage into India by Europeans, [1] is a European word connected with our series: Latin *caputium*, Italian *capuccio*, Portuguese *couve*, French *cabus*, *caboche*, English *cabbage*.

In central and eastern Tibet the name *pe-tse* or *pi-tsi* is used. This, as already observed by Jäschke, is Chinese *pai ts'ai* 白菜 (*Brassica chinensis*), a colloquial term for the species *sun* 菘. The Tibetan word, likewise in oral use only, was apparently conveyed through the medium of a Sze-ch'uan dialect, as shown by the vocalization *pe*; and the tenuis instead of the aspirate proves that we have not a rigorous transcription of the written language before us. The term *pai ts'ai* itself is not old, but makes its first appearance in the *P'i ya* 埤雅 of Lu Tien 陸佃 (1042—1102). [2] The conditions of the Tibetan language, accordingly, bear out the fact that Central Asia, on the one hand, participated in the European variety or varieties of cabbage (and A. DE CANDOLLE [3] has well demonstrated the European origin of this species, *Brassica oleracea*), and, on the other hand, received (probably as imported product only) another species anciently cultivated in China. [4] The Si-hia word *liu-na* bears no relation either to Chinese or to any other Indo-Chinese language (compare, for instance, Lo-lo *o-za*, *o-lai-ma*; Lepcha *bi-bum*), and it is therefore justifiable, as in the case of Tibetan *kram*, to seek its origin in a western language. The Turkish word *lahana* كهنه suggests itself. [5] This is derived from Greek λάχανον, which passed into Arabic as *lahana* كهنه. The Turkish, Arabic, and Si-hia ending -*na* is due to the Greek plural λάχανα; the word (from λαχαίνω, "to dig") was chiefly used in the plural form in the sense of

[1] WATT, *Dictionary of the Economic Products of India*, Vol. I, p 533

[2] *Pèa ts'ao kang mu*, Ch. 26, p. 10 b.

[3] *Origin of Cultivated Plants*, pp. 83—86.

[4] BRETSCHNEIDER, *Bot. Sin.*, pt 3, No 245. G A. STUART, *Chinese Materia Medica*, p. 73.

[5] RADLOFF, *Wörterbuch der Turk-Dialecte*, Vol. IV, col. 731.

"garden-herbs." Again, it seems to me that also in this case the Nestorians may be responsible for the transplanting of both the name and the object.

NOTE ON SI-HIA PLANTS.

The preceding observations show that the Si-hia names of cultivated plants are of particular interest, and augur a peculiar position of Si-hia culture in Central Asia. It is therefore appropriate to elucidate this subject to some extent from an historical point of view; and an attempt in this direction is supported by a list of plants growing in the Si-hia country, and recorded in the Chinese Annals. [1]

The only cultivated plant the name of which is common to Si-hia and Tibetan, as far as we can judge at present from our fragmentary material, is barley: Si-hia *na*, Tibetan *na-s* (No. 116). The translation of this word given by Mr. Ivanov is "cereal," but the standard cereal of Tibetan tribes has at all times been barley. Barley is expressly mentioned in the Annals, and occupies the first place among the plants cultivated in the Si-hia country. The food-plants raised by the ancient Tibetans are enumerated in the *Kiu Tᶜang shu* (Ch. 196 A, p. 1 b) as barley, a certain species of beans, [2] wheat, and buckwheat. [3] The *Sin Tᶜang shu* gives the same in the

---

[1] *Liao shi*, Ch. 115, p. 3.

[2] *Lao tou* 營豆, according to BRETSCHNEIDER (*Bot. Sin.*, pt. 2, No. 96), "a climbing leguminous plant, wild-growing, used as a vegetable; the small black seeds, which resemble pepper, are edible" (see also G. A. STUART, *Chinese Materia Medica*, p. 378), identified with *Rhynchosia volubilis*. The proper mode of writing is 勞 *lao* ("to weed"), the plant being a weed growing in wheat-fields. There can be no doubt that this is not the plant intended in the above passage of the T'ang Annals, where a cultivated plant is in question. In the Polyglot Dictionary (Ch. 29, p. 11) we find the term *lao tou*, written 澇豆, with the following equivalents: Tibetan *sran-čᶜuṅ*, Manchu *laifa*, Mongol *khosingur*. Tibetan *sran-čᶜuṅ* means "small bean or pea," and appears in the Mahāvyutpatti as rendering of Sanskrit *masūra* ("lentil;" see A. DE CANDOLLE, *Origin of Cultivated Plants*, p. 323); but it is not known to me that lentils are grown in Tibet.

[3] Tibetan *bra-bo*, Purig *bro*, Jyaruṅ *dru*.

order wheat, barley, buckwheat, and *lao* beans. The Chinese term for "barley" employed in this passage is *ts'iñ k'o* 青稞, which in meaning answers to Tibetan *na-s*.[1] The word *na, na-s*, does not seem to cover much ground in Tibeto-Burman.

Barley is among the most ancient cultivated plants, and many varieties have been brought into existence through the process of cultivation. The word *nas* refers at the present time to the beardless variety of barley (*Hordeum gymnodistichum*), which has only two rows of spikelets, and further presents the curious feature of having the flower-scales non-adherent to the grains. These scales drop in threshing, leaving the grains naked like those of the wheat. Three sub-varieties are said to be largely cultivated in Tibet,—a dull green, a white, and a dark or chocolate brown. It was recently (1886) introduced into India by seed obtained in Tibet.[2] It is known in the Indian vernaculars as *paigambari, rasuli*. In China it is called *kuñ mai* 穬麥,[3] and *ts'iñ k'o* 青稞.

The term *nas*, however, has a wider application in literature; for in the Mahāvyutpatti, translated into Tibetan in the ninth century, it is identified with Sanskrit *yava*. The latter term, as conclusively shown in particular by J. Hoops,[4] referred to the barley in the earliest period of Indian history. Moreover, as *Hordeum hexastichon* (the six-rowed barley) is almost the only cultivated form, the barley par excellence, of India,[5] we are justified in identifying with it both the terms *yava* and *nas*.

---

[1] Polyglot Dictionary, Ch. 28, p 45. BRETSCHNEIDER (*Chinese Recorder*, Vol. IV, 1871, p. 225) is inclined to think that *ts'in k'o* in the above text refers to oats. This, however, is not correct; *ts'in k'o* denotes the so-called naked barley (*Hordeum gymnodistichum*).

[2] WATT, *Dictionary of the Economic Products of India*, Vol. IV, p. 274.

[3] BRETSCHNEIDER, *Bot. Sin*, pt. 2, No. 32; pt. 3, No. 220.

[4] *Waldbäume und Kulturpflanzen*, pp. 344, 358—359.

[5] WATT, *l c*, p. 275.

Another variety of barley is styled in Tibetan *so-ba, so-wa*; Ladākhī *sóa* or *swa*. This word is widely diffused in Tibeto-Burman: Jyaruṅ *sui*, Nyi Lo-lo *ze-ma* (*ša-za*, "oats"), A-hi Lo-lo *e-sa*, Lo-lo of Tᶜung River (BABER) *zo*, Mi-ñag *mu-dza* (the element *mu* is apparently identical with Burmese *mu-yau*, "barley," and Chinese *mou* 牟 or 麰, "barley"), Nepal *to-sa*; Bunan *za-d*. The primeval form appears to be **za, zo*; we may derive from it also the Tibetan word *ka-rtsam, ka-sam, rtsam-pa, tsam-pa, tsam-ba* ("roasted flour from barley or oats"). The barley here in question is the common, four-rowed variety (*Hordeum vulgare*). An interesting identification of Tibetan *so-ba* is made in Kᶜien-lung's Polyglot Dictionary (Ch. 38, p. 45), where it corresponds to Manchu *arfa* and Mongol *arbai*. Both these words are identical with, and presumably derived from, Turkish *arpa* (in some dialects *arba*, Salar *arfa*), whence the Hungarians received their *árpa*. [1]

Another Tibetan word that belongs to this group is *yu-gu, yu-kᶜu*, or *yug-po* (Ladākhī *ug-pa*), which relates principally to oats (*Avena sativa*), but is locally employed also for barley (compare Burmese *mu-yau*, "barley"). Tibetan *yo-s* ("roasted corn") and Bunan *yu-ši* ("flour") are derived from the same base, that we have also in Chinese *yu* 㢈 ("oats"). The wild oat (*Avena fatua*), from which the cultivated species is now generally believed to have been obtained, occurs spontaneously in the Himalaya up to 9500 and 11500 feet, [2] everywhere in Eastern Tibet, [3] and in several parts

---

[1] Z. GOMBOCZ, *Die bulgarisch-türkischen Lehnwörter*, p. 30. Other Manchu words for "barley" are *muji* (corresponding to *ta mai* 大麥) and *murfa* (corresponding to Tibetan *nas*). The Chinese equivalent for Tibetan *so-ba* in the Polyglot Dictionary is *liṅ-taṅ mai* 鈴鐺麥 ("bell wheat"), a Peking colloquial term explained by BRETSCHNEIDER (*Chinese Recorder*, Vol. IV, 1871, p. 225) as "oats;" the meaning "barley" seems more appropriate. In Mongol dialects we meet the following variations: *arpa, irpei, arpai*, and *χarba* (POTANIN, *Tibeto Tangutan Border-Land of China*, in Russian, Vol. II, p. 398).

[2] WATT, *l. c.*, Vol. I, p. 355.

[3] According to the observations of the writer who secured specimens in the field.

of China. [1] Though no investigations as to the relation of the wild to the cultivated species in Tibet have as yet been made, it is very likely that the latter has sprung from the former. As to the oat of Europe, the same relation has been assumed, but the origin of European oat-cultivation is a problem as yet unsolved. It has been carried somewhat vaguely into Central Asia or Turkistan by some investigators; [2] but I have no doubt that Tibet, where we find the wild and the cultivated species side by side, must be regarded as the home of oat culture. The *Liao* Annals mention for the Si-hia country no other cereal than barley (*ta mai*), which probably includes also oats. The Si-hia people, accordingly, like the Tibetans, must chiefly have been barley and oat eaters.

The designation for "beans", *tu* (No. 96), separates the Si-hia from the Tibetans, [3] and draws them near the Chinese; for *tu* is apparently akin to Chinese *tou* 荳 ("beans, pulse"). [4] A special kind of black beans (*tu hei*) is recorded in our Si-hia glossary. In the language of the Chuñ-kia we meet *lok-tu*, in Miao-tse *ka-tu*, *lañ-tao*, and *tu*. [5] The Lepcha terms for different kinds of beans (*tuk-byit*, *tuñ-ki*, *tuñ-kuñ*, etc.) do not belong to this series; Lepcha *tuk* is a distinct word, that is independent of Chinese *tou*. Lo-lo-p°o *no*, A-hi Lo-lo *a-nu*, Nyi Lo-lo *a-nu-ma*, present likewise a separate group. Mo-so *beber* is a mysterious word. Considering the numerous varieties of beans, the diversity of words is not surprising. The

---

[1] FORBES and HEMSLEY, *Journal Linnean Soc.*, Botany, Vol. XXXVI. p. 401. In Turkistan we have *Avena sterilis* and *A. desertorum* (S KORŽINSKI, *Sketches on the Vegetation of Turkistan*, in Russian, pp. 20, 66, 76).

[2] See HOOPS, *l. c.*, p. 405. The information of this author as to China and Central Asia is certainly insufficient.

[3] Tibetan *sran-ma*, *srad-ma*; Yün-nan Tibetan *se-mer*; East Tibetan (so-called Si-fan) dialects *se-mer*, *se-ma*.

[4] The case of Manchu *turi* ("bean"), derived from Chinese *tou*, might favor the assumption that Si-hia *tu* is likewise a Chinese loan-word.

[5] D'OLLONE, *Langues des peuples non chinois de la Chine*, p 51

cultivation and consumption of beans (*pi tou* 畢豆) in the Si-hia country is testified to by the Annals.

As imports from China we meet fruits like apricots, *kaki*, plums, *lungan*, and oranges. Pears (No. 90) were possibly cultivated.

The relationship of the Si-hia designation for rice (*k*°*o*) to the T°ai languages (No. 84) is curious. It is independent of the Tibetan (*abras*) and Chinese terms, and probably points to the fact that the inhabitants of the Si-hia kingdom received the T°ai word from Miao-tse and Mo-so tribes, which likewise possess it, and also that they may have traded this staple from their southern and south-eastern neighbors. It is likewise interesting that the nomenclature for "field, forest, tree, wood, fruit" (Nos. 85 — 87) exhibits decidedly T°ai affinities; and this may hint at a certain degree of influence exerted by the T°ai on Si-hia agriculture.

Among the plants enumerated in the *Liao shi*, we find the fruits of the *p°uñ* of the salty soil 鹹地蓬實, which may be the equivalent of *kien p°uñ*, identified with *Salsola asparagoides* (family *Chenopodiaceae*).[1] FORBES and HEMSLEY[2] enumerate three Chinese species, — *Salsola collina*, *S. kali*, and *S. soda*, — and state that several others are in the Kew Herbarium. These desert shrubs grow everywhere in Persia, Tibet, Mongolia, and Turkistan.[3]

The *Liao shi*, further, mentions sprouts of *ts°uñ yuñ* 苁蓉苗. T°ao Hung-king indicated that the best ones came from Lung-si (Kan-su).[4] The plant is common in southern Siberia, Dsungaria, and Mongolia, and belongs to the family *Orobanchaceae*. The species

---

[1] BRETSCHNEIDER, *Bot. Sin*, pt. 2, p. 254.

[2] *Journal Linnean Soc.*, Botany, Vol. XXVI, p. 330.

[3] WATT, *l. c*, Vol. VI, pt. 2, p. 392. The Mongol names will be found in POTANIN, *Sketches of North-Western Mongolia*, Vol. IV, p. 152; and *Tanguto-Tibetan Border-Land of China*, Vol. II, p. 404 (both in Russian); the Turkish names in GRUM-GRŽIMAILO, *Description of a Journey in Western China* (in Russian), Vol. III, p. 498.

[4] BRETSCHNEIDER, *Bot. Sin.*, pt. 3, p. 38.

in question is probably *Cistanche salsa*, which occurs in Kan-su and Siberia. [1]

The Si-hia people turned out mats from the bark-fibre of the small *wu-i* tree 小蕪荑席. Two kinds of *wu-i* are distinguished, a larger and a smaller one. The former has been identified with *Ulmus macrocarpa* (family *Urticaceae*). [2] The latter presumably is *Ulmus parvifolia* or some other kind of elm. [3] The bark of *Ulmus* contains a strong fibre suitable for the manufacture of cordage, sandals, and mats.

Finally we meet the following wild-growing plants in the Annals of the Liao:

Leaves from *ti huan* 地黃葉, *Rehmannia glutinosa*. [4] The dried leaves of this plant furnish digitalis.

Kü hui t'iao 拒灰蓧, presumably identical with *hui t'iao* 灰滌, *Limnanthemum peltatum* (family *Gentianaceae*); [5] applied to the Kan-su region, it may refer as well to one of the ten species of *Gentiana* occurring there. [6]

Pai hao 白蒿, *Artemisia stelleriana vesiculosa*. [7] Root and

---

[1] FORBES and HEMSLEY, *J. Linnean Soc.*, Botany, Vol. 26, p. 222. The above designation of the species is the same as *Phelipaea salsa* given by G. A. STUART (*Chinese Materia Medica*, p. 61). Compare the interesting descriptions of this and the preceding plants by S. KORŽINSKI, *Sketches on the Vegetation of Turkistan* (in Russian), pp. 5—6.

[2] BRETSCHNEIDER, *Bot. Sin.*, pt. 3, No. 330. G. A. STUART, *Chinese Materia Medica*, p. 448. The discrimination of the two kinds is attributed to Li Shi-chên by Bretschneider and Stuart; it is first made, however, in the *Pên ts'ao yen i* of the year 1116 by K'ou Tsung-shi. There the smaller species is said to be identical with elm-seeds 榆莢. In fact, as follows from our above text, it must have designated a particular species.

[3] Eight species of *Ulmus* are known from China (FORBES and HEMSLEY, *l. c*, p. 448); the genus comprises about sixteen species. The elm of Mongolia is *Ulmus campestris*.

[4] BRETSCHNEIDER, *l. c.*, No. 100. G. A. STUART, *l. c.*, p. 150. FORBES and HEMSLEY, *l. c.*, p. 193.

[5] G. A. STUART, p. 241. FORBES and HEMSLEY (*l. c.*, p. 142) enumerate the two species *L. cristatum* and *L. nymphoides*, both of which occur likewise in India (WATT, *l. c.*, Vol. IV, p. 641), where stems, fruit, and leaves of *L. cristatum* are eaten in certain localities, *L. nymphoides* being largely employed as fodder in Kashmir.

[6] GRUM-GRŽIMAILO, *l. c.*, Vol. III, pp. 492—494.

[7] G. A. STUART, *l. c.*, p. 52

leaves are used as food; a decoction is employed as a wash in ulcerous skin affections.

Fruits of *kien ti suñ* 鹹地松寶, pine-nuts of a particular species growing in salty soil. The tree in question is perhaps identical with the *lo* or *ro* of the Si-hia glossary (No. 143). At least eleven species of *Pinus* are known from China. [1]

The following plants recorded for the Si-hia country in the *Liao shi*, as far as I know, have not yet been identified: *ts⁽iñ lo mi-tse* 青稞床子, *ku-tse man* 古子蔓, *ki ts⁽ao-tse* 雞草子, and *tŏñ siañ ts⁽ao* 登廂草. The plants previously mentioned, may have been employed partially by the Si-hia as food-stuffs, and partially by the Chinese as medicines. The Si-hia, if we may depend on the Chinese annalist, in case of sickness, did not resort to physicians and drugs, but summoned their shamans for the exorcism of the devils causing the complaint.

In our Si-hia glossary, the name of a flower is given as *t⁽o-lŭ* 托綠, and through the Chinese translation *mu-tan* is identified by Mr. Ivanov with *Paeonia chinensis*. It is more probable, however, that this is not the Chinese, but an indigenous Si-hia species. Presumably it is *Paeonia anomala*, found in Kan-su and Mongolia. According to POTANIN, [2] it is styled by the Tangut *tombu-tuglan*; and the Si-hia name may bear some relation to this word *tuglan*.

Likewise the Si-hia term *fu-sai* (No. 136), alleged to mean "lotus," in all likelihood refers to an autochthonous plant of the Kan-su and Amdo regions, where we know of four species of *Iris* (*bungei, dichotoma, ensata,* and *gracilis*). [3]

It is regrettable that the Si-hia vocabulary does not impart the words for "onion" and "garlic." It is related in the Annals of the

---

[1] FORBES and HEMSLEY, *l. c.*, pp. 549—553.

[2] *Tanguto-Tibetan Border-Land of China*, Vol. II, p. 401.

[3] GRUM-GRŽIMAILO, *Description of a Journey in Western China*, Vol. III, p. 500.

Liao that the Si-hia country produced two wild species, called *ša
tsᶜuṅ* 沙葱 ("sand-onion") and *ye kiu* 野韭 ("wild-growing leek").
The former species is not mentioned in the *Pên tsᶜao kang mu*;
but it is explained in the continuation of this work, the *Pên tsᶜao
kang mu shi i*,[1] according to which *ša tsᶜuṅ* is a Kan-su name, and
relates to a wild-growing onion occurring everywhere in Mongolia,
the leaves being the same as those of the cultivated variety. It is
styled "sand-onion," because it thrives in sandy places; and the
Mohammedans of Kan-su especially relish it. The term is listed
in the Polyglot Dictionary (Ch. 27, p. 27) with the literal Tibetan
translation *bye-tsoṅ*, Manchu *engule*, Mongol *günggel*. Eight wild
species of *Allium* have become known from Kan-su.[2]

The alliaceous plants belong to the oldest cultivated within the
dominion of the Indo-Chinese family. We have the following
interesting coincidences of names:

1. Chinese *tsᶜuṅ* (even upper tone) 葱 (general term for alliaceous
plants, as onions, garlic, leek), Korean *čᶜoṅ*, Japanese *sō*. Tibetan
b-tsoṅ (high tone), Tromowa, Sikkim, and Lepcha *o-tsoṅ*, eastern
Tibetan (so-called Si-fan) *a-čᶜuṅ*, Chö-ko *a-suṅ*, Nyi Lo-lo *a-tsᶜe*.
Primary form presumably **dzuṅ, juṅ*.

2. Chinese *suan* (sinking upper tone) 蒜, "garlic" (*Allium sativum*),
Cantonese *süṅ*, Hakka *son*, Fukien *sauṅ*, Korean and Japanese *san*,
in north-western Chinese dialects *suai*.[3] Lepcha *suṅ-gu*, A-hi Lo-lo *šo*,
Lo-lo-pᶜo *šu*, Nyi Lo-lo *še-ma*; Burmese *krak-swan*, So. Chin *kwet-šon*.

3. Chinese *kiu* (rising upper tone) 韭 or 韭 ("leeks, scallions,"
Allium odorum), Cantonese *kau*. Tibetan *s-goy(-pa)* (high tone),
Yün-nan Tibetan *yau-pa*, Suṅ-pan Tibetan *čon-grog*;[4] Burmese

---

[1] Ch. 8, p. 16 b (see *T'oung Pao*, 1913, p. 326).

[2] Enumerated by GRUM-GRŽIMAILO, *l. c.*, Vol. III, pp. 500—501.

[3] POTANIN, *Sketches of North-Western Mongolia* (in Russian), Vol. IV, p. 146.

[4] POTANIN, *Tanguto-Tibetan Border-Land of China* (in Russian), Vol. II, p. 395.
Li-žii gur kᶜaṅ (fol. 15) has also *keu* in the sense of wild garlic (*ri sgog*).

krak-swan, Southern Chin *kwet-šon*. The primary form seems to be **gau, gou*. The suffix *-g, -k*, is peculiar to Tibeto-Burman. Presumably also Tibetan *ske-tse* ("wild onion") belongs to the same base; perhaps even Tibetan *kwon-doñ* or *kon-toñ* [1] and *kiu-ljoñ* ("wild onion," = Mongol-Manchu *suduli*).

4. The element *doñ, toñ*, in the Tibetan compounds *kwon-doñ*, *kon-toñ*, and the element *čon* in Suñ-pan Tibetan *čon-grog*, may be anciently related to Chinese **diem, tiem* 藏, now *tsʿien* and *tʿien* ("wild garlic or onion").

Chinese *tsʿuñ* and *suan* (*sun, suñ*), with their corresponding equations in the other languages, apparently are allied words; and, what is still more interesting, appear to be historically connected in some manner with Turkish (Cumanian, Chuvaš, and Osmanli) *soɣan* ("onion"), [2] Uigur *soɣun*, Baskir *ϑugan*, Djagatai *soiɣan*, Mongol *soñɣina* (in dialects also *soñɣinok*), Shirongol *soñgnyk*, Teleutic *soɣono*, Altaic *sōno*, Manchu *suñgina*, Mongol and Manchu *suduli* ("wild garlic"). From a purely philological standpoint it is difficult to decide which side is the borrower, and which is the recipient. A further interrelation seems to prevail as to the Indo-Chinese base **gau, gou, gok, ko*, and Shirongol *gogo*, Ordos Mongol *kogut, kogyt* (*y* = Russian ы); also as to Tibetan *kwon, kon*, previously mentioned, and Mongol *günggel*. The Turkish-Mongol-Tungusian series bears the genuine imprint of Altaic words, both as to their phonetic structure and particularly as to their endings. On the other hand, the Chinese words, as shown by documentary evidence, point to a great antiquity, and this conclusion is corroborated by their wide diffusion in the cognate languages. The philologist,

---

[1] Polyglot Dictionary, Ch. 29, p. 24. The former is equalized with Chinese *ye suan miao* 野蒜苗 ("sprouts of wild garlic"), Manchu *sejulen*, Mongol *khaliyar*; the latter, with *siao ye kiu* 小野韭 ("small wild leek"). *Li-šii* writes *kun-doñ*.

[2] SCHRADER (in Hehn, *Kulturpflanzen*, 8th ed, p. 208) compares with this Lithuanian *swogũnas*, a word that stands alone in Indo-European languages.

7

however, cannot decide two important botanical questions without
which the solution of the problem is hopeless: and these are whether
the Indo-Chinese and Turkish-Mongol names in their origin refer
to a wild or to a cultivated species, and where the home of the wild
and cultivated species is to be sought. BRETSCHNEIDER [1] informs us
that the *ts'uñ* of North China is *Allium fistulosum*, a native of
Siberia, Dauria, and northern Mongolia, and that the ancient
dictionary *Erh ya* does not mention the cultivated *ts'uñ*, but notices
only the mountain or wild onion. Since the researches of E. REGEL, [2]
who found *Allium sativum* growing wild in the Kirgiz steppe,
botanists are agreed that this region should be regarded as the
original habitat of garlic. In Egypt, the cultivation of garlic and
onion is very ancient, and traceable at least to about 1200 B.C. [3]
Yet their spontaneous origin in Egypt cannot be proved. [4] The
ancient Semitic name (Assyrian *šūmu*, Hebrew *šūm*, Punic *σουμ*,
Arabic *ṭūm*) is probably not correlated with the Turkish term.
Relying on the botanical evidence, the assumption would be possible
that a certain species of garlic or onion was first cultivated by
Turkish tribes and handed on by them to the Chinese and their
neighbors in Central Asia; this transmission, as borne out by the
linguistic evidence, must have taken place in a very remote, prehistoric period. [5] It is of great interest also that Mongol and Turkish

---

[1] *Bot. Sin.*, pt. 2, p. 169.

[2] *Alliorum adhuc cognitorum monographia*, p. 44 (St. Petersburg, 1875).

[3] V. LORET, *L'ail chez les anciens Égyptiens* (*Sphinx*, Vol. VIII, 1904, pp. 135—147).

[4] F. WOENIG, *Pflanzen im alten Aegypten*, pp. 198, 199.

[5] This prehistoric transmission must not be confounded with the late historical introduction in the Han period of another species of *Allium* (*Allium scorodoprasum*), *ta suan*
大蒜, *hu* 胡 *suan*, or *hu* 葫. HIRTH (*T'oung Pao*, Vol. VI, 1895, p. 439) has
represented the matter as though garlic had been introduced into China for the first time
under the Han from Fergana by General Chang K'ien. The attribution of the introduction
to Chang K'ien, however, is not an historical fact; this event is not reported in his
authentic biography inserted in the Han Annals, but it is on record only in that spurious
and untrustworthy Taoist production, the *Po wu chi* (see BRETSCHNEIDER, *Bot. Sin.*, pt. 2,

possess a common word for a wild species of leek or garlic thriving in the steppes (*Allium senescens*): Mongol *mangin*, *mangir*, Kalmuk *mangirsun*, Buryat *manehun*, *manehan*; [1] Teleutic *manyr*. [2] The popularity of this plant is borne out by the fact that the Buryat designate June the "month of leek" (*manehan hara*). In view of the situation on which the problem rests, and in view of two wild species of garlic and leek utilized by the Si-hia, we readily see that it would be important to know the Si-hia terms for the latter, which would possibly shed light on the subject.

PHONOLOGY OF SI-HIA.

A complete and positively assured phonology of Si-hia can naturally not be based on the limited material available for the present. Some characteristic phonetic traits, however, may be pointed out.

1. The most striking phenomenon in the vocalic system of Si-hia is that in a number of cases the vowel *i* corresponds to Tibetan *a*:

> *li*, moon = Tib. *z-la* (No. 12).
> *rni*, ear (corresponding to Chinese *ni*, *ni*) =
> Tib. *rna* (No. 10).
> *ni*, nose = Tib. *s-na* (No. 11).
> *ci*, flesh = Tib. *sa* (No. 48).

p. 171). F. P. SMITH (*Contributions toward the Materia Medica*, etc., p 8; and repeated in the new edition of G. A. STUART, p. 28) says that according to the *Pên ts'ao* Chang K'ien introduced the plant; but Li Shi-chên, the editor of that work (Ch. 26, p. 6 b) gives as his own opinion only that "people of the Han dynasty obtained the *hu suan* from the Western Regions," while he cites the dictionary *T'ang yün* of Sun Mien (published in 750) to the effect that Chang K'ien was the first to bring it back from his expedition to Central Asia It therefore was in the T'ang period that this opinion prevailed, but, as far as I know, there is no evidence thereof accruing from a contemporaneous source of the Han. The connection of the plant with the name of the great general is purely traditional, as he was famed for having introduced a number of other useful plants; and all we may assert safely is that it is possible that a new alliaceous species was received by the Chinese during the Han era from inner Asia.

[1] CASTRÉN, *Burjätische Sprachlehre*, p. 172.
[2] RADLOFF, *Wörterbuch*, Vol. IV, col. 2007.

čᶜi-ma, orange = Tib. *tsᶜa-lum-pa* (No. 44).
tsᶜi-ñu, salt = Tib. *tsᶜa* (No. 13).
si, zi, grass (Mo-so *zi*) = Tib. *tsa* (No. 42).
niñ, morning = Tib. *nañ* (No. 120).
yi, hundred = Tib. *b-r-g-ya*, Burmese *ta-ya* (p. 14).
yiñ, light (in weight) = Tib. *yañ* (No. 120).

2. If, however, the Tibetan word containing the vowel *a* is closed by final *s* or *g*, which are eliminated in Si-hia, [1] the vowel *a* is preserved in the latter language:

na, barley = Tib. *nas* (pronounced *na*).
la, hand = Tib. *lag*.

Interchange of *a* and *i* occurs within the pale of the Tibetan written language: *r-mañ-lam* ("dream;" *lam*, "road") and *r-mi-lam* (*mañ* [2] = Chinese *moñ* 夢, "dream"). In this case we may safely assert that the form *r-mañ* is older than *r-mi*. Ladākhī *lčä* corresponds to common Tibetan *lči* ("dung"). The word *rmañ* ("ground, foundation") has a provincial form *rmiñ*; both *tᶜag* and *tᶜig* mean "cord." The same alternation is met between Tibetan and other Indo-Chinese languages: Tib. *šig* ("louse") — Lepcha *šak*; Tib. *byañ* ("pure") — Newārī *bᶜiñ* ("good, fine"), but Chepañ *bᶜañ-to*; Tib. *za-ba* ("to eat") — Chepañ *jᶜi-sa*; Tib. *ña* ("fish") — Mo-so *ñi*. In the word for "eye" (No. 53) we noted the forms *mik* and *mak*.

3. In other cases, Tibetan *a* in an open syllable changes in Si-hia into *o* or *u*:

Tib. *na*, sick, disease = Si-hia *no* (corresponding to Ahi Lo-lo *no*).
Tib. *ba*, cow = Si-hia *wu*.

---

[1] In this, as well as in the following paragraphs, where the elimination of finals is mentioned, it should be understood that elimination merely refers to an existing fact, but is not intended to convey any notion of genetic development. It is more than doubtful whether Si-hia (like other Indo-Chinese languages) has ever possessed such finals (see Conclusions).

[2] This base occurs also in the Karen languages: *ner-mañ, mye-mañ, biñ-mañ, mi-mañ* (*Gazetteer*, etc., p. 654).

4. Tibetan *a* followed by a final labial or guttural consonant is transformed into *o* in Si-hia, the labial or guttural being eliminated:

>Tib. *r-gyab*, back = Si-hia *ṅo*.
>Tib. *pᶜag*, hog = Si-hia *vo* (Ahi Lo-lo *vye*).

5. Si-hia *u* corresponds to Tibetan *u*:

>*lu*, body = Tib. *lus*.
>*wu*, head = Tib. *wu* (*dbu*).
>*wu*, centre = Tib. *dbus* (*wu*).

6. Si-hia *u* corresponds to Tibetan *o*:

>*lu*, season = Tib. *lo*, year.

7. Si-hia *u* corresponds to Tibetan *a*:

>*wu*, cow = Tib. *ba*.
>*wu*, father = Tib. *pᶜa*.
>*k-ṅu*, five = Tib. *l-ṅa*.
>*k-ṅum*, heaven = Tib. *g-nam* (*d-guṅ*).
>*ẓu*, fish = Tib. *ña*.

8. Si-hia *u* corresponds to Chinese *ou*:

>*tu*, bean = Chin. *tou*.

9. Si-hia *o* corresponds to Tibetan *u*:

>*lo*, sheep = Tib. *lug*.
>*mo*, worm = Tib. *ạbu*.

10. Si-hia *o* corresponds to Tibetan *ö* (*e*):

>*mo*, fire = Tib. *mȯ* (*me*).

11. Si-hia *ö* corresponds to Tibetan *ȯ* developed from *o*:

>*tsö*, *Pantholops hodgsoni* = Tib. *tsȯ* (*gtsod*, *gtso*).

12. Si-hia *i*, usually in a closed syllable, corresponds to Tibetan *i*:

>*niṅ*, heart = Tib. *s-ñiṅ*.
>*miṅ*, man = Tib. *mi*.
>*liṅ* or *riṅ*, great = Tib. *riṅ*, long.
>*čᶜi* (*kᶜi*), gall = Tib. *m-kᶜri-s* (*-pa*).

13. Si-hia *i* in an open syllable corresponds to Tibetan *i* and Chinese *ę*:

si, to die = Tib. *śi*, Chin. *sę*, Mo-so *śe*.

14. Nasalized vowels seem to occur in the following examples:

mĩ, man = Tib. *mi*.
mĩ, not = Tib. *mi*.
rĩ, bear = A-hi Lo-lo *ru-mo*, Lo-lo-pʻo *vu-mo*,
          Tib. *d-re-d*.
čʻu-lĩ, plum = Chin. *li* (No. 184).
ko (go)-čẽ, lion (No. 198).
nõ, two (see p. 13), beside *nöñ*.
nõ, spring (No. 73).
čʻĩ-ni, summer (No. 74).

15. Diphthongs occur with comparative frequency. The one most characteristic of the language is *ou*. The words in which it is found have only a simple vowel in the allied languages: *šou* (No. 6), *kou* or *gou* (No. 18), *mou* (No. 21), *kou* (No. 70), *čʻou-na* ("coin"), and *lou* ("to stew"). The same diphthong is characteristic also of Jyaruṅ, a Tibetan dialect with which Si-hia shares other features: *smou* ("medicine") = Tibetan *sman*; *šou* ("paper") = Tib. *šog*.

16. Likewise the diphthong *ei* thrives in Si-hia where the allied languages have a plain vowel: *mei* ("eye"), *zei* ("panther"), *jei* ("water"), *šañ-wei* ("under-garment"), *wei* ("snow"), *wei* ("year"), *dzei* ("south"), *wei* ("monkey"), *pei* ("present, current"). In *wei* ("to do") the diphthong agrees with Chinese *wei*.

17. The diphthong *ai* is met with in *tsai-šu* ("rat") and *nai* (two of the cyclical signs, see No. 176). It will be noticed that, with the single exception of *tʻien-čo*, all diphthongs close the syllable, and are never followed by a consonant.

18. The diphthong *ao* is found in the Chinese loan-words *yao* ("furnace") and *maoʻr* ("cat"); and in the indigenous words *yao* ("day"), and *yao* ("bird") in the compound *wo-yao* ("domestic fowl").

19. The diphthong *ui* appears after palatals in *č'ui* ("tooth"), where no diphthong is encountered in related idioms, *ṅo-čui-ni* ("chair"), and *č'ui* ("nit"), The diphthong *iu* occurs only in the loan-word *liu-na* ("cabbage"); *ie,* in *t'ien-čo* ("loam-house") and *sie* ("following, next").

20. The triphthong *iao* occurs in *liao* ("blood") and *siao* ("to be born"). The existence of the triphthong *uai* seems to me doubtful (see No. 196).

21. The consonantal system of Si-hia is as follows:

k	*k'*	*g*			*ṅ*
č	*č'*	*j*			*ñ*
t	*t'*	*d*			*n*
p	*p'*	*f*	*b*	*w*	*m*
ts	*ts'*	*dz*			
ž	*z*	*ś*			*s*
y	*r*	*l*			*h*

This system coincides with Tibetan save the fricative *f*, which does not exist in Tibetan. All consonants occur as initials. No double consonants are found in our material. Only *ṅ*, *n*, *m*, and *r* are utilized as finals. A Si-hia syllable or word is therefore composed of initial consonant + vowel + *ṅ*, *n*, *m*, or *r*. Phonetic groups consisting solely of a vowel or diphthong or initial vowel + final consonant seem to be scarce.

22. Initial gutturals and palatals are capable of palatalization: *k'ä* or *g'ä* ("gold"), *k'ä-i* or *č'ä-i* ("true"), *gu-g'ä* ("phœnix").

23. Consonantal prefixes can be pointed out only in four cases: *k-ñä* or *k-ñu* ("five"), *k-ñum* ("heaven"), *r-ni* ("ear"), and possibly in *m-ru*, *m-lu* ("worm," No. 22).

24. Si-hia is destitute of final *g*, *l*, and *s*, as compared with Tibetan and other languages:

> *la,* hand = Tib. *lag.*
> *lo,* sheep = Tib. *lug.*
> *dzei, zei,* panther = Tib. *g-zig.*

·

wo, swine = Tib. *pᶜag.*
ńo, silver = Tib. *d-ńul,* Mo-so *ńo.*
na, barley = Tib. *nas.*
wu, centre = Tib. *dbus* (*wu*).
lu, body = Tib. *lus.*

25. Final *ń* is frequent in Si-hia: *č'öń* ("mother"), *lińn, rińn* ("west"), *šań-wei* ("shirt"), *yińn* ("star"), *tań-lańn* ("quail"), *mońn-tsi* ("fly"), *tańn* ("to set," of the sun).

26. Si-hia final *ń* corresponds to Tibetan final *ń*:

yińn, light (in weight) = Tib. *yańn.*
lińn or *rińn,* great = Tib. *rińn,* long
nińn, heart = Tib. *s-ñiń.*

27. Si-hia final *ń* corresponds to Tib. final *g*:

šańn, iron = Tib. *l-čag-s.*

28. Si-hia final *ń* corresponds to Chinese and Tibetan final *m*:

sińn, heart, mind = Chin. *sim,* Tib. *sem(s).*

29. Final *n* appears in Si-hia *tsan* ("lungs"), *tsan* ("autumn"), *tan* ("to be").

30. Final *m* occurs in *rom* ("spring of water") = Tib. *k'ron-pa, k'rom-pa*; *kńum* ("heaven") = Tib. *gnam.*

31. There is only one instance of final *r* in Si-hia: *nir, n'ir* ("relatives") = Tib. *g-ńer, g-ńen, ńen, ńe.*

32. Si-hia *kᶜ* corresponds to Tibetan *sk* in:

kᶜia, magpie = Tib. *skya.*

33. *č* and *š* alternate in Si-hia and Tibetan:

či, flesh = Tib. *ša.*
šuo, water = Tib. *čᶜu.*
šańn, iron = Tib. *l-čag-s,* Lepcha *ču.*

34. Si-hia *m* answers to Tibetan *m*:

m̃i, man = Tib. *mi.*
mo, fire = Tib. *mo, me.*

35. Si-hia *m* corresponds to Chinese *m*:

　　moṅ-tsi, fly = Chin. *moṅ* 虻 , gadfly.

36. Si-hia *ts* answers to Tibetan *ts*:

　　tsö, Pantholops hodgsoni = Tib. *tsö, g-tso(-d)*.

37. Si-hia *z* corresponds to Tibetan *ts*:

　　zi, grass = Tib. *tsa* (*r-tsa*).

38. Si-hia *y* answers to Chinese *y* (from *ṅ*) and Tibetan *ṅ*:

　　ya, goose = Chin. *yen*, Cantonese and Hakka *ṅan*,
　　Fukien *ṅaṅ*; Tib. *ṅaṅ*, Burmese *ṅan*.

39. Si-hia *w* corresponds to Tibetan *b*, *p'*:

　　wu, cow = Tib. *ba*.
　　wo, stomach = Tib. *p'o*.

40. Si-hia *w* corresponds to Tibetan and Chinese *ṅ*:

　　wen, bad = Tib. *ṅan* (in Central Tibet *ṅem-pa*),
　　Tromowa *ñen-po*, Chin. *ṅan* ("false, fraudulent").
　　wu, cow = Cantonese *ṅau*, Siamese *ṅoa* (*ṅ-woa*),
　　but Tib. *ba*.

41. Si-hia *w* is the equivalent of Chinese *f*:

　　wu, father = Chin. *fu*.

MORPHOLOGICAL TRAITS.

1. The syllable *ni* appears at the end of several stems, so that it may be regarded as a suffix, used similarly to Tibetan *-pa*: *tsu-ni* ("rain"), *k'iṅ-ni* ("summer"), *tsu-ni* ("man"), *ṅo-čui-ni* ("chair"), *ku-ni* ("flour"), *tsö-ni, jö-ni* ("town"). It occurs with a certain preference in the names of animals: *čui-ni* ("fox"), *tsö-ni* or *jö-ni* ("wild goose"), *ko-ni* ("owl"), *mo-ni* ("cuckoo"), *wu-ni* ("wild animals"), *č'aṅ-ni* ("birds"). In the phrase *tsiṅ-ni* ("in time") it seems to have an adverbial function.

2. A suffix -*i* appears in *k'o-i* ("boot"), derived from *k'o* ("foot"); and in *ro-i* ("saddle"), from *ro* ("horse"). Thus it seems to imply the meaning of a covering. [1] Perhaps also *wu-i* ("backbone") may belong here, if the element *wu* bear any relation to *ňo, o* ("back").

3. The diminutive suffix *bu*, in the same manner as in Tibetan (from *bu*, "son"), seems to be employed in *pa-bu* ("daddy"). In *pan-bu, ban-bu* ("butterfly"), however, the element *bu* (= Tibetan *ạbu*) means "insect."

SYNTACTICAL TRAITS.

I have nothing new to add to the observations of my predecessors along this line. Our vocabulary imparts only a few brief sentences which allow of no far-reaching inferences as to syntax. The main point in the construction of the Si-hia sentence, as already remarked by M. Morisse, is that the verb concludes the phrase, while it is preceded by both the direct and the indirect object. This feature is in striking agreement with Tibetan, and presents additional proof for the fact that Si-hia belongs to the Tibeto-Burman group. The position of the attribute, which may precede or follow the noun, is also in harmony with Tibetan. M. Morisse is presumably right in observing that the place of the adjective may be regulated according to certain rules of euphony analogous to those of French. The glossary offers certain fixed terms, as *pei wei* ("current year"), *sie wei* ("next year"), *yi wei* ("past year"), *sie niň* ("after to-morrow"), but *wei saň* ("sparrow," No. 149), *yaň hei* ("raven"), *tu hei* ("black bean"). In regard to the last, Mr. Ivanov remarks that the attribute is placed after the noun if terms of Tibetan origin are involved. I do not see why *tu hei* should be of Tibetan origin; both *tu* and *hei* are probably Chinese words (Nos. 96 and 182).

---

[1] A suffix with analogous significance occurs in Mongol *-bči*: *ebčigin* ("chest") — *ebčigü-bči* ("thorax"); *dalu* ("shoulder-blade") — *dala-bči* ("collar"); *bǔlgǎgusǔ-bči* ("girdle").

CONCLUSIONS.

The preceding analysis leaves no doubt that the Si-hia language belongs to the Tibeto-Burman group of the Indo-Chinese family. It is more difficult for the present to assign to it an exact position within that group. It appears as a certainty that Si-hia is not a mere dialect of Tibetan, Tibetan being taken in the strict ethnographical sense. It is as distinct from Tibetan as is Lo-lo or Mo-so, and has peculiar characteristics by which it is clearly set off from Tibetan proper. To these belong the prevalence of the vowel *i* corresponding to Tibetan *a*, the predominance of diphthongs, particularly of *ou*, and the lack of final explosive consonants. In the latter trait, Si-hia agrees with Lo-lo and Mo-so. Further, it has a common basis with these two languages, as shown by the large number of coincident words. While the Si-hia vocabulary displays certain affinities with Chinese, on the one hand, and with Tibetan, on the other, there are numerous words that are not related to Chinese or Tibetan, but that closely agree with Lo-lo and Mo-so. A goodly proportion of these (sixteen, Nos. 67—82) meet with no counterpart in Chinese and Tibetan, and must be designated as formations peculiar to the Si-hia, Lo-lo, and Mo-so stock; while those words of the same stock, which are traceable to Chinese and Tibetan, exhibit a closer degree of relationship with one another than with Chinese and Tibetan. It must therefore be conceded that we are entitled to the uniting of Si-hia with Lo-lo and Mo-so into a well-defined group of Indo-Chinese languages, which for brevity's sake might be termed the Si-lo-mo group (by choosing the first syllable of each name). How this curious fact is to be explained from the standpoint of history is a question that is not yet capable of a satisfactory solution. Si-hia is a dead language, and the remains discussed on these pages come down from the end of the twelfth century; Lo-lo and Mo-so,

however, are known to us only in their present state. The missing
links between Lo-lo and Mo-so on the one hand, and Si-hia on the
other, must have existed in the territory of Sze-ch'uan (or may still
survive there), but little of the aboriginal languages spoken in that
region has come to our knowledge. One trail leads from the Si-lo-mo
group to the Burmese, Assam, and Nepal languages; and another
trail takes us to the T'ai group. Unsuspected relations between the
various groups of languages are revealed; these may partially be
of ancient date, partially may point also to a lively interchange of
ideas in historical times. Among the characteristic traits of the
Si-lo-mo branch, two are prominent: in distinction from Tibetan,
a very limited number of consonantal prefixes; and in distinction
from Chinese and Tibetan, the lack of final explosives. These two
features may be illustrated somewhat more in detail.

The significance of the preceding investigation for a study of the
historical grammar of Tibetan (and a language can be properly
comprehended only if we grasp its genetic growth) is self-evident.
In former days it was permissible, for instance, to compare such
words as Tibetan *č̓u* and Chinese *šwi* ("water"); but now we
recognize that the two words, though certainly interrelated, are not
directly comparable, each representing a different phase of development from a common root (see No. 37). The most signal result
of our study is that many monosyllabic phenomena of the Tibetan
language, which at first sight appear as indivisible stem-words and
have indeed been taken as such by previous scholars, now turn out
to be compounds contracted from two, three, and even four bases:
compare *bya* (No. 32), *k̓yim* and *lam* (No. 47), *sbrul* (No. 22),
rṅa-boṅ (No. 7), *rta* (No. 8), *gaṅs* (No. 38), *rluṅ* (No. 39), *rdzoṅ*
(No. 47), *groṅ* and *k̓rom* (No. 47), *lče* and *ljags* (No. 49), *ayro*
(No. 51), *mt̓o* (No. 65), *stag* (No. 67). These examples demonstrate that it is our primary task to ascertain the history of Tibetan

words through careful analysis of their components, based upon comparative methods, before venturing direct comparisons of the word in question with the corresponding notions of cognate forms of speech. It is not good method, for example, to correlate Tibetan *spreu* ("monkey") with Burmese *myok* (*myauk*). [1] The Tibetan word is a secondary formation contracted from *spra-bu* (*bu* being a diminutive ending which affects the stem-vowel after the elision of *b*; *bya-bu* becomes *byeu*, *bye*; *lo*, "leaf" — *löu*, *lö*, "section of a book;" *mda*, "arrow" — *mdöu*, *mdö*, "arrowhead"). [2] Judging from experience, the word *spra* certainly is not a plain stem-word, but, on the contrary, is a double or even a triple composition. Theoretically it should be analyzed into **sa-pa-ra* or *sa-pa-la*, and in fact we meet these single components in the Lepcha designation of the monkey, *sa-hu-pa-lá-p*. It is therefore improbable that Tibetan *spra* bears any relation to Burmese *myok*, which, on the contrary, is an independent word. In the same manner, Tibetan *skra* ("hair") has been evolved from **sa-ka-ra*, for in Aka we have *sa-kå* ("hair") and in Kachin *ka-ra* ("hair"); the element *sa* further appears in Kuki-Chin *sa-m* and Burmese *sa-n*. Tibetan *gru* ("boat") is developed from **ge-ru*, *ge-lu*: the two elements are transposed in Mi-ñag *lo-ge*; and the base **lu* occurs in Mo-so *lu*, *lö* ("boat"), Lo-lo of T'ung River *lo*, Lo-lo-p'o *li*, A-hi Lo-lo *li* and *li-zo*, Nyi Lo-lo *š-li*; Aka *lü*, Burmese *hle*, Kachin *hli*. In the languages of the illiterate tribes we naturally find older forms preserved; this is the case also in many Tibetan and Chinese dialects, as contrasted with the written languages and the standard colloquial forms now in use. When we compare Tibetan *bya-wañ* or *p'a-wañ* ("bat") with Chinese *pien-fu* 蝙蝠, a word that refers to the same animal, no relationship between the two words is apparent; but if we fall back on the dialectic forms,

---

[1] E. HUBER, *Bull. de l'Ecole française*, Vol. V, 1905, p. 326.

[2] See also A. SCHIEFNER, *Mélanges asiatiques*, Vol. I, pp. 357, 358.

Tromowa *p'o-loṅ-da*, Sikkim *p'yo-loṅ-da*, and Hakka *p'o-fuk*, we begin to realize the identity of the first element in the Tibetan and Chinese compounds (compare also Sino-Annamese *bien* with Lepcha *bryan*).

Tibetan *r-mi-ba* ("to dream"), past tense *r-mi-s*, at first sight, seems to be a widely different formation from Chinese *moṅ* 夢 ("dream, to dream"); but the Tibetan noun *r-moṅ-lam* ("dream;" *lam*, "road;" of rare occurrence for *r-mi-lam*) at once prepares the connection. We further have Lepcha *moṅ* (*moṅ myon*, "to dream a dream") and Burmese *mak*. The final guttural surd in Burmese in lieu of the guttural nasal shows us the variability of the finals, and the Tibetan stem *r-mi* may raise the question whether the finals *ṅ* or *k* in their origin formed really part of the stem, or rather present subsequent formative additions. He who has carefully gone over the analysis of Si-hia words must have noticed that this problem presents itself in more than one instance. It has been said that Lo-lo has dropped all final consonants with the exception of all nasals;[1] it has been asserted also, and is generally assumed, that the final explosives *k*, *p*, *t*, in Chinese are inherent in the stem, and have been eliminated in the northern dialects. These suppositions, however, are by no means borne out by careful observation of the facts. In a number of cases, it is true, Chinese and Tibetan, as well as other cognate languages, agree as to their finals; for instance:

Chin. *ciuk* 竹, bamboo; Tib. *s-ñug*; Magar *huk*.

Chin. *dzak* 賊, brigand; Tib. *jag*.

Chin. *tuk* 毒, poison; Tib. *dug*.

Chin. *muk* 目, eye; Tib. *mig*.

Chin. *luk* 六, six; Tib. *drug*.

---

[1] E. HUBER, *Bull. de l'Ecole française*, Vol. V, 1905, p. 325.

Chin. *čik* 織, to weave: Tib. *t'ag*.

Chin. *swiet* 說, to speak; Tib. *b-šad*.

Chin. *šat* 殺, to kill; Tib. *b-sad*.

Chin. *kiap* 甲, armor: Tib. *k'rab*.

In other instances, however, Tibetan is destitute of a final, whereas Chinese is provided with it:

Tib. *agro*, Si-hia *k'o*, Chin. *gu-k* (No. 51).

Tib. *ñi*, sun; Chin. *žit* 日.

Tib. *ko*, hide, leather; Chin. *kok* 革.

Tib. *za-ba* (*zo*, *b-zo-s*), to eat; Shanghai *zok* 食.

Tib. *l-če*, tongue; Chin. *dziet* 舌 (see No. 49).

Tib. *g-la*, Si-hia *la*, stag; Chin. *luk* (No. 31). [1]

Again, in other cases, Tibetan has a final, while Chinese is devoid of it:

Tib. *s-gog*, leek; Chin. *kiu* (see p. 96).

Tib. *zug-pa*, pain, *gdzug-pa*, to sting, prick; Chin. *dz'u* 愁, to grieve.

Or, the finals differ in Chinese and Tibetan:

Chin. *šik* (Hakka *šit*) 識, to know; Tib. *šes*.

Chin. *dat* 達, to penetrate; Tib. *dar*.

Chin. *šit* 蝨, louse; Tib. *šig*.

Chin. *džak*, *dzak* 作, to make; Tib. *m-dzad*. [2]

---

[1] Compare also Tib. *ro*, taste; Siamese *rot*, taste.

[2] Final *m*, *n*, and *ñ* are inherent in the stem:

Tib. *sem-s*; Chin. *sim* 心 (No. 9).

Tib. *šam*; Chin. *šam* 衫 (No. 100).

Tib. *ram-s*; Chin. *lam* 藍, indigo (Korean *ram*, Manchu *lam-un*).

Tib. *žon-pa*, to mount, ride; Chin. *džien* 乘.

Tib. *mañ*, many, *dmañ-s*, people, multitude; Chin. *moñ* 氓, people, subjects.

Tib. *r-moñ-ba*, to be obscured, stultified, *mun-pa*, obscurity; Chin. *muñ* 蒙 and 懜, dull, stupid. Compare, however, *lu-ñ*, dragon (No. 22).

The final consonants of Tibetan are by no means stable. SCHIEFNER [1] has already called attention to two phenomena, — parallel words with and without final consonants, and confusion of final consonants. This, however, is not an explanation of the phenomenon. Certain final consonants, as I endeavored to show on a former occasion, [2] were added as formative elements with a specific significance, or with a grammatical function which became lost in course of time. From *ko-ba* ("hide, leather") is formed *ko-g-pa*, with the meaning "shell, peel, rind." The word *kug* ("crooked"), identical with Chinese *k^cuk* 曲 and Burmese *kok*, appears also as *kum* and *koň*, while in Ahom it is *kut*, in Siamese *koň*. [3] In Ahom we have *lat* ("to speak"), in Tibetan *lab*. More examples will be found in the Appendix.

Within the Chinese language we find variations of the finals in Cantonese, Hakka, and Fukien:

Cantonese *yět* 一, one; Fukien *eik*.

Cant. *pat* 八, eight; Fukien *paik*.

Cant. *yět* 日, sun; Fukien *nik*.

Cant. *lip* 立, to stand; Fukien *lik*.

Cant. *lik* 力, strength; Hakka *lit*; Swatow *lat*.

Cant. *šik* 食, to eat; Hakka *šit*; Amoy *sit*; Shanghai *zok*.

Cant. *fat* 法, law; Hakka *fap*; Swatow *hwap*.

Cant. *tsat* 疾, disease; Swatow *čit*; Amoy *ček*; Fukien *čik*.

Cant. *šat* 室, room; Fukien *sek*; Shanghai *sak*.

Comparison shows us that Fukien has in some cases preserved

---

[1] *Mélanges asiatiques*, Vol. I, pp. 346—348.

[2] *T'oung Pao*, 1915, p. 424.

[3] The primary root is **gug*, which is preserved in the Tibetan verb *a-gug-s* ("to make crooked, to bend"), corresponding to Chinese **guk, kuk* 局 ("crooked;" analogous to Tibetan *kug-kug*), and in Bunan *gur-gur*. The final *t* appears also in Milčan *kuta* ("crooked"), beside *k'uň-šim*.

the ancient final *k*, whereas Hakka and Cantonese have exchanged it for *t*: Fukien *saik* ("louse") agrees with Tibetan *šig* in the final, where Cantonese has *šĕt* and Hakka *sit* (No. 6).

If we now take Indo-Chinese philology in its widest range, we observe that vast tracts of its domain are occupied by languages which are destitute of any finals, and that the lack of finals even covers a larger geographical area than the area where they occur. The word for "tongue" (No. 49), for instance, possesses a final only in the South-Chinese dialects (*še-t, šu-t, sie-k*), while such is absent in all other languages of the family, notably in Tibetan (*l-če*). Examining the word *la* (No. 50), we note that Si-hia, Mo-so, Lo-lo, Kachin, and Nepalese are equally devoid of a final, while Tibetan has one in *la-g*; but Tromowa and Sikkim *la-ko* demonstrates that the Tibetan final *g* is not inherent in the stem, but the survival of a syllable *ko*, which was contracted with the base *la*. Hence it is reasonable to conclude that *la* presents the primary root-base, and that *la-g* denotes a secondary development. It is likewise obvious from the facts cited under No. 53 that the base of the word for "eye" is **me, mi*, not *mik* or *mit*. Not only is the number of languages without a final in this case very large, but also derivatives like Tibetan *s-mi-n-ma* and Chinese *mou, mei, mi* (No. 54) uphold this point of view.

The secondary character of final Tibetan *l* has been demonstrated in the case of *dñul* (No. 1), where *-l* corresponds to Chinese *-n* or *-ñ*, and where the base is *ñu, ñū, ño*; likewise in the case of *sbrul* (No. 22). The same holds good for *sbal* ("frog"), for we meet in Mo-so *pa*, in Burmese *p'a*, in Nyi Lo-lo *a-pa-ma*, in Chinese *wa* 蛙. In *spre-l* ("monkey," from *spre*), *-l* is secondary.

There is accordingly no valid reason to regard the final consonants as the prior event and to construe a theory of their gradual elision in the light of a posterior move. It is not only conceivable, but it

8

is borne out by the data of many languages, particularly by the branch which interests us here, Si-hia, Lo-lo, and Mo-so, that there are Indo-Chinese idioms without final explosives. To my mind, these did not originally belong to the stem, but were subsequent formative elements. In this respect, the relationship between Chinese and Tibetan is closer than between these languages and the Si-hia, Lo-lo and Mo-so branch, which is essentially characterized by the lack of final explosives. As to Mo-so, compare, for instance, *bo* ("pig") -- Tib. *p'ag*, *da* ("to weave") —Tib. *t'ag*, *nau* ("forest") — Tib. *nags*, *še-še* ("paper") — Tib. *šog*, *miö* ("eye") —Tib. *mig*, *du* ("poison") —Tib. *dug*, *be* ("to do") — Tib. *byed*, *p'e-le* ("butterfly") —Tib. *p'ye-ma-leb*, *ma* ("oil") —Tib. *mar*. As to Si-hia, see Phonology, § 24 (p. 103).

A similar observation holds good for the prefixes. There is no basis for the preconceived assumption that prefixes should once have been general in Indo-Chinese, that in view of an abundance of prefixes in some languages those with scanty or no prefixes should have lost them, and that the type of prefix-language is older than that devoid of prefixes. It seems certain that Chinese roots have never had any consonantal prefixes (the only instance of a Chinese prefix is the vocalic *a*). [1] In general we might say that those idioms which are destitute of prefixes at the present time were likewise so in the past. Si-hia sides with Lo-lo and Mo-so in the very limited number of prefixes; [2] and the mere fact that the prefixes of the

---

[1] See the writer's *The prefix A- in the Indo-Chinese Languages* (*J.R.A.S.*, 1915, pp. 757—780).

[2] The following examples of Mo-so words show lack of prefix, as compared with Tibetan: *k'o* ("door")—Tib. *sgo*, *pa* ("wolf")—Tib *spyaṅ*, *kö* ("star")—Tib *skar*, *pa* ("frog")—Tib. *sbal*, *ts'ö* ("lake")—Tib. *mts'o*, *dsü* ("to do")—Tib. *mdsad*, *ba* ("goitre")—Tib. *lba*, *či* ("tongue")—Tib. *lče*, *ṅö* ("silver")—Tib. *dṅul*.

same words are variable within the Tibetan dialects, and again are at variance with the prefixes in the corresponding words of cognate languages, is apt to show that the prefixes represent a secondary stage of development. [1]

Prefixes as well as final consonants, therefore, are of minor importance in the comparative study of Indo-Chinese languages. The same may be said about the stem-vowels, which are likewise vacillating, although we may arrive at certain laws in course of time. The staff of comparison remains the initial consonant and the tone conditioned by it: these form the backbone of the word and the basis of all investigation. The tendency of the sonants to change into surds and aspirates has gradually modified the face of many root-words and obscured the mutual relationship of Chinese and Tibetan (compare, for instance, No. 143). The manner and degree of relationship in the Indo-Chinese camp, therefore, do not lie so clearly at the surface as in the Indo-European field. Strictly speaking, we first ought to elaborate the historical grammar of Chinese and Tibetan, taken individually, before attempting the comparative study of the two languages. Nevertheless, the one cannot well be accomplished without the other, and both efforts undertaken simultaneously and leading in the same direction will yield results to benefit and to advance both historical and comparative research of these languages.

---

[1] Compare the prefix *d* in Jyarun in cases where common Tibetan is without a prefix (*T'oung Pao*, 1914, p. 108).

APPENDIX.

In connection with the preceding observations, the following Chinese and Tibetan concordances may be of interest; these demonstrate at the same time that the relationship between the two languages is much closer than assumed heretofore. The first group here selected shows Tibetan words, most of them being provided with prefixes. The fact that these are of secondary character becomes evident from their absence in Chinese.

1. *kan* 乾, dry. Tib. *s-kam, s-kem,* dry. (Compare also **go, ko, k'o* 頷, chin; Tib. *ko-ko, ko-sko, kos-ko,* chin.)

2. *gaṅ 岡, ridge of a hill, mound. Tib. *s-gaṅ,* a projecting hill.

3. *kaṅ* 槓, trunk, box. Tib. *s-gam,* chest, trunk, box.

4. *dzaṅ 藏, granary. Tib. *r-dzaṅ,* storage-chest (West Tib. *zem,* box, chest).

5. **džam, dzam, dam* 參, to counsel, advise. Tib. *g-dam, ạ-dam,* to advise, exhort.

6. **džam, dzam* 慚, to feel shame. Tib. *ạ-dzem,* to feel ashamed.

7. **džiet, diet* 節, knots or joints of plants. Tib. *m-dud,* knot; *m-dze-r,* knot in wood; wart.

8. **džiet, diet* (Japanese *šitsu*) 櫛, a comb, to comb the hair; *šwat 刷, brush, to brush. Tib. *šad, g-šad, g-šod,* to comb, curry, brush, stroke; *zed,* brush.

9. **džaṅ, dzaṅ, zaṅ* 曾, past, done, finished. Tib. *ạ-dzaṅ-s, za-d,* spent, consumed, exhausted.

10. **džaṅ, dzaṅ, zaṅ* 層 (Shanghai *dzəṅ, zəṅ*), story of a building, layer. Tib. *b-zaṅ-s* (also *k'aṅ b-zaṅ*), a two-storied building; *ạ-dzeṅ,* to project, to jut out.

11. **džan, dzan, zan* 粲, viands; 餐, to eat, a meal. Tib. *zan, b-zan,* food, porridge; *za,* to eat.

12. *zan 娿, beautiful. Tib. b-zaṅ, good, beautiful.

13. *džai, dzai 材, stuff, materials; 財, property, wealth. Tib. r-dza-s, objects, materials, goods, property, treasures, jewels.

14. *dzuṅ, duṅ 同, together, the same as, alike; 雙, a match for, a peer. Tib. m-tsʻuṅ-s, similar, like, equal; a match for.

15. *dzuṅ 聰, quick of apprehension, clever; 憁, intelligent. Tib. m-dzaṅ-s, wise. Regarding the change of vowel compare *dzuṅ, duṅ 銅, copper, and Tib. zaṅ-s.

16. *dzin 盡, exhausted. Tib. zin, exhausted.

17. *dže, dze, de 砥, whetstone. Tib. a-dze-ṅ, whetstone.

18. *dzuṅ, duṅ 鏦, javelin, to stab with a spear. Tib. m-duṅ, spear.

19. *džuṅ 中, middle. Tib. g-žuṅ, middle.

20. *džu, du 聚, to collect, to assemble. Tib. a-du, to come together, to assemble, to unite, to join one another.

21. *džwi, dzwi, zwi 醉, intoxicated. Tib. zi, b-zi, g-zi, intoxication.

22. *džie (Wen-chou zi, Ningpo zie) 射, to shoot, to aim at, archery. Tib. g-žu, bow.

23. *džie (Hakka sa) 麝, musk-deer. Tib. ša, šwa, stag; šwi, new-born fawn.

24. šö (Hakka ša, Fukien sia) 社, personification of Earth. Tib. sa, earth.

25. *bun 糞, ordure, dung. Tib. brun, dung, excrements.

26. *boṅ 胖 or 肨, fat. Tib. boṅ, size, bulk; s-bom, thick, stout, coarse.

27. *bun 氛, vapor, miasma, poisonous exhalations. Tib. bem, dead matter, a pestilential disease.

28. *bu 舞, dance. Tib. bro, dance.

29. *bat, pat 拔, lofty, eminent. Tib. d-pal, glory, splendor.

30. *bun 墳, grave, mound. Tib. a-bum, tomb, sepulchre.

31. pa 疤, scar, birth-mark. Tib. r-ma, wound; r-me, d-me, s-me, natural mark, spot, mole, birth-mark.

32. *pa 波, wave; Tib. d-ba, wave.

33. pa 霸, to rule by force; might, leadership. Tib. d-ba-ṅ, might, power, rule.

34. pa 笆, a kind of bamboo. Tib. s-pa, s-ba, cane; Lepcha po; Burmese wā; Kačāri oá; Ahom bai, cane, rattan; Shan wai; Thādo wo, go.

35. *wat 曰, to say. Tib. a-bar, to talk.

36. *dam 毯, rug, carpet. Tib. g-dan, bolster, seat of several quilts; s-tan, mat, carpet, cushion.

37. *dam 潭, pool, lake. Tib. l-teṅ, pool, pond.

38. *duṅ 洞, hole, cave, grotto, ravine. Tib. doṅ, deep hole, pit, pitch.

39. *duṅ 筒, tube, pipe, duct. Tib. doṅ, l-doṅ, tube; a hollow, cylindrical vessel.

40. *duṅ 疼, pain, soreness, ache. 痌 or 恫, to moan with pain. Tib. g-duṅ, to desire, long for; to feel pain, to be tormented, afflicted.

41. *diṅ 梃, stalk or grain; 梃, stalk, staff, cudgel; 莛, stalk of grasses and plants. Tib. s-doṅ, stalk of a plant, trunk of a tree, tree (Mūrmi d'oṅ, tree).

42. *diṅ 靛, indigo. Tib. m-t'iṅ, indigo.

43. *do 砣, heavy stone, stone roller, weight. Tib. r-do, stone, weight.

44. *duk 督, to superintend, to direct, to rule. Tib. b-dog, to possess, to own; b-dag, master, lord, self; b-dag byed-pa, to reign, to possess.

45. *džek, zek, šek, sek 石, stone. Tib. g-seg, small stones; šag, pebbles, gravel.

46. t'o, t'u 唾, to spit, saliva. Tib. t'u, t'o-le, to spit.

47. leu 簍, basket. Tib. s-le, le, basket. (Compare also lei 儽, lazy, and Tib. le-lo, lazy.)

48. luṅ 惏, stupid. Tib. b-lun, dull, stupid.

49. *dži, dzi, dza 瓷 磁, pottery. Tib. r-dza (West-Tib. za), clay.

50. *tsi* 漆, lacquer. Tib. *r-tsi*, varnish, paints.

51. **hap* 呷, to swallow. Tib. *hab*, a mouthful; *hab-hab za-ba*, to devour greedily.

Chinese final explosives correspond to a Tibetan liquida:

52. **g'ap* 俠, heroic, bold; **dž'ap, dz'ap* 捷, to gain a victory in battle. Tib. *r-gyal*, to be victorious, victory, king.

53. **g'ap* 夾, to squeeze, to press. Tib. *b-čer, b-čir*, to squeeze, to press.

54. **get* 訐, to accuse. Tib. *a-gel, r-gol*, to accuse.

55. **džo, dzo, džak, dzok* 錯, to err, to make a mistake. Tib. *a-dzol*, fault, error, mistake; *r-dzu*, delusion; *r-dzu-n*, falsehood.

56. **džik, džek* (Cantou *ts'ek*) 莿, thorn, to prick. Tib. *ts'er*, thorn, prick, brier.

57. **wat* 曰, to say. Tib. *a-bar*, to talk.

*džiet, diet 節, knot. Tib. *m-dzer*, knot in wood (see No. 7).

58. **dat* 達, to pass through, to penetrate. Tib. *dar*, to be diffused, to spread. See also No. 29.

Final explosives in Chinese where Tibetan has none:

59. **dži-p, dzi-p* (Ningpo *dzi*) 睫, eye-lashes. Tib. *r-dzi*, eye-lashes.

60. **džu-k, dzu-k, zu-k* 畜, to feed, nourish, rear. Tib. *a-ts'o, g-so*, to feed, nourish, rear, cure.

61. **ku-t* 掘, to dig out, excavate. Tib. *r-ko*, to dig, dig out, hoe, engrave.

Final explosives or liquidæ in Tibetan where Chinese has none:

62. **giao* 膠, glue. Tib. *r-gya-g*, glue.

63. **giao* 交, to unite, friendship. Tib. *gro-g-s*, friend; *a-gro-g-s*, to be associated.

64. **giao* 絞, to bind, to twist. Tib. *s-gro-g*, cord, rope, feathers.

65. **giao* 叫, to call out. Tib. *s-gro-g*, to call out, to proclaim.

66. *kiao* 皎, the bright white moon; effulgent. Tib. *b-kra-g*, brightness, lustre.

67. *džo, dzo, zo (Shanghai zu) 坐, to sit. Tib. s-do-d, b-žu-g-s,
     to sit.

68. *džai, dzai, zai 纔, just now, then. Tib. g-zo-d, now, this moment.

69. *zai, ts'ai 猜, to guess (猜謎, to guess a riddle). Tib. ts'o-d,
     estimation, guess; ts'od šes, ts'od bya, riddle.

70. *tš'ai, ts'ai 菜, vegetables. Tib. ts'o-d, vegetables.

71. *džai, dzai, dzoi, čai 栽, to cut off (but čöi 截). Tib. g-čo-d, to cut.

72. *giai 解, to loosen; to explain; to get free from. Tib. ɑ-gro-l,
     to be released from; to loose, untie, release; to explain;
     s-gro-l, to rescue, deliver. As to the initial guttural with
     following r or l, compare Tib. sgro, large feather for or-
     namenting arrows, as a charm, etc., with Chinese *gio 翹,
     long-tail feathers used as ornaments; Tib. sgro, to elevate,
     exalt, sgrob, haughtiness, pride, with Chinese *gio 驕, high,
     elevated, 喬, high, stately, proud; and Tib. glen, to talk, to
     preach, with Chinese *gian 講, to talk, to preach.

73. *giai, kai 偕, to accompany. Tib. s-kye-l, to accompany.

74. *gia, ga, ge 跏, to sit cross-legged. Tib. s-kyi-l, to sit cross-legged.

Chinese final nasals correspond to Tibetan liquidæ:

75. *zan 燦, bright, glittering. Tib. g-zer, ray, beam; zil, brightness;
     ɑ-ts'er, to shine, glitter.

76. *džam, dzam 慘, grieved, sad; 憯, to be sorrowful. Tib. g-zer,
     pain, ache, illness; to feel pain; ɑ-ts'er, to grieve, grief, sorrow.

77. *džam, dzum, dziem, zam 鏨, to pierce, cut, chisel out, engrave;
     鑽, to bore a hole, to pierce; Tib. g-zer, to bore into; zer,
     to drive in nails; g-zer, nail; g-zon, chisel, graving-tool.

78. *kwon 勸, to admonish. Tib. s-kul, to admonish.

79. *kwon 圈, circle. Tib. s-kor, ɑ-k'or, s-gor, circle.

The following comparisons are instructive both as to the finals
     and initials of the words:

80. *džien, dzien 剪, to cut with scissors or shears. Tib. čem-tse,

scissors; *ts'em-pa*, tailor; *ts'em-po*, seam; *a-ts'em*, *b-tsem-s*, etc., to sew.

81. **k'em*, *k'im* 衾, coverlet, quilt. Tib. *k'eb-s*, *k'yeb-s*, cover.

82. **zuṅ*, *suṅ* 雙, a pair, couple. Tib. *zuṅ*, a pair, couple.

83. **dza-k* 作, to make. Tib. *m-dza-d*, to make.

84. **džuk*, *dzuk*, *dzok* 足, to be sufficient. Tib. *č'og*, to be sufficient.

85. **bik*, *p'ak* 甓, tile. Tib. *pag*, brick.

86. **dzok*, *zok* 鑿, to chisel out, to bore into. Tib. *a-dzug-s*, *zug*, to prick, sting, pierce, bore.

87. **džiek* 食, to eat. Tib. *b-žes*, *b-za*, food.

88. **džan* 殘, to injure, to destroy. Tib. *a-jom-s*, to conquer, to destroy.

89. **giao* 狡, crafty, clever. Tib. *s-gam*, *s-grin*, clever.

90. **giao* 較, to compare. Tib. *s-grun*, to compare.

91. **l'äp* (Canton *lip*) 獵, to hunt. Tib. *liṅ-s* (probably from **lim-s*), hunting, chase. Lepcha *lyŭm mat*, to hunt (*lyŭm*, the god of hunting).

92. **džai*, *dzai*, *dzoi* 彩, color, gay-colored, ornamented. Tib. *b-tso*, *a-ts'o-d*, to dye; *ts'o-s*, *ts'o-n*, paint, dye; *b-tsa-g*, red ochre, earths of different color.

93. **džok*, *džak*, *dzak* 昨, yesterday. Tib. (*k'a*)-*r-tsaṅ*, *m-daṅ*, yesterday.

94. *yaṅ* 颺, to be tossed about, as by wind or waves. Tib. *g-yeṅ*, to be moved by the water to and fro.

95. *yaṅ* 揚, to raise, to hold up, to praise. Tib. *g-yaṅ*, happiness, blessing. Shan *yâṅ*, to praise; Ahom *jâṅ*, fame, glory.

96. **lok* 烙, to burn, roast. Tib. *s-reg* (West-Tib. *š-rag*), to burn, to roast. The Tibetan stem is **ra*, *re*, *ro*: compare *s-ro*, heat, to make warm; *d-ro*, hot time of the day; *d-ro-ba*, *d-ro-n-ma*, warm, *d-ro-d*, warmth; Bunan *ko-s-ra*, hot; Ahom *rau*, *raw*, *râ-n*, heat.

INDICES.

(The figures refer to the numbers of the words.)

miů, man 107
miů, not 108
mu 131
mei, eye 53
mei, virtuous 164
mo, drum 158
mo, eyebrows 54
mo, fire 35
mo, forest 85
mo, goat 173
mo, heaven 34
mo, lip 57
mo, not 108
mo, ox 172
mo-jei 189
mo-ni 148
mo-niů-liů 119
mo-wo 194
mo yiů 192
mo-lu 22
mou-šu 21
moů-tsi 20

tsai-šu 29
tsan, autumn 126
tsan, Chinese 127
tsan, lungs 126
tsiů 163
tsu, drum 158
tsu, winter 72
tsu-ni, man 45
tsu-ni, rain 123
tsei 37
tsò, hare 169
tsò, mountain 79
tsö-ni, town 148
tsö-ni, wild goose 47
tso-wei 63

tsʻi-ňu 13
tsʻüan-ni-na 184 no.
tsʻun 187

dsai-šu 29
dsiů 163
dsei, panther 26
dsei, south 80
dsei, water 37
dsò, hare 169
dsò, mountain 79

žu 5

zi 42
zei 26

ya 97
yao, bird 32
yao, day 145
yao, furnace 180
yaů-hei 148
yi, charcoal 151
yi, ladle 156
yi, woman 125
yiů, light 120
yiů, mark 162
yiů, star 124, 144
yiů-na 124
yü, hog 176
yü, silk 160
ye 68

riů, bear 24
riů, great 119
riů-lo 8
ro, copper 199
ro, fir 143
ro, wind 39
ro, wolf 25
ro-i 8
rom 114
b-ru 21

la, hand 50
la, north 81
la, stag 31

la, tongue 49
la-nu 146
laů-to 152
laů-nòn 7
li 12
liao 83
liu 168
liu-na 202
liů, great 119
liů, tortoise 133
liů, west 82 no.
liů-lo 8
lu, body 102
lu, season 115
lu, stone 41
lu-ni 10
lu-yi 41
le, lo, earth, field 40
le, lo, heavy 64
lo, copper 199
lo, fir 143
lo, tiger 67
lò, lo, wind 39
lo-i 8
lo-wo 30
lom 114
lo-tsei 112
m-lu, m-ru 22

šaů 3
šaů-wei 100
šwi-ma 184
šu-kʻuai 196
šou 6
šoů 188

saů 149
si, to die 16
si, grass 42
si, liver 55
si-na 201
si yiů 193
sie 165

sie-niů 120
siů 9
siů-le 9
se 138

hei 182
hi 171
hiů 183

a-ko 141
i 125
u-yi, u-i 61
o 101
o-diů 106

2. Tibetan.

kwaů-gul 94
kram 202
kluůs 40
lkog 106
skya-ka 111
skrag-pa 47

kʻa-ba 38
kʻug 134
kʻyi 28
kʻyim 47
kʻyuů 33
kʻro 199
kʻrom 47
mkʻris 56
ąkʻar 199

gaůs 38
gwa-nu 99
gi-liů 8
go-bi 202
go-bo 33
goů-ma 46
grog-ma 18
groů 47
gla 31

3. English.

action 117
agate 124
ancients, the 122
animal 95
ant 18
antelope 112, 113
apricot 183
autumn 126

back 60
backbone 61
bad 17
barley 116
bean 96
bear 24
bee 21
bird 32
black 182
blood 83
body 102
boot 52
brain 154
brother 141, 152
butterfly 19

cabbage 202
camel 7
cat 181
cattle 4
centre 104
chair 130
charcoal 41, 151
Chinese 127
coal 41
coin 161
copper 199
coral 196
cow 4, 172
cuckoo 148

day 145

day after to-
  morrow 120
deed 117
to die 16
dipper 146
disease 62
dog 28, 175
dragon 23
drum 158
duck 89

ear 10
earth 40
east 82
egg 32
egg-plant 184 no.
elephant 131
eye 53
eyebrows 54

father 91
female 150
field 40
fir-tree 143
fire 35
fish 5
flea 6
flesh 48
flour 134
flower 43
fly 20
foot 51
forest 85
form 118
fowl 32, 170
fox 99
fruit 87, 121
furnace 180

galaxy 189
gall 56
genuine 77
goat 173

gold 2
goose 97, 148
grass 42
great 119

hair 92
hand 50
hare 30, 169
hatchet 63
head 103
heart 9, 105
heaven 34
heavy 64
high 65
hog 27, 176
holy 188
honey 21
horse 8
house 47

I 14
inch 187
insect 21
iron 3

Jupiter 193

Kaki 184
kidney 76
king 46

ladle 156
lake 101
lamp 178
law 163
light, in weight 120
lion 198
lip 57
liver 55
loam house 155
Longan 195
lotus 136
louse 6

low 137
lungs 126

magpie 111
male 149
man 45, 107
mandarin duck
  194
mark 162
Mars 192
mind 9
monkey 174
moon 12
morning 120
mosquito 21
mother 122, 153
mountain 79
mouth 75
mule 69
mustard 201
mutton 68

neck 94, 106
at night 66
nit 132
north 81
nose 11
not 108

orange 44
owl 147
ox 4, 172

palm 93
panther 26
peacock 197
pear 90
phœnix 33
pig 27
pigeon 148
plum 185

quail 148

radish 200	silk 160	swallow 148	water 37
rain 123	silver 1	swine 27	well 114
rat 29, 171	snake 22		west 82 no.
raven 148	snow 38	throat 106	white 135
relatives 122	south 80	tiger 67	wind 39
ribbon 140	sparrow 149	tongue 49	winter 72
rice 84	spider 159	tooth 58	wise 165
river 109	spleen 129	tortoise 133	wolf 25
road 47	spoon 156	town 47	woman 125
	spring 73	tree 86	word 167
saddle 8	spring (water) 114	true 77, 166	world 168
salt 13	stag 31		worm 22
serpent 98	star 124, 144	undergarment 100	
sheep 68	stomach 59		year 70, 71, 115
shoe 179	stone 41	Venus 190	Yellow River 109
shoulder 128	summer 74	vinegar 186	

Addenda. — Two valuable studies of Chinese scholars have meanwhile reached me, — a Dissertation on the national writing of the Si-hia 西夏國書略說 by Lo Fu-ch'ang 羅福萇 (accompanied by good reproductions of Si-hia Antiquities) and an Investigation on the Si-hia translation of the Saddharma-pundarika-sūtra (p. 4) 西夏譯蓮華經考釋 by Lo Fu-ch'êng 羅福成.

P. 12. In the language of the Min-kia, *a* is the numeral 1 (LIÉTARD, *Anthropos*, Vol. VII, 1912, pp. 678, 692).

P. 112. The following may be added to the base *gug* ("crooked, bent"): *giok* 角 ("horn"), *g'uk* 毬 ("leather ball"), *gou* 句 ("crooked"), *gou* 痀, "hunchback" (corresponding to Tibetan *gye-gu*, "hunch, hump"). Compare also CONRADY, *Causativbildung*, p. 168.

127

海象和独角鲸牙补说

SUPPLEMENTARY NOTES ON
WALRUS AND NARWHAL IVORY.

BY

BERTHOLD LAUFER.

———◆◁▷◆———

The following notes are intended to supplement my essay published in *T'oung Pao*, 1913 (pp. 315—364), and accompanied by additional notes of M. PELLIOT (pp. 365—370). [1] Page references given without further specification pertain to that article.

---

[1] These notes were written in the beginning of 1914, but their publication has been delayed owing to circumstances beyond the writer's control. I avail myself of this opportunity to express to my esteemed friend, M. Pelliot, my sincere thanks for his generous co-operation and his valuable additional notes. The present article contains also several interesting contributions from his pen. — Further bibliographical references may find place here. The *T'u shu tsi ch'êng*, following the procedure of Li Shi-chên, has placed that author's text on *ku-tu-si* in the section on "Snakes" (XIX, 181, *hui k'ao* II, p. 15), without adding any further matter, while the text of the *Yün yen kuo yen lu* is inserted among the miscellaneous notes on the "Rhinoceros" (XIX, 69, *tsa lu*, p. 3). This shows that no scholarly investigation of the subject was made in the Manchu period. Likewise it is worthy of note that the editors of the great cyclopædia, in the same manner as Li Shi-chên, overlooked the fundamental definition of the term *ku-tu-si* in the *Liao shi*, and its employment in the Annals of the Liao and Kin Dynasties. This, as well as other instances, bears out the fact that the encyclopædic works of the Chinese, with all their vast accumulation of material, are far from being complete or perfect. — The passage quoted on p. 327, note 1, after *P'ei wên yün fu*, is in *T'ang shu*, Ch. 40, p. 8 b; also in *Ta t'ang leu tien* 大唐六典, Ch. 3, p 6 b (ed. of *Kuang ya shu ku*, 1895). — Ohtere's account (p. 337) is easily accessible in H. SWEET's *Anglo-Saxon Reader* (pp. 17—23, Oxford, 1908); and in English translation in C R. BEAZLEY, *Texts and Versions of Carpini and Rubruquis* (p. 8, London, Hakluyt Society, 1903)

In regard to the modern trade with China in marine ivory, S. WELLS WILLIAMS [1] had stated,—

"Seahorse teeth, 海馬牙 *hai ma ya*,[2] are brought from California, Sitka, and other parts of western America, and are used by the Chinese in the same manner as ivory. Under this term are also included the teeth and tushes of the walrus, sperm whale, and other cetaceous and phocine animals; but with the cessation of the whale fishery, the importation has dwindled to almost nothing."

On September 6, 1913, the "Daily Consular Trade Reports" published by the Department of Commerce and Labor of Washington contained the following, written by Consul-General F. D. CHESHIRE, Canton (p. 1356):

---

[1] *Chinese Commercial Guide*, 5th ed., p. 102 (Hongkong, 1863).

[2] The adoption of the word *hai ma* in the sense of "walrus" seems to be of recent origin. The Manchu word corresponding in sense to Chinese *hai ma* or given as its equivalent in the Manchu-Chinese dictionaries is *malta* which, judging from the native definitions, seems to denote a kind of seal (see AMIOT, *Eloge de la ville de Moukden*, p. 289, Paris, 1770; and SACHAROV, *Manchu-Russian Dictionary*, p. 872). The *Polyglot Dictionary* of K'ien-lung (Ch. 32, p. 47) adds to this equation Tibetan *mts'o srin* and Mongol *aramana*. A. KIRCHER (*La Chine illustrée*, p. 259, Amsterdam, 1670) figures a (somewhat grotesque) hippopotamus with the legend "l'Hippopotame ou Cheval-Marin appellé Hayme par les Siriens." On the cut we read *haý mà*, accompanied by Chinese characters. Kircher gives a somewhat confused description of the animal after Boim, who asserts that the Chinese make chaplets, crosses, and images of saints from its teeth, "et on assure qu'il n'y a rien de si salutaire pour empêcher le flux de sang, que de porter quelqu'une de ses pieces sur soy." See also O. DAPPER, *Beschryving des Keizerryks van Taising of Sina*, p. 241 (Amsterdam, 1670). The ancient authors, however, understand by the term *hai ma* a kind of shrimp (鰕類) having the shape of a horse, — according to Ch'ên Ts'ang-k'i of the T'ang period five to six inches long, according to K'ou Tsung-shi of the Sung period two to three inches long, — and occurring in the southern sea (*T'u shu tsi ch'éng*, XIX, 164). The Chinese naturalists seem to be ignorant of the term *hai ma*, as applied by the archæologists with reference to the lion-like animals displayed on a certain class of metal mirrors. — The word "sea-horse" was used in English in the same sense. The *Century Dictionary* credits it with the meaning "morse of walrus", and gives sea-horse tooth as "the ivory-yielding tooth of the walrus or of the hippopotamus." The famous English naturalist John Ray (1627—1705), who wrote under the Latinized name Raius, has the following observation in his work *Synopsis methodica animalium quadrupedum* (p. 193, Londini, 1693): "Vidi etiam penem eiusdem animalis [i. e. rosmari, walrus] osseum rotundum, cubitum et amplius longum, crassum, ponderosum ac solidum, in fine prope glandem longe crassiorem et rotundiorem. Huius pulvere ad pellendum calculum Moscovitae utuntur. Dentes hi nostratibus Equi marini dentes *Sea-horses Teeth* appellantur."

"Before the revolution, about eighteen months ago, there was considerable trade in the manufacture from walrus ivory tusks of tobacco-pipe mouth-pieces, handles of fans, thumb-rings, and peacock-feather tubes for mandarin hats. These articles were sent to Peking, where they were dyed a green color, resembling the color of jade, but since the revolution there has been very little activity in the manufacture of such goods from walrus tusks. The demand has fallen off considerably, and the trade is confined to making cigarette holders, tooth brushes, and chopsticks. The value of walrus tusks is $ 280 to $ 400 Hongkong currency per picul (133½ pounds). Elephant tusks are worth $ 700 to $ 1,200 Hongkong currency per picul. The elephant tusks are more serviceable and at the same time more valuable."

On the same page, Consul-General G. E. ANDERSON, Hongkong, reports that inquiry among local importing and exporting firms and dealers in ivory of Hongkong failed to locate any importations of walrus ivory, but that elephant ivory is imported in large quantities, and is shipped mostly to Canton.

An inquiry regarding the trade in walrus and narwhal ivory from Alaska, addressed to the Department of Commerce and Labor, Washington, elicited the information that during the fiscal year 1913 a quantity of ivory, and manufactures thereof, amounting to $ 2,475, was received from Alaska, but that no figures were known there concerning the export of these articles to China. It was therefore intimated to me to communicate with the United States collectors of customs at Juneau (Alaska), San Francisco (California), and Seattle (Washington) for further information. The collector of customs, San Francisco, wrote as follows: —

"There are no statistics kept at this office, from which the desired information can be furnished. I have made several inquiries regarding this matter, but can find no one that can furnish the requested information."

The collector of customs, Seattle, reported, —

"I regret to advise you that no record is kept by this office of the ivory, or other products, received in this District from Alaska."

The following positive information was received from the collector of customs, Juneau, Alaska:—

"TREASURY DEPARTMENT,
UNITED STATES CUSTOMS SERVICE,
PORT OF JUNEAU, ALASKA,
December 15, 1913.

"Replying to your letter of the first instant relative to exportation from this district of walrus and narwhal ivory, I have to state that there was during the present year exported direct from Alaska to China 4,000 lbs. of walrus ivory, value $1,200, and from Alaska to the United States 7,763 lbs. of foreign walrus ivory, value $2,717. The destination of the latter quantity is unknown to this office, but it is believed that the bulk of this ivory is exported to Japan and China."

To a further inquiry as to the route or line upon which direct exportation of ivory from Alaska to China is undertaken, the collector of customs of Juneau was good enough to reply on January 20, 1914. that this shipment was made by the Norwegian tramp steamer "Kit" from Nome to Japan, that there is no regular transportation line direct from the Alaskan coast to the Orient, but that occasionally tramp steamers call at different ports, bound for the Orient.

As the fact of a direct Alaskan-Chinese ivory trade was now established, and as, according to the report of our Consul-General at Canton, the material is handled and wrought there by the Chinese, it seemed to me an essential point to inquire if the ancient name *ku-tu-si* is still known to the Cantonese. The Consul-General of Canton, in a letter of January 16, 1914, favored me with the following reply:—

"I beg to state that I have made many inquiries, and find that the name by which walrus ivory is commonly known in Canton is *hai ma ya* 海馬牙. The term for walrus ivory which you state in your paper was common in ancient Chinese literature,—*ku-tu-si*,— I find is not known or used at present in Canton."

Simultaneously my old friend P. P. Schmidt, Professor of the Chinese language at the Oriental Institute of Vladivostok, whom I had interested in the problem because of the importation of walrus ivory from Gishiginsk and Baron Korff's Bay to that port (p. 335), was good enough to send me the following note:—

"The word *ku-tu-si* is not known here. The tusks are called in Chinese *hai ma ya* 海馬牙, in Japanese *kaiba no kiba*. [1] In Korean the walrus is styled *yung sê* 靈犀 (Chinese *ling si*). Since 1909 the firm Tshurin has taken the northern trade into its hands, and annually receives from forty to fifty *pud* of walrus ivory, which is transported *via* Moscow to London. This article was formerly handled by a merchant from St. Petersburg. It has always been disposed of, however, in London, not in China, Japan, or Russia. The bulk of walrus ivory is collected by American smugglers, and exchanged for alcohol. Such a ship has recently brought together

[1] This means likewise "sea-horse teeth." *Kai-ba* is the Sinico-Japanese reading of *hai ma*. The Japanese dictionaries of Hepburn, Gubbins, Nitobe and Takakusu, also assign to *kai-ba* the meaning "walrus." Among the temple treasures of Nikko, Japan, a narwhal-tooth is still preserved in the temple of Iyemitsu. It is figured in the little guidebook *Nikkō-zan rin-ō-ji giu-hōmotsu zu-kai* 日光山輪王寺御寶物圖解 (p. 5, Tokyō, 1896). Here we find the Portuguese name *unikōrn* (p. 349) written in Kana ウニカウゥ (Portuguese *unicorne, unicornio*); and the rectangular box in which the tusk is kept is inscribed with the words *Ban-kaku is-shi* 蠻角一枝, "a horn of the Barbarians," *Ban* (Chinese *Man*) being a Japanese designation of Europeans, in particular of Portuguese and Hollanders. Dr. O. Nachod writes me that according to *Gwai kō shi kō* (p. 706) Holland presented this tusk in 1671. J. Dautremer (*Nikkō passé et présent, guide historique*, p. 103, Tokyō, 1894), in an enumeration of the treasures, terms it "une dent d'espadon;" but the illustration mentioned leaves no doubt that it is a narwhal-tusk — An interesting allusion to the trade of the Dutch in narwhal ivory to Japan is contained in the work of Ch. P. Thunberg (*Voyage en Afrique et en Asie, principalement au Japon, pendant les années 1770—1779*, traduit du suédois, p. 296, Paris, 1794), who has the following: "Aussitot que les marchandises qui composent la cargaison des vaisseaux hollandais, sont déposées dans les magasins de la Compagnie, le gouverneur fait annoncer cette nouvelle aux négocians qui se rendent alors chez lui, pour examiner les échantillons des marchandises dont la vente se fait dans un encan public, ou Kambang. Les offres se font en Mas, dont dix font un Thail. La corne de Narval se payait cette année assez cher; c'était autrefois un objet de contrebande sur lequel les Hollandais gagnaient immensément; les Japonais qui attribuent à cette production animale, semblable à l'ivoire, toutes les vertus médicinales que les adeptes vantent de la pierre philosophale, la payaient à des prix exorbitaus."

about three hundred walruses. It is not known here where this merchandise is sold. Japanese smugglers have not been noticed in the high north among the Chukchi. This is all I am able to learn here."

The term *ku-tu-si*, accordingly, is now extinct in China, and this is exactly what we should expect; and S. Wells Williams was correct in applying the term *hai ma ya* to walrus ivory.

G. CAHEN [1] calls our attention to the fact that in the first part of the eighteenth century the Russians bartered with the Chinese two articles,—seal-skins and walrus-tusks, called "bones of the walrus-tooth" (кости моржевого зуба). The Russians, consequently, appear twice in the history of this trade with China,—first in the Mongol period, when Russian walrus ivory, through the medium of the Mongols, reached Turkistan (p. 338); and again in recent times, as direct traders of the article in northern China.

The *Pa hung yi shi* (p. 321), the preface of which is dated 1683, has the following account of Russia (Ch. 2, p. 1 b): "Russia 阿路索 [2] is situated north-west from China, and in the north-east of Europe. The country has walled cities. As to apparel, sable coats are most highly esteemed. Men and women themselves settle their marriage affairs. They live in blockhouses. Vassal states are numerous. The population is sparse compared with the extent of the area. The climate is exceedingly cold, and the soil of the far-off corners is frozen up during six months. During the Ming dynasty no intercourse was as yet established with China. It was only at the time of K'ang-hi of the present dynasty that they first presented black sables 黑貂, *fish-teeth* 魚牙, gyr-falcons 海青, [3]

---

[1] *Le Livre de Comptes de la caravane russe à Pékin en 1727—1728*, p. 104, note 1 (Paris, 1911). M. Cahen's statement is based on L. Lang, *Journal de la résidence du sieur Lange, agent de sa Majesté impériale de la Grande Russie à la cour de Chine dans les années 1721 et 1722* (Leiden, 1726); LANG speaks of "les dents de loups-marins."

[2] BRETSCHNEIDER, *Mediæval Researches*, Vol. II, p. 71.

[3] In the *Polyglot Dictionary* (Ch. 30, p. 10) this term corresponds to Tibetan *k'ra č'en*, Mongol *šongkor*, Manchu *šongkon* (see also AMIOT, *Éloge de Moukden*, p. 265).

a striking-clock, glass mirrors, and other objects. The speech of this people differs from the other languages of Europe, but as to writing they use the European letters. The western scholar 西儒 Nan Huai-jên 南懷仁 [1] understood their language."

Since the Russians styled walrus-tusks "fish-teeth" (p. 337), and, according to their own documents, imported them into China during the K'ang-hi period (1662—1722), the fact is established that the term *yü ya* 魚牙 of the Chinese text quoted refers to walrus-tusks. It was not that Russian influence of recent date, however, which caused this Chinese phrase to assume this specific meaning. Prior to the arrival of the Russians, the expression existed, and is clearly enough defined to leave no doubt that it denotes marine ivory. The *Wu li siao shi* (p. 328), Ch. 8, p. 21, speaking of wrought objects of elephant ivory 象牙器, has a notice to this effect: "That kind of ivory styled *shu kio* [p. 343, note] is marine ivory (or the tooth of the sea-elephant). [2] The tooth of the red boar is in color like oysters and jujubes, and is veined like elephant ivory. The fish-teeth are like elephant's teeth" (其曰殊角者海象牙也。紅猪牙如蚌棗色如象牙紋。魚牙如象齒). It is unambiguously expressed in this passage that "fish-

---

E. D. Ross (*Polyglot List of Birds*, No. 68) is quite right in assigning to it the meaning "gyr-falcon," which, by the way, had already been established by W. Schott (*Über die ächten Kirgizen, A.B.A.W.*, 1865, p. 449). The Russian name is кречетъ. S. von Herberstein (*Rerum Moscoviticarum Commentarii*, translated by R. H. Major, Hakluyt Society, 1852, Vol. II, p. 35) emphasizes the large number of gyr-falcons in Muscovy, trained for taking swans, cranes, and other birds of that kind.

[1] Ferdinand Verbiest (1623—88).

[2] D. Crantz (*History of Greenland*, Vol I, p. 125, London, 1767) describes the walrus under the name "sea-cow" (adding German *wallross*, Latin *rosmarus*, and French *vache marine*), saying, "Their bodies resemble a seal, but their heads are very different; for the head of this is not long, but stubbed and broad, and therefore it might be called a sea-lion, or perhaps elephant, on account of the two long tusks it has." The *Persian-English Dictionary* by Johnson and Richardson (ed. Steingass, p. 945) quotes *fīl-i daryā* فیلی دریا "sea-elephant" as a name for walrus.

teeth" are a product of the nature of ivory; and for this reason I am disposed to conclude also that the "fish-tooth silks" mentioned in the T'ang Annals as tribute of Sinra and the Mo-ho (pp 338 — 342) [1] were so named from peculiar patterns woven in these stuffs, and resembling the natural designs occurring in walrus ivory. In fact, nothing else could be intended by this expression. Such designs as might be imitated in textiles are peculiar to walrus ivory, and to this kind of ivory only, which in its cross-sections exhibits designs of the character of grained wood, and along the sides is intersected by fine yellow lines, or overstrewn with larger yellow flamed spots.

In 1518 the prince of T'ien-fang (Arabia), Sie-yi-pa-la-k'o 寫亦把剌克, sent an envoy to the Chinese Court offering as tribute horses, camels, knives made of fish-teeth 魚牙刀, and other objects; and he received for his sovereign precious garments, silk-stuffs, musk, etc. [2]

It is by no means striking that the term 魚牙 "fish-tooth" assumed the specific significance of "walrus-tooth." This development is quite in harmony with the genius of the Chinese language. From remote times the word yü 魚 has denoted not only "fish," but also "sea-mammals." In the *Shi king* (II, ɪ, VII, 6; and II, ɪɪɪ, IV, 1) [3] we find twice the compound yü fu 魚服 rendered

---

[1] M. PELLIOT (p. 366) is doubtless correct in assuming that 納 in the text of the *Kiu T'ang shu* is a misprint for 紬 (this reading, indeed, is given in the *Pien tse lei pien*, Ch. 221, p. 1 b); this emendation, however, does not affect my conclusion. In the *T'ang hui yao* 唐會要 (Ch 95, pp 16 b, 17 b; ed of *Kiang-su shu kü*) 納 appears twice after yü ya. A curious passage extracted from a Manchu work by AMIOT (*Éloge de Moukden*, p. 290) deserves mention in this connection: "Le *lekerhi* est un animal aquatique, dont la peau blanche et noire ressemble à une très belle étoffe." The Manchu word *lekerhi* refers to a marine mammal.

[2] According to the Annals of the Ming Dynasty, as already indicated by BRETSCHNEIDER (*China Review*, Vol. V, 1876, p. 173), without explaining what these knives are.

[3] LEGGE, *Chinese Classics*, Vol. IV, pp 261, 285; S. COUVREUR, *Cheu king*, p. 186.

by LEGGE "seal-skin quiver," and by COUVREUR "carquois de peau
de veau marin" (*phocae pellis* in his Latin translation).   The word
"fish" in this instance is explained by Lu Ki 陸璣 (of the
T'ang dynasty) as "the name of an animal like a pig, found in the
eastern sea, spotted on the back and dark (純青) underneath." [1]
Medhurst identifies it with a seal; Legge is inclined to think that
a porpoise may be meant.   If we remember that the Gilyak, Ainu,
and other North-Pacific tribes, still turn out quivers of seal-skin,
we move on the basis of reality.   Elephant ivory was perfectly
well known in the early days of Chinese antiquity: combs of ivory
象之搞 are twice alluded to in the *Shi king*, [2] and the ends of
the bow were tipped with ivory 象弭. [3]   In the *Shu king* (Tribute
of Yü) ivory is simply designated as "teeth" (*ch'i* 齒), [4] and in
the same manner the word is employed in the *Chou li*. [5]   As the
word *ya* enters into the compound *siang ya* 象牙 "elephant's tooth,"
the conditions were given in the language that *yü ya*, as used in
the T'ang period, could easily assume the significance "tooth or
ivory of a sea-mammal."   A fresh impetus was received during the
Mongol period, when walrus-tusks were transmitted to the Mongols
by the Russians under the name "fish-teeth," and when the latter
designation with its specific meaning, no doubt under Russian
influence, was revived in the East.   There is a piece of evidence
to this effect in the tradition of the Mongols.

The Armenian King Haithon, who reigned 1224—69 and died

---

[1] *Mao shi ts'ao mu ch'ung yü su*, Ch. 下, p. 9 (ed. of *T'ang Sung ts'ung shu*)..
Under the category "fish" a marine mammal styled *kien-t'ung* 建同 and occurring
around the littoral of Cambodja is described in the *Sui shu* (Ch. 82, p. 3 b).

[2] LEGGE, *l. c.*, pp. 77, 164.

[3] LEGGE, *l. c.*, p. 261.

[4] LEGGE, *Chinese Classics*, Vol. III, pp. 111, 115; COUVREUR, *Chou king*, pp. 71, 73.
It must be noted, however, that *ya* 牙 properly means "canine tooth," "tusk," and *ch'i*
齒 "front or molar tooth."

[5] HIRTH, *Ancient History of China*, p. 121.

in 1271, iu the narrative of his journey to the Mongols written
by Kirakos of Gandsak, [1] "told many marvelous things about the
barbarous peoples whom he had seen and heard." "He asserted
that beyond Cathay there was a country where the women have
human shape and are endowed with reason, and where the men
are without reason, big and hairy. These dogs do not allow anybody to penetrate into their territory, they go ahunting and subsist,
together with the women, on the game which they seize. From
the union of the dogs with the women are born boys having the
shape of dogs, and girls of the shape of women. There is also a
sandy island there where is found a precious bone in the form of
a tree, called fish-tooth (*dent de poisson*). When it is cut, another
bone will shoot forth at the same spot, in the manner of deer's
antlers." [2] As shown by KLAPROTH, who was the first to make this
document known, the fable of the Country of Dogs was generally
known among the Mongols in the thirteenth century. As to the
Chinese sources of the story, KLAPROTH refers only to *San ts'ai t'u
hui*, [3] but, as is well known, the earliest records of it are contained
in the *Liang shu* (Ch. 54, p. 12), *Nan shi* (Ch. 79, p. 3 b), and
Wu Tai shi (Ch. 73, p. 4). [4] Haithon's story closely approaches
these Chinese traditions. We dwell here merely on the point which

---

[1] BRETSCHNEIDER, *Mediæval Researches,* Vol. I, p 164.

[2] KLAPROTH, *Journal asiatique,* 1833, p 288; and DULAURIER, *ibid.,* 1858, p. 472.
Neither of these authors explains what the fish-tooth is.

[3] The passage is translated, with a reproduction of the illustration, also by F. DE
MÉLY (*Revue archéologique,* Vol. III, 1897, p 359).

[4] Compare W. W. ROCKHILL, *Journey of William of Rubruck,* pp. 12, 36; C. R
BEAZLEY, *Texts and Versions of John de Plano Carpini,* p 117; and chiefly G SCHLEGEL
(*T'oung Pao,* Vol III, 1892, pp. 495 *et seq*), who has made a special study of this legend,
but has overlooked the fact that the substance of the story must have been borrowed by
the Chinese from western sources, and that it is only localized by them on the far-off islands
in the north-eastern ocean. From this point of view the subject has been treated by me in
a preliminary article published in the Anniversary Volume in honor of E. Kuhn. See,
further, CHAVANNES, *Journal asiatique,* 1897, mai-juin, p 408.

is of interest with reference to our subject. The Country of the
Dogs or Dog-Headed (*kou kuo* 狗國) is vaguely defined as an
island in the eastern ocean;[1] the Kingdom of Women (*nü kuo*
女國), which must be identical with it, is first mentioned in the
Hou Han shu (Ch. 115, p. 4 b) as situated in the ocean off the
coast of Korea, and is stated by the çramaṇa Huei Shên 慧深
to be distant a thousand *li* east from the country of Fu-sang 扶
桑; while Fu-sang was alleged to be twenty thousand *li* east of
Ta-han 大漢, the latter over five thousand *li* east of Wên-shên
文身, and Wên-shên over seven thousand *li* north-east of Japan.
All this, of course, is not real geography, but geographical myth
and literary reconstruction, in which a curious medley of Taoist
speculations is blended with western fables and possibly with a
certain substratum of traditions coming down from the tribes of the
North-Pacific area, and presumably conveyed through the medium
of Japanese and Chinese mariners. Haithon's country of dog-headed
men with women of human shape, in the belief of the Chinese,
was located in an island of the northern Pacific; and there, according
to Haithon, was the home of the fish-tooth. This feature of the
story, as far as I know, has not yet been pointed out in any
Chinese version; but doubtless Haithon appears to have received
the report from the Mongols, who, on their part, had picked it up
from the Chinese. There is no other possibility than that Haithon's
fish-tooth relates to walrus and narwhal tusks.[2] The origin of the

---

[1] According to the *Wu Tai shi* (*l. c.*), it was situated on the mainland north of the
Shi-wei 室韋.

[2] The northern Woman Country has another curious relation to marine mammals.
In the *Tu yang tsa pien* 杜陽雜編 (Ch 下, p. 15, ed. of *Pai hai*) we read that
"in the period Ta-chung 大中 (847—860) the Woman Kingdom 女王國
sent as tribute whale-blubber stuffs 龍油綾 and seal-blubber stuffs 魚油錦
with very queer, manicolored patterns; when placed in water, they did not become wet,

legend of the fish-tooth growing in the manner of trees is not far to seek, and is implied by Haithon's own words. The tusk, as previously demonstrated, was regarded as a horn; and as the stag sheds his antlers, so also the 'horns" of the marine mammals were believed to become detached from the animal, and to grow again.

The fact that the notion of the tree-like character of the horn is not a personal fancy of Haithon, but a tradition which really obtained in the East, is well attested by the peculiar Khitan writing of the word *ku-tu* 榾柮, where the classifier "tree" appears in either element. It will be seen below that the earliest writing of this word, as it occurs in the T'ang Annals, is 骨咄, where the first element *ku* ("bone"), and the use of the classifier "bone" in the second element, very appropriately indicate a product consisting of bone or ivory. The tradition of *ku-tu* being in structure or appearance like wood seems to have originated in the Liao period, and, as demonstrated by Haithon, was perpetuated down to the age of the Mongols. Nevertheless the writing of the word *ku-tu* with the classifier "tree" is a peculiar characteristic of the Liao only,

---

the cause thereof being attributed to the whale and seal-blubber (皆入水不濡濕云有龍油魚油故也); the best were made in the Woman Kingdom" What Woman Kingdom is meant here, follows from the comment added, which refers us to the chapter on the "Eastern Barbarians" in the *Hou Han shu* (Ch. 115, p. 4 b) as saying that in the ocean there is the Woman Kingdom, in which the women become pregnant by looking into a well. This passage contains a curious reference to the blubber of marine animals, in this case utilized to render certain kinds of textiles water-proof. The earliest allusion to seal-blubber is made in the *Shi ki* (CHAVANNES, *Mémoires historiques de Se-ma Ts'ien*, Vol. II, p. 195), where torches or lamps fed by the grease of "the human fish" (p. 351) in the tomb of Ts'in Shi Huang-ti are mentioned. The employment of seal-blubber (海牛脂) for purposes of illumination in Shan-tung is recorded as early as the fifth century in the *Ts'i ti chi* 齊地志 of Fu Ch'ên 伏琛. With the exception of this single reference, Li Shi-chên does not speak about blubber. Chao Hio-min, in his *Pên ts'ao kang mu shi i* (p. 326; Ch. 9, p. 11), points to this omission, and devotes a brief notice to *hai kou yu* 海狗油 "seal-blubber," describing the life of the animals and the mode of their capture off the coast of Shan-tung.

and was abandoned by the Kiu (p. 366) as well as by the Mongols (p. 320). Under the word *ku* 榾 the Dictionary of K'ang-hi cites the *T'u king pén ts'ao* of the Sung period to the effect that this word refers to the trunk of a tree which is white like bone, and hence receives its name, and that the southerners make from it very fine utensils. The term *ku-tu* does not relate to any specific tree, but denotes the burls or knotty excrescences on the trunks of various trees which in diverse parts of the world, owing to their fine veneer, are chosen with a predilection for carvings, particularly of bowls. Every one who has been in China has observed these fist-like knots on the mulberry-trees, called *sang ku-tu* 桑榾柮, which, according to *Pén ts'ao kang mu shi i* (Ch. 6, p. 36), are employed as a remedy for pleurisy (治膈症). The most clever artists in burl-carved work known to me are the Tibetans, whose eating-bowls justly evoke also the admiration of the Chinese. The burls used by them, as was established by the botanist J. D. HOOKER, [1] are produced on the roots of oaks, maples, and other mountain-forest trees, by a parasitical plant known as *balanophora*. These bowls have two peculiar features in common with *ku-tu-si*: many of them are white and yellow, and, with their peculiar veins, offer a somewhat ivory-like appearance; and some of them are believed by the Tibetans to be capable of detecting poison. [2] This observation may possibly account for some Chinese writers ascribing *ku-tu-si* to Tibet (pp. 320—321) by confounding the ivory *ku-tu* with the vegetal *ku-tu* 榾柮. On the other hand, we thus obtain a clew as to the reasoning of the Khitan in choosing the latter characters for the purpose of writing the *ku-tu* 骨䯏 of the T'ang.

The account of S. VON HERBERSTEIN (p. 337), [3] who was am-

---

[1] *Himalayan Journals*, p. 91 (London, 1893).

[2] HOOKER, *l c.*, p. 90; ROCKHILL, *J.R.A.S.*, 1891, p 274.

[3] *Notes upon Russia: being a Translation of the Earliest Account of that Country*,

bassador to the Grand Prince Vasiliy Ivanovich in the years 1517 and 1526, is as follows: "The articles of merchandise which are exported from Russia ... into Lithuania and Turkey, are leather, skins, and the long white teeth of animals which they call *mors*, and which inhabit the northern ocean, out of which the Turks are accustomed very skilfully to make the handles of daggers; our people think they are the teeth of fish, and call them so." "The ocean which lies about the mouths of the river Petchora, to the right of the mouths of the Dwina, is said to contain animals of great size. Amongst others, there is one animal of the size of an ox, which the people of the country call *mors*. It has short feet, like those of a beaver; a chest rather broad and deep compared to the rest of its body; and two tusks in the upper jaw protruding to a considerable length. ... The hunters pursue these animals only for the tusks, of which the Russians, the Tartars, and especially the Turks, skilfully make handles for their swords and daggers, rather for ornament than for inflicting a heavier blow, as has been incorrectly stated. These tusks are sold by weight, and are described as fishes' teeth." Von Herberstein, accordingly, identifies the commercial label "fish-teeth" with the zoölogical term "morse;" that is, the walrus. [1]

---

entitled *Rerum Moscoviticarum Commentarii*, translated by R H. MAJOR, Vol. I, p. 112; Vol. II, p. 111 (Hakluyt Society, 1851, 1852)

[1] The origin of the Russian word *morž* (морҗъ) is still obscure. Certain it is that it is not Slavic (Polish *mors* is derived from French *morse*), but its source is not yet traced. The derivation from Russian *more* (море), "sea," as given in the *Century Dictionary* (see *morse*), is impossible. The relation to Lapp *morša, moršša*, and Finnish *mursu*, is not clear (E. BERNEKER, *Slavisches etymol. Wörterb.*, Vol. II, p. 80). The chances are that these may be based upon the Russian words as well. No lesser Finnish scholar than LÖNNROT (*Finskt-Svenskt Lexicon*, Vol. I, p 1094) traces Finnish *mursu* to Russian *morž*; and KNUD LEEM (*Lexicon lapponicum*, Vol. I, p. 825, Nidrosiae, 1753; and *Beskrivelse over Finmarkens Lapper*, p. 216, Copenhague, 1767) records the Lapp word only in the form *morsh*, which would thus point to a Russian source. R. MECKELEIN (*Finnisch-Ugrische Elemente im Russischen*, 1914) does not cite the word *morž* among the Finno-Ugrian

A most interesting reference to the employment on the part
of the Turks of knife-hilts of walrus ivory is made in 1553 by
PIERRE BELON [1] as follows: "Les Turcs sont quasi aussi grande
despense en leur endroict en l'orfeuerie, que nous: et ce qu'ils font,
est de fort bonne matiere. Ils aiment à porter des anneaux, et
veulent que leurs cousteaux soyët bien façonnez: et les pendent à
vne chaine d'argent, dont la gaine est enrichie de quelques belles
garnitures d'or ou d'argët. C'est vne coustume commune aux Turcs
comme aux Grecs de porter les cousteaux aux pëdants à la ceinc-
ture: et sont cõmunement forgez en Hongrie, ayants le mãche moult
long: mais quand les merciers de Turquie les ont achetez, lors ils
les baillent aux ouuriers pour leur mettre vn bout, qui est com-
munement de dent de Rohart, [2] dont y en à de deux sortes. L'vne

---

loan-words in Russian Other Finno-Ugrian and the Samoyed languages have different
words; for instance, Ostyak peŋk-voi, "tooth-animal" = walrus (A. AHLQUIST, Sprache der
Nord-Ostjaken, p. 120), and Samoyed tewote, tiutei (CASTRÉN, Wörterverzeichnisse, pp 27,
300; regarding the Samoyed's relation to the walrus see V. KRESTININ in Klaproth's
Magasin asiatique, Vol. II, pp. 56, 74, 76). According to the new Oxford English
Dictionary the earliest occurrences of the word morse in English literature are in CAXTON,
Chron. Eng. of 1482 ("This yere were take four grete fisshes between Erethe and london,
that one was callyd mors marine") and in CHANCELOUR (circa 1553) in Hakluyt's Voyages
of 1599 ("There are also a fishes teeth, which fish is called a Morsse"). — It may be
added that according to Dal' "morse-eaters" (моржеѣды) is a nickname for the inhabitants
of Archangel, and that рогозубъ ("horn-tooth") is a synonyme of the narwhal.

   [1] Les Observations de plusieurs singularitez et choses memorables, trouuées en Grece,
Asie, Iudée, Egypte, Arabie, et autres pays estranges, p. 298 (Anvers, 1555). The first
edition was published in Paris, 1553. Belon (1518—64) was a prominent traveller and
naturalist. "L'amour de la vérité, un désir avide d'acquérir des connaissances, un courage
infatigable, l'art d'observer et l'esprit d'analyze, en firent un savant distingué, et on le
place au nombre de ceux qui contribuèrent puissamment au progrès des sciences dans le
XVIᵉ siècle. On peut se fier à l'exactitude de ses observations et à la véracité de ses
récits" (Biographie universelle, Vol IV, 1811). The author's spelling is retained in the
above quotation.

   [2] Explained by E. LITTRÉ (Dictionnaire de la langue française) as "ivoire des morses,
de l'hippopotame. HIST. XIVᵉ siècle. Un coustel à un vieil manche de rohart, DE LA-
BORDE, Emaux, p. 486. XVᵉ siècle. L'ivire et le rochal et les pierres precieuses, DU
CANGE, rohautum (où il interprète, probablement à tort, rochal par cristal de roche).
XVIᵉ siècle Par quoy luy en faut adapter d'autres [dents] d'os ou ivoire, ou de dents

est droictement blanche compacte, ressemblant à la Licorne: [1] et est si dure que l'acier à peine y peut mordre, s'il n'est bien trempé. L'autre dent de Rohart est courbée comme celle d'vn Sanglier: qu'eussions creu estre dent d'Hippopotame, n'eust esté qu'auõs veu des Hippopotames en vie, qui n'en auoyent pas de telles." [2] In the Latin translation of Belon's work, [3] the name "morse" for the animal has been added: "ut manubrio ex dente beluae marinae *Mors* quibusdam dictae, Gallis *Rohart*, adornent." [4] Von Herberstein's and Belon's accounts are coeval with the Turkish source indicated by Jacob (p. 317).

An important contribution to the subject is furnished by the Jesuit father AVRIL, [5] who in the latter part of the seventeenth century gathered the following information from the Russians: "Besides furs of all sorts, which they fetch from all quarters, ... they have discovered a sort of ivory, which is whiter and smoother than that which comes from the Indies. Not that they have any elephants that furnish them with this commodity (for the northern

---

de rohart, qui sont excellentes pour cest effet, PARÉ, XVII, 3. ÉTYM. Probablement, corruption de *rorqual*" The latter word refers to a species of whale and is explained, after Cuvier, from Swedish *rær* ("tube") and *qval* ("whale"), "baleine à tuyaux, à cause des plis de la peau sous la gorge et la poitrine." In the *Supplément* is added Bugge's etymology from *rohal, roshal*, Norse *hrosshval*, which is the more probable of the two.

[1] The narwhal.

[2] The walrus has frequently been confounded with the hippopotamus (compare below the quotation from Avril). The new *Oxford English Dictionary* also states that the term "morse" has been erroneously applied to the hippopotamus. Belon's observation shows conclusively that hippopotamus' teeth are not involved.

[3] PETRI BELLONII CENOMANI, *Plurimarum singularium et memorabilium rerum in Graecia, Asia, Aegypto, Iudaea, Arabia, aliisque provinciis ab ipso conspectarum observationes, tribus libris expressae* C. Clusius Atrebas e Gallicis Latinas faciebat, p. 395 (Antverpiae, 1589).

[4] In the same manner the word *morse* (*morsse*) has been interpolated in the early English translations of Ohthere's Anglo-Saxon account, where only the term *hors-hwæl* ("horse-whale" = walrus) occurs.

[5] *Travels into Divers Parts of Europe and Asia, undertaken by the French King's Order to discover a New Way by Land into China,* done out of French, p. 175 (London, 1693). M Pelliot has been good enough to call my attention to this source.

24

countries are too cold for those sort of creatures that naturally love heat), but other amphibious animals, which they call by the name of *Behemot*,[1] which are usually found in the River Lena, or upon the shores of the Tartarian Sea. Several teeth of this monster were shewn us at Moskow, which were ten inches long, and two at the diameter at the root: nor are the elephant's teeth comparable to them, either for beauty or whiteness, besides that they have a peculiar property to stanch blood, being carried about a person subject to bleeding. The Persians and Turks who buy them up put a high value upon them, and prefer a scimiter or a dagger haft of this precious ivory before a handle of massy gold or silver. But certainly nobody better understands the price of this ivory than they who first brought it into request; considering how they venture their lives in attacking the creature that produces it, which is as big and as dangerous as a crocodile." Farther on, Avril quotes a

---

[1] *Behemōth* בְּהֵמוֹת, the Hebrew word used in the Old Testament (Job, XL, 10) for the hippopotamus of the Nile, and presumably derived from Egyptian *p-ehe-mau* ("water-ox"). In Russian it is бегемóтъ. Vladimir Dal' (*Complete Dictionary of the Live Great-Russian Language*, in Russian, Vol. I. col. 144) attributes to it only the meaning "hippopotamus," but does not state that it is used with reference to the walrus. It will be seen below (p 367) that Sir George Watt employs the term "hippopotamus ivory" as synonymous with "sea-horse [narwhal] ivory." See also above, p. 363. On the other hand, *behemōth* was applied in Russia also to the mammoth (Russian *mammont*; C. Witsen, *Noord en Oost Tatarye*, p. 742), and P. J. von Strahlenberg (*Nord- und Östliche Theil von Europa und Asia*, p. 394, Stockholm, 1720) derived the word *mammoth* from Hebrew *behemōth* through the medium of an Arabic *mehemoth*. H. H Howorth (*The Mammoth and the Flood*, p. 49) therefore thinks that Avril has possibly confounded mammoth ivory with the teeth of walrus or narwhal. In my opinion it is not necessary to assume such a confusion, as Avril plainly describes the hunting of the walrus and nought else; and the term *behemoth* was used rather flexibly, being referred to any large and strange beast, for instance, also to the rhinoceros (*Chinese Clay Figures*, p. 83, note 4). The *Oxford English Dictionary* says that *behemoth* is used in modern English literature as a general expression for one of the largest and strongest animals Von Strahlenberg's etymology, moreover, is doubtful, and is not accepted by Russian lexicographers It is more probable that the word *mammoth* is traceable to some language of Siberia, but this is not the occasion for a discussion of this subject.

story told him by the Voyevoda of Smolensk about an island at the mouth of the great River Kawoina, beyond the Obi, that discharges itself into the Frozeu Sea. "This island is spacious and very well peopled, and is no less considerable for hunting the *Behemot*, an amphibious animal, whose teeth arc in great esteem. The inhabitants go frequently upon the side of the Frozen Sea to hunt this monster; and because it requires great labor and assiduity, they carry their families usually along with them." Avril, accordingly, confirms the fact that the Russians hunted the walrus along the shores of the Arctic Sea, and that the animal's tusks were conveyed to Moscow and traded to the Persians and Turks.

The term "fish-tooth" covers still more ground than Russia, Turkey, and China; it advanced also to Persia and India along with the importation of the article. E. WIEDEMANN [1] has pointed out as a Persian name for "fish-tooth," *dandān māhi* دندان ماهی, occurring in a Turkish work on precious stones by al-Gaffārī, written in 1511—12 and partially translated from the Persian *Tansūq nāmeh ilkhāni* ("the Ilkhan Book of Precious Objects"). In this work, the substances occurring in nature are enumerated as fish-tooth from which combs and knife-hilts are turned out, ivory, ebony, *khutū*, etc. But even more than that, the Persians actually possessed fish-teeth [2] and sent them on to India, as demonstrated by H. BEVERIDGE in a highly interesting article [3] suggested by my previous study. In the second volume of his Memoirs, Miss BEVERIDGE tells us, the Emperor Jahangir describes how delighted he was when he received from Persia a dagger whose hilt was made of a fish-tooth. He was so much impressed by the

---

[1] *Zur Mineralogie im Islam*, p. 210.

[2] Termed also ماهی شیر *šir māhi* "lion-fish," translated "dent de morse" by A. BERGÉ (*Dict. persan-français*, col. 246).

[3] *The Emperor Jahangir's Treasures of Walrus and Narwhal Ivory* (*Indian Magazine*, February, 1914, pp. 37—39)

hilt that he despatched skilful men to search for other specimens
in Persia and Transoxania. Their instructions were to bring fish-
teeth from anywhere, and from any person, and at any cost. A
little later a fine specimen was picked up in the bazar of his own
capital of Agra, and was brought to him by his son, Shah Jahan.
Jahangir had the tooth made into dagger-hilts, and gave one of the
craftsmen an elephant as a reward, and bestowed on the other
increase of pay and a jewelled bracelet. Miss BEVERIDGE adds further,
"The idea that this ivory was an antidote to poison, and also reduced
swellings, added greatly to its value. From a statement in the history
of Akbar the Great, known as the *Akbarnāma*, it appears that
about 1569 a Rajah in Malabar, who probably was the Rajah of
Cochin, sent Akbar a knife which had the property of reducing or
removing swellings, and that Akbar told his secretary that it had
been successfully applied in more than two hundred cases. Probably
this knife was made, wholly or in part, of walrus or morse ivory,
which could easily have been brought to Cochin by sea."

The following interesting notes on the subject are contributed
by Sir GEORGE WATT:[1]

> "Ivory is in Indian as in European commerce spoken of as the
> 'elephant tooth' but a second substance is called the 'fish tooth'
> (*mahlīka-dant*). This is always of a dirty (oily) yellow color with
> the texture looking as if crystallized into patches. The significance
> of being called in every language and dialect of India 'fish tooth' at
> once suggests a common and, most probably, foreign origin for the
> material. Upon inquiry it was found that it was more highly valued
> for sword and dagger hafts and more extensively used for these pur-
> poses than is ivory. It is put through an elaborate and protracted
> process of curing before being worked up. The crude 'fish tooth' is
> wrapped up in a certain mixture (*masala*) and retained in that
> condition for various periods, the finer samples for as long as fifty years.
> The advantages are its greater strength, finer and smoother surface,

---

[1] *Indian Art at Delhi, 1903, being the official Catalogue of the Delhi Exhibition,*
p. 173 (Calcutta, 1903).

and greater resistance (less liability to slip in the hand) than is the case with ivory.

"So far as the writer has been able to discover, the 'fish tooth' of Indian trade is mainly, if not entirely, the so-called fossil ivory of Siberia — the ivory of the mammoth — a substance that has lain for countless ages in the frost-bound drifts of Liakoff and New Siberia. It is also possible that a fair amount of hippopotamus or 'sea-horse ivory' and even of the 'walrus ivory' finds its way to India by passing like the Siberian ivory by land routes to India. And from the antiquity of some of the swords, found in the armories of the princes of India with 'fish tooth' hafts, it would seem possible that there has existed for centuries a traffic in carrying this material to India."

The chain of evidence thus seems to me to be complete: fish-teeth in the sense of walrus ivory were known to the Russians, Bulgar, Turks, Arabs, Persians, Hindu, Mongols, and Chinese; and we may now confidently state that, in whatever European and Asiatic languages and documents the term may still come to the fore, it will invariably refer to walrus ivory. [1]

Reference has been made to the Mo-ho as possibly having been acquainted with walrus ivory during the T'ang period (pp. 324, 340). First of all, the passage quoted in the *Man-chou yüan liu k'ao* from the *T'ang hui yao* 唐會要, completed by Wang P'u 王溥 in 961, indeed occurs in this work. [2] The addition *kio* ("horn") to the term *ku-tu* is rather suggestive, and it now appears certain that this word was known in the age of the T'ang. Further evidence to this effect will be given below. There are now further reasons that strengthen this belief. The Mo-ho were in close proximity

---

[1] The Russian designation "fish-tooth" seems to have survived to at least the end of the eighteenth century. In S. KRASHENINNIKOV's *Beschreibung des Landes Kamtschatka* (p. 148, Lemgo, 1766), translated from the Russian, we read in the description of the walrus, "Their teeth are what is commonly called fish-bones." The term "horn" is still employed in Russian: VLADIMIR DAL' (*Complete Dictionary of the Live Great-Russian Language*, in Russian, Vol. III, col. 1696) says, "The horn of the narwhal grows out of its jaw-bone, and hence is a tooth."

[2] Ch. 96, p 10 b (ed. of *Kiang-su shu kü*, 1896). The text is as follows: 土多貂鼠皮尾骨咄角白莬白鷹等.

and intercourse with the Liu-kuei 流鬼, a people briefly described in the *T'ang shu*.[1] SCHLEGEL[2] has made a special study of this tribe, and we may agree with him in his main result, — that the geographical position of the country of the Liu-kuei is clearly enough defined to lead us to Kamchatka;[3] and that the culture of this people, as characterized by the Chinese, plainly reveals a type that is still found in the North-Pacific area. These cultural traits are, absence of agriculture, economy essentially based on the maintenance of numerous dogs, subterranean habitations, utilization of furs as winter costume, employment of fish-skins as clothing in the summer, and transportation on snowshoes. The Mo-ho entertained a lucrative commerce with the Liu-kuei by way of the sea, the voyage lasting fifteen days; and when the latter in 640 sent a mission to China, it travelled over the Mo-ho country. One of the three interpreters with whom they arrived at the Chinese Court appears to have been a Mo-ho, and the extract in the Annals is doubtless derived from a report made by the Mo-ho. The Mo-ho, accordingly, were in intimate touch with a people that had the walrus and its product within easy reach; and from the descriptions of Steller and Krasheninnikov, on which our knowledge of the ancient Kamchadal or Itelmen now almost extinct is based, we know surely enough that these tribes hunted the walrus and utilized its ivory for industrial work.[4]

---

[1] Ch. 220, p. 11 b. The text has been published by SCHLEGEL (*T'oung Pao*, Vol. IV, p. 338) from the *Pien i tien*; it agrees with *T'ang shu*, except that the latter reads 少海之北 in lieu of 北海.

[2] *T'oung Pao*, Vol. IV, 1893, pp. 335—343.

[3] SCHOTT (*Kirgisen, l. c.*, p. 448) was the first, though somewhat hesitatingly, to connect the Liu-kuei with Kamchatka.

[4] G. W. STELLER, *Beschreibung von dem Lande Kamtschatka*, p. 106; S. KRASHENINNIKOV, *Beschreibung des Landes Kamtschatka*, p. 147; V. MARGARITOV, *Kamchatka and Its Inhabitants*, p. 82 (in Russian, Chabarovsk, 1899). According to J. G. GMELIN (*Sibirische Reisen*, Vol. III, p. 164) walrus were numerous in the sea of Kamchatka; and on the shore

Aside from this evidence, actual proof of the occurrence of the word *ku-tu* in the Annals of the T'ang Dynasty may now be offered. In the *T'ang shu* (Ch. 39, p. 9 b) the tribute of Ying chou 營州[1] in Liao-tung is stated to have consisted of ginseng 人葠, musk,[2] leopard tails[3] and skins 豹尾皮, and *ku-tu* 骨骴. The reading of the latter character is certain, being explained in the *T'ang shu shi yin* (Ch. 4, p. 2 b) as 都骨; that is, *tu*. A definition of the term, unfortunately, is not added; and K'ang-hi's Dictionary, which quotes the same passage under 骴, tells no more. Nevertheless the form in which the term is written is very suggestive: the word *ku* means "bone," and the syllable *tu* is written with a character formed by means of the classifier "bone" (compare p. 319, note);

---

they found great numbers of discarded teeth that were much larger and heavier than the Greenland teeth, weighing ten, twenty, and thirty pounds each.

[1] It was formerly the district of Liao-si 遼西郡, established under the Ts'in, being 3300 *li* north-east of Lo-yang (*Hou Han shu*, Ch. 33, p. 6 b). It was captured by the Khitan in 696. In 699 it was connected with the administration of Yü yang 漁陽; in 717, with the administration of Liu ch'êng 柳城. In 742 the name was altered into the latter. See chiefly *T'ai p'ing huan yü ki*, Ch 71, pp 4 b *et seq.* This work enumerates as products of Ying chou only leopard tails, musk, pongee 絹, and domestic animals like cattle, horses, sheep, and swine, but does not mention *ku-tu*. Compare also CHAVANNES, *Journal asiatique*, 1898, mai-juin, p. 398.

[2] Old Pallas had already observed that the musk-deer is distributed over the entire Amur region as far as the shores of the Pacific; and L. v. SCHRENCK (*Reisen und Forschungen im Amur-Lande*, Vol. I, Säugetiere, p. 161) has more accurately defined the localities of its occurrence, inclusive of the island of Saghalin. All the Amur and Saghalin tribes are well acquainted with the animal, and, as has been shown by the writer, it is frequently represented in their decorative art. This point is especially mentioned here, as J. MARQUART (*Osteuropäische und ostasiatische Streifzüge*, p. 82), on the authority of Schlözer (eighteenth century), believes that the musk-deer does not inhabit the northern countries, and doubts the identification of Khitayān خنبان with Kitai or Khitan, for the reason that excellent musk is mentioned in their country. As far as this point is concerned, the identification is all right. The musk-deer is ubiquitous in Nepal, Tibet, Kukunōr, and the high mountains of Sze-ch'uan, Kan-su, and Shen-si.

[3] These were extensively utilized by the Chinese as pendants for their spears and as decorations of chariots.

so that this *ku-tu* apparently refers to a product of osseous nature. [1] It comes from the region of the Khitan, or, in general, from the domiciles of Tungusian tribes, [2] and this feature brings us in immediate contact with the *ku-tu-si* observed by Hung Hao among the Khitan (p. 318); so that the *ku-tu* of the T'ang Annals may be affiliated with the latter, and in all probability refers to walrus ivory.

The word *ku-tu* in the spelling 骨咄, further, appears in the T'ang Annals [3] as the designation of a wild animal living in the country of the Kirgiz (Kie-kia-se 黠戛斯). [4] The passage runs thus: "Among the animals of this country, there are wild horses, [5]

---

[1] Turkish *kümik* means "bone" and "ivory" (RADLOFF, *Worterbuch der Türk-Dialecte*, Vol. I, col. 1208).

[2] According to the *T'ai p'ing huan yü ki* (*l. c.*), the Shi-wei, Mo-ho, and other tribes were settled to the north-east of Liao-si at a distance varying from two to six thousand *li*.

[3] *T'ang shu*, Ch. 217 下, p. 8. Professor Hirth has been kind enough to call my attention to this passage.

[4] I do not think that this name is to be read Hia-kia-se, as maintained by KLAPROTH (*Mémoires relatifs à l'Asie*, Vol. I, p. 164). Since the name Kyrkyz appears in the Orkhon inscriptions (W. RADLOFF, *Die alttürkischen Inschriften*, p. 425), it seems to me that the Chinese is a regular transcription of this name; 黠 was formerly possessed of initial *k* and final *t* (Japanese *katsu*). Also the writing Kie-ku 結骨 (*ket* or *kyt kut*) is doubtless intended for Kyrkyz. The older name Kien-kun 堅昆 seems to go back to the same original. — [M. Laufer a mille fois raison de considérer Kie-kia-sseu comme la transcription directe du nom des Kirgiz, et non de "Hakas, ancien nom des Kirgiz", comme on le dit généralement. Les prétendus Hakas et les Dulgha, dont M. Pozdnéev maintient la tradition, sont deux *idola libri* dont nous sommes redevables au P. Hyacinthe Biĉurin. Le P. Hyacinthe avait en réalité forgé ces noms sur les transcriptions chinoises Kie-kia-sseu et T'ou-kiue; mais ces transcriptions représentent simplement Kirgiz et Türk; les Hakas et les Dulgha n'ont jamais existé. Kien-kouen paraît transcrit sur une forme *Qirqun, singulier répondant au pluriel Qirγud qu'on trouve dans les textes mongols; Qirqız (Kirgiz), qui se trouve déjà dans les inscriptions de l'Orkhon, le χερχίς de Ménandre, n'est lui-même vraisemblablement qu'une autre forme du même pluriel. — P. PELLIOT.]

[5] The numerous references to wild horses, wild asses, and wild camels, in the older Chinese records, are of great scientific significance; but the Chinese terms, unfortunately, are equivocal. The Chinese do not make that fundamental distinction established by our science between wild horses and feral horses; that is, horses descended from domesticated stocks and subsequently reverting to a wild state. This problem of reversion in animals

ku-tu, yellow sheep, *Ovis ammon*, [1] deer, and black-tails resembling the species of deer styled *chang* 麠 but their tails being larger and black." [2] The Glossary of the *T'ang shu*, unfortunately, gives no

---

and plants which have run wild was first clearly set forth by DARWIN (*Variation of Animals and Plants under Domestication*, Vol II, p. 6, Murray's ed. of 1905), and then further developed by Geoffroy Saint-Hilaire and E. HAHN (*Haustiere*, p 20). It is difficult, if not impossible, to decide in every instance with certainty whether the Chinese, who do not give us descriptions, refer to wild or to feral animals. In a few cases the matter is certain: thus, the "wild" horses of Kan-su are feral horses (HAHN, p 193). The general rule may be laid down, that, the nearer the locality of the Chinese report to the present habitats of the wild equidae (*Equus hemionus, E h. kiang*, and *E. onager*), the greater the probability that genuine wild horses are to be understood; the farther removed from that centre, the less likely is it to be the case. In a tribe of horsemen like the Kirgiz, whose great wealth of horses is emphasized, it is most unlikely that wild horses still occurred, and the term *ye ma* 野馬 in the above passage would rather seem to mean "feral horses." As to the "wild" horses sent as tribute to the imperial Court (for example, *T'ang shu*, Ch. 37, p. 7 b, where three instances are mentioned), it seems out of the question that wild horses could be meant, nor is there much sense in the assumption that feral horses would be presentable gifts. In this case, another consideration must be made. In the northern steppes of Tibet there are still numerous half-domesticated horses, now called by the Chinese *ts'ao ti ma* 草地馬 "horses of the steppe." These horses are kept and ridden by men, but are not yet accustomed to grain fodder or stall-feeding, and subsist exclusively on grass. While travelling in Tibet, I had several such horses in my caravan; and even when brought to Chinese territory and sheltered in stables, they refused to take any grain The process of domestication naturally was one of long-continued development, running through many stages; and among the Tibetan nomads we still find horses in the savage or uncivilized state, little cared for by man, and looking for their own means of subsistence. Such-like horses, I venture to presume, were the "wild" tribute horses, and perhaps also other "wild" horses of the Chinese. MARCO POLO (ed. of YULE and CORDIER, Vol. I, p. 260) mentions these horses as peculiar to the Mongols, saying, "Their horses also will subsist entirely on the grass of the plains, so that there is no need to carry store of barley or straw or oats."

[1] 猨貾 *nguan ti* (see *T'oung Pao*, 1914, p. 71).

[2] W. SCHOTT (*Über die ächten Kirgisen*, *A. B. A. W.*, 1865, p. 433) omits the *ku-tu* (so does E H PARKER, *A Thousand Years of the Tartars*, p 255) and enumerates as wild animals of the Kirgiz only wild horses, wild goats, and various species of birds-of-prey. KLAPROTH (*Tableaux historiques de l'Asie*, p. 171) has dodged the wild animals entirely; but VISDELOU (in d'Herbelot, *Bibliothèque orientale*, Vol IV, p 174) has given the series complete (he transcribes *khou-thou*) For the black tailed deer the *T'ai p'ing huan yü ki* of Yo Shi gives the native word *se-mu* 巴没, which SCHOTT (p 471) identifies with *syn* or *sin* ("stag") in the languages of the Koibal and Soyot; this is not very plausible. The Chinese characters point to a word *se-mut*, *se-mur*; and *samur* is a well-known Turkish

explanation of the word *ku-tu*, the spelling of which coincides with
that of the *Cho keng lu* (p. 322) and *Yün yen kuo yen lu* (p. 358).
For this reason it may appear as justifiable at first sight to link
the *ku-tu* of the Kirgiz with the *ku-tu* of the Khitan region. We
remember al-Bērūnī's words, that it is asserted in regard to *khutū*
(*chutww*) ختو that it is the frontal bone of a bull living in the
country of the Kirgiz (p. 316); and when we recall the commercial
relations of the Arabs with the Kirgiz,[1] the whole question seems
to assume a new turn. It is possible, as stated (p. 354), that al-
Bērūnī's bull furnishing ivory may be an allusion to the mammoth;
or rather it may have grown out of a tradition that the ivory was
derived from a "marine bull" (sea-cow, seal). The *ku-tu* of the
above Chinese text, however, cannot refer to a mammoth or a walrus,
for *ku-tu* is plainly spoken of as a live animal (and the mammoth

---

word for the sable (RADLOFF, *Wörterbuch der Turk-Dialecte*, Vol. IV, col. 511), used
likewise by the Arabs, *semmūr* (for instance, in Dimeški's Cosmography, also with reference
to marten-like and weasel-like animals; see also G. JACOB, *Handelsartikel*, p. 31). —
[J'ai vérifié le texte dans le *T'ai p'ing houan yu ki*, ch. 199, fol 15 v.—16 r.: de l'éd.
de Nankin, 1882 (sur laquelle cf. *B.E.F.E.-O*, II, 338—339); il est bien conforme aux
indications de Schott. Il est clair que phonétiquement la restitution de M. Laufer serait
très satisfaisante, au lieu que celle de Schott est inadmissible. Je me demande seulement
si on peut songer à la zibeline quand le texte parle d'un daim. Il ne faut pas oublier
qu'un élément d'incertitude provient de la confusion constante dans les textes de 己 *ki*,
巳 *yi* et 已 *sseu*; l'analogie de la transcription *bāš-ai*, "premier mois", par 茂師
袞 *mao-che-ngouai* dans le même texte et l'emploi de 沒 pour rendre *bol* (*bolmiš*) sous
les T'ang permettent en outre de songer à une forme en *b* au moins autant qu'à une forme
en *m*. — P. PELLIOT.] M. Pelliot is right in his contention. In agreement with the indication in the above Chinese text, the Kirgiz employ the term "Black-Tailed" with reference to *Antilope subgutturosa*; PALLAS (*Zoographia rosso-asiatica*, p. 252) records this term
in the form *kara-keuruk* and translates this "nigri caudata;" POTANIN (*Sketches of North-
western Mongolia*, in Russian, Vol IV, p. 156) writes it *karaguiruk* (there are several
species of antelopes with black tail). POTANIN (p. 157), further, gives an animal name *bur*
as peculiar to the dialect of the Uryankhayans on Lake Terkul and referring to *Cervus alces*.
This *bur* may be sought in Chinese 沒, but I do not believe that the animal intended in
our text is the elk.

[1] KLAPROTH, *l. c*, p 172; SCHOTT, p. 451.

certainly was as extinct in the T'ang period as it is now); and, further, Chinese traditions regarding mammoth and *ku-tu* or *ku-tu-si* are not interrelated, but entirely distinct and individual matters (p. 329).

Li Shi-chên, in his discourse on the seal (*wu-nu shou* 膃 肭 獸), has this interesting passage: "According to a statement in the T'ang Annals, the animal *ku-nu* has its habitat in Ying chou in Liao-si and in the country of the Kirgiz" (按唐書云骨貀獸出遼西營州及結骨國). HIRTH [1] has accepted this passage at its face value; but it is evident that in this form it is not contained in the *T'ang shu*, which, as has been shown, with reference to Ying chou as well as to the Kirgiz, speaks of *ku-tu*, not of *ku-nu*: and it is on these two texts that Li Shi-chên's opinion is apparently based. Li Shi-chên, accordingly, makes two points: he combines the *ku-tu* 骨咄 of Ying chou with the *ku-tu* 骨咄 of the Kirgiz, and identifies both with the animal *ku-nu* 骨貀. [2] This view seems rather sensible, as the first elements of the two forms are identical, and the elements *tu* and *nu* are phonetically interrelated. This matter is not pursued here any further, as it has no relation to the subject under review, but bears on numerous other problems of great complexity. These will be taken up in a special monograph in which the Siberian fauna known to the Chinese will be discussed in detail. [3] Suffice it for the present to

---

[1] HIRTH and ROCKHILL, *Chau Ju-kua*, p 234, line 21.

[2] This reading may have existed in some editions of the *T'ang shu*; for K'ang-hi gives it in this manner, quoting it under the word 貀. PALLADIUS (*Dictionary*, Vol. I, p. 436) writes 骨豽, and states, "name of an animal from the dominion of the Kien-kun (Kirgiz)." His entry is probably based on K'ang-hi. Both *ku-tu* and *ku-nu* seem to be correct in this passage, and merely appear as phonetic variants or transcriptions of the same foreign word.

[3] In co-operation with M. Pelliot. Our manuscripts were ready in 1914 and would have been published long ago, if the world conflict had not interfered.

remark that the *ku-tu* or *ku-nu* ascribed to the country of the Kirgiz in all probability denotes the beaver. [1]

As to the Khitan word *t'u-hu*, M. PELLIOT (p. 367) has pointed out a text in the *Liao shi*, and another in the *Kin shi*. The fundamental passage, however, is *Kin shi*, Ch. 43, p. 7 (其束帶曰吐鶻 "the girdles worn by them [that is, the Kin] are styled *t'u-hu*"), where these girdles, with their accessory ornaments, are minutely described. Jade ranked as the supreme material for them; while gold, rhinoceros-horn, ivory, bone, and horn followed suit. The substances employed for *t'u-hu* are noteworthy. If, accordingly, the Emperor T'ien Tsu had a *t'u-hu* made of *ku-tu-si* (p. 359), this was an exceptional case, which simultaneously bears out the fact that *ku-tu-si* cannot have been rhinoceros-horn or elephant-ivory, which were the common materials for *t'u-hu*. There is another piece of evidence to the effect that *ku-tu-si* is neither elephant-ivory nor rhinoceros-horn. M. PELLIOT (p. 366) has happily discovered the term *ku-tu* in the Annals of the Kin Dynasty, from which it appears that the Niüchi perpetuated the word inherited by their predecessors, the Khitan. The Niüchi language, however, possessed particular terms for both elephant-ivory and rhinoceros-horn,— *sufa weihe* and *si uihe* respectively; [2] and these terms, most certainly, are not connected with *ku-tu-si* or *tu-na-si*. The same condition of affairs is reflected in the Annals of the Liao and Kin Dynasties, where elephant-ivory and rhinoceros-horn are frequently mentioned, and are surely distinct from *ku-tu-si*. The former play a prominent

---

[1] According to the *T'ang hui yao* (Ch. 98, p. 16 b) the animal *ku-tu* 骨吐, together with panthers and rodents, occurred also in the country of the Yü-che 俞折, while the *T'ang shu* (Ch. 217 下, p. 7 b) locates there only an abundance of sables and rodents. The Yü-che territory was situated fifteen days' journey eastward from the country Kü 鞠 (identical with the Wu-huan 烏丸), the latter six days' journey north-east of the Pa-ye-ku 拔野古 (the Bayirku of the Orkhon inscriptions).

[2] See W. GRUBE, *Sprache und Schrift der Jučen*, pp. 31, 98.

part in official and ceremonial costume, and were perfectly known to both Liao and Kin; while the latter, being a rare article of import, does not.

In regard to the term *pi-si* 碧㷊 employed by the *Ko ku yao lun* for the definition of *ku-tu-si* (p. 325), M. PELLIOT (p. 365) is quite right in maintaining that it cannot be credited in this passage with the meaning "rhinoceros-horn." The *Pien tse lei pien* (under *pi-si*) cites the text in question as the only instance of the occurrence of the term. T. WADA,[1] the eminent Japanese mineralogist, who is well acquainted with the nomenclature of Chinese mineralogy, observes, "Transparent jewels are much utilized at present in China and more highly esteemed than jade. In distinction from *yŭ* 玉, the Chinese designate those *pi-si* 壁璽 (that is, precious stone)."

The term *kuo hia ma* 果下馬,[2] listed as a Khitan word (p. 359), after all, may be purely Chinese. As stated previously, it is not traceable in Korean. Furthermore, it is not applied exclusively to the dwarf ponies of Korea, but also to those of a South-Chinese breed. Fan Ch'eng-ta 范成大, in his work *Kuei hai yŭ heng chi* 桂海虞衡志, the preface of which is dated 1175, in the chapter dealing with the animals of southern China, makes reference to *kuo hia ma* as being bred in Lung-shui 瀧水 in the prefecture (now *chou*) of Tê-k'ing 德慶 in Kuang-tung Province, where the highest ones are produced; the fine ones, which do not exceed three feet, have two backbones, and are therefore styled also "double-ridge horses" (*shuang chi ma* 雙脊馬), which are robust and fond of walking. In the Ming period these *kuo hia* horses appear among the taxes sent by the prefecture of Chao-k'ing 肇慶, Lung-shui being given as the place of their provenience.[3]

---

[1] *Beiträge zur Mineralogie von Japan*, No. 1, p. 19 (Tōkyō, 1905).

[2] First explained by KLAPROTH, *Aperçu général des trois royaumes*, p 162.

[3] *T'u shu tsi ch'êng*, XXVII, 191, *hui k'ao* 9, p. 6.

There is a term of the Chinese language, *k'uei* or *k'ui* 夔,[1] for which the translation "walrus" has been proposed.[2] Besides this meaning, GILES gives the definition "a one-legged creature," and explains the term *k'uei lung* as "one of the varieties of the dragon." Giles's quotation, "the walrus said to the centipede, 'I hop about on one leg,'" is taken from the philosopher Chuang-tse, and occurs on p. 211 of Giles's translation of this work. At the outset it is difficult to see how Chuang-tse, who lived in the fourth and third centuries B.C., and his contemporaries could have had any knowledge of an arctic animal like the walrus, how the walrus came to be credited with a single leg, and how walrus and centipede could occur in the same geographical area. The rendering "walrus" is conjectural and does not result from the definitions of the word *k'uei* found in early Chinese sources. Thus PALLADIUS[3] interprets it in the sense of a "spirit resembling a dragon, with a single foot." COUVREUR states that it is a demon in the shape of a dragon with a single paw, that occurs in the mountains, and cites from the work *Lu yü* that *k'uei* is a strange apparition in the midst of trees or rocks. L. WIEGER[4] defines *k'uei* as a fabulous animal. The *P'i ya* of the eleventh century says that "*k'uei* is a beast in the eastern sea, having the appearance of an ox, with blue body, without horn, and with a single foot; when entering or leaving the water, there is storm and rain, and its voice is like thunder." An allusion to a marine mammal looms up in this definition, but I hardly believe that it can be referred to the walrus. It is certainly possible that vague descriptions of this creature might have reached the Chinese

---

[1] Written in various other forms that are recorded in COUVREUR's *Dictionnaire chinois-français*, p. 439.

[2] GILES, No 6507.

[3] *Chinese-Russian Dictionary*, Vol. I, p. 287.

[4] *Pères du système taoiste*, p. 343.

through the medium of the Su-shên, Mo-ho, or other northern tribes. [1]

The snake-horn of the Khitan tradition (p. 318) [2] was revived in a curious manner during the eighteenth century. The *Pén ts'ao kang mu shi i* (Ch. 2, p. 4) contains a lengthy dissertation on a stone called *hi tu shi* 吸毒石 ("poison-attracting stone") not yet mentioned in the *Pén ts'ao kang mu*. From the various quotations given, it becomes clear that this article was introduced into China by the Spaniards 小西洋, and that it is identical with the snake-stone well known in the west, which was believed to originate in the head of a snake, and, when placed on a wound caused by a snake-bite, to draw the poisonous matter out of the body. A certain Mr. Hiao-lan 曉嵐先生, in his work *Luan yang siao hia lu* 灤陽消夏錄, [3] tells a very fantastic story anent a huge serpent once seen in Urumtsi with a single horn over a foot in length on its head. A flock of pheasants passed above it, and, attracted by the horn, with fluttering wings fell to the ground like arrows into a jar in the game of pitch-pot (如矢投壺). The horn of this snake is poisonous, and can neutralize poison, and Hiao-lan

---

[1] In the journal *Kuo sui hio pao* 國粹學報 (Vol. IV, No 2) there is a series of illustrations of animals after European models, identified with their Chinese names. The animal *k'uei* is here illustrated by the figure of a walrus of European origin This modern attempt, of course, proves nothing.

[2] A point which the Arabists did not touch upon in the discussion of *khutū* is that this tradition appears to have spread to Persia; for JOHNSON and RICHARDSON (*Persian-English Dictionary*, ed STEINGASS, p 448) assign to خُتُو *khutū* the significance "the tooth or bones of a viper" (besides, "the horn of a Chinese bovine animal, the horn of a rhinoceros"). I trust that some one will be able to point out the Arabic or Persian source on which this explanation is founded.

[3] [Hiao-lan est le *tseu* de 紀昀 Ki Yun, un lettré de la deuxième moitié du XVIIIe siècle, et l'un des principaux rédacteurs des notices critiques du *Catalogue impérial* (cf. Giles, *Biogr. Dict*, no. 301) Ki Yun avait été exilé à Urumtsi et a même laissé une suite de poésies sur cette ville. Son *Louan yang siao hia lou* est l'une des œuvres qui composent son 閱微草堂筆記 *Yue wei ts'ao t'ang pi ki* J'ai lu jadis cet ouvrage, dont il y a peu à tirer, tant l'auteur s'y montre accessible aux contes les moins vraisemblables. — P. PELLIOT.]

rashly identifies it with the newly introduced poison-attracting stone. The editor of the work, Chao Hio-min, justly opposes this view as incorrect by referring to the article *ku-tu-si* in the *Pén ts'ao*, where the passages quoted agree in stating that the snake-horn cures poison, but they do not state that it attracts poison. [1]

---

[1] The notion of stones encountered in the heads of serpents and curing snake-bites doubtless originated in India. Three kinds of stones — arising in the heads of man, the serpent, and the frog, respectively — are distinguished in the *Agastimata* (posterior to the sixth century); and according to Varāhamihira (A.D. 505—587), a very brilliant blue stone is formed in the head of the serpent (L. Finot, *Lapidaires indiens*, p. xx). Yule and Burnell (*Hobson-Jobson*, pp 847—849) have devoted an elaborate note to the subject, without pointing out, however, any Indian, Arabic, or Persian references The earliest testimony mentioned by them is that of the European travellers to India at the end of the seventeenth century. The Arabic views have crystallized in Qazwīnī (J. Ruska, *Steinbuch aus der Kosmographie des al-Kazwīnī*, p. 15), who describes the snake-stone thus: "This is a stone called in Persian *muhreh-i-mār* مهره ٴ مار [Vullers, *Lexicon persico-latinum*: "lapis qui in occipite serpentum reperitur"], being of the size of a small nut, and being found in the heads of many snakes. It has the special effect that, when thrown into curdled milk or hot water in which the bitten organ is placed, it will stick to that spot and suck up the poison." Entirely distinct from the Indian notion is the ophites of Pliny (*Nat. hist.*, XXXVI, 11, § 56), a kind of marble occurring in two varieties, worn as an amulet, and regarded as a cure for headache and for wounds inflicted by serpents (dicuntur ambo capitis dolores sedare adalligati et serpentium ictus). This belief is solely inspired by the name of the stone (from ὄφις, "serpent"), caused by its being marked with streaks resembling serpents in appearance (serpentium maculis simile, unde et nomen accepit; compare our term "serpentine") Pliny's text was adopted by Dioscorides (F. de Mély, *Lapidaires grecs*, p. 25) and Ibn al-Baitār (L. Leclerc, *Traité des simples*, Vol. I, p. 412), whose "snake-stone," accordingly, is different from that of Qazwīnī. The classical and Indian notions are amalgamated in the Armenian *lapidarium* (K. P. Patkanov, *Precious Stones according to the Notions of the Armenians in the Seventeenth Century*, p. 41, in Russian, St. Petersburg, 1873), which first describes the Plinian "serpentine" and its application to snake-bites, and then joins to it the Indian snake-stone; curiously enough, the latter is characterized as belonging to the species mother-of-pearl, entirely white, round, convex on one side, and smooth on the other, which is bordered by a fine, black edge resembling a coiled snake; the stone is placed on the wound, is rubbed in with honey, and allowed to remain there for eight days. Again, the so-called snake-bezoar is a substance distinct from snake-stone. Bezoar was sometimes designated "snake-stone," and believed to be found in the head of a snake; Pseudo-Aristotle (J. Ruska, *Steinbuch des Aristoteles*, pp. 147—149, and L. Leclerc, *Traité des simples*, Vol. I, p 196) gives the best account of it. According to him, bezoar is powdered and administered internally to him who is poisoned; it drives the poison, by means of perspiration, out of the veins of his body. Certainly the theory

Finally I may be allowed to offer some additional comment on

---

based on this process is different from that upheld in India regarding snake-stone. But Pseudo-Aristotle, in his further discussion of the subject, also reverts to the Indian practice by saying that bezoar, if pulverized and strewn on the sting, attracts the poison and heals the wound; and this is thus far the earliest account traceable as to the conception of bezoar attracting and absorbing poison (吸 毒) that agitates a human body (repeated by Qazwinī [RUSKA, *Steinbuch aus der Kosmographie,* p. 29] and al-Akfānī [WIEDEMANN, *Zur Mineralogie im Islam,* p. 228]). I am not inclined to believe, however, that this conception arose in the west. There is nothing to this effect in the classical authors, particularly in Theophrastus, Pliny, and Dioscorides. I presume that this notion was developed in India, and has migrated westward. Indian mineralogical ideas occur as early as the Alexandrian epoch in the Physiologus, which speaks of "the Indian stone" (F. LAUCHERT, *Geschichte des Physiologus,* p. 37). This stone is described as having the specific quality of sucking up the diseased matter of a dropsical person to whose body it is bound; when the stone is exposed to the sun for three hours, it emits the water and is cleansed. This is exactly the same idea as that expressed by Pseudo-Aristotle in regard to bezoar being an antidote of snake bites. Lauchert emphasizes the fact that exactly corresponding earlier testimony (*scil.* from classical literature) is not known. Since the stone itself is designated "Indian," it is more than probable that also the tradition accompanying it was derived from India. Whether the transfer of the notion concerning the dropsy-stone to bezoar was effected by Pseudo-Aristotle himself or by an earlier source utilized by him, I do not know. Certain it is that the influence of the Physiologus, whether direct or indirect, is apparent in Pseudo-Aristotle. Thus the legend of the parturition-stone (RUSKA, p. 165), localized in India, is found in all versions of the Physiologus, and was already booked by HOMMEL (*Aethiopische Übersetzung des Physiologus,* p. xv) as one of the instances of Indian influence on the Greek work. The name *huan,* mentioned for this stone in the Syriac and Arabic translations of the Physiologus (E. PETERS, *Der griechische Physiologus,* p. 99), should be traceable, after all, to an Indian language. The snake bezoar is known also in Chinese pharmacology under the name *shĕ huang* 蛇 黃 (a counterpart of *niu huang* 牛 黃), "snake yellow." The statement of the *Pén ts'ao* (F. DE MÉLY, *Lapidaires chinois,* p. 133) that it is formed in the belly of snakes, clearly sets it off from the Indian snake-stone which is found in the head of snakes (regarding the various names of bezoar in China, compare the interesting notes of M. PELLIOT, *T'oung Pao,* 1912, pp. 437—438, note). — The word "snake-stone" (*pedra de cobra*) was introduced into Europe by the Portuguese, as we are informed by E. KAEMPFER *Amoenitates exoticae,* p. 395): ita dictus lapis, vocabulo a Lusitanis imposito, adversus viperarum morsus praestat auxilium, externe applicatus. As the Portuguese imported the article into Europe, it is very likely also that they and the Spaniards brought it to China, as stated in the above Chinese text. A. KIRCHER (*La Chine illustrée,* p. 108) has a lengthy discussion of the snake-stone with an illustration of it and the snake supposed to yield it and called by the Portuguese *cobra de capelos;* he says also that it is partially artificial. All competent informants are agreed that the snake-stone was not a stone, but an artifact, — an opinion shared by Yule, and confirmed by the descriptions of Thevenot, Tavernier,

25

the Arabic accounts.[1]   In his study "Zur Mineralogie im Islam,"[2] which contains a translation of the mineralogical treatise of Ibn al-Akfānī, E. WIEDEMANN has returned to the question of *khutū* (*chutww* خَتُو), and is inclined to regard it as rhinoceros-horn, because al-Akfānī, besides the word *chutww*, avails himself of the word *chartūt*, and because his informants in Egypt tell him that *chartīt* or *chirtīt* is still the name for the African rhinoceros.[3] This may very well be the case, but it cannot be construed to mean that *khutū* as a product is identical with *khartūt*. A critical and historical attitude toward the subject is indispensable. Al-Akfānī is a late author, who died in 1347—48, and who depended entirely, in his statement regarding *khutū*, on his predecessor al-Bērūnī (973—1048). For this reason al-Bērūnī remains the oldest and the

---

Kaempfer, Tennent, and others. — [C'est le P. Verbiest (1623—1688) qui paraît avoir popularisé en Chine la "pierre qui attire le poison"; il a en effet écrit un 吸毒石 原由用法 *Hi tou che yuan yeou yong fa*, "Origine et emploi de la pierre qui attire le poison", dont deux exemplaires se trouvent à la Bibliothèque Nationale (Courant, *Catalogue des livres chinois*, nos. 5321, 5322). Je tiens d'autant plus à attirer l'attention sur ce curieux opuscule qu'il n'en est fait mention ni dans l'*Imprimerie sino-européenne en Chine* de M. Cordier, ni dans les travaux consacrés récemment aux œuvres chinoises de Verbiest, l'un dû au P. Louis Van Hée, *Ferdinand Verbiest, écrivain chinois*, Société d'émulation de Bruges, Mélanges, VII, Bruges, 1913, in-8°, et l'autre au P. H. Bosmans, *Les écrits chinois de Verbiest*, dans la *Revue des questions scientifiques* de juillet 1913. Quant au bezoar, il est absolument certain que c'est là la "pierre *jada*" des Mongols, et il est hors de question de considérer le *jada* comme du jade, ainsi que le fait M. Blochet, dans *J.R.A.S.*, 1914, p. 168. — P. PELLIOT.]

[1] The results of my previous study have meanwhile been acknowledged by G. FERRAND (*Textes relatifs à l'Extrême-Orient*, Vol. II, p. 679), J. RUSKA (*Der Islam*, Vol. V, 1914, p. 239), and Miss A. BEVERIDGE (in the article previously cited). E. WIEDEMANN has been good enough to write me that he sides with my opinion, but that *chutww* and *charṭūṭ* are also frequently confounded. E. LITTMANN, the well-known Arabist, in a letter kindly addressed to me, says that he has accepted my result, and remarks in regard to the transcription that it is preferable to write simply *chutū* or at best *chutūw*.

[2] Published in *SB. P. M. S. Erlg.*, Vol XLIV, 1912, pp. 205—256. The manuscript of my former article on the subject was sent to press in May, 1913; a copy of Wiedemann's work, I received, thanks to the courtesy of the author, on June 26, 1913.

[3] It is worthy of note that Wiedemann's inquiries have failed to trace the word *khutū* in modern Arabic. It seems to have shared the fate of Chinese *ku-tu-si*.

pre-eminent Arabic authority on the question. Al-Akfānī simply copies him; and the additions which he makes are merely fanciful, and show that *khutū* as an object of reality was foreign to him. It is necessary to discriminate between actual conditions or realities, and purely literary or bookish reconstructions. Al-Bērūnī does not offer the term *khartūt*, which is plainly an addition peculiar to al-Akfānī, but only the terms *khutū* and "fish-teeth," — the latter, as now conclusively shown, strictly referring to walrus ivory. Nor does he let drop in this connection a word about the rhinoceros or its horn; and this silence is conclusive, as al-Bērūnī, in his account of India, shows himself closely familiar with the rhinoceros of India as well as with that of Africa. [1] Indeed, he is the Mohammedan author who has furnished the best and clearest description of the animal, founded on keen observation. He noticed correctly that it possesses three hoofs on each foot, and that the horn is placed on the top of its nose. [2] In this description, however, he does not allude to *khutū*, nor in his notice of the latter to rhinoceros. [3] It is impossible to assume that a keen observer of his type who correctly described what he saw should have mistaken rhinoceros-horn for any kind of ivory, and it can be stated most positively that such a confusion is out of the question for any one who has ever seen and examined the two. [4] Only a scribbler or copyist who

---

[1] E C. SACHAU, *Alberuni's India*, Vol. I, pp. 203—204.

[2] See the writer's *Chinese Clay Figures*, p. 95, note 6.

[3] Regarding the Arabic names of the rhinoceros, see M. REINAUD (*Relation des voyages faits par les Arabes et les Persans*, Vol. II, p. 66), F. HOMMEL (*Namen der Säugetiere bei den südsemitischen Völkern*, pp. 332, 382, 395), and G. FERRAND (*Textes relatifs à l'Extrême-Orient*, p. 675). No Arabic author has ever used the word *khutū* with reference to the rhinoceros.

[4] Ivory and rhinoceros-horn are substances of radically distinct biological origin and structure, which do not have a single trait in common. Rhinoceros-horn is an epidermal formation composed of a solid mass of agglutinated hairs or bristles, and has no firm attachment to the bones of the skull, which are merely roughened so as to fit into the concave base of the horn. Ivory is a tooth-substance which in transverse sections displays lines of different colors running in circular arcs.

lacks in critical faculties and is not in touch with life, but with books only, is capable of such confusion; and such a figure is al-Akfānī. This is evidenced by his interpolation that *khutū*, according to some, is derived from the forehead of a large bird which falls on some of these islands (p. 316), — a product naturally different from *khutū* and rhinoceros-horn. It must be conceded that the man who confounds a bird's beak with mammal-tusks is fairly ripe for confusion of the latter with rhinoceros-horn. But such extravagances of an erratic mind cannot really mean that *khutū* designates rhinoceros-horn. The passage of al-Akfānī, in my opinion, is a literary concoction of no value for the whole question.

As regards the horn from the forehead of a large bird, which al-Akfānī has interpolated in the text of al-Bērūnī, Wiedemann has offered no explanation for it; and our Arabists, as far as I know, have not yet discussed this matter, which, however, is well known to students of China. GROENEVELDT [1] appears to have been the first to call attention to it in studying a text of the *Ying yai shêng lan* of 1416 relative to Palembang on Sumatra, where a bird called "crane-crest" *hao ting* 鶴頂 is described as being larger than a duck, with black feathers and a long neck, the bones of its cranium being over an inch thick, outside red and inside yellow, used by the natives for the handles and scabbards of their swords, and for other different purposes. Groeneveldt, in general, has given a correct

---

[1] *Notes on the Malay Archipelago*, p. 198 (in *Miscellaneous Papers relating to Indo-China*, Vol. I, London, 1887). Some inadvertencies in his translation have been rectified from the original. — "Crane-crests" are mentioned as tribute from Borneo and Malacca (*Ming shi*, Ch 325, pp 1, 4 b); also from Bengal (*Si yang ch'ao kung tien lu*, Ch. B, p. 11, ed. of *Pie hia chai ts'ung shu*). The *Tung si yang k'ao* 東西洋考 of 1618 (Ch. 2, p. 19) says after the *Hua i k'ao* 華夷考 that the heads of the birds were sold by the native hunters directly to the Chinese trading-junks for exportation into Fu-kien and Kuang-tung, and that their value equalled that of gold and jade.

identification of the bird by observing that it is not a crane,[1] but
the buceros, characterized by a large beak, with an excrescence on
the top of it, which is usually hollow, but in some species solid;
and that even now it is much used in Canton, where brooches and
other ornaments[2] are cut out of it. The current name of this
bird is now *siang* 象鳥. The *Bucerotidæ* form a large family of
tropical birds, distributed over India, Ceylon, south-eastern Asia,
and the Archipelago, and characterized by the extraordinary development of a horn-like excrescence or protuberance of the upper mandible.
The species eagerly sought by the carver is chiefly *Rhinoplax vigil*,
the solid-billed or helmeted hornbill, inhabiting the Malay Peninsula,
Sumatra, and Borneo, being a shy bird of the highest forest-trees.
It has a nearly straight, sharp-pointed bill, the casque being high
and in its anterior part a dense and solid mass. The front portions
of the bill and the casque are yellow, while the remainder of the
latter and the basal portion of the bill are crimson.[3] The Chinese

---

[1] The Chinese Maritime Customs were led into error by distilling from the term
hao ting "a sort of yellow substance much resembling amber," and believing in its being
"the upper part of the beak of a crane" (see the writer's *Jottings on Amber in Asia*,
p. 243). No wonder that among the private collectors of America the most fantastic
notions are current concerning the character of this product which is usually confounded
with amber. In a private collection of Chicago, I once came across a snuff-bottle carved
from this material, and labelled by a dealer in China "egret's head" (!). Such errors are
interesting to note, because they bring us nearer to the psychology of the Arabs.

[2] He adds, "for the European market." It may be doubted whether these carvings
ever had a large demand in Europe, or were especially made to fill foreign orders. The
good specimens, of the K'ien-lung period, are snuff-bottles and girdle-buckles of thoroughly
Chinese style, and of such exquisite technical execution that they cannot be suspected of
any foreign odor.

[3] Compare F. H. KNOWLTON, *Birds of the World*, p. 507 (New York, 1909); *Encyclopædie van Nederlandsch-Indië*, Vol. III, p. 15; and chiefly the fine monograph, illustrated
by colored plates, of D. G. ELLIOT (*A Monograph of the Bucerotidæ or Hornbills*, 1882).
Collectors of Chinese specimens, suspecting them of being hornbill carvings, may compare
the colors of these with the birds on those plates, the colors of which are very exactly
reproduced. A section of the bird's cranium is figured in H. O. FORBES (*A Naturalist's
Wanderings in the Eastern Archipelago*, p. 155, London, 1885).

carvers display great skill in utilizing to best advantage these two very beautiful colors; in a snuff-bottle, for instance, the yellow portions forming the two large surfaces, and the crimson parts the two narrow sides. The natives of Borneo sometimes carve the hard substance of the beak of the helmeted hornbill into an ear-ornament having the form of the canine tooth of the tiger-cat, a pair of these being worn by elderly men, or men who have captured heads. [1]

Following is the information in regard to the bird given in the *T'u shu tsi ch'eng* (XIX, 42). The earliest text quoted there is the *Nan Yüe chi* 南越志, [2] where the bird is designated "bird of the King of Yüe" (*Yüe wang niao* 越王鳥). [3] "The bird of the King of Yüe is shaped like a kite (*yüan* 鳶), and has on its upper mandible an excrescence (吻末) which has a capacity of over two pints (升). The southerners make wine-vessels out of it,

---

[1] C. HOSE and W. MCDOUGALL, *Pagan Tribes of Borneo*, Vol. II, p. 60.

[2] A description of southern China by Shên Huai-yüan 沈懷遠 of the fifth century (BRETSCHNEIDER, *Bot. Sin.*, pt. 1, p. 177).

[3] This term is still used in the *Ming shi*, where the bird is listed among the products and taxes of the prefecture of Chao-k'ing 肇慶府 in Kuang-tung Province, and where it is identified with the *mung-tiao* 蒙雕. This indicates that the bird extended (and probably still extends) into southern China. *Yüe niao* 越鳥 "bird of Yüe" is one of the epithets of the peacock (*T'u shu tsi ch'êng*, XIX, 41, p. 1). According to SCHLEGEL (*T'oung Pao*, Vol. X, 1899, p. 461) "birds of the King of Yüe" are attributed to the country Tan-tan 丹丹 in the *T'ung tien* 通典 of Tu Yu 杜佑 (regarding this country compare PELLIOT, *Bull. de l'Ecole française*, Vol IV, 1904, p. 284). The *Polyglot Dictionary* of K'ien-lung (Ch. 30, p. 36) knows a swallow of Yüe 越燕. The shell of the coconut (*ye-tse* 椰子) is known under the term "head of the King of Yüe" (*Yüe wang t'ou* 越王頭), because the latter, in a feud with the King of Lin-yi 林邑, was assassinated at the instigation of his adversary; and his head, hung from a tree, became metamorphosed into a coconut with two eyes on the shell (*Nan fang ts'ao mu chuang* 南方草木狀, Ch. B, p. 2, ed. of *Han Wei ts'ung shu*; and BRETSCHNEIDER, *Chinese Recorder*, Vol. III, 1871, p. 244). In a study on *Le coco du roi de Yueh et l'arbre aux enfants* (*Transactions of the Ninth Congress of Orientalists*, London, 1893, Vol. II, pp 897—905), TERRIEN DE LACOUPERIE has endeavored to correlate the story regarding the head of the King of Yüe with the well-known tale of the tree of the Wāqwāq.

which are highly prized, like conch-shells. The bird does not tread upon the ground, nor drink out of rivers and lakes, nor feed on herbage, nor swallow vermin or fish, but subsists exclusively on the leaves of trees. [1] Its guano resembles the incense *hiūn-lu* 薰陸香, [2] and the southerners get hold of it to prepare incense (or perfume) from it. It is also a curative for various kinds of ulcers."

The *Yu yang tsa tsu* [3] terms the bird *mung tiao* 蒙鵰, [4] and states that "its bill is large and the excrescence a foot in length, red and yellow in color, and of a capacity of two pints, and that the southerners make it into wine-cups."

---

[1] ELLIOT (*l. c.*, p. 3) states, "As a rule the food of *Buceros bicornis* is strictly fruits — certainly so, says Hodgson, at certain seasons, as in the months of January and February, when he found the stomachs contained nothing but the fruit of the Pipal tree. Tickell states that it eats lizards readily, not only from the hand, but will search for them and seize them. With this exception, authors generally agree in regarding fruit as the sole food of this bird." O. BECCARI (*Wanderings in the Great Forests of Borneo*, p. 117) observes that the hornbill subsists mostly on the fruits of various species of *ficus*; these the bird easily plucks with its bill, but it is then obliged to throw each fruit high up in the air, and catch it with open mandibles and a clever jerk. Likewise P. J. VETH (*Java*, Vol. III, p. 282): "Zij leven voornamelijk van vruchten, inzonderheid van die der vijge-boomen." This diet may account for the scent of the bird's dung.

[2] See the interesting discussion of P. PELLIOT in *T'oung Pao*, 1912, pp. 475—479.

[3] Ch. 16, p. 8 (edition of *Pai hai*). According to P. PELLIOT (*T'oung Pao*, 1912, p. 375), this work was written about A.D. 860. It is an excellent source for many questions of Chinese zoölogy and animal-lore.

[4] Chinese *mung* and *mung-tung* (of the *Kiao chou ki*) are apparently the reproductions of a Malayan name. In the language of the Dayak of Borneo, the bird is called *bungai* and *tingang* (A. HARDELAND, *Dajaksch-deutsches Wörterbuch*, pp. 78, 604). Neither the Malayan word for the hornbill, *enggang* (F. A. SWETTENHAM, *Vocabulary of the English and Malay Languages*, p. 28), nor the Javanese name *rangkok* (VETH, *Java*, Vol. III, p. 282), furnish the foundations of the Chinese transcription. As the phonetic element 冡 was anciently **buṅ* (Japanese *bō*), the word 鵬 may very well have reproduced the syllable *buṅ* of some Malayan term of the type of Dayak *bungai* (*buṅai*). There are presumably other Malayan links of the word not known to me. Beaks of the hornbill were sent as tribute from Borneo to China in 1370 (GROENEVELDT, *l. c.*, p. 231; *Ming shi*, Ch. 325, p. 1). — The phrase *mung tiao* would mean "hornbill carvings;" but 雕 is presumably to be corrected to 鵰 "eagle," as shown by K'ang-hi's 雘鵰, thus written likewise in the *Yu yang tsa tsu*.

Li Shi-chên, in his *Pên ts'ao kang mu*, annotates, "According to the *Kiao chou ki* 交州記, by Liu Hin-k'i 劉欣期,[1] the bird *mung-tung* 鸏䳍 is identical with the bird of the King of Yüe, and is a water-bird. Its habitat is in Kiu-chên 九眞 and Kiao-chi 交趾. It is as large as a peacock; its bill is over a foot long and tinged yellow, white, and black, being lustrous like lacquer. The southerners make it into drinking-vessels. According to the *Lo fou shan su* 羅浮山疏,[2] the bird of the King of Yüe is shaped like a black kite, with long feet, and has on its upper mandible an excrescence like a cap, which may hold over two pints, and which is made into wine-vessels. These are extremely strong and solid. [Then follows the passage on the food and guano of the bird, as in the *Nan Yüe chi*.] Yang Shên 楊槇 (1488—1559), in his *Tan k'ien lu* 丹鉛錄,[3] states that the bird *mung-tung* is identical with the one now styled *hao ting* 鶴頂." The *Chéng tse t'ung* 正字通 does not contain much that is new. It quotes the *T'ung ya* 通雅[4] to the effect that the bird *hao ting* is the *mung-tung*, but that in fact it is not the crest of a crane; there is also the designation *siang tiao* 象鳥[5]雕, referring to the large size of the bird, and this is now the character in vogue for it.[6]

---

[1] According to BRETSCHNEIDER (*l. c.*, p. 159), "probably fourth or fifth century."

[2] The second character is omitted in the text; but there is no doubt that the above work is meant, as it is so quoted in the same passage by K'ang-hi (under *mung*), with the addition of the author's name, Chu Chên 竺眞 (see BRETSCHNEIDER, *l. c.*, p. 172). The Lo-fou Mountains in Kuang-tung Province, famed for their flora, fauna, and temples, have been described by R. C. HENRY (*Ling-nam*, p. 307, London, 1886) and F. S. A. BOURNE (*The Lo-fou Mountains, an Excursion*, Hongkong, 1895, — a brief but interesting pamphlet of 48 pages).

[3] Published in 1554 (WYLIE, *Notes*, p. 162). The term *hao ting*, accordingly, is no older than the Ming period.

[4] A collection of miscellaneous notes by Fang I-chi 方以智, who lived under the last emperor of the Ming dynasty.

[5] Even without Chinese comment, it may be inferred that this character is a recent, artificial formation, intended to convey the meaning "elephant-bird" (compare our "rhinoceros-bird").

[6] The *T'u shu tsi ch'éng* has added an illustration of the bird which has not the

It is certain that the Chinese accounts unequivocally describe the hornbill which is peculiar to the Indo-Malayan culture-area, and which plays a significant rôle in the religious beliefs of the Malayans. It is no less certain that al-Akfānī's product, derived from the forehead of a large bird, presents an allusion to the same matter. His color descriptions — "changing from yellow into red, and apricot-colored" — are indeed very appropriate. But how is

---

slightest resemblance to a hornbill The Peking draughtsman, quite naturally, had never had occasion to see the tropical bird, and pieced his picture together from the scraps which he encountered in the definitions of the text. He consequently sketched a kite on the wing, and added a sort of fantastic wine-vessel above the neck and skull! In a similar manner, the Chinese book-illustrations of the rhinoceros are not based on realities, but on the definitions of the dictionaries Consequently identifications of such animals cannot be founded on the illustrations, as has been done, but only on the texts. The text is the key to the how and why of the illustration. — It is assumed also that Pliny has made reference to the hornbill. Speaking of the birds called *pegasi* and *grypae*, he states that he looks upon them as fabulous, and proceeds to say that the same is his opinion in regard to the *tragopana*, of which several assert that it is larger than an eagle, has curved horns on the temples, and a plumage of iron color, while only the head is purple (Equidem et tragopana, de qua plures adfirmant, maiorem aquila, cornua in temporibus curvata habentem, ferruginei coloris, tantum capite phoeniceo. — *Nat. hist.*, x, 49, § 70) The identification of the *tragopana* (which literally means the "goat Pan") with the hornbill is still given in the last edition of the *Encyclopædia Britannica*, but this is erroneous. G. CUVIER (*Le règne animal*, Vol. I, p. 479, Paris, 1829) was doubtless correct in terming *tragopan* the napaul or horned pheasant (*faisan cornu*) first described under this name by his great predecessor BUFFON (*Histoire naturelle des oiseaux*, Vol. II, p. 381, Paris, 1772). This bird lives in the Himālayas, chiefly in Nepāl and Sikkim, and is, in the words of Cuvier, "l'un des oiseaux dont la tête est dans le mâle le plus bizarrement ornée; presque nue, elle a derrière chaque œil une petite corne grêle." These lateral fleshy protuberances perfectly answer Pliny's description. The bird is now termed *Ceriornis satyra*; and colored illustrations of it may be seen in D. G. ELLIOT (*Monograph of the Phasianidae*, Vol. I, Plate XXII, New York, 1872), or in GOULD (*Birds of Asia*, Vol. VII, Plate XX, 1868). It is apparently this bird which is hinted at in K'ien-lung's *Polyglot Dictionary* (Appendix, Ch. 4, p 86) by the name *kio ki* 角鷄 "horn pheasant," accompanied by the Tibetan name *bya p'o p'od-čan*; the proper Tibetan designation seems to be *,oṅ-log* (SANDBERG, *Handbook of Colloquial Tibetan*, p. 171), which in our dictionaries is rendered "ptarmigan" on the ground of a surmise of I. J. Schmidt. The Lepcha know this species as *ta-ryok-fo* (MAINWARING-GRÜNWEDEL, *Dictionary of the Lepcha Language*, p. 119), and are likewise acquainted with two species of hornbill, — *ka-hlet-fo* and *ka-groṅ-fo* (*ibid.*, p. 468).

the mystery that al-Akfānī places the hornbill on the same level as *khutū*, and possibly rhinoceros-horn, to be explained? This problem is as follows: the Malayans entertain the belief that horn-shavings of the hornbill, placed in a suspected beverage or food, color it blood-red, in case poison has been added. [1] The hornbill substance thus ranked among the poison-detecting remedies, and was easily associated with *khutū* and rhinoceros-horn. This result is instructive, in that my former conclusions as to the development of the beliefs in the virtues of *ku-tu-si* are signally confirmed by it. Nobody could assert that *khutū* originally designated the horn of the buceros bird: this is impossible, for the reason that the Chinese *ku-tu-si* is not linked with this bird, and that the Chinese traditions regarding the latter are a chapter radically distinct and independent of the former. The anti-poisonous property ascribed to hornbill is

---

[1] H. V. STEVENS, *Materialien zur Kenntnis der wilden Stämme auf der Halbinsel Malāka*, II, p. 134, note 3 (*Ver. Mus. Völk.*, Vol. III, 1894). This explanation of the matter renders intelligible also the following entry under خُتُ *khutū* in JOHNSON's and RICHARDSON's *Persian-English Dictionary* (p. 448): "A Chinese bird, of whose bones they make handles to knives, which, being dipped into any victuals suspected to be poisoned, are said to have the virtue of immediately discovering it" He who is not satisfied with my explanation of al-Akfānī's bird may fall back on another theory. Several Siberian tribes conceive the skulls of the fossil rhinoceros as "birds," and term the horn "bird's claws" (see chiefly H. H. HOWORTH, *The Mammoth and the Flood*, pp. 6 *et seq.*). This tradition, however, is not known to the Mohammedan writers, whereas fables of the buceros were current among them. The Persian allusion to *khutū* as a "Chinese bird" presents one example. Further we read in Damīrī (translation of JAYAKAR, Vol. I, p. 667) about the bird *al-khaṭaq* الْخَطَق as follows: "Aristotle states that it is a certain large bird found in China, Babylon, and the land of the Turks. Nobody has seen it alive, for nobody is able to catch it in that state. One of its peculiarities is that when it smells a poison, it becomes benumbed or paralyzed, perspires, and loses its senses. Another authority states that on the way to its winter and summer quarters, there are many poisons on the road, and that when it smells one of them, it becomes benumbed and drops dead: its body is then taken, and vessels and knife-handles are made from it. If its bone perceives the smell of a poison, it breaks out into perspiration, by which means poisoned food may be detected. The marrow of the bones of this bird is a poison to all kinds of animals, and the serpent flees away from its bones, so much so that it cannot be overtaken." We apparently meet here an allusion to the buceros based on oral traditions, and it seems preferable to think that it is this bird which al-Akfānī had in mind.

a purely Malayan idea, and has apparently been handed on from that quarter to the Arabs during their commercial relations with Malayan tribes. Our Arabists will presumably be able to tell us more about the trade in this article. It is a wholly secondary development that the hornbill was classified in the same category with *khutū*, — a notion absent in al-Bērūnī, who does not know rhinoceros-horn, either, in this connection. The latter is as secondary as the hornbill; and consequently the *khutū* of al-Bērūnī can be neither the one nor the other, but only walrus ivory. His identification of *khutū* with the fish-teeth brought by the Bulgar from the northern sea renders this conclusion quite certain.

Addenda. — P. 349, note 2. The dictionary *Cheng tse t'ung* defines the term *hai ma* as the designation of a fish or seal (*yü*) with teeth as strong and bright as bone and adorned with designs as fine as silk, — workable into implements. Evidently this is the walrus.

P. 375. Regarding the *kuo hia ma* of southern China see also *Ling wai tai ta* (Ch. 9, p. 5; ed. of *Chi pu tsu chai ts'ung shu*).

P. 385. Also the *Ling piao lu i* 嶺表錄異 (Ch. 中, p. 7b; ed. of *Wu ying tien*) has a brief note on the bird of the King of Yüe, and speaks of a yellow cap on its head in the shape of a cup; this cap is so solid that it can be wrought into wine-cups.

128

斯调

MÉLANGES.

SE-TIAO.

In speaking of the country Se-tiao (*Tʻoung Pao*, 1915, pp. 351, 373) I had overlooked the fact that M. CHAVANNES had drawn attention to this locality with reference to an interesting text of the *Lo yang kia lan ki* (*Journal asiatique*, 1903, nov.-déc., p. 531), adding that the information given by the çramaṇa Pʻu-tʻi-pa-tʻo 菩提拔陁 (Bodhibhadra) in regard to this country would merit a special examination. Bodhibhadra, after reaching the capital Lo-yang in 509, was interviewed by the Buddhist clergy of this place as to the customs of the southern countries and stated, "Formerly there was the country Nu-tiao, where four-wheeled carts drawn by horses were employed as means of conveyance. The country Se-tiao produces asbestine cloth made from the bark of a tree; this tree when exposed to a fire will not be consumed" (古有奴調國乘四輪馬爲車。斯調國出火浣布以樹皮爲之。其樹入火不燃。 *Lo yang kia lan ki*, Ch. 4, p. 15 b, ed. of *Han Wei tsʻung shu*). This is the passage to which I incidentally referred on p. 353, unfortunately relying on the *Tʻu shu tsi chʻéng*, which has the wrong reading 車斯國 (my remark on this alleged country Kü-se in note 4 must accordingly be discarded).

I avail myself of this opportunity to make a small addition to the notice regarding the animal *ki-ku* (p. 342). The text of the *Sung chi* is in the main derived from the older work *Yu yang tsa tsu* (Ch. 16, p. 15, ed. of *Pai hai*), where it is said: 獵得者斫剌不傷積薪焚之不死、乃大杖擊之骨碎乃死. The animal *fêng li* (p. 343), however, is not given as a synonyme of *ki-ku* in this work, as wrongly stated in the *Pén tsʻao kang mu*, but is treated there as a separate subject (Ch. 15, p. 8 b). See also *Tʻai pʻing huan yü ki*, Ch. 177, p. 8 b.

B. LAUFER.